Personality

IN NATURE, SOCIETY,

AND CULTURE

EDITED BY

Clyde Kluckhohn AND *Henry A. Murray*

with the collaboration of *David M. Schneider*

Personality

IN NATURE, SOCIETY,

AND CULTURE

SECOND EDITION, REVISED AND ENLARGED

NEW YORK: ALFRED·A·KNOPF

1961

L. C. Catalog Card Number 52–12410

THIS IS A BORZOI BOOK
PUBLISHED BY ALFRED A. KNOPF, INC.

Originally published in 1948
Reprinted four times
Second Edition, revised and enlarged, published February 1953
Reprinted 1954, 1955, 1956, 1959, 1961

TO

Gordon Allport

O. Hobart Mowrer

Talcott Parsons

With whom it was our privilege to collaborate in planning the Department of Social Relations at Harvard which is concerned both with personalities and with the sociocultural matrix in which they develop

The Contributors

Franz Alexander, M.D., Director of the Chicago Institute for Psychoanalysis and Clinical Professor of Psychiatry, University of Illinois

Gordon W. Allport, Professor of Psychology, Harvard University

Raymond A. Bauer, Lecturer on Social Psychology in Social Relations and Research Associate of the Russian Research Center, Harvard University

* Ruth Benedict, formerly of Columbia University

J. S. Bruner, Professor of Psychology, Harvard University

Bingham Dai, Professor, Departments of Psychiatry and Psychology, Duke University

Allison Davis, Associate Professor of Education, University of Chicago

Rose Davis, Mt. Sinai Hospital, New York City

John Dollard, Professor of Social Psychology, Yale University

Dorothy Eggan, Chicago, Illinois

Erik Homburger Erikson, Senior Staff, Austen Riggs Foundation and Professor of Psychology in the Department of Psychiatry, University of Pittsburgh

Lawrence K. Frank, New York City, New York

Erich Fromm, Bennington College and William Alanson White Institute of Psychiatry

Frieda Goldman-Eisler, Institute of Psychiatry, Maudsley Hospital, London

Geoffrey Gorer, Haywards Heath, Sussex, England.

Phyllis Greenacre, M.D., Professor of Clinical Psychiatry, Cornell University Medical College, New York City

A. Irving Hallowell, Professor of Anthropology and Curator of Social Anthropology at the University Museum, University of Pennsylvania

Robert J. Havighurst, Professor of Education, University of Chicago

Jules Henry, Associate Professor of Anthropology, Washington University, St. Louis

* Deceased.

Zunia Henry, St. Louis, Missouri

Donald Horton, Research Associate in Sociology, University of Chicago

J. McV. Hunt, Professor of Psychology, University of Illinois

Alex Inkeles, Lecturer on Sociology in Social Relations and Senior Research Fellow, Russian Research Center, Harvard University

* E. M. Jandorf, formerly of Harvard University

Hudson Jost, Professor of Psychology, University of Georgia

Franz J. Kallmann, M.D., Principal Research Scientist (Medical Genetics), New York State Psychiatric Institute; Assistant Clinical Professor of Psychiatry, Columbia University

Florence Rockwood Kluckhohn, Lecturer on Sociology and Research Associate in the Laboratory of Social Relations, Harvard University

Dorothy Lee, Professor of Anthropology, Vassar College

Nathan Leites, Social Science Division, the Rand Corporation, and Visiting Lecturer, Department of Political Science, Yale University

David M. Levy, M.D., Research Professor of Psychiatry, Yale University

Margaret Mead, Associate Curator of Ethnology, American Museum of Natural History

Robert K. Merton, Professor of Sociology, Columbia University

Benjamin D. Paul, Lecturer on Social Anthropology, Harvard School of Public Health, Harvard University

Talcott Parsons, Professor of Sociology, Harvard University

Hortense Powdermaker, Associate Professor of Anthropology, Queens College

Jurgen Ruesch, M.D., Associate Professor of Psychiatry, University of California School of Medicine

Saul Rosenzweig, Professor of Psychology, Washington University, St. Louis

R. Nevitt Sanford, Associate Professor of Psychology, University of California

David M. Schneider, Lecturer on Social Anthropology in Social Relations, Harvard University

Lester W. Sontag, M.D., Director, Fels Research Institute, Antioch College

* Deceased.

Alfred M. Tozzer, Professor of Anthropology, Emeritus, Harvard University

Robert Waelder, Lecturer, Philadelphia Psychoanalytic Institute

Henry J. Wegrocki, M.D., Chief of Psychiatric Service, California Hospital, Los Angeles

John W. M. Whiting, Lecturer in Education and Research Associate in the Laboratory of Human Development, Harvard University

* Ira Wile, M.D., former president, American Orthopsychiatric Association

* Deceased.

ACKNOWLEDGMENT

(REVISED EDITION)

OUR GRATITUDE to authors and publishers for permission to reprint is recorded at the beginning of each selection. We wish to express special appreciation to Drs. Raymond A. Bauer, Alex Inkeles, and Jurgen Ruesch, who prepared papers specifically for this volume, and to Drs. Erikson, Henry, Horton, Florence Kluckhohn, and Schneider, who made substantial revisions for our needs.

INTRODUCTION

to the Second Edition, Revised

IT IS ALWAYS DIFFICULT to keep roots in the past and still keep up with the future. We have tried, in this revision, to include as much new material as will keep it abreast of the rapid developments in the field, yet retain from the previous edition those papers which seem to us the best in their particular fields—at least so far as the less easily accessible literature is concerned.

The first essay, Part One, *A Conception of Personality*, has been entirely rewritten. The second has had only minor additions and revisions. Although limitations of space preclude the extended treatment we had hoped to give to the concept of values, the new papers by Drs. Dorothy Lee, Florence Kluckhohn, Erik H. Erikson, and Raymond A. Bauer should give the serious student a good start in this field, which cannot be ignored by the student of personality. The papers by Drs. Bauer and Inkeles were written especially for this volume and are not available elsewhere. Drs. Florence Kluckhohn, Erik H. Erikson, and David M. Schneider rewrote and substantially reworked their papers especially for this volume. The papers by Drs. Leites and Rosenzweig each treat of an aspect of literature, a problem neglected in the first edition of this book.

We are enormously grateful to our collaborator, Dr. David M. Schneider, upon whom the major burden of reviewing the newer literature and editing this second edition has fallen. The genuine strengthening of the book which we feel to have been accomplished is largely the result of his hard work and sensitive judgment.

C. K.

H. A. M.

INTRODUCTION
to the First Edition

IN RECENT YEARS, psychiatrists, psychologists, sociologists, social workers, students of education, biologists, and anthropologists have been preoccupied with the relation of the individual to his society. Is there a connection between the mental illnesses characteristic of a given group and the social norms that this group enforces with special severity? Does a people's way of bringing up their children make a particular type of personality unusually common in that society? Why do certain kinds of delinquency appear most frequently among the children of certain social classes? How much of an individual's personality is fixed by his biological constitution? How much is personal life style influenced by the society's traditional designs for living (its "culture")? Why are ways of behaving that are punished by one group rewarded, or at any rate tolerated, by another? Is individual deviance culturally defined, or are there absolute standards of psychological abnormality? What is the role of outstanding personalities and of other individuals in social change?

These are some of the questions that have been asked—and partly answered. Unfortunately for the undergraduate and the general reader, many of the most interesting and significant contributions have been published in professional journals. The articles are scattered among papers on quite different topics, and many of the journals are not available save in the largest libraries. The purpose of this book is to make readily accessible a representative selection of the more important publications that have appeared in the periodical literature. They are arranged in such a way as to give a somewhat orderly presentation of the theory and of the empirical generalizations, and the linkage between each paper and those that have gone before is sketched in brief transitional passages.

The approach of the editors is a "field" approach. That is, we regard the conventional separation of the "organism and his environment," the drama of "the individual *versus* his society," the bi-polarity between "personality and culture" as false or at least misleading in some important senses. Knowledge of a society or a culture must rest upon knowledge of the individuals who are in that society or share that culture. But the converse is equally true. Personal figures get their definition only when seen against the social and cultural background in which they have their being. An individual who has no part in any group and who has not been influenced by some traditional way of behaving cannot be found in real life. Although those who study culture and society are primarily interested in the similarities in personality and those who practice psychotherapy and investigate individual psychology have their focus of attention upon the differences, the two sets of facts are inextricably interwoven. One defines the other. In actual experience, ⌐individuals and societies constitute a single field⌐ Only by abstraction can one be dealt with apart from the other, although in scientific analysis there can be a legitimate alternation of focus of interest between individuals and cultures or groups.

The inquiry into the interdependence between the personal characteristics of a people and their culture and the environment in which they live is not, as some contemporary enthusiasts for the "culture and personality" viewpoint have implied, something new under the sun. On the contrary, this interest is at least as old as the fifth-century-B.-C. Greek physician, Hippocrates:

> . . . The same reasoning applies also to character. In such a climate arise wildness, unsociability, and spirit. For the frequent shocks to the mind impart wildness, destroying tameness and gentleness. For this reason, I think, Europeans are also more courageous than Asiatics. For uniformity engenders slackness, while variation fosters endurance in both body and soul; rest and slackness are food for cowardice, endurance and exertion for bravery. Wherefore Europeans are more warlike, and also because of their institutions, not being under kings as are Asiatics. For, as I said above, where there are kings, there must be the greatest cowards. For men's souls are enslaved, they refuse to run risks readily and recklessly to increase the power of somebody else. But independent people, taking risks on their own behalf and not on behalf of others, are willing and eager to go into danger, for they themselves enjoy the prize of victory. So institutions contribute a great deal to the formation of courageousness.

Hippocrates, however, gives us merely the speculations of an acute clinician, based upon his general observations. The modern enquirer can draw upon the systematic investigations of biologists, physicians, geographers, anthropologists, sociologists, and psychologists. There are laboratory studies, case histories of social workers, clinical interviews of psychiatrists and clinical psychologists, tests and experiments, the observations of sociologists in our own society, and the field studies of geographers and anthropologists among remote lands and people.

In 1925, Dr. Leslie A. White published an article called "Personality and Culture" in *The Open Court*. Dr. White was then a student of the linguist-anthropologist, Professor Edward Sapir, and in 1932–3 Professor Sapir, with Dr. John Dollard, conducted a seminar in culture and personality at Yale University for a group of fellows of the Rockefeller Foundation chosen from various European countries. Mr. Lawrence Frank, then of the Rockefeller Foundation, was instrumental in promoting and planning this seminar. During the same period, Dr. W. I. Thomas, also much influenced by the ideas of Edward Sapir, was surveying this field for the Social Science Research Council. In April 1933, Dr. Thomas submitted his report "On the Organization of a Program in the Field of Personality and Culture." This document, which was supported by ninety-seven appendices (many of which have since been published), forms a major landmark in the growth of organized research on the relationship between individual development and the biological, social, and cultural matrix in which it occurs. Thomas' insistence upon the necessity for a multi-dimensional attack is worth recalling at a time when so many publications are set in a framework that in fact is purely biological, or social, or cultural—however much verbal hat-tipping there may be to the other dimensions.

Since the early thirties, when Dr. Sapir initiated formal instruction in this field and Dr. Thomas first systematically mapped the research territory, interest has steadily grown. The editors have examined more than a thousand pertinent articles and discovered several hundred which it was necessary to consider very seriously for inclusion in this volume. The final selection was exceedingly difficult and had to be based upon somewhat arbitrary principles. For example, there are a number of excellent discussions which use Pueblo Indian data, but we restricted ourselves to one of these in order to secure the widest possible variety of geographical and cultural representation. When choice had to be made between articles dealing with the same area and of approximately equal interest, preference has been given to a paper that appeared in a journal not ordinarily in the smaller college libraries. Similarly, we rigorously excluded all articles, even though of the highest quality, which are or will shortly be available in book form. A volume on "culture and personality" without Sapir seems like *Hamlet* without Hamlet, but the University of California Press will soon make available all of Sapir's writings in this field.[1] The essays of Lawrence Frank [2] and of T. H. Pear are also about to be published in collected form. Nor is Harry Stack Sullivan represented except indirectly by contributors influenced by his teachings, since his lectures are available in one volume.[3] After making the initial selections for this volume, we saw the table of contents for *Readings in Social Psychology* [4] and found some duplications. This fact we took as an invitation to make accessible other materials of equal significance.

The same criterion of accessibility decided us against any attempt to

---

[1] *Selected Writings in Language, Culture and Personality*, David Mandelbaum, editor (1948).

[2] *Society As the Patient*, Rutgers University Press (1948).

[3] *Conceptions of Modern Psychiatry*, William Alanson White Psychiatric Foundation (1946).

[4] Edited by the Society for the Psychological Study of Social Issues, Theodore M. Newcomb and Eugene L. Hartley, chairmen of editors (1947).

include passages from important books that are readily available.[5] By assigning material from one or more of the books already available and the present volume, the teacher can select reading that will cover the subject extremely well without recourse to periodicals that need to be preserved for the use of graduate students and for research.

The deliberate intention of the editors is to illustrate the range and variety of work published in this field. They do not necessarily agree with all points made in the selections. Indeed it is obvious that, while the overlap of agreements is impressive, there are some disagreements in approach and contradictory findings in some details. Our aim is to see that various important points of view are represented. It is equally obvious that there is some repetition. It seems to us highly desirable for pedagogical reasons that certain basic themes should be reasserted in different languages and in different factual contexts.

In general, this source book is directed to an audience of college Juniors and Seniors and of educated laymen. For this purpose, authors of the original articles have been generous in allowing the editors to delete most footnotes and to omit highly technical passages. However, for completeness of coverage it has been thought wise to include a few selections that demand some technical language. These have been marked with an asterisk, and they may be omitted in undergraduate courses without a serious break in the general argument of the book. The teacher of such courses is also advised to assign only the first three sections of Part One.

Most of the book is given over to the topic of personality formation because it is here that one sees most clearly how intricate and manifold are the connections between an individual's physique and the press of his total environment (ecological, biological, social, and cultural). The selections in Part Three indicate the range of implications for contemporary issues offered by studies of personality in nature, society, and culture. The most basic contribution of these studies is precisely that of helping us to understand other peoples—and ourselves. Insofar as one gains some knowledge of the variety of personality types in the world and of universal psychological processes, cross-cultural tolerance and genuine communication become possible. All research in this field is in the last analysis directly or indirectly oriented to one central type of question: What makes an Englishman an Englishman? an American an American? a Russian a Russian?

[5] Such as Ralph Linton, *The Cultural Background of Personality;* Cora DuBois, *The People of Alor;* Erich Fromm, *Escape from Freedom;* and Abram Kardiner, and collaborators, *The Individual and His Society* and *The Psychological Frontiers of Society.*

CONTENTS

Part Three

SOME APPLICATIONS TO MODERN PROBLEMS

Contents

TABLES AND CHARTS

[xxiii

Part One

A CONCEPTION OF
PERSONALITY

1

Henry A. Murray
AND *Clyde Kluckhohn*

OUTLINE OF A CONCEPTION
OF PERSONALITY

I. HISTORICAL INTRODUCTION[1]

ORIGINALLY, adult man's perceptions and apperceptions of his physical and social environment were, like the child's, elaborated by a great variety of dramatic, and often moral, imaginations. Animism was in power: everything was viewed in a framework of supernaturalism. In the West, the first strategic inroads of scientific objectivity were made in the domains of astronomy and of physics. The result was a different universe. Subsequent advances in chemistry, geology, biology, and palaeontology eventually brought about a complete revolution in man's beliefs as to his origin and present place in nature. More recently the science of personality has emerged to transform man's notions of his own dispositions and powers.

An initial attempt to use the primitive bio-chemical concept of "humors" as basis for a fundamental typology of personalities was largely abortive. Then from biology, Darwin largely, and from medicine, came the concepts of instinct and of constitution. The former was later redefined as animal drive or human need, the overt instrumentalities of which were proved to be learned through experiences of "success" and "failure." Anthropological observations also indicated that action patterns are not

---

[1] Certain ideas and phrases in this and later sections derive from an unpublished paper by O. H. Mowrer and C. Kluckhohn. The writers thank Dr. Mowrer for his permission to use these excerpts.

innate, as was originally supposed, but in large measure culturally determined. Complementary to these studies was Pavlov's work which showed
how the nervous system becomes organized by experience to "foresee" the
most probable immediate future and to act accordingly. Another important strand of facts and theories came concurrently from the medical
psychologists, some from Janet and others, but many more from Freud
and the followers of Freud. A human being does not grow up in a vacuum:
his development is determined not only by the physical environment as the
biologists proved, and by the family environment as Freud demonstrated,
but, as the massive data collected by the cultural anthropologists showed,
by the larger societal and cultural institutions that are extolled, preached,
and practiced not only by parent "carriers" but by the leading minority
(authority figures), if not by the majority, of the group in which the individual is reared.

The resultant of these and other investigations and conceptualizations is
a re-structured discipline, more comprehensive and sophisticated, with its
own sphere of concern, its appropriate methods, concepts, terms, and
working hypotheses. This new approach does not deny the biological basis
of human nature nor the physio-chemical basis of biology. It accepts the
axiom that no postulate of a science of behavior may be incongruent with
facts or principles accredited in other spheres. But it holds to the proposition that there are different "levels" of observation, analysis, and formulation, and that the behaviors of human personalities are on a different level
from physiological phenomena, and hence should be studied and conceptualized in their own right without waiting for more "basic" sciences
to provide a complete foundation.

Social science has one incomparable advantage over other sciences: the
objects of its attention, human beings, in contrast to stars, chemical com-
pounds, and insects, are capable of remembering and communicating—
often with sufficient accuracy—a great deal of critical information about
themselves and their associates. (No April warbler, for instance, can tell us
whence he came and whither he is going, and who his mother was and
how she treated him.) Like every blessing, this unique advantage is offset
to some extent by the armaments of man's *amour propre*, his desire to be
respected by himself and others, and hence his disposition to dwell on his
virtues and successes and to repress, falsify, or deny his vices and his failures. For these reasons, his spontaneous accounts of himself are habitually
warped, and, when exposed to the cool eye of inquiry, he bristles in self-
defense. It is no accident that the first deep penetration of the guarded
fortress was made by a physician (Dr. Sigmund Freud), since, as a rule,
only miserably sick people in need of help can be encouraged to abandon
their defenses one by one and eventually reveal their innermost secrets,
and, then, only after they have become convinced that their cure depends
upon this sacrifice. Thus, the underlying mental processes, without a
knowledge of which no complete science of human nature can be built,

are generally exhibited only to those whose role it is to administer relief. The psychiatrist, in short, has become an essential member of the social scientific fraternity.

From Freud came the first relatively comprehensive dynamic theory of personality—psychoanalysis. Like all great scientific contributions, Freud's most fundamental discovery, though essentially simple, subsumed under a common formula a host of superficially diverse phenomena. Slips of the tongue and pen, acts of misreading and misinterpreting, play, humor, and dreams, artistic and literary creations, myths and rituals, the long and varied gamut of neurotic symptoms and sexual perversions, and the bizarre manifestations of the psychoses—all these phenomena Freud explained *in part* as more or less well disguised expressions and gratifications of needs which the individual is afraid to express and to gratify openly.

Every society makes many demands upon the persons who become acculturated by it. When and how renunciations are demanded differ in various societies and in the different social classes of the same society. A time-minded culture like that of the United States stresses more distant rewards than does, say, Mexican culture. Cultures also offer many potential enrichments of experience to the individual, but these are less apparent to the child than are the more direct advantages of social, as contrasted with solitary, existence. Although psychotherapists have probably stressed the repressive and renunciatory demands of culture too much and the benefits too little, most children seem to feel these demands as more or less arbitrary and hence repugnant. Even the adult rarely appreciates the "wisdom of the culture." Instead, culture is represented to him, first, by a succession of imperious requirements set by external authority and, later, by a rigid, unreasoning part of his own personality which must be obeyed implicitly or hazardously defied. But, it is probably the phase of culture in which we are now living that has disposed people to overlook the phases in which religionists and ethical philosophers were enthusiastically engaged in building up, rationalizing, and propagating the moral edifice against which recent generations have rebelled.

In this connection it might be well to mention one special impediment to the advance of impartial objectivity in the realm of social science: the limitations and distortions of perception, apperception, and conceptualization, the blind spots, rigidities, and defensive rejections in the personality of the social scientist himself and in the members of the society in which he must work. To adults who have painfully achieved some degree of integration of character, an honest examination of the processes involved is as distressing as the reopening of an old wound. For many persons it is precisely the "givenness" of personality that is the bulwark of their security system. Students of personality and culture are themselves victims of their own personalities and products of their culture. Only recently has a cross-cultural perspective provided some emancipation from those values that are not broadly "human" but merely local, in both space and time.

Only gradually is our culture altering so as to permit social scientists to work and think with something approaching objectivity. Problems of personality have, in general, been deeply embedded in moral issues; and the student who did not wish to be punished by society was forced to take sides in disputes over the "goodness" and the "badness" of this or that disposition, instead of being free to ask how such dispositions develop, and why some individuals manifest them more than others.

2. A FEW BASIC DEFINITIONS AND ASSUMPTIONS[2]

1. *Persons.* The psychologist's objects of concern are individual human organisms, or persons, rather than groups (the sociologist's objects of concern).

2. *Person in and of environments.* A person is an emergent entity *of* and *in* a certain physical, social, and cultural milieu. He cannot be properly represented in isolation from his locale, or from the culture of the group of which he is a member, or from his status (role) in the structure of that group. Basically, every person is a social person, an interdependent part of a system of human interactions.

3. *Seat of the personality of a person.* The superordinate governing institution of the human organism is the *personality*, or mind. Its main physiological and neurological basis lies in the head.

According to this view, the *establishments* and *processes* which constitute personality are out of sight, but their characteristics, relations, and operations can be defined and conceptualized on the basis of the subject's verbal reports (memories of events, introspective judgments, and avowals) and on the basis of observations of his overt behaviors, physical and verbal. Thus the psychologist is directly concerned only with the manifestations of personality, *the facts.* The personality is something that must be inferred from the facts. Hence, in actual practice, the personality is an abstract formulation composed by the psychologist.

Much evidence goes to show that the brain is the locus of the integrations which are manifested in all co-ordinated and effective actions of the total organism. It is surely the locus of the feelings which evaluate situations as they appear and discriminate goals for action. It is the seat of consciousness, of streams of thought, of conflict, and of decision. It is also the repository of all *establishments* of personality: traces of past experiences, images, symbols, concepts, beliefs, attached emotions, ideologies, plans, commitments, resolutions, and expectations. Thus, both the enduring components and the ceaseless, kinetic processes of personality are—physically and concretely—located in the head.

The operations of personality are markedly affected by brain injuries and lesions, by the operations of lobotomy, lobectomy, and topectomy,

---

[2] We are purposely repeating here statements that one or both of us have published elsewhere, most fully in Talcott Parsons and Edward A. Shils (eds.), *Toward a General Theory of Action* (Cambridge, Harvard University Press, 1951).

by infections of the brain, by a narrowing of the cranial blood vessels, and by changes in the composition of the circulating blood (its internal physiological environment).

Personality may be greatly modified, temporarily or permanently, by happenings in other parts of the body. The glands of internal secretion, in fact, are so influential in determining energy level, temperament, emotion, and drive that these might well be included as outlying organs of the personality, contributing, say, to what psychoanalysts have termed the "id." But other considerations militate against this notion: the operations of personality are even more dependent upon atmospheric oxygen; they are modified by low oxygen tension just as they may be modified by drugs or by happenings in distant regions of the globe (e.g. death of a beloved person). It seems preferable, then, to regard the glands of internal secretion as important chemical or constitutional *determinants*, rather than anatomical centers, of personality.

The *establishments* of the personality, such as a person's knowledge, value-system, or vocational goal, and the *processes* of personality, such as a transient perception, emotion, or intention, have been located physically in the brain region because there is no other possible place for them—except as abstractions. Furthermore, the day may come when advances in neuro-physiology and electronics will reveal some of the physical correlates of psychic processes. But it should be understood that according to our present conception no components of the personality are identified with any particular structures or processes of the brain as we know it, morphologically and physiologically, today. The physico-chemical brain is not the personality, but one of its constitutional determinants. For the present at least, most psychologists will be content to forget the brain and restrict themselves to the psychical-behavioral level of analysis and formulation.

4. *Conscious and unconscious regnant processes.* For generations, psychology was considered to be the science of the "psyche" or the conscious mind: psychic processes were conscious processes. Since Freud, however, this assumption has been untenable, and today there is much evidence to support the accepted view that many conscious events and many behavioral events can be sufficiently explained only by inferring the occurrence of more processes than the subject was aware of at the time, the inferred processes being no different in their effects from the processes with which introspection is familiar. The semantic problem presented by the discovery of unconscious "conscious" processes can be readily solved by using the adjective "regnant" to distinguish the momentarily governing processes at the superordinate (regnant) level of integration in the brainfield. Many, but not all, of the regnant processes have the property of consciousness at the moment of their occurrence, or soon afterwards if recalled by retrospection. According to this way of stating things, every conscious process is regnant, but not every regnant process is conscious; or, with the double-aspect hypothesis in mind, one might say that most regnant

processes have both a conscious (subjective, inner) aspect and a physical aspect, the nature of which is still unknown.

In designating the different regnant processes there is nothing better than the traditional terminology derived from introspection, terms such as perception, apperception, emotion, evaluation, intellectual, expectation, and so forth.

The same term is used whether the regnant process is conscious or unconscious. It is quite proper to speak of an unconscious perception, unconscious emotion, or unconscious expectation. If each kind of process is defined operationally by reference to objective criteria as well as subjectively in the terms of a verbal report, the same variables can be used in explaining the behaviors of lower animals and infants as are used in formulating the activities of human beings capable of speech.

One reason for starting this way, dwelling on conscious and unconscious regnant processes at the "center" of personality, rather than on the activities at its "periphery," is that many American psychologists today—influenced, quite naturally, by the successful methods of physical science, by mechanistic conceptions, and by the creed of behaviorism—are disposed to neglect, or even to rule out, subjective processes, and we believe this trend should be balanced by its opposite.

Furthermore, if the psychologist's aim is to arrive at a structured formulation of a personality, rather than a mere list of overt action patterns, he must refer everything back to the only place where superordinate structures can exist or be created. The brain is the locus of over-all integrations, because it is the *only* place where afferent (sensory) processes from the whole body terminate and efferent (motor) processes to the whole body originate.

5. *Personality in process*. The personality is participant in a long succession of activities from birth to death, regularly punctuated by periods of sleep. From one point of view the history of the personality *is* the personality. It cannot be properly represented as a fixed structure. Personality is an on-going manifold of structured processes.

6. *Proceedings of personality, internal*. The waking hours of a person's life may, for convenience, be divided into a sequence of temporal segments, or *proceedings*, some of which are mostly *internal*, some mostly *external*. An *internal proceeding* is a temporal segment during which the person, abstracted from his environment, is attending to his feelings, memories, evaluations, fantasies, artistic or theoretical creations, plans, or expectations of the future. Knowledge of a person cannot be restricted to the observation of his overt behaviors. Reported memories of internal proceedings as well as verbalizations of concurrent streams of thought are indispensable to the psychologist.

7. *External proceedings*. An *external proceeding* is a stretch of time during which the person is overtly engaged in coping with some part of his immediate physical or social environment. Such a proceeding consists of the concrete chronological activities of a concrete chronological person

in a concrete chronological situation. This is the psychologist's simplest *real entity*, the thing he should observe, analyze, try to reconstruct and represent, if possible, with a model, and thus explain; it is the thing he attempts to predict, and against which he tests the adequacy of his formulations and hypotheses. Real entities have duration: to perceive and interpret them the psychologist must be present and alert from start to finish. Furthermore, they are complex, occur rapidly, and are gone, never to be repeated exactly as before. Hence for detailed and precise studies, a moving picture film with sound track, which can be examined over and over again by any number of experts, is of the greatest utility.

8. *Situations*. Since the environmental situation from moment to moment is an integral part of every external proceeding (real entity), a description or symbolic representation of the more relevant portions of it is required. A person's behavior cannot be adequately characterized by a list of adjectives; it is necessary to record the nature of the successive situations, imposed or selected, in relation to which the person's activity has been oriented. According to this notion, then, a human life, as objectively viewed, consists of a chronology of distinguishable person-situation transactions.

9. *Interpersonal proceedings*. An interpersonal or dyadic proceeding (a discussion, say, between parent and child, two siblings, two enemies, two friends, or two lovers) is the psychologist's most significant type of real entity, or transaction. Here the object (alter, second person) constitutes the ever-changing situation for the subject (first person), and the subject constitutes the ever-changing situation for the object. Consequently, both sides of the equation must be given equal analytical consideration, that is to say, the psychologist's reconstruction of the episode should include as much formulation of object's status, sentiments, thought, and speech as of the subject's status, sentiments, thought, and speech. For, among the things a psychologist wants to know about a subject is how he behaves in the presence of this and that kind of person, say, a shy stranger of lower status, a boastful person of the opposite sex, a teacher who makes frequent dogmatic statements, etc.

10. *Fields*. The course or result of a proceeding, the total effect, say, of the subject's and the object's behavior, can be roughly indicated by representing the structure of the *field* (defined as the instantaneous total situation) at the beginning and at the end of the transaction. For a more exact understanding, however, it is necessary to conceive of a proceeding as a succession of instantaneous fields, each of which determines the action that ensues at that moment. According to Lewin's definition the *field* at an instant includes both the external situation (say, the smile of a certain acquaintance who wants to borrow money) and the internal situation (say, fatigue and a feeling of depression) and *both* are within the head of the subject, because for him external reality is what *he* perceives and apperceives out there, no more, no less. Since this formulation abolishes the important difference between normal verifiable perceptions and appercep-

tions, on the one hand, and illusions and delusions, on the other,—not to speak of the many extravagant projections of normal people—it has seemed advisable to call the external situation as it actually exists (insofar as this can be determined by careful inquiry) the *alpha situation*, and to call the external situation as the subject apperceives it the *beta situation*. Although the subject's response to a given situation provides a fairly reliable clue to the nature of the beta situation, the latter should be ascertained, whenever possible, by direct inquiry.

11. *Uniqueness of proceedings.* Every proceeding leaves behind it some trace of its occurrence—a new fact, the germ of an idea, a re-evaluation of something, a more affectionate attachment to some person, a slight improvement of skill, a renewal of hope, another reason for despondency. Thus, slowly, by scarcely perceptible gradations—though sometimes suddenly by a leap forward or a slide backward—the person changes from day to day. Since his familiar associates also change, it can be said that every time he meets with one of them, both are different. In short, every proceeding is in some respects unique. As a rule, the psychologist looks for uniformities among events, but if he attends exclusively to these he will overlook those exceptional occasions which leave a person, let us say, with an incurable psychic injury, or start him off on a fresh path, perhaps a new vocation.

12. *Serials.* Many actions, though temporally discreet, are by no means functionally discreet; they are continuations of a shorter or longer series of preceding actions and are performed in the expectation of further actions of a similar sort in the future. Of this nature are skill-learning activities and behaviors which are oriented toward some distant goal, a goal which cannot be reached without months or years of effort. Also to be included here are the behaviors which form part of an enduring friendship or marriage. Such an intermittent series of proceedings, each of which is dynamically related to the last and yet separated from it by an interval of time for recuperation and the exercise of other functions, may be termed a *serial.*

Most people are in the midst of several on-going serials which occupy their minds whenever they are not forced by circumstance to attend to more pressing matters. Men who are intensely interested in constructing something—a house, a relationship, a political group, a scientific book—return eagerly every day to the next step or stage of their endeavor. Their behavior is so different from what is denoted by the old S-R formula—since there is no distinguishable stimulus which evokes a response—that it calls for a differentiating word. I will suggest, then, that the term *proaction* (in contrast to *reaction*) be used to designate an action that is not initiated by the confronting external situation, but spontaneously from within. An action of this sort is likely to be part of a serial program, one that is guided by some directional force, which, in turn, may be subsidiary to a more comprehensive aim. As a rule, a *proaction* is not merely homeostatic, in the sense that it serves to restore a previously experienced physical, social, or cultural condition. If successful, a proaction may be said to be *superstatic*, inasmuch as it results in the acquisition or in the production of something new, small though it may be, over and above the previous condition—a little physical growth, more strength, more property, a new instrument, a more cohesive friendship, the conception of an offspring, another chapter of a novel, or the statement of a proposition. The integrates of serials, of plans, strategies, and intended proactions directed toward distal goals constitute a large portion of the ego system, the establishment of personality which inhibits irrelevant impulses and renounces courses of action that interfere with progress along the elected paths of life.

Since a substantial part of means-end learning, skill learning, value learning, as well as what has been termed emotional development, occurs during the course of *serials*, representative samples of the major on-going serials should be carefully studied by the psychologist in order to determine the kinds and rates of transformations which are characteristic for a person. In predicting future behaviors, values, and achievements it is not enough to know a subject's more consistent current dispositions, habits, and abilities; in addition one must have evidences necessary to an estimation both of his capacity and his determination to develop further.

13. *Cathected objects.* The study of serials teaches us that certain specific objects, or kinds of objects, may become more or less enduringly attractive or repellent to the subject. In such cases, we say that the object has acquired a *positive value,* or *positive cathexis,* that is, the power to evoke affection and to attract; or that it has acquired a *negative value* or *negative cathexis,* that is, the power to evoke dislike (either antagonism, fear, or revulsion). When the object is both liked (valued) and disliked (disvalued) it is said to be *ambivalent.* A positively cathected object is said to have final, or intrinsic, value when it is enjoyed for its own sake, and instrumental, or extrinsic, value when it merely serves as an efficient means of attaining a final value. In the development of the child the most significant objects are persons—the mother (or mother-surrogate), the father (or father-surrogate), the siblings, and, to a lesser degree, early playmates. Since the interactive patterns, or themas, that recur and become established in transactions with these significant persons are apt to be repeated later when dealing with other authorities or with peers, these are the interpersonal serials which must be most thoroughly understood. A binding cathection which prevents a person from cathecting some other ("more appropriate") object is called a *fixation.*

14. *Transformations of personality.* Personality is both the ground and the product of activity—physical, social, and mental. It is a hive of imaginations, conceptions, hypotheses, and expectations, which influence its own perceptions and interpretations of the world, and these in turn influence behavior. Man invents a heaven and a hell and modifies his ways accordingly. He dreams of rape and murder and develops neurotic symptoms. He makes tools and becomes skillful in re-structuring the physical

environment which, then, in turn, re-structures him. With people he is both affecting and affected. He apperceives them as benevolent, malicious, or mere robots, and treating them as such, is treated so by them. He anticipates conflict, amasses weapons, provokes resentment, precipitates war. Thus, does he shape, in some measure, the situations to which he must respond, the situations which shape him.

Through trials and failures and renewed efforts a capable youth will gradually approximate reality in his feelings, thoughts, and actions. By ranging experiments and much suffering he learns what satisfies him most and how to re-experience it. But there are always rival interests and claims, and renunciation is not easy. A unifying trend is commonly observable, though, whatever its degree, unity is not easily achieved. It is born out of conflicts and resolutions of conflicts, the course of life being punctuated by repeated choices to be made between alternative or opposing percepts, concepts, needs, goals, loyalties, philosophies, tactics, or modes of expression. Thus, the nature of a personality is revealed in the sequence of its major dilemmas as well as in its wholeness. Here again, since time is an integral element of personality, the psychologist may have to wait, sometimes for years, until a transformation has been completed, before he can formulate the events which led to it.

During any given period, the development of a personality—by assimilations, differentiations and integrations (of perceptions, apperceptions, valuations, strategies, language, or motor patterns)—is usually confined to one or a few serials or regions of special concern (some region of technical skill, social relations, artistic appreciation, philosophy, etc.). In less cathected regions the personality is apt to be merely "holding its own," conserving by exercise (repetition) what was previously acquired, or possibly losing some elements by disuse. Even along the frontier of creativeness, most developments occur slowly by hardly perceptible degrees, although in some instances the change is sudden and striking. There are reasons to believe, however, that a dramatic transformation (conversion) is commonly preceded by a long period of incubation, that is, by a series of unconscious reconstructions. Anyhow, personality seems to be an ever-changing institution. Precise observations teach us that both its establishments and functional operations are different at different points in its career.

15. *Periods of personality.* Since the establishments of a personality change with time, there are not many things that can be said about them which are applicable to all stages of a person's career from infancy to senility. It is customary, therefore, to divide each life, more or less arbitrarily, into a number of *long periods,* such as infancy, childhood, adolescence, and so on, each of which is composed of an indefinite sequence of proceedings, many of them being parts of concurrent serials. In order to gain some knowledge of the design, or life-style, of the personality (the ways its different activities have been temporally organized or scheduled) in any one of these long periods, it is advisable to examine, in

some detail, reports of selected *short periods*—a day, a week—which are judged to be representative. A man's schedule of activities is no less than his philosophy in practice.

16. *Forms of activity, process activity.* Aside from the simpler segmental reflexes, there appear to be three major types of activity: 1), *ungoverned, or process, activity*, and 2), *governed activity*, of which there are two forms: a), *modal, or formal, activity*, and b), *directional, or instrumental, activity.* The latter, *directional activity*, differs from the other two, insofar as it is oriented towards a supposedly satisfying end state, or goal. Both *process activity* and *modal activity* are enjoyed in their own right from start to finish, and hence are intrinsically, rather than extrinsically, satisfying. *Modal activity*, however, differs from mere *process activity*, insofar as it is governed by some effort to appreciate or to achieve a certain degree of sensuous, dramatic, humorous, or purely technical excellence for its own sake.

Process activity is both effortless and aimless. The personality is not an institution that remains inert unless excited by extracranial stimuli (as the S–R formula suggests). It is marked from first to last by a continuous flow of on-going activities. These unabatable processes are fundamental "givens," the most elementary characteristics of the mind—random perceptions, random sequences, and combinations of images, words, and sentences (symbols), random moods and feelings, random vocalizations (hummings) and verbalizations (babblings), random gestures and movements (manipulations and locomotions). When the energy level is high, such spontaneities are rapid, abundant, and, as a rule, highly pleasurable *per se* (process pleasure). They are sluggish during periods of exhaustion or staleness.

In their more elementary forms, as observed in infancy, these mental processes are so unco-ordinated, impersistent, and seemingly ineffective that they can hardly be called motivated in the ordinary sense, even though they do occasionally arrive (as if by accident) at something which gratifies a latent need.

Whenever some strong viscerogenic need, such as hunger, or a social need is activated, these self-same mental processes become the integrated instruments of its fulfillment. But, in the dependent and irresponsible days of childhood, these needful moments are periodic and relatively brief. Furthermore, they require, in the last analysis, hardly more than one instrumental act, that of crying. The mother does the rest. Thus, because of the protective parent's constant ministrations, the child's mind is not by necessity held down (as a young animal's mind is held down) to external realities and practical adjustments. For long periods the mind is free to exercise its own processes in its own way. What these mental processes do, naturally and without prompting, is to *transform* percepts into images and also, later, into words (symbols), and out of these images and words to *construct* dreams and fantasies. At one extreme these imaginations consist of elementary animistic conceptions of environmental events (world pictures), and, at the other, dramas in which invented selves play leading roles, tragic or triumphant.

17. *Modal activity.* The neglect of modal activity (which is more closely related to *being* than to *doing*) by American psychologists may perhaps be correlated with our addiction to competitive strivings and tangible rewards. The chief difference between modal activity and directional activity is this: in the former, satisfaction is normally concurrent with the activity, whereas in the latter satisfaction is linked with the ultimate effect of the activity. The activity itself may be tedious, boring, or even painful. The two types, of course, are often compounded: a person may enjoy the exercise of his functions as well as the results which they achieve. Furthermore, it is not at all unlikely that the satisfaction which has been so commonly associated with the attainment of a final state (reduction of tension, satiation) is, in fact, more closely correlated with the exciting modal activities (gustatory sensations in eating, and genital sensations in sexual intercourse) which precede the end point.

The most obvious illustrations of modal activity are expressions and receptions of sensuous, dramatic, or comic patterns: ballet dancing and witnessing a ballet, playing music and listening to music, acting in a drama and watching a drama, telling a funny story and laughing at it, exhibitionism and voyeurism. Whether other elements (such as vicarious identification with the hero in a play, or catharsis of aggression in certain types of humor) are present or not, the essence of the presentation lies in its form or style. For example, a joke that belittles a disliked person will not evoke laughter if it is ineptly told. Considering the number of hours that millions of Americans spend at the movies or before a television set, it is strange that neither this type of activity nor its complement, the activity of the actors, has been given a place in catalogues of behavior. Perhaps one reason for this omission is that we have been slow in realizing that the mind is not always bent on serving some somatic or social need, but has its own peculiar occupations and enjoyments (see 19. *Mental needs*).

18. *Directional activity.* Most dynamic psychologists have assumed, explicitly or implicitly, that the most important thing to observe, to define, and to represent in formulating any single proceeding or any single serial in the career of a person is the superordinate, or major, *directionality* of its activity, whether this activity be chiefly physical, verbal, or mental. Instead of *directionality* one might say the *effect* of the activity, if it were not that we often wish to characterize the behavior of a person when no significant effect has been produced. Either 1), the man is "on his way" to some effect but has not had time to achieve it, or he has not achieved it 2), because of lack of relevant knowledge or ability, or 3), because it was a "competitive effect" and his efforts were surpassed by one or more rivals.

Since directional activities vary in respect to their intensity and duration, their degree of co-ordination and focality (definiteness of aim), it has been judged necessary to conceptualize a directing superordinate "force" in the brain region (merely analogous to a physical force). It has been variously named—tendency, drive, need, propensity, instinct, urge, impulse, desire, wish, purpose, motive, conation, resolution, intention, etc. The number of different, yet more or less synonymous, terms which have been used to designate this hypothetical force suggest confusion of thought and/or the absence in our language of the precisely definitive word. Anyhow, the conception is that of a vector (directional magnitude) which guides mental, verbal and/or physical processes along a certain course.

Conforming with Lewin and many others, we may use the term "need" or "need disposition" to refer to the roughly measurable "force" in the personality which is co-ordinating activities in the direction of a roughly definable goal; and we may use the term "aim" to refer to this need's *specific* goal (to be achieved perhaps in association with a *specific* object in a *specific* place at a *specific* time). For example, a man may be motivated by a general *need* for dominance (power, authority, leadership, a decision-making role, an administrative position, etc.), but his *aim* at a particular time may be to persuade residents of Bordeaux to elect him mayor of that city.

In a need of this class the satisfaction comes at the end with the attainment of the goal (effect pleasure), and perhaps also along the way from anticipations of the goal (anticipatory effect pleasure).

Thus, "directionality" means the *trend* of the activity and this can be defined only in terms of its *aim*, or *goal*, or *intended effect*. Since Freud we have not been troubled by the absence of a *conscious* aim in the actor's (subject's) mind. It is now scientifically respectable to speak of an *unconscious* aim, or goal, or intention if sufficient evidence for it is at hand. It is only necessary to exclude those cases where the effect was clearly a mistake, or accident.

In America it is customary for psychologists to take the hunger drive (food-seeking and eating behavior) and the sex drive (mate-seeking and copulating behavior) in animals (rats, monkeys) as basic illustrations of motivated activity. This practice has helped to clarify certain issues, but by restricting reflections to two kinds of relatively simple tendencies, it has served to discourage efforts to conceptualize other kinds of activities, especially "higher order" activities. These last will have to be investigated before much progress can be made towards a satisfactory classification of human tendencies.

19. *Mental needs*. The infant's mind is not acting most of the time as the instrument of some urgent animal drive (as many psychologists assume), but is preoccupied with *gratifying itself*. Among these purely mental gratifications of early life is the need to dramatize everything, to invent another world that is more exciting and hence more "real" than the reality of sheer perception. This inner world of childhood provides the basic stuff out of which are shaped forms of play, myths and religious rituals, songs and dances, folk-tales and poetry, and even, at a later date, philosophies,

scientific theories, and technical inventions. When the child is not fabricating its own stories, it is listening with rapt attention to stories told by others, or reading the comics, or watching television. In childhood, furthermore, these dreams are influential in determining what is perceived in the external world and how it is interpreted. Figures of the imagination are constantly projected into objects of the environment, into animals, into sun, moon, and stars, into trees and streams, the deeps of the sea, and the expanse of heaven.

All this sustains the proposition expounded years ago by Santayana: in the human being it is *not* perception that is fundamental, but imagination. It also conforms to the views advanced more recently by Suzanne Langer: the human mind is inherently a transforming, creating, and representing organ; its function is to make symbols for things, to combine and re-combine these symbols incessantly, and to communicate the most interesting of these combinations in a variety of languages, discursive (referential, scientific) and expressive (emotive, artistic).

In short, the mind lives in two worlds—the world of the imagination (religion, art, and science) and the world of perception and practical action. In some areas the two may coalesce; in other areas they may be poles apart. Of these two worlds, it is the first—the inner world of heaven and hell, of literature and music, of philosophies and scientific speculations (cultural productions)—which chiefly distinguishes human beings from the predominantly practical lower organisms. By satisfying the "higher" mental needs, the internal world may be more compelling than the external world in which the prepotent viscerogenic and socio-relational needs are gratified. The mental needs are a good deal more important in the lives of intellectuals—scientists, artists, philosophers, religionists—all of whom (positivists included) are more attached to their abstractions, theories, fables, poems, or mystical visions than they are to any facts.

Furthermore, the mind's imaginations—especially in their religious and artistic forms—not infrequently provide fantasied satisfactions of viscerogenic and socio-relational needs (in addition to the real satisfactions of mental needs).

In insanity a person does not "go out of his mind," but wholly "in his mind" (as Santayana said) and out of the external world. At the other end of the detachment-attachment continuum are the people who apply their imaginations to the solution of environmental problems. Creative mental needs are peculiar in having no clearly envisaged target or goal, the goal being something that has never existed (new machine or gadget, new social group, new political constitution, new dramatic epic, new scientific hypothesis or theory, new philosophy). The goal is something that must be constructed step by step. This might be compared to nest building in animals.

20. *Diagnosis of needs*. Some critics have objected to the need theory on the grounds that one cannot immediately tell them which need is being exhibited by a given person at a given time. But if anything should be clear

it is that needs are not discernible facts. A need is an intervening variable, hidden in the head, the operation of which can only be inferred on the basis of certain criteria. Hence, the task of identifying an active need is not that of labeling the kind of behavior that is observed, but of making a diagnosis. Sometimes the diagnosis is easy: the need is almost as obvious as the movements and the words. But, often, it is impossible to decide, even after hours of investigation, whether or not this or that disposition was operating during the observed event.

21. *Dissatisfaction, satisfaction.* One of the strangest, least interpretable symptoms of our time is the neglect by psychologists of the problem of happiness, the inner state which Plato, Aristotle, and almost all succeeding thinkers of the first rank assumed to be "the highest of all goods achievable by action." Although the crucial role of dissatisfaction and of satisfaction is implicit in much that is said about motivation, activity, and reinforcement, psychologists are generally disposed to shun these terms as well as all their synonyms.

Actually, satisfaction is an affective state which is likely to manifest itself *objectively* as well as *subjectively*. It is no more difficult to diagnose than anxiety or anger. It should be systematically investigated, because it is the most refined sign that we have of whether need processes are being obstructed, advancing without friction, or attaining their aim, or, after cessation of action, whether the effect produced did in fact appease the need.

It might be helpful to conceive of a physical variable (H) in the brain (possibly the thalamus) the concentration of which varies along a continuum, just as the hydrogen ion concentration of the blood varies along the acid-alkaline continuum. At the lowest end of the continuum H manifests itself *subjectively* as a feeling of extreme dissatisfaction (dejection, depression) and *objectively* by certain readily distinguishable postures, grimaces, gestures, and expressive sounds. At the highest end of the continuum we get extreme satisfaction (elation, joy) combined with its objective symptoms. At the middle of the continuum would be the zone of indifference. If we call H "hedone," and its continuum, the "hedonic scale," we can say that every occurrence which moves H (hedone) *down* the scale is *hedonically negative* and every occurrence which moves H (hedone) *up* the scale is *hedonically positive*. With this terminological framework we can go on to study the kinds of events that are hedonically negative and the kinds that are hedonically positive.

There is not room here to discuss the pain-pleasure principle. All we can do is reiterate the antique proposition (which research over the years has verified as often as any other psychological hypothesis or dictum), namely that the aim of all needs is hedonically positive (in the imagination). That is to say, the imaged goal of all activity is *believed* to be associated with more hedone (less dissatisfaction or more satisfaction) than the field in which the individual finds himself at the moment. This belief is no more than a prediction. The man may be mistaken. After attaining his goal the expected satisfaction may not occur, which means among other things that the discrimination of goals (values) is something that must be learned through experience.

This is not psychological hedonism in the sense that behavior can be understood as riddance of pain and pursuit of pleasure, because we are never dealing with pain *in general* or pleasure *in general*. The most that we can say is that behaviors tend to reduce or are calculated to reduce a *kind* of dissatisfaction (thirst, hunger pangs, sexual tensions, destitution, loneliness, inferiority feelings, disgust, and so forth), and that they tend to attain and to prolong a certain *kind* of satisfaction. In other words, a certain kind of dissatisfaction (e.g. fullness of the bladder) cannot be relieved by *any* kind of satisfaction (e.g. congenial discourse).

Need, then, is the fundamental variable, and degree of satisfaction (hedone) the best indicator of its state or progress.

22. *Social commitments, role actions.* As the sociologists have shown us, the concept of role is strategic to the integration of the two levels of theoretical analysis, psychological and sociological. To make this clear we might stretch this concept, for the moment, beyond its intended limits and say that every self-and-body, in order to develop, maintain, express, and reproduce itself, must perform a number of *individual* roles (functions) such as respiration, ingestion of food, construction of new tissue, excretion, defense against assault and disease, expression of emotions and sentiments, copulation, and so forth. Likewise, it can be said, that every group (social system), in order to develop, maintain, express, and reproduce itself, must perform a number of *social* roles, such as recruitment and training of new members, hierarchical organization of functions, elimination of incorrigible members, defense against attack by rival groups, expansion by reproduction (formation of similar groups in other parts of the world), and so forth. Also, both persons and social systems are devoted to the accomplishment of one or more further purposes, such as the manufacture and exchange of utilities, the acquisition and communication of knowledge, the creation and performance of plays (on the stage, through moving pictures), correction of delinquents, the subjugation of enemies, and so forth. Finally, both persons and social systems (each taken as a consensus of intentions) are desirous of improvement, of living up to their ideals, of deserving recognition and prestige. In the personality it is the governing ego system which assumes responsibility for the integration of *individual* roles and the actualization of plans. In the group it is the leader or government (system of legislators and administrators) that assumes responsibility for the structuring of *social* roles and the carrying out of policies (domestic and foreign). The id of the personality is somewhat comparable to the disaffected low status members of a social system, the "unwashed masses," including the "creative minority" (Toynbee), the radical reformers and fanatics, as well as the criminals and psychotics. Every structured ego "holds a lunatic in leash" (Santayana).

Thus, by extending the concept of role (social role) to include personal roles, a personality action system and a social action system can be represented as roughly homologous, at least in certain respects.

Furthermore, all social roles require the execution of one or more kinds of actions, that is, the habitual production of one or more kinds of effects, and these effects (goals) can be classified in the same manner as needs are classified. Indeed, the need-aim and the role-aim may exactly correspond. For example, a gifted actor with a need for artistic expression may be asked by a theater manager to play Hamlet (the very part which has long excited his ambition). Thus a man may *want* to do exactly what he is expected to do. But, so happy a congruence of "want" and "must" is, in the lives of most people, more of an ideal than an actual daily occurrence.

The chief reason for the frequent discrepancy between desire and obligation is the differentiation of society into sub-systems, and the differentiation of sub-systems into specialized and *temporally integrated* role functions, and, finally, the necessity of *committing* men to the scheduled performance of these functions. It is not so much that a man is obliged (expected) to do certain things, but that he is obliged (in order to integrate his actions with others) to do them at a *fixed time*. Consequently, it may happen that a man eats when he is not hungry, converses when he feels unsociable, administers justice when he has a hangover, makes a speech when his head is bereft of enlivening ideas, goes to the theater when he wants to sleep. Thus, in many, many cases, a need is not the initiator of action, but the hands of the clock. Spontaneity is lost, and will-power (an unpermitted concept) must be constantly exerted to get through the days.

The point is that here every action is an instrumental one, not satisfying in itself. Instrumental to what? This varies from individual to individual: the need for sheer survival and hence the need for money, the need for upward mobility, or for fellowship, or for authority, or for prestige; but, more generally and more closely, the *need for roleship*, that is, the need to become and to remain an accepted and respected, differentiated and integrated, part of a congenial, functioning group, the collective purposes of which are congruent with the individual's ideals. So long as the individual feels this way about the group that he has joined, he will try to abide, as best he can, by its *schedule* of role functions.

23. *Generality and specificity: diffuse and focal needs.* A need is a general disposition which commonly becomes associated (through "focalization," or "canalization," as Murphy would say) with a number of *specific* entities (e.g. a certain doll, or dog, or person, or group, or town, or theory, or work of art, or religion, etc.), and (through "generalization") with a number of *kinds* of (*semi-specific*) entities (e.g. French wines, or horses, or women, or music, or novels, or philosophies, etc.).

These focalizations (specificities and semi-specificities) rarely exhaust the possibilities of need activity. Unless the structure of the disposition has become rigid and fixated, it is always capable of becoming attached to a new object—new kind of food, new place, new acquaintance, new organization, new kind of art, new ideology. Indeed, the development of personality can be partially represented by listing, in chronological order, the attachments it has acquired and outgrown in the course of its career (e.g. a series of material objects, such as a rattle, toy truck, mechano set, bicycle, motorcycle, automobile, airplane; or a series of aesthetic forms, such as nursery tales, adventure stories, Dickens, Tolstoy, Sophocles). Attachments (sentiments, attitudes), then, may be more or less enduring.

A permanent attachment to a nurturant person (e.g. mother fixation) dating from infancy is regarded by psychoanalysts as a sign of emotional immaturity. Equally indicative of retardation is its apparent opposite: lack of enduring attachments, that is, the inability to remain loyally committed to anybody or anything. Here we might speak of a *diffuse* need which is sensitive to a large number or variety of objects (e.g. free-floating sociability, free-floating anxiety, free-floating irritability) in contrast to a *focal* need which is enduringly centered on one object (e.g. satisfying marriage, specific phobia, canalized revenge).

As we have noted earlier (see 13. *Cathected objects*) an entity (material object, person, group, political policy, philosophy) to which one or more needs have become attached is said to have value, or *cathexis* (power to excite). A *liked* entity which attracts the subject is said to have *positive* cathexis whereas a *disliked* entity is said to have *negative* cathexis. An entity with negative cathexis (negatively cathected object) may evoke avoidant reactions, defensive reactions, or destructive reactions. Thus a goal object may be either a positively cathected (loved) person or a negatively cathected (hated) person whom the subject wants to subdue, injure, insult, or murder. Entities (places, animals, persons, topics of conversation) which the subject wishes to avoid (withdraw or flee from) may be called *noal* objects. The distribution in any personality of these different kinds of cathections (evaluations, sentiments, attitudes) is correlated with the relative potency of the three vectorial dispositions so well described by Horney: moving toward people (positive goal objects), moving against people (negative goal objects), and moving away from people (noal objects).

In a developing individual, positively cathected goals and goal objects (kinds of values) are of two classes: 1), satisfying goals which have been experienced more or less regularly week after week, and 2), goals which have not yet been attained, but which sway the imagination and orient the planning processes of the mind. In sanguine temperaments, visions of future goals (grass on the other side of the fence) are likely to be given higher *value* than goals which are attainable every day. This applies especially to the creative needs, the goals of which are always novel, never-yet-constructed entities.

We can affirm that goals of the above-defined first class are learned, but can we say this about the more greatly esteemed goals of the second class? Perhaps it can be said that these are learned by "trial and error" in the

imagination. Many of them, we know, become established through identification with some exemplar, in conformity with cultural expectations,
and are therefore to be subsumed under the heading of social imitation or
mimesis. In any event, they are not innately given.

The fact that the word "learning" (when used by an American psychologist) refers almost always to the process of acquiring effective instrumental action patterns, might, I suppose, be cited as another illustration of how a prevailing ideology (e.g. the high valuation of technical
skills, "How to Make Friends and Influence People," etc.) can influence
the course of our supposedly chaste science.

More important than means-end learning, of course, is goal and goal-
object learning, that is, the process whereby an individual comes to some
conclusions as to the relative *values* of different possible goals and goal
objects, or, looking at it from a developmental or educational standpoint,
the process whereby he learns to enjoy (and so discovers for himself) the
kinds of goals and goal-objects which are worth striving for.

24. *Attitudes, interests, values.* Social psychologists looking in at themselves and looking out at others concluded some time back that a great
deal of human thought and behavior is value-oriented. It seemed, for example, that the psychologist's own intellectual preoccupations, conversations, and activities could be largely explained by stating that he was
interested in theories of personality and social behavior, *interested* in research, *interested* in educational procedures, *interested* in political policies,
and so forth. These regions of thought and action were highly *valued*. He
had a *positive attitude* towards them. More specifically, his attitude was
positive towards *certain* concepts (e.g. traits), certain *kinds* of research
(e.g. opinion polling), certain *kinds* of educational procedures (e.g. small
research seminars), certain *kinds* of political policies (e.g. labor legislation), and so forth. "Attitude," as defined by Allport and others, became
the social psychologist's key concept. A man's personality was conceived
as a more or less integrated system of attitudes, each of which is a relatively permanent disposition to evaluate some entity negatively or positively, and, as a rule, to support this evaluation with reasons, or arguments.
The general or specific entity (object of the attitude) *was a value*, positive
or negative, or *had* value (power to attract or repel).

(Here it might be said parenthetically that "attitude" and "sentiment"
are synonyms, both of them referring to a more or less lasting disposition
in the personality; also that "value" in one sense is synonymous with
"cathexis" and "valence," and, in the other sense, with "cathected entity"
and "object with valence." There is always a representation (image) of
the valued, or cathected, entity *in* the personality.)

In the judgment of most social psychologists there is a big difference or
gap between attitudes (vaguely defined as dynamical forces) and needs
(or drives), probably because the latter are fixedly associated in their
minds with hunger and sex. They are willing to concede that there are
such "pushes from the rear" as hunger and sex, but these are said to be

"segmental" and "lower." Attitudes, on the contrary, are consciously and rationally selected "pulls from the front." The principle of functional autonomy is usually invoked to sever whatever associations might have existed in the subject's past between needs and attitudes.

Assuming that this greatly telescoped account of the concept of attitude conforms roughly to the conventional notion, we submit that two modifications are required before it can be assigned an important place in a theoretical system. First, its scope must be extended to include all affect-invoking entities from faeces to the Fallen Angel. There is no reason to limit the term to dispositions towards "higher" entities. We must be able to speak of attitudes towards the mother's breast, the mother herself, siblings, spinach, mechanical toys, play group, comics, fairy tales, and so forth. Second, the concept must be logically connected with directional activity, that is to say, with the concept of need. Knowledge that a certain person has, let us say, a positive attitude towards music does not tell us what he *does* about it, if anything. Perhaps he values music very highly (as checked on a questionnaire), but is too busy to listen to it. Or, to take the opposite extreme, he may be a composer of music. If we are not told, the man does not become alive. We cannot guess how he spends his day, or predict what he would do at a certain choice-point. Does he keep music alive in his head by humming it? Does he discuss music with other appreciators and defend the excellence of his favorite concertos? Does he play some instrument, privately for his own satisfaction, or publicly for the satisfaction of others? Is playing in an orchestra his professional role, his path to money and eminence? He may be a music critic, a writer of books on music, or a singing teacher. Or, is he merely an enjoyer of music as it comes over the radio? If a high evaluation of music is linked with one or more vectorial dispositions (such as reception, construction, expression), we can represent what a given appreciator *does* with music; we can picture many proceedings in his life. Otherwise, one has a dangling abstraction.

Now, having pointed to the beam in the eye of the attitude construct, it is time for us to acknowledge the mote in the conception of need: *not all needs have been defined as dispositions operating in the service of a certain kind of value.* Take, for example, anger and aggression (the goal of which is partial or complete destruction of some object). This is usually defined as a general drive or need which is likely to be provoked by frustration. In the discussion of a case, a psychologist will usually mention the object (e.g. mother, father, sibling, enemy) towards which the aggression is directed, but rarely, if ever, do we find any reference to the value in the service of which aggression has operated. Since observation and experience testify to the fact that aggression, as well as every other kind of action, has an effect (function) which can be best defined in terms of some valued entity (its construction, conservation, expression, or reproduction), the naming of the valued entity in conjunction with the named activity should contribute a good deal to our understanding of the dynamics of the behavior. Aggression, for example, may serve to defend the *body* from injury, to establish an *erotic relationship* (when obstructed by a rival), to acquire more *property*, to maintain *authority*, to increase one's area of *freedom*, to depreciate the over-all values of others in order to make way for the acceptance (reproduction) of one's own *ideology*, to revenge an insult in order to restore one's self-respect, or *prestige*, and so forth.

This means that both valued entities and action tendencies must be classified. Following Lewin and Erikson we are calling the action tendencies *vectors*, each of which is a physical or psychological *direction* of activity, such as 1), *rejection*, 2), *reception*, 3), *acquisition*, 4), *construction*, 5), *conservation*, 6), *expression*, 7), *transmission*, 8), *elimination*, 9), *destruction*, 10), *defendance*, and 11), *avoidance*. The classification of the valued entities presents more difficulties. Merely as an illustration the following might be listed: A), Body (physical well-being), B), Property (useful objects, wealth), C), Authority (decision-making power), D), Affiliation (interpersonal affection), E), Knowledge (facts and theories, science, history), F), Aesthetic Form (beauty, art), G), Ideology (system of values, philosophy, religion). These are the six well-known Spranger values, with the addition of Body (which includes sex and reproduction as well as all the other viscerogenic needs). Others should be included, but this number is sufficient to indicate how each *kind of value* and each *kind of vector* can be combined to form a manageable number of *value-vectors*. For example, a man might serve his knowledge, E (valued entity), by 1), *rejecting* irrelevant or inaccurate observations, 2), passively *receiving* reliable information, 3), actively seeking and *acquiring* more facts, 4), *constructing* an adequate theory, 5), *conserving* (remembering, recording, and utilizing) the knowledge he has acquired, 6), *expressing* his thoughts, 7), *transmitting* (implanting) his theories into the minds of others, 8), *eliminating* (from his mind) fallacious notions, 9), *destroying* rival (erroneous) theories, 10), *defending* his conceptions, or 11), *avoiding* harmful criticism by concealing his ideas or by escaping from the field of discourse (changing the topic). According to this conception each value-vector is a need.

Such a scheme can be used to distinguish the action trend (*need-aim*) in the subject as well as the action trend (*press-aim*) in the object. The two in sequence might be said to constitute the *thema* of a proceeding. For example, the proaction of the subject could be that of expressing (6) an idea (E) and the reaction of the object might be that of rejecting (1) the idea (E): Thema: sE6→oE1. For a more complete statement of the thema other variables (cathexis, status, conformity to role) must be added to each side of the formula.

This scheme is not yet ready to be applied to the topics which will be discussed next. It is still in process of construction, tentative and incomplete.

25. *Id, Ego.* Up to the time of Janet and Freud the human mind was al-

most wholly equated with consciousness, with rationality and voluntary action. Psychology was restricted to "ego psychology." But the discovery of dissociated subconscious complexes in patients with hysteria, and, then, of primitive unconscious establishments and processes in all people, normal as well as abnormal, called for a greatly expanded conception of the mind.

Freud's first distinction was between the conscious parts and the unconscious parts of the personality. Neuroses were products of conflict between the former and the latter. But this distinction was abandoned by Freud as soon as he concluded that the unconscious processes of repression and resistance operated in the service of the conscious parts of the personality. This led to a new distinction, that between the *ego* and the *id*, the latter consisting of a medley of unconscious pleasure-seeking processes that were unacceptable to the ego, and hence subject to repression. The ego was the conscious, objective, rational, and reality-oriented part of the personality. It was supported by a variety of unconscious "defense-mechanisms."

In due time it became apparent to other analysts, if not to Freud, that the concept of id could not be limited to unacceptable dispositions. In infancy, for example, when the ego system is non-existent or at best very rudimentary, the mind is a hive of involuntary spontaneities, emotions, and needs, many of which are not only acceptable to the child and its mother during these early years, but continue to be acceptable and, what is more, culturally encouraged throughout life. It would not be proper to say that respiration, ingestion of food, defecation, expressions of affection, endeavors to master the environment, and so forth, had their sources in the ego. Also, as Jung, Rivers, and others have pointed out, the id is evidently the breeding ground of love and worship, as well as of the novel imaginations which are eventually applauded, instituted, and cherished by society. For these and other reasons, it seems best to think of the id as consisting of all the basic energies, emotions, and needs (value-vectors) of the personality, some of which are wholly acceptable and some wholly unacceptable, but most of which are acceptable when expressed in a culturally approved form, towards a culturally approved object, in a culturally approved place, at a culturally approved time. Thus, the function of the ego is not so much to suppress instinctual needs as to govern them by moderating their intensities and determining the modes and times of their fulfillment.

Some of the criteria of ego structure, or ego strength, are the following:

A. *Perception and apperception.*
  1. *External objectivity:* the ability to perceive human actions and events without distortion, to analyze and interpret them realistically, to predict the behavior of others.
  2. *Internal objectivity:* the capacity for self-detachment and self-analysis; insight into one's own motives, evaluations, and emo-

tional reactions; also, the entertainment of a goal of personal development and accomplishment which is suited to one's own circumstances and capacities.

3. *Long apperceptive span:* the habit of making causal connections between events that are not temporally contiguous in experience; the ability to foresee broad or distant consequences of one's actions (time-binding power or long time-perspective).

B. *Intellection.*

4. *Concentration, directionality:* the ability to apply one's mind to an assigned or selected topic, to direct one's thoughts along a chosen path, to persist when bored, to inhibit day-dreaming.

5. *Conjunctivity of thought and speech:* the ability to think, speak, and write clearly, coherently, and logically, to inhibit irrelevant ideas.

6. *Referentiality of thought and speech:* the habit of using concepts and words which refer to real things, events, and experiences; the absence of vague, undefined, and essentially meaningless terms and expressions.

C. *Conation.*

7. *Will-power:* the ability to do what one resolves to do and is capable of doing, to persist in the face of difficulties, to complete a prescribed or elected course of action; also, to re-strive after failure (counteraction).

8. *Conjunctivity of action:* the ability to schedule and organize one's activities, to make a plan and follow it, to live an ordered life.

9. *Resolution of conflicts:* the ability to choose between alternative courses of action. The absence of protracted periods of hesitation, indecision, vacillation, or perplexity.

10. *Selection of impulses:* the power to repress temporarily, inhibit, or modify unacceptable emotions or tendencies, to resist "temptations"; also, the habit of selecting and expressing, without qualms or conflict, impulses which are intrinsically enjoyable or extrinsically rewarding; absence of disturbing worries or anxieties.

11. *Selection of social pressures and influences:* the ability to choose among the demands, claims, enticements, and suggestions that are made by other people, to comply with those that are acceptable and reject those that are not; especially the power to resist intolerable coercions from society, but to submit if there is no way out; power "to will the obligatory."

12. *Initiative and self-sufficiency:* the ability to decide for oneself and act without waiting to be stimulated, urged, or encouraged. The habit of trusting one's own nature, of having reasonable confidence in one's own decisions (self-reliance). Also, the ability to stand alone, to do and finish things alone, without help; to endure solitude and to tolerate misfortune without appealing for sympathy; absence of marked dependence on others.

13. *Responsibility for collective action:* the willingness and ability to take responsibility and effectively organize and direct the be-

haviors of others; the experience of feeling secure in a position of authority, rather than being threatened, worried, and on the defensive.

14. *Adherence to resolutions and agreements:* the disposition and ability to abide by long-term decisions and commitments, to keep a promise or pledge.

15. *Absence of pathological symptoms:* freedom from incapacitating neurotic or psychotic symptoms.

Since the ego system is the differentiated governing establishment of the personality, it is not possible to estimate its power (relative to other ego systems) without some knowledge of the strength of the id forces with which it has to cope. Some "egos" are sitting in the saddle of a docile Shetland pony, others are astride a wild bronco of the plains.

The chief criteria of id strength are these: 1), intense energy, zest, spontaneity, enthusiasm; 2), intense needs and appetites; 3), intense emotions; 4), abundant, vivid imaginations. As stated above, some of these needs, emotions, and imaginations are accepted, enjoyed, and cherished; others are unacceptable and, if possible, repressed. Of the latter, Freud has stressed the need for sex (deviant forms of it especially) and the need for aggression, that is, lust and wrath, two of the Catholic Church's seven Cardinal Sins. But equally important in American culture is the need for acquisition (greed) and the need for superiority or prestige, manifesting itself as ruthless competitiveness, vainglorious self-display, extravagant pride, boasting, and envy of the accomplishments or statuses of others. All of these tendencies must be inhibited, appropriately moderated, or modified.

A weak id is conducive to another of the seven sins, sloth or passivity. In extreme cases one speaks of neurasthenia or psychasthenia.

26. *Superego, ego ideal.* According to Freud's conception, the *superego,* or conscience, is a part of the personality which interiorly represents and enforces the moral imperatives of the society insofar as these have been implanted by the parents and parent-surrogates during childhood. It is an establishment which is gradually constructed by the internalization of the system of rewards for "good" behavior and punishments for "bad" behavior that is practiced by each of the succession of respected authorities—parents, nurses, teachers, priests, etc.—who assume responsibility for the education of the child's character. If socialization has been completed, the superego will privately reward and punish the self in conformity with cultural standards and in this way replace the authorities who inculcated these standards. Thus, the concept of the superego provides the necessary link between the laws and customs of a society as a whole and the conforming behavior of its individual members. Each socialized person bears in his superego the value system of his society, and in terms of this system he is disposed to judge others and to discipline his children. This over-simplified account of the superego must, of course, be radically qualified and revised.

In the first place, the superego develops by stages, stratum upon stratum. In its most primitive form it consists of no more than projections of the child's own oral acquisitive and aggressive tendencies, represented, say, by a vampire, werewolf, or gigantic biting animal, such as Moby Dick himself, or, possibly, by a child-devouring witch or deity such as Cronos. In the child's imagination "totem" animal figures are gradually replaced, as they were in the evolution of religions, by human figures, first, perhaps, by an affrighting Jehovah image, and then usually by a more kindly symbol of moral authority. These are "father" figures, for the most part, but in some cultures, and in some personalities in our culture, the superego is more mother-oriented, represented sometimes by an androgynic figure. Here we are speaking largely of the lower layers of superego formation, marked by archaic, semi-human, and mostly punishing images. As socialization proceeds in Western cultures these strata, overlaid by others, become unconscious, and in later life are manifested only in dreams, artistic productions, and insane delusions.

Contemporaneous with the development of these imaginative, figurative, mythological, or religious aspects of the superego, one finds the gradual structuration of behavior according to certain abstract moral imperatives, transmitted by parental example, suggestion, persuasion, promises, rewards, threats, or punishments. At one extreme, these imperatives amount to scarcely more than a medley of arbitrary and inconsistent injunctions or aggregate of heterogeneous apothegms; at the other, they constitute a fairly rational ordering of ethical principles. But even if the system of rewards and punishments practiced by the parents is reasonably clear and coherent in the child's mind, there are bound to be serious conflicts *within* the superego as soon as the boy or girl goes to a school or becomes a member of a clique in which a somewhat different code of conduct is respected and enforced.

The reciprocative and co-operative interactions among members of a peer group constitute, as Piaget has pointed out, the second major source of superego constituents. No doubt there have always been antagonisms between parental standards and peer group standards, but in recent years there has been an almost world-wide exacerbation of the conflict, resulting in a step by step compliance, if not surrender, by each parental pair to the mass demands of the adolescent peer group for an earlier emancipation and a greater range of freedom, especially for young girls in the sphere of sex. Since in America the youth culture is unanimous only in its opposition to authority, lacks a unifying ideal, and is split in several ways by rivalries, the superegos of its members are, for the most part, in a state of flux, if not of *anomie*.

One more class of superego determinants should be mentioned—literature. An intelligent youth who reads widely will be exposed to a great variety of contrasting moral judgments: the Bible, Aristotle, Epicurus, Seneca, Montaigne, Spinoza, Hobbes, Rousseau, Jefferson, Nietzsche, Whitman, Tolstoy—each may contribute something to his maturing superego. Since the majority of mortals are incapable of making order out of such diversity, and since among Protestants and Jews there is no body of philosophers officially delegated to create and re-create such order, it is only in members of a totalitarian system—Catholic, Communist, Fascist—that one finds considerable agreement in respect to basic assumptions and rationalized derivations.

As a result of these several factors the disposition to adjust one's behavior to accord with the evaluations of one's peers, fitful as they may be, is gaining ground in America, whereas, the strength of the elevated superego with most of its manifestations—inwardness, moral aspiration, righteous indignation, guilt, remorse, penitence, reformation, and so forth—is in process of decline.

This recent weakening of conscience is all of a piece with Freud's conception of it as a chiefly prohibitive, repressive, life-denying agency. In short, the superego, or puritan conscience, of the Western world lost its "charm" (to use Toynbee's apt term) sometime ago and became an establishment of bearded threats, a purely negative Maginot Line which was destined to succumb in due course to the onslaughts of freedom-loving moral revolutionists.

The deterioration of the religious and ethical superego becomes apparent when it is compared to the intellectual or scientific superego. The latter is not only highly evaluated but it invites continuous creative efforts, attempts at re-definition, refinement, elaboration, and re-construction. In brief, instead of demanding more leniency from their superegos, scientists are generally vying with each other to see who can define and then conform to ever more exacting standards.

In childhood the principal conflicts are between insurgent id tendencies and parental standards. Later, if socialization advances, many conflicts are between id and superego (internalized standards). During these phases, the ego is hardly more than a battle ground, or possibly a puppet, dominated sometimes by the id and sometimes by the superego. Eventually, however, guided by an exemplar, or an envisaged ideal self (ego ideal), the ego becomes more differentiated and integrated, more self-conscious and self-reliant, until eventually it acquires enough power to arbitrate to some extent between emotional impulses from the id and superego imperatives.

At best, the superego is an inviting conception of an ideal future world, or at least a better world, to be constructed by stages, and its imperatives define those modes of behavior which are most conducive to that end. The ego ideal, on the other hand, is an inviting conception of an ideal future self, or at least a more able or better self, also to be attained by stages. If integrated with the superego, the ego ideal will portray an individual who is promoting in some small or large way the development of a better world. If not so integrated, the ego ideal may consist of images of the Master Criminal or of any other wholly egotistical "successful" person.

3. FORMULATIONS OF PERSONALITY

A. *Subjective and objective facts for the formulation of a personality.*
Since we cannot observe the transient regnant processes in the brain, these
must be distinguished (whenever necessary) by retrospection and reported to the psychologist or communicated as they pass in the stream of
consciousness. But since it has been proved that not all regnant processes
have the attribute of consciousness, those that are not conscious must be
inferred from what the subject says and does. The data, then, consist of
subjective facts reported by the individual, and of *objective facts* observed
by the psychologist or by others.

The obtainable subjective facts are many and various. Besides the
above-mentioned communications of current mental processes, the subject
is capable of reporting (with more or less accuracy): countless memories
of past proceedings (internal and external); his failures and successes; estimates of his past and present valuations and attachments (attitudes and
sentiments); his past and present memberships and commitments; his past
and present fantasies, plans, hopes, and expectations; estimates of his major
needs (motivations); traits, and abilities of significant figures (persons who
have affected the course of his development); and a multiplicity of other
impressions and self-assessments.

Instead of crippling himself by renouncing this source of invaluable
information, limiting his data (as some scientists advise) to the behaviors
which he himself can observe and faithfully record, the psychologist
should discover the principal kinds and sources of errors in subjective
communications and, as far as possible, correct for these in each case
studied.

The obtainable objective facts consist largely of observations of daily
uncontrolled proceedings, of lifelike (though somewhat artificial) situational tests, of laboratory reactions to standard stimuli, of so-called "projective" tests, and of tests of aptitude and ability. Since the observation
and interpretation of these behaviors are liable to all the distortions that
can occur in the human mind, the perfection of the psychologist as "in
strument of precision" is, if anything, more important than the perfection
of technical devices.

Furnished with a sufficient supply of the above-listed kinds of data, the
psychologist attempts to discriminate the most significant uniformities of
action and of accomplishment during each period of the subject's (or patient's) development. He will also take note of exceptional past events and
attainments: those which were followed by modifications of behavior and
those which exhibit the subject's range of variability, his so-far-reached
limits of failure and success. The bulk of the data, however, will inevitably
consist of observations of present-time actions and achievements, some
of which, to the psychologist's discomfort, may be more exceptional than

habitual. In any event, the psychologist, reviewing his selected facts, will then attempt to arrive at an integrated conception of the on-going structure of the subject's personality.

B. *Purposes of formulation.* The term "personality" has been reserved for the hypothetical structure of the mind, the consistent establishments and processes of which are manifested over and over again (together with some unique or novel elements) in the internal and external proceedings which constitute a person's life. Thus personality is not a series of biographical facts but something more general and enduring that is inferred from the facts. A biographer might describe every significant thing that a man felt and thought and said and did from the start to the finish of his life without conceptualizing a single part of his personality. The subjective and objective activities of a personality continue without cessation from birth to death, and, were it possible to observe and formulate them all, the massive result would be as undesirable as it was unmanageable. It would be necessary to construct a *sufficient formulation* of the complete formulation. We might compare a complete formulation to the score of a more or less integrated series of musical pieces which took a lifetime to play. If it were possible, as in music, to represent with appropriate symbols twenty-four hours of activity on 2,400 pages of paper, making five compact volumes per day, there would be 1,825 for each year, and over 125,000 for a Biblical lifetime. Here, surely, filling a large library, would be the facts about one person, a truth that was too huge to serve. Thus we must accept a sufficient formulation as the proper goal; and this provokes the question: sufficient for what purpose?

In answer, we would say, and this is our next point, that the general purposes of formulation are three: 1) to *order and explain* past and present events; 2) to *predict* future events (the conditions being specified); and 3) to serve, if required, as a basis for the selection of effective measures of *control*. The validity of a formulation is to be found in the degree to which these purposes are fulfilled. But this statement is still much too broad. A formulation which would explain *every* occurrence in the past and permit predictions of behavior in *every* conceivable situation in the future, after exposure to *every* possible series of controls, would have to be a complete formulation, unwieldy and wasteful. In actual practice, one formulates a personality in accordance with a purpose that is more specific —to explain, say, a given set of facts, to predict effectiveness in this or that vocation, to foretell the success of marriage with a given person, to rectify a maladjustment, to eliminate delinquent behavior, to cure a cluster of neurotic symptoms, and so forth. Thus one is always dealing with a *partial* formulation that is meant to be sufficient for a special purpose. This fact accounts for some of the misunderstandings which occur among various specialists. The formulation of the personality of Mr. X arrived at by a psychoanalyst who is paid to cure his melancholia will necessarily be different from that reached by a vocational psychologist after administering some tests of aptitude, or by a social psychologist interested in political sentiments.

C. *Personality-as-a-whole.* As we have said, the psychologist never aims at a "complete formulation" of the "total" personality, but he usually does attempt to represent, in some way, the principal components of the "personality-as-a-whole." This commonly used term means the unity of emotional involvement and of action that is manifested 1), through one *proceeding;* or 2), through a long *serial* (a directionally progressive series of proceedings each separated from the last by an interval of time); or 3), through a *period* of life marked by a variety of endeavors and reactions.

Through one proceeding the criterion of unity is the effective co-ordination of muscles and/or words (coherence of speech). Through a serial, the criteria are continuity of purpose and co-ordination of the successive discrete proceedings. Through a period the criterion is the harmony and balance (absence of confusion, friction, discord, and conflict) that is manifested in the sequence of different kinds of behaviors.

Personality is not always "a whole"; that is, it is seldom perfectly integrated (completely unified). Since the course of life is punctuated by countless occasions when some choice must be made (between alternative, if not opposing, needs, goals, goal-objects, concepts, tactics, or modes of expression), indecisions and conflicts are common and final resolutions of conflicts are rare. Consequently, the psychologist is well-advised to include some account of his subject's major dilemmas and conflicts during critical periods of his life.

The most useful concept for representing the major components of the several on-going proactive systems of the personality-as-a-whole might prove to be that of need (value-vector), which gives both the kind of entity that is valued (e.g. material possession, knowledge, interpersonal relationship), and the kind of activity which is manifested in behalf of that valued entity. This information should, as far as feasible, be supplemented by characterizations of the cathected objects of each class of values toward which the needs are specifically oriented. In addition, some description of the temporal organization of the different on-going proactive systems is necessary in order to reveal their relative strengths as well as the harmonies and discords which prevail among them.

A representation of the consciously directed superordinate proaction systems would be tantamount to a description of the chief constituents of the ego system. One large proactive system, comprising several sub-systems, let us say, would be concerned with the fullfillment of the subject's vocational roles and aspirations—economic, political, legal, artistic, medical, or what-not. Another system might be concerned with his marriage, with a sub-system for each child; another with a cluster of friendships; still another with recreational pursuits or hobbies. Together, those would constitute the more or less differentiated and integrated activities and aims which engage the energies of the subject, the things which preoccupy him, which he daydreams about, plans for, and eagerly looks forward to. But these systems would not always be inviting and agreeable. The subject might be concerned with abandoning his vocation and looking for another job, divorcing his wife, revenging an insult received from a friend, or working over-time to pay a debt.

Also, there are the reactive systems which are evoked by unselected though fairly frequent situations (press), some of which are situations that the subject would like to forestall or avoid if possible. As a rule the psychologist limits his account of reactions to those which depart from the culturally expected norm, either in intensity or in frequency.

An account of the proactive and reactive systems would correspond reasonably well to the subject's physical, social, and mental behaviors and their conscious motivations, including the general and specific values which are furthered by these behaviors.

Next to be considered would be the unacceptable id dispositions—the anxieties, resentments, envies, revulsions, and depressions—which interfere with the harmonious and effective fulfillment of the needs. Experience has shown that these have their origin in unconscious complexes generated by infantile experiences and cannot be properly understood until the latter have been discovered. The system of infantile complexes—capable of explaining so many otherwise unintelligible dispositions and reactions in later life—has been called the *unity thema*. The schematic representation of the unity thema provides the depth dimension that is necessary to any comprehensive formulation of the personality.

Then, there are the id-bred fantasies and imaginations indicative of emergent (as-yet-unestablished) dispositions, of novel developments or endeavors in the possible future. These constitute the pioneering frontier of the expanding personality.

Finally a dynamic description of a personality would have to include a record of its typical effects over the years—its principal successes and failures in attempting to achieve goals and the enduring constructions which it has created. The effects produced by a personality are not, to be sure, the personality itself, any more than a newspaper is the printing press; but, just as a description of a machine that gave no account of what the machine could do would be judged deficient, so should any formulation of a personality be judged, which omits a report of its effects—attitudes and affections generated in other people, conservative and creative achievements, lasting contributions to culture, etc.

D. *Formal attributes*. Distinct from dynamic analysis and reconstruction, which sets forth the predictable actions of the personality under different specified conditions, is that type of analysis and reconstruction the aim of which is to characterize regnant configurations in a general way by representing the *formal* properties of typical units of activity. Although the *formal* variables and the *dynamic* variables of any temporal segment of behavior are mutually dependent, it is convenient to distinguish them, because the first, the formal variables, define the consistent properties of the interrelated regnant processes (such as perception, interpretation, affection, evaluation, decision, planning, expectation) which operate in every event regardless of its nature, whereas the second, the dynamic variables, define the kind of situation that is perceived and how it is apperceived, and how felt and evaluated, and how adjusted to or manipulated, and with what expectations of success, and so forth. Thus a dynamic analysis deals with constituents of the personality—such as certain needs (value-vectors) and emotions, certain parcels of knowledge, certain valued concepts, certain plans and expectations—which are not always operating in consciousness, but which consistently belong among the potential establishments or latent resources of the personality.

A *formal* analysis calls for the observation of such attributes as the speed, coherence, and congruity (application to reality) of verbal and motor activity. Also included in a formal analysis are such attributes as the typical duration of a functional unit (persistence of trends), harmony and coherence between consecutive units, ratio of sameness and novelty of units, flexibility or rigidity of patterns, ratio of internal (mental) over external (social) units of activity, ratio of directed purposive action to free, random, and expressive movements, degree of conformity to cultural patterns. Some of these qualities depend on such variables as energy-level, temperament, the subjectivity-objectivity balance, the introversion-extraversion balance, general intelligence, and so forth—many of which, though modified by environmental influences, are, in large measure, constitutionally determined.

4. MAJOR FUNCTIONS OF PERSONALITY

What are the functions of the personality? In physiology, the study of processes from this point of view has been justified by its fruitfulness, by repeated proofs that almost all the occurrences in the body *are* functional, in the sense that they contribute in some way to the maintenance, development, or reproduction of the organism. Cannon's account of the physiological structure of an emotional event, illustrative of the principle of homeostasis, is a case in point.

Unlike the physiologists, however, most psychologists are extremely wary of the concept of function. One reason for this is that many of the regnant processes which the psychologist studies, if judged by their actual effects, are not functional in the conventional sense; that is, they do not lead to psychological well-being, satisfaction, happiness, survival, but, instead, to pain and misery, and in some desperate people, to suicide.

Here we might remind ourselves that when the physiologist finds processes which interfere with the harmonious functioning of the body, or, as a result of the atrophy of some organ, notes the absence of certain processes which are necessary for health, he introduces the concept of malfunction. The psychologist might do likewise. The brain functions in an extremely complex, ever changing, and often hurtful, environment, and it rarely acquires the ability to do this in a manner completely satisfactory to the organism except, say, in certain rather simple uniform situations. Therefore, the psychologist has as much to do with processes which fail to achieve "beneficial" results and with processes which lead to "hurtful" results as he has to do with "successful" processes. It is because the other organs of the body "know" how to perform their relatively simple functions in a relatively stable internal environment that the physiologist finds that most of the processes he studies take a beneficial course. The personality has a more difficult assignment: it must learn how to improvise new and socially respected ways to deal with novel situations, and, perhaps, to deal with them more effectively than other people do, or, if not this, to lower its acquired craving for prestige. Because this is all so difficult, the psychologist finds that many governmental processes lead nowhere and the result of others is a hell on earth. This line of reasoning might help to explain many of the failures to function adequately, but it does not solve the psychologist's problem, because the notion of malfunction depends on an adequate definition of function, and the latter, in turn, depends on some formulation of an end state (such as equilibrium, or homeostasis) or of a temporal pattern of states towards which processes tend to take their course.

Another reason why the psychologist has not arrived at an acceptable conception of an end state is that, if there is such a thing, it is difficult to measure. Aristotle's assertion that the only rational goal of goals is happiness has never been successfully refuted as far as we know, but, as yet, no scientist has ventured to break ground for a psychology of happiness. In place of happiness, or, rather, recurrent satisfaction, psychologists have proposed a number of concepts, such as reduction of tension, but none is any easier to estimate in a human being than the amount of satisfaction. In the hope of solving this problem in an objective fashion, animal psychologists have sought some physical fact (such as diminished activity after feeding or copulation) which could be accepted as an invariable correlate of satisfaction, or of tension-reduction. This practice has proved expedient and might be extended to include the objective manifestations of these states in human beings, but its limitations are self-defeating. Objective facts are valuable because they compel unanimous acceptance and so fall readily into science, but a psychologist who confines his attention to such irrefutable data will have to exclude from his sphere of concern a wide range of phenomena and thus prevent himself from arriving at an adequate conception of personality.

A third reason why some psychologists have fought shy of the concept of function is that it is teleological (according to one definition of this word), and, so, raises from its grave the horrific image of final causes and of supernatural design. A few psychologists are still unconsciously so close to their possessive mother, philosophy, and their dogmatic grandfather, theology, that they are compelled to assert their autonomy and self-sufficiency in the most radical possible manner, by adopting the extreme position of nineteenth-century mechanism (as represented by the stimulus-response formula).

1. *Reduction of need tension*. Perhaps the nearest thing to an all-embracing principle, which avoids reference to final causes, and which, though difficult to measure is currently accepted, in one form or another, by many social scientists, is the concept of need, drive, or vectorial force. It avoids final causes because it points to an *existing* state of tension, a compelling uneasiness or dissatisfaction, a hypothetical disequilibrium, within the organism as the action-initiating state. This is a *present* "push" rather than a "pull" from a non-existent future. Faced by a novel situation, the tense or disequilibrated organism will be disposed to proceed by trials and discriminations—which may take place in the imagination or overtly in behavior—and finally, perhaps, after some exertion, by intelligence or luck, an effect (goal) will be attained which will re-establish equilibrium (reduce the tension, appease the need). The reduction of tension will be attended by a feeling of satisfaction. After one or more experiences of this sort, the object (person, thing), or kind of object, which was dynamically connected with the satisfaction becomes valued (cathected) as a goal-object; the habitual location of the object may also become cathected as the goal-place and the road to it as the pathway. Furthermore, with repetition, unsuccessful patterns of action will tend to be eliminated and successful patterns, including the agencies (things, persons) that were of service, will be conserved. Thus a simple need-integrate (a compound of need, affect, goal, one or more goal-objects, one or more goal-places, and pathways, a variety of action patterns, and, perhaps, one or more agencies) will become established. From then on, either a percept or a mental image of the goal-object, or, in fact, the image of any component of the system may arouse the need and thus initiate activity.

This oversimplified and over-rigid conception of the organization of a positive need-integrate (e.g., hunger system, sex system, etc.) requires some modification before it can be applied to a negative need-integrate, the goal of which is a withdrawal from, an avoidance of, or a defense against, a threatening situation, rather than a "going towards" something. But a discussion of these details, important as they are, is not essential to this exposition. Suffice it to say that practically all action patterns and goal-objects are acquired by learning and, hence, in large measure, are culturally determined; and the same applies to many of the psychological needs which, like the need for morphine, we may assume, are potentially present in all people but are developed only if certain satisfactions are experienced, and, in most cases, only if the enjoyment of these satisfactions is permitted by society. It is generally agreed that the somatic needs (for oxygen, water, food, excretion, sex, etc.) are constitutionally determined.

Anyhow, we seem to have arrived at a general formula applicable to a large number of needs: tension $\rightarrow$ reduction of tension; and so, as a first approximation, we might say that one function of regnant processes is the periodic appeasement of different needs, or more generally, the satisfying reduction of tension. The word "tension" is being used here in the most general sense to include impulse, wish, desire, and so forth.

In this conception, the chief point of reference is the initiating state, the discomforting tension rather than the end state, the satisfying reduction. Thus we are provided with an explanation of suicide and of certain other apparently anti-biological effects as so many forms of riddance of intolerable suffering. Suicide does not have *adaptive* (survival) value but it does have *adjustive* value for the organism. Suicide is *functional* because it abolishes painful tension. But since goal-images are likely to arise as soon as a need is incited, and since the action patterns which ensue will be integrated directionally towards the chosen goal, it is the latter, the so-called "aim" of the activity, which is the outstanding conscious fact. A person is capable of telling us, at any moment, what he is doing, what he is trying to do, or what he intends to do, though often he does not recognize the need that prompted his resolution. Usually several needs are operating and the person will be conscious only of those which are acceptable to him. In any case, the goal, or aim, towards which he strives is founded on nothing more solid than a faith, conscious or unconscious, that its realization will serve to appease the existing tension. Actually the effective goal may be something else. Thus it is often difficult to determine either by questioning or by observation precisely which needs are operating at a given moment.

Satisfaction involves a cultural as well as a constitutional determinant. The eating of grasshoppers (raw, or roasted) will not ordinarily reduce the hunger tension of an American. One must know the culture (as well as the individual's specific situation and life history) before one can hope to predict whether a given tension, say, mental anguish, will be resolved by suicide, by recourse to psychotherapy, by withdrawal from social life, by change of occupation, or by some other alternative response. In short, the concept of need-tension by no means answers all questions about motivation. It helps us to understand why a person acts in the first place, but it does not solve the problem of "why this act and not that one?" or "why this object and not that one?"

2. *Generation of tension.* It is important to note that it is not a tensionless state, as Freud supposed, which is generally most satisfying to a healthy organism, but the *process* of reducing tension, and, often factors being equal, the degree of satisfaction is roughly proportional to the amount of tension that is reduced per unit of time. The hungrier a man is, the more he will enjoy his dinner; the lonelier he is, the more pleasure he will experience in meeting a congenial friend, and so forth. Hence, some people will exercise in order to "work up" an appetite or use absence as a means of revivifying their affections. A tensionless state is sometimes the ideal of those who suffer from chronic anxiety or resentment or a frustrated sex drive; but, as a rule, the absence of positive need-tensions—no appetite, no curiosity, no desire for fellowship, no zest—is very distressing. This calls our attention to the fact that the formula, tension → reduction of tension, takes account of only one side of the metabolic cycle. It covers

catabolism, but not anabolism (which is the synthetic growth process by which tissues and potential energies are not only restored but, during youth, actually increased). The principle of homeostasis represents conservation but not construction. Anabolism occurs unconsciously, mostly during the night, but its results are apparent each morning in the renewal of physical and mental energies and of need-tensions, and, every so often, in the emergence of new and ever more challenging goal-images which serve as incitements to the further development of certain needs.

Most people take the enjoyable consequences of the anabolic phase of metabolism for granted, and only when it fails to recur do they discover how much they depended on it. More sleep, recreation, a holiday, tonics, glandular injections, vitamins, and stimulants are some of the means used to recapture the lost tensions. Excitement-evoking objects and situations are often deliberately sought (e.g., ghost stories and roller coasters). These considerations lead us to submit tentatively a more inclusive formula: generation of tension → reduction of tension. This formula represents a temporal pattern of states instead of an end state, a way of life rather than a goal; but it applies only to the positive need systems. The conservative systems that are directed towards withdrawals, avoidances, defenses, and preventions, are adequately covered by the reduction-of-tension formula.

Since some positive need systems generate images of difficult and distant goals, one of the important functions of personality is the temporal arrangement of the sub-goals which must be attained in order to arrive at each of these destinations. These *serial programs* projected into months and years of the anticipated future constitute one of the major determinants of development. One young man wants to become a senator or a great naturalist, another wishes to compose symphonies or write novels, another to go to Mecca or achieve the state of Nirvana; but whatever his imagined end, it cannot be reached except by a long series of steps. Human beings are taught to anticipate these steps and, so, to establish in their personalities serial programs of action which are more or less coherent, strategically and tactically, and more or less congruent with the realities to be encountered. Some are pure fantasies.

3. *Self-expression.* More basic and elementary than integrated goal-directed activities are the somewhat anarchic, unco-ordinated medley of tentative, short-lived mental processes which characterize the stream of consciousness during periods of rest and day dreaming, at one extreme, and during periods of intense emotional excitement or lunacy, at the other. For these spontaneous, random, ungoverned, but yet expressive cacophonies of energy we have proposed the term "process activity." This is pure Being, a state in which the mind moves in its own inherent manner for its own intrinsic pleasure. We have only to conceptualize a free irresponsible and playful release of vitality, enjoyed for its own sake. These movements (rather than actions) of the mind and body are functional exercises often of an imaginative or experimental sort which at any moment may be suddenly integrated to produce some effect. In this way small fragments of

thinking patterns and action patterns are formed and retained, ready for future use.

When these random processes become shaped into effective or aesthetic patterns—telling gestures, ritual dances, songs, dramas—which are more perfectly expressive of emotion or of valuations, we may speak of "modal activities" and "modal pleasures," in contrast to mere "process pleasures," or to the "end pleasures" that are concomitant with tension reductions.

It is the process activities—the incessant spontaneities, the seizures of novelty, the new combinations and transformations, and the bizarre inventions—that constitute the substrata out of which all the more orderly, rational, and effective patterns of mental action are sooner or later formed. One of the many incongruent representations of reality which have resulted from our fixation on the old stimulus-response formula has been the notion of a personality as a more or less inert aggregate of reaction patterns, which requires a stimulus (often an external stimulus from the experimenter) to start it going, instead of conceiving of a matrix of incessant functional processes (the catabolic phase which spontaneously succeeds the anabolic phase of sleep), a brain seething with fantasies, programs, and projects, with tentative goals, expectations, and hopes, with fears and dreads, which lead on to actions which *select* certain regions and constituents of the environment and which apprehensively *avoid* others.

4. *Scheduling.* Besides the play of functional processes, self-expressions, and reductions of specific tensions, the personality is almost continuously involved in deciding between alternative or conflicting tendencies or elements. Among a great host of possibilities, a person will ask himself: Which shall I attend to? Which is the right interpretation? Which is the best hypothesis? Which objective is the best? Am I capable of it? Whom should I marry? Which is the most effective course of action? Which is the best word to convey my meaning? and so forth. The alternatives are fewer and the choice less difficult in non-literate and folk societies, but always and everywhere choices must be made. Of these different types of questions, the most pressing and determining are conflicts between different purposes or aims. Since aims derive their energies from needs, one could speak here of conflicts between needs, if it were not for the fact that many critical conflicts are not between needs but between alternative goal-objects which might serve to satisfy a single need (or a synthesis of needs), or between alternative goal-places where one or more needs might find their valued objects. Therefore, it seems better to refer to aims, for these are specific in respect to goal-place or goal-object.

Personalities construct schedules which permit the execution of as many aims as possible, one after the other. A schedule should be distinguished from a serial program, which, as we have said, consists of an imagined sequence of sub-goals, or steps, leading to a single distant goal. A schedule is the division of a period time—day, month, year—into a temporal sequence of differentiated and often unrelated activities. It is by creating a schedule of this sort that a man will shape his day so that certain recurrent needs will be successively appeased (washing himself, dressing, stoking the furnace, eating, drinking, defecating, urinating, smoking, ventilating the room, seeing his friends, exercising, feasting his eyes on the trees in the park, relaxing, copulating, sleeping, and so forth). Time will also be set aside for the carrying forward of one or more serial programs. The man will, perhaps, work for eight hours on a job for which he will be paid a certain sum at the end of the week; he will acquire some usable information, buy two tickets to the theatre, purchase a rare stamp to add to his collection, advance his standing in the mind of an important official, work a little on his ship-model, and so forth. By designing orderly and efficient schedules and serial programs for activities of such kinds a man will allow for the expression and appeasement of many diverse needs, and, in this way, diminish the number of aim conflicts and attain some measure of harmony and balance.

Were a person to live by organic or subjective time, doing nothing except when he was prompted by a strong impulse, he would become clearly aware of each need as it arose. A feeling of loneliness, say, would lead him to seek the company of a friend; an intense curiosity would start him searching for a certain solution, and so forth. But under the conditions that generally prevail today, especially in highly-integrated urban communities, a man lives by a clock-determined schedule. By committing himself to an unsuitable schedule, some of a man's needs will be stilled before they arise, and perhaps be oversatiated, and others will be kept waiting or entirely defeated. As a result, he will not only be scarcely conscious of the existence of some needs but will miss the enjoyment of reducing high tensions. Also, he will experience discomfort from the frustration of other needs. The designing of subjectively suitable schedules which do not seriously impair social relations is one of personality's most important assignments.

In summary, then, we might say that the *actual* order of events is determined to a varying degree by social conventions, by schedules of organizations to which the individual belongs, by the strength of his disposition to conform to these conventions and schedules, by the programs of his own proactive systems, by the unexpected upsurge of new impulses and ideas, and by luck, a hundred and one unpredictable occurrences and encounters.

So far, then, we have concluded that the general functions of personality are to exercise its processes, to express itself, to learn to generate and reduce insistent need-tensions, to form serial programs for the attainment of distant goals, and, finally, to lessen or resolve conflicts by forming schedules which more nearly permit the frictionless appeasement of its major needs.

5. *Adjusting of aspiration levels.* For a variety of reasons—lack of ability, defects of character, lack of money, deficient environment, social barriers—very few individuals are capable of fully satisfying their needs. They fail, let us say, to realize their material ambitions; or they are un-

successful in love; or they fail to win friends; or they are defeated in the mad race for power and glory; in any event, their efforts lead to frustration and dissatisfaction, time after time. As a result, many people learn to lower their levels of aspiration or learn to accept substitute goals, so that eventually their needs become realizable. Thus the ratio of achievements to anticipations, which James selected as the simplest index of happiness, comes closer to 1.0. If we accept the hypothesis that personality is generally disposed to rid itself of dissatisfactions and to prevent their recurrence, the process of lowering aspirations to realizable levels may be considered functional.

The yogi, whose discipline leads him to the attainment of an inwardly concentrated passive state, devoid of all overt needs and claims, whose goal is to have no goals in the social world, would be an extreme case in point. Possibly one of the determinants of this philosophy was an environment in which the ratio of gratification to frustration was, for many people, very low. By voluntarily attaining the state of Nirvana, a man deprives the environment of its power to move or frustrate him. This aspect of the state of Nirvana would constitute the defensive triumph of an introvert living in a repellent world. The positive triumph would be a feeling of Oneness with the All. All religions and almost all philosophers have advocated the moderation, curtailment, or introversion of needs normally directed towards the environment. But in the last hundred years, interest in this direction of effort has given way to ideals of oneness with the social all and a collective will to power. Democratic ideologies, following a different course, have encouraged a high level of extravert aspiration for every individual (e.g., in the United States: a large fortune, leading to privilege and prestige) and, thus, have opposed the natural tendency to reduce the level after repeated failures. The over-all result, in the United States, has been an extraordinary degree of material progress with a high standard of living, on the one hand, and an equally extraordinary degree of discontent (griping about the lack of material "necessities") on the other. The ideology, in other words, prevents many individuals from achieving happiness.

One of the important establishments of a personality is the ego ideal, an integrate of images which portrays the person "at his future best," realizing all his ambitions. More specifically, it is a set of serial programs, each of which has a different level of aspiration. Ego ideals run all the way from the Master Criminal to the Serene Sage. They are imaginatively created and recreated in the course of development in response to patterns offered by the environment—mythological, historical, or living exemplars. Thus the history of the ideal self may be depicted as a series of imaginative identifications, of heroes and their worship. A typology of these exemplars, based on the personalities and accomplishments of the real or fictitious characters who have acted as magnets to the imagination, in this and that culture, at this and that period, is not too difficult. Certain figures recur so regularly that they have been called "archetypes."

6. *Conforming to social expectations.* Last to be considered here is the most difficult and painful (though also, in some ways, rewarding) function of personality, that of accommodating its expressions, needs, choice of goal-objects, methods, and time-programs to the patterns that are conventionally sanctioned by society. In most cases this is a matter of learning to conform. Only rarely and to a limited extent is an individual able to modify the culture patterns of his group so that they will correspond more closely to his own inclinations and persuasions.

Many acute and sensitive students of human behavior feel an understandable discontent with descriptions of human life as lived in groups when these descriptions use cultural patterns as a conceptual framework. It is perfectly true, of course, that a cultural pattern is one thing; the utilization of pattern by individuals quite another. It is studies of participation and the utilization of the concepts of status and role which bridge the gap between the abstracted culture patterns and specific individuals.

If one is studying a particular personality, questions of the following kind are useful to ask: In what patterns does this person participate? Are they those which are deemed by other members of the society to be appropriate to his statuses? To what extent does his behavior fall outside the permitted range of variation of the given patterns? What patterns are flatly rejected by him? Pattern transgression is invariably a useful clue in the investigation of personality formation because doing the same as others do is regularly rewarded. Hence, when someone characteristically—in at least certain sectors of his behavior—does not do as others do, the investigator is always led to ask specific questions about accidents of the socialization process affecting this particular individual.

Likewise, one should examine very carefully the extent to which the individual's habits conform to the standards of his group. Does he profess full acceptance of group standards? If a given individual does not seek the culturally selected ends by socially approved means, other individuals will express disapproval or, indeed, endeavor to force conformity by various sorts of deprivations. Thus, as we study differential participation in patterns, we gain invaluable clues as to the personal habit systems which characterize specific individuals. Similarly, patterns often give the key to certain type conflicts which beset individuals of particular ages or statuses. In our society one may instance the constant pressure of the conflict between age-group standards and parental standards upon youngsters and adolescents. This is without doubt a type conflict of our culture. The problem of the mutual adjustment of patterns is a complicated and interesting one—largely arising out of the circumstance that it is both the business of culture to check unacceptable impulses of all kinds, and then again to check the aggressions mobilized by the first check.

Sometimes, however, comments upon pattern conflicts and the inconsistency of the total system of patterns are superficial and misleading. The thing which so many theorists miss is the segmentation of behaviors. There are various groups of patterns within a culture which are superimposed upon each other like a stencil. If viewed as somewhat separate structural entities shaping various rather separate sectors of the total culture, these patterns are not inconsistent at all.

The ideal patterns defining statuses are in most cultures so adjusted that the segmentation of behaviors by roles does not create personality disorganization. However, there is another type of segmentation, peculiarly prevalent in large societies, which subjects the individual to various critical press. In a social structure as complex as our American social structure today, there are great cultural variations—differences between regions, between economic and social classes. Likewise, in a culture which is changing as rapidly as ours, there are often major discrepancies between the ideal patterns of various age groups. The youngster soon discovers that one type of response is rewarded by his own age group, quite another by his parents and their age group. The hostility between adults and adolescents, which is so frequent among us, is due only in part to a real conflict of interests; in large part it is also due to conflicting culture patterns which arise and are transmitted at the adolescent level. The situation is further complicated by the circumstance that the adolescent also finds that the standards or ideal patterns which prevail in his family are quite different from those which prevail in the families of some of his friends—perhaps greatly admired friends, friends whose families have greater prestige, and so forth.

The influence of cultural patterns upon individuals has certain characteristic differences in literate and non-literate societies. In non-literate societies the emphasis is upon patterns as binding individuals and upon the satisfactions individuals find in the fulfillment of intricately involuted patterns. In literate societies the emphasis must be placed more upon pattern conflict and upon the relationship this bears to personality disorganization. These are, of course, not black-and-white distinctions. In literate cultures also individuals often find much gratification in the fulfillment of patterns. That most behavior is culturally patterned means not only economy of energy and thought for the individual; there is a positively-toned affective side too. Every time an individual in our society feels that he *knows* what the appropriate pattern for his behavior in a given situation is, he (speaking statistically) gets a relief from anxiety, a sense of security from carrying it out.

This is not the place to discuss the genesis of culture patterns. The origin and rationale of many of them are a mystery. Often it is impossible to decide whether a particular pattern is an expression of the profound "wisdom of the culture," or merely, like man's coccyx, a vestigial fixation. Suffice it to say that although some culture patterns, especially the class-linked aesthetic forms, are subject to periodic changes dictated by the autocrats of fashion, in most cases it is long usage which sanctifies them. Anyhow, it is chiefly the traditional patterns which parents and educators, policemen and judges, and, indeed, all socially responsible adults, as carriers of the culture, are *expected* to uphold and to teach by example, by persuasion, and by an accepted, culturally defined system of rewards and punishments. The process of inculcating and learning these patterns, until they become "second nature," is termed "socialization." In part, this process is rewarding to the child, for he learns how to do many things toward which he is groping.

However, a child is also confronted by a multiplicity of Don'ts, each connected with some activity, or place, or objects towards which he is naturally disposed, and a large number of Do's connected with actions towards which he is not disposed, at least at the time or in the manner indicated. If he is to avoid punishment and enjoy the many rewards which adults have to offer, he must learn to inhibit or redirect certain insistent impulses, temporarily or permanently, as well as learn to force himself to perform certain other actions which at the time are repugnant to his feelings. After countless protestations and rebellions, the average child, with great reluctance, learns to do these things, to the extent of conforming to most of the patterns which are considered normal for his age. It seems that the ability to suppress a prohibited inclination is acquired by associating it with the images of punishment suggested by the parent and, later, those suggested by the sanctions of the youngster's own peer group. These images arouse anxiety to a degree that is greater or more painful than the tension of the impulse and, consequently, the child is led to reduce the former rather than the latter. This is achieved by turning to other activities and thus side-tracking, or inhibiting, the disturbing inclination.

Although, by a little juggling, the tension-reduction formula can be used to represent the described course of events, the type of learning that occurs here is different, in one significant respect, from that usually studied by animal psychologists. In the conventional means-effect learning, the organism is wholly concerned with the reduction of one kind of tension, and none of the ineffective action patterns which, after some practice, it succeeds in eliminating or inhibiting have any power to reduce this or any other kind of tension. But in "moral" learning, which forms the core of the socialization process, the child must acquire the ability to suppress action patterns which, like masturbation or knocking down a younger brother, are highly effective in reducing strong tensions. The former is a matter of solving slight conflicts between means; the latter calls for the suppression, rather than the appeasement, of powerful needs and emotions which return, over and over again, to disquiet, plague, or mortify the self.

Under the term "rewards and punishments" are included all the satisfactions and dissatisfactions which other people are capable of producing in a person. Of chief importance to the child is the assurance of being the object of the love of the mother or mother surrogate, the significant person, and, so, of enjoying all the benefits that she is able to provide. Any impulse which leads to an apparent weakening of her affection, to anger or to rejection, to verbal or physical aggression, to deprivation of customary benefits, or, even, to her sorrowful disappointment, will become associated with anxiety-invoking (or pity-invoking) images of one or more of these consequences. The same applies, at a somewhat later age and in less degree, to the father and other significant persons. In the Western world the parental sanctions were, for centuries, tremendously strengthened by constant references to a supernatural system of rewards and punishments—by promises of everlasting bliss in heaven and by threats of eternal punishment in hell. In many non-literate societies, parents refuse, as it were, to take the blame for frustrating their children and displace the sanctions onto supernaturals or society in general ("people will make fun of you if you act that way").

During the course of development, figures of authority, capable of rewarding and punishing—the father, the mother, father- and mother-surrogates, God, admired exemplars, priests and teachers, magistrates and policeman—associated with a heterogeneous system of moral principles, laws, and conventions, become incorporated in the personality as a distinguishable establishment which is termed the "superego." Depending on the ideology that has been inculcated as part of the socialization process, the superego can represent one or usually several of the following: "voice of God," the "eternal laws of conduct," the "right," "moral truth," the "commandments of society," the "good of the whole," the "will of the majority," or merely "cultural conventions." But once the superego has been accepted with an emotional attitude compounded of fear, respect, and love, the individual becomes a true carrier of the moral culture of his society, who inhibits many of his delinquent impulses, not so much out of fear of external retributions as out of dread of superego disapproval, the secret pains of guilt and remorse. This, at least, is the picture in the Judaic-Christian world. In some cultures, "shame" (the fear of being discovered in non-conformity by others) appears more powerful than "guilt" (self-punishment) in inducing inhibition.

In all cultures there develops in the child, as a supplementary source of motivation towards "good" behavior, a dread of insult and personal humiliation, of the loss of self-respect, combined, perhaps, with a positive desire for moral prestige or righteousness. This works very well so long as the parents retain the respect of the child and so long as the latter's pride is connected with membership in his family and with his recently won powers of self-control, and so long as no unjust or degrading punishments are actually administered. For, if the child does lose respect for his family or is unjustly, humiliatingly, and repeatedly punished, he may become asocial with a vengeance, in his sentiments or in pursuing a life of crime. Most boys in our society pass through a partly accepted phase of this sort when they go to school and join a gang, one of the avowed or unavowed aims of which is to break as many laws as possible without getting caught. Status in some gangs, and hence the pride of its most respected members, is connected with degree of courage and cleverness required to "get away with" minor crimes. But here the youth is still dependent on social approval and hence is susceptible to influence. It is the lone wolf, in feelings or in actions, whose pride has become connected with the maintenance of his own self-sufficient sovereignty and with revenge, who is difficult to touch.

Beginning in the nursery, the process of socialization continues throughout life. Among other things, what must be learned are: the power to inhibit, or to moderate, the expression of unacceptable needs; the ability to transfer cathexis from a prohibited goal-object to an acceptable substitute; the habitual and automatic use of a large number of approved action patterns (methods, manners, and emotional attitudes); and the ability to adapt to schedules (to do things at the proper time, keep appointments, etc.). It is assumed that, having acquired these abilities, the average person will be capable of establishing satisfactory interpersonal relations within the legal and conventional framework of society. When the child begins to behave in a predictable, expectable manner it is well on the road to being socialized. But degree of socialization cannot be estimated solely on the basis of objective evidence. The everlasting practice of suppression, demanded by some societies, may be extremely painful, and the not infrequent result of these repeated renunciations is a deep-seated cumulative resentment, conscious or unconscious, against cultural restrictions. Held under control, this resentment manifests itself in tensions, dissatisfactions, irritabilities, recurrent complaints, periods of dejection, cynicism, pessimism, and, in more extreme cases, in neurosis or psychosis. A high degree of repressed resentment is indicative of a fundamental emotional maladjustment, and, hence, of the partial failure of the socialization process, regardless of how successful the individual may be in "making friends and influencing people." The goal of the socialization process is an unreluctant emotional identification with the developing ethos of the society, but rarely is this attained: some degree of conflict is inevitable.

Greater behavioral conformity often leads to an accumulation of resentment, and thus to an increase in emotional incompatibility. It is as if the individual's hatred of society rose with each renunciation that was involved in the process of adjustment. It might seem that in such cases we would have to say that any process is functional which results in a greater disposition or willingness to conform to the standards of society and, so, to a diminution of emotional incompatibility.

History, as well as everyday experience, teaches us that, in some instances, the unacceptable patterns which an individual has expressed or wished to express in action have turned out to be more functional, more conducive to social as well as to individual well-being, than the patterns which were supported by his society. Consequently, in dealing with a situation of emotional incompatibility, one must consider the fitness of the culture for individual development. This is particularly true when the personality considered is strongly identified with the welfare of his society and wants to participate in the developing forms which he feels are valuable. If a man of this sort were to try to reduce strain by learning to like

the culture patterns to which he is opposed, he would create another conflict which could be solved only by renouncing his identification with the
creative forces of society. Hence, for those who cannot easily abandon an
ideal, it is usually more functional to attempt to modify the culture patterns which are most objectionable to them than it is to attempt to abolish
their objections. In doing this, the individual will usually be identifying
with others who, like himself, have suffered from the hurtful culture patterns.

The actual fact is that some culture patterns are enduring whereas
others are changing from year to year, and, although it is often impossible
to agree in our appraisals of the different trends, it is certain that some of
them are constructive of new social values, others are conservative of the
old, others forcefully destructive, and still others diffusely and wastefully
disintegrating. A man who is emotionally identified with his culture *in toto*
would be one who accepts and exemplifies in his actions the great majority
of these different trends. But, actually, such heterogeneity in one personality is rarely, if ever, seen. Perhaps most men are "culture carriers," or
conservers; but there are also culture creators, destroyers, and disintegra-
tors. Adopting a broad functional standpoint, we might say now that the
goal of the socialization process is the disposition and the ability to reciprocate and co-operate with members of the society who are conserving
its most valuable patterns as well as with those who are endeavoring to
improve them. The path to this goal is marked by a long and varied series
of compromise formations which represent modes of mediation of conflict
between different types of personalities and their societies; and, although
most people stop at one or another of these way-stations, a psychologist
might be justified in assuming hypothetically that one function of personality is to transform itself as far as possible in the direction of identification with both the conserving and creative forces of humanity.

Human personality is a compromise formation, a dynamic resultant of
the conflict between the individual's own impulses (as given by biology
and modified by culture and by specific situations) and the demands, interests, and impulses of other individuals. The compromise is attained in a
great variety of ways. An individual may be over-socialized in one sector
of his behavior, adequately flexible in others, inadequately socialized in
another behavioral area. Conflicts arising from the varying demands upon
personality in different roles the individual must play may be solved by
compartmentalization—the personality adopts habitual strategies in one
area of life that are not carried over to another. Rationalizations are also
invoked, many of these being supplied ready-made by the cultural ideology. Conformity may be accompanied by overt release of hostility
against the institutions, or the required performance of conventions may
be subtly distorted. In the effective reactions that attend both the learning
of cultural patterns and their transmission to others inhere most fertile
sources of culture change.

The socially deviant forms of personality may arise as a result of various dynamisms and their interactions. The study of concrete cases shows that the processes are often excessively complex. But let us first state the matter over simply. If, as a result of special combinations of factors operating in the life of one individual, the primary biological impulses remain unchecked, such an individual will be variously known as "anti-social," "criminal," or "id-dominated." If, on the other hand, an individual emerges from his socialization with his impulse life unusually inhibited and with an exaggerated need for and dependency upon the approval of others, this type of person may be designated as "neurotic," "conscience-ridden," or "superego-dominated." But if, as a result of a more fortunate admixture and integration of both biological and cultural demands, an individual emerges who represents a balanced type, we speak of him as "normal," "rational," or "ego-controlled."

While the under-socialized, the over-socialized, and the adequately-socialized individual may well stand as three types representing central tendencies in a range of dispersion, this classification is too clear-cut and schematic to serve as a wholly satisfactory model of the observed facts. The psychotics, for example, do not fit easily into such a schema. Some, but by no means all, individuals having "functional" psychoses may fall into a category of persons whose socialization was so severe and so anxiety-laden that (given also constitutional predispositions) they reacted by rejecting alike cultural goals and the culturally approved means of attaining them. The lawbreaker is not always just a person who has not learned to express his impulses in socially acceptable ways. Criminal behavior is sometimes rejection of existing institutions, sometimes the active assertion of socially prohibited responses.

The total cultural structure or certain special pressures within a society (economics, class, etc.) may increase the frequency of certain types of deviation. There can be little doubt but that certain constitutional types find certain cultural constraints more intolerable than others do. The manner in which the learning process is structured by cultural and idiosyncratic forces can lead to certain familiar forms of personal adjustment.

If the child is socialized in such a way that both cultural goals and institutionalized means are affectively toned with a strong threat of punishment for their violation, his personality may have a conformist [3] coloring. He may also be a conformist if he has accepted goals and means with little anxiety—provided these goals give adequate satisfaction to the needs conditioned by his particular organic equipment and the other idiosyncratic components of his personality. But such a person's conformity will lack compulsive quality, so that if cultural goals and means change rapidly in accord with alterations in the technological, economic, scientific, or political environment, he will readjust with little difficulty. But some persons who have been socialized with a minimum of anxiety are still unable

---

[3] The argument and terminology of the following four paragraphs draw much from Robert K. Merton, "Social Structure and Anomie," *American Sociological Review*, Vol. 3 (1938) pp. 672–82.

to accept emotionally the definition of themselves provided by the culture. Such persons will probably be innovators or recluses.

Innovating or radical personalities will also develop when accidents of the socialization process cause certain institutionalized means, or both means and goals, to be "tabooed" for the individual. Note that many radicals develop in socially and economically advantaged families and that many workmen are stubborn conservatives. Barnett has shown how even in a relatively homogeneous non-literate society personal conflict is the primary motivation for invention, and how "the disgruntled, the maladjusted, the frustrated, and the incompetent are pre-eminently the accepters of cultural innovation and change." [4] The criminal is often one who accepts goals (such as wealth and power) but who cannot attain, or rejects, the approved means of achieving them.

When socialization (perhaps especially "wrong" identification) or constitutional temperament enforce a rejection of some or many cultural goals but acceptance of means, the personality tends towards ritualism. It isn't the objectives which seem important but a fixed manner of performance. When both goals and means are too heavily fraught with anxiety or are otherwise rejected, the type of adjustment may be characterized as "retreatism." Such personalities may be hobos or hermits or romantic escapists in our culture.

Still other discriminations need to be made. For example, the innovator (action constructively toned) should be distinguished from the rebel (action destructively toned). But a dominant or a partial source for all such variations may be found in the way in which the culture (some societies are far more characterized by conservatives than others) or adventitious events of the individual's life history influence the process of socialization. The fact that an organism can be over-socialized also requires further elaboration. Inasmuch as the development of harmonious social living, with its innumerable gains and advantages, has made progressively greater demands for the suppression of the animal side of his nature, man has historically extolled the virtues of conscience and subservience to the interests of others. Because of the painful difficulty of making the transition from asocial to social existence, one prevalent view has been that man is naturally all evil and that the more completely and utterly he can be forced to give up his "original sinfulness" and accept the socially approved ways of life, the better it is for him and all concerned. Hence the tendency for society to press the individual to the utmost for renunciation of "self-interest" and acceptance of "sacrifice and service." By definition, a highly socialized person is just like the majority, the average. Constructive social innovations and intellectual and artistic creations will hardly come from such individuals. Moreover, too great socialization tends to eliminate that spontaneity of response which is a great value in ordinary social intercourse. Whatever else they may be or may become, human beings are and

---

[4] H. G. Barnett, "Personal Conflicts and Cultural Change," *Social Forces*, Vol. 20 (1941), p. 171.

must always remain animals; and this unalterable fact sets definite limits on the extent to which suppression of biologically given needs and inclinations can go, with benefit either to the individual or to the group of which he is a member.

5. SUMMARY

Personality is the continuity of functional forms and forces manifested through sequences of organized regnant processes and overt behaviors from birth to death. The functions of personality are: to allow for the periodic regeneration of energies by sleep; to exercise its processes; to express its feelings and valuations; to reduce successive need-tensions; to design serial programs for the attainments of distant goals; to reduce conflicts between needs by following schedules which result in an harmonious way of life; to rid itself of certain persisting tensions by restricting the number and lowering the levels of goals to be attained; and, finally, to reduce conflicts between personal dispositions and social sanctions, between the vagaries of antisocial impulses and the dictates of the superego by successive compromise formations, the trend of which is towards a wholehearted emotional identification with both the conserving and creative forces of society.

Part Two

THE FORMATION OF
PERSONALITY

SECTION I

THE DETERMINANTS OF
PERSONALITY FORMATION

CHAPTER

2

Clyde Kluckhohn AND
Henry A. Murray

PERSONALITY FORMATION:

THE DETERMINANTS

EVERY MAN is in certain respects
    a. like all other men,
    b. like some other men,
    c. like no other man.

He is like all other men because some of the determinants of his personality are universal to the species. That is to say, there are common features in the biological endowments of all men, in the physical environments they inhabit, and in the societies and cultures in which they develop. It is the very obviousness of this fact which makes restatements of it expedient, since, like other people, we students of personality are naturally disposed to be attracted by what is unusual, by the qualities which distinguish individuals, environments, and societies, and so to overlook the common heritage and lot of man. It is possible that the most important of the undiscovered determinants of personality and culture are only to be revealed by close attention to the commonplace. Every man experiences birth and must learn to move about and explore his environment, to protect himself against extremes of temperature and to avoid serious injuries; every man experiences sexual tensions and other importunate needs and must learn to find ways of appeasing them; every man grows

NOTE: This paper represents a complete revision of an earlier scheme published by C. Kluckhohn and O. H. Mowrer, "Culture and Personality: A Conceptual Scheme," *American Anthropologist*, Vol. 46 (1944), pp. 1-29. The present writers gratefully acknowledge their indebtedness to Dr. Mowrer.

in stature, matures, and dies; and he does all this and much more, from
first to last, as a member of a society. These characteristics he shares with
the majority of herd animals, but others are unique to him. Only with
those of his own kind does he enjoy an erect posture, hands that grasp,
three-dimensional and color vision, and a nervous system that permits
elaborate speech and learning processes of the highest order.

Any one personality is like all others, also, because, as social animals,
men must adjust to a condition of interdependence with other members
of their society and of groups within it, and, as cultural animals, they
must adjust to traditionally defined expectations. All men are born helpless into an inanimate and impersonal world which presents countless
threats to survival; the human species would die out if social life were
abandoned. Human adaptation to the external environment depends upon
that mutual support which is social life; and, in addition, it depends upon
culture. Many types of insects live socially yet have no culture. Their
capacity to survive resides in action patterns which are inherited via the
germ plasm. Higher organisms have less rigid habits and can learn more
from experience. Human beings, however, learn not only from experience
but also from each other. All human societies rely greatly for their survival upon accumulated learning (culture). Culture is a great storehouse
of ready-made solutions to problems which human animals are wont to
encounter. This storehouse is man's substitute for instinct. It is filled not
merely with the pooled learning of the living members of the society,
but also with the learning of men long dead and of men belonging to other
societies.

Human personalities are similar, furthermore, insofar as they all experience both gratifications and deprivations. They are frustrated by the
impersonal environment (weather, physical obstacles, etc.) and by physiological conditions within their own bodies (physical incapacities, illnesses, etc.). Likewise, social life means some sacrifice of autonomy,
subordination, and the responsibities of superordination. The pleasure
and pain men experience depend also upon what culture has taught them
to expect from one another. Anticipations of pain and pleasure are internalized through punishment and reward.

These universalities of human life produce comparable effects upon
the developing personalities of men of all times, places, and races. But they
are seldom explicitly observed or commented upon. They tend to remain
background phenomena—taken for granted like the air we breathe.

Frequently remarked, however, are the similarities in personality traits
among members of groups or in specific individuals from different groups.
In certain features of personality, most men are "like some other men."
The similarity may be to other members of the same socio-cultural unit.
The statistical prediction can safely be made that a hundred Americans,
for example, will display certain defined characteristics more frequently
than will a hundred Englishmen comparably distributed as to age, sex,
social class, and vocation.

But being "like some men" is by no means limited to members of social units like nations, tribes, and classes. Seafaring people, regardless of the communities from which they come, tend to manifest similar qualities. The same may be said for desert folk. Intellectuals and athletes the world over have something in common; so have those who were born to wealth or poverty. Persons who have exercised authority over large groups for many years develop parallel reaction systems, in spite of culturally tailored differences in the details of their behaviors. Probably tyrannical fathers leave a detectably similar imprint upon their children, though the uniformity may be superficially obscured by local manners. Certainly the hyperpituitary type is equally recognizable among Europeans, African Negroes, and American Indians. Also, even where organic causes are unknown or doubtful, certain neurotic and psychotic syndromes in persons of one society remind us of other individuals belonging to very different societies.

Finally, there is the inescapable fact that a man is in many respects like no other man. Each individual's modes of perceiving, feeling, needing, and behaving have characteristic patterns which are not precisely duplicated by those of any other individual. This is traceable, in part, to the unique combination of biological materials which the person has received from his parents. More exactly, the ultimate uniqueness of each personality is the product of countless and successive interactions between the maturing constitution and different environing situations from birth onward. An identical sequence of such determining influences is never reproduced. In this connection it is necessary to emphasize the importance of "accidents," that is, of events that are not predictable for any given individual on the basis of generalized knowledge of his physical, social, and cultural environments. A child gets lost in the woods and suffers from exposure and hunger. Another child is nearly drowned by a sudden flood in a canyon. Another loses his mother and is reared by an aged grandmother, or his father remarries and his education is entrusted to a stepmother with a psychopathic personality. Although the personalities of children who have experienced a trauma of the same type will often resemble each other in certain respects, the differences between them may be even more apparent, partly because the traumatic situation in each case had certain unique features, and partly because at the time of the trauma the personality of each child, being already unique, responded in a unique manner. Thus there is uniqueness in each inheritance and uniqueness in each environment, but, more particularly, uniqueness in the number, kinds, and temporal order of critically determining situations encountered in the course of life.

In personal relations, in psychotherapy, and in the arts, this uniqueness of personality usually is, and should be, accented. But for general scientific purposes the observation of uniformities, uniformities of elements and uniformities of patterns, is of first importance. This is so because without the discovery of uniformities there can be no concepts, no classifications, no formulations, no principles, no laws; and without these no science can exist.

The writers suggest that clear and orderly thinking about personality formation will be facilitated if four classes of determinants (and their interactions) are distinguished: *constitutional, group-membership, role,* and *situational.* These will help us to understand in what ways every man is "like all other men," "like some other men," "like no other man."

I. CONSTITUTIONAL DETERMINANTS

The old problem of "heredity *or* environment" is essentially meaningless. The two sets of determinants can rarely be completely disentangled once the environment has begun to operate. All geneticists are agreed today that traits are not inherited in any simple sense. The observed characters of organisms are, at any given point in time, the product of a long series of complex interactions between biologically-inherited potentialities and environmental forces. The outcome of each interaction is a modification of the personality. The only pertinent questions therefore are: (1) which of the various genetic potentialities will be actualized as a consequence of a particular series of life-events in a given physical, social, and cultural environment? and (2) what limits to the development of this personality are set by genetic constitution?

Because there are only a few extreme cases in which an individual is definitely committed by his germ plasm to particular personality traits, we use the term "constitutional" rather than "hereditary." "Constitution" refers to the total physiological make-up of an individual at a given time. This is a product of influences emanating from the germ plasm and influences derived from the environment (diet, drugs, etc.).

Since most human beings (including scientists) crave simple solutions and tend to feel that because simple questions can be asked there must be simple answers, there are numberless examples both of overestimation and of underestimation of constitutional factors in theories of personality formation. Under the spell of the spectacular success of Darwinian biology and the medicine of the last hundred years, it has often been assumed that personality was no less definitely "given" at birth than was physique. At most, it was granted that a personality "unfolded" as the result of a strictly biological process of maturation.

On the other hand, certain psychiatrists, sociologists, and anthropologists have recently tended to neglect constitutional factors almost completely. Their assumptions are understandable in terms of common human motivations. Excited by discovering the effectiveness of certain determinants, people are inclined to make these explain everything instead of something. Moreover, it is much more cheerful and reassuring to believe that environmental factors (which can be manipulated) are all important, and that hereditary factors (which can't be changed) are comparatively inconsequential. Finally, the psychiatrists, one suspects, are consciously or unconsciously defending their livelihood when they minimize the constitutional side of personality.

The writers recognize the enormous importance of biological events and event patterns in molding the different forms which personalities assume. In fact, in the last chapter personality was defined as "the entire sequence of organized governmental processes in the brain from birth to death." They also insist that biological inheritance provides the stuff from which personality is fashioned and, as manifested in the physique at a given time-point, determines trends and sets limits within which variation is constrained. There are substantial reasons for believing that different genetic structures carry with them varying potentialities for learning, for reaction time, for energy level, for frustration tolerance. Different people appear to have different biological rhythms: of growth, of menstrual cycle, of activity, of depression and exaltation. The various biologically inherited malfunctions certainly have implications for personality development, though there are wide variations among those who share the same physical handicap (deafness, for example).

Sex and age must be regarded as among the more striking constitutional determinants of personality. Personality is also shaped through such traits of physique as stature, pigmentation, strength, conformity of features to the culturally fashionable type, etc. Such characteristics influence a man's needs and expectations. The kind of world he finds about him is to a considerable extent determined by the way other people react to his appearance and physical capacities. Occasionally a physically weak youth, such as Theodore Roosevelt was, may be driven to achieve feats of physical prowess as a form of over-compensation, but usually a man will learn to accept the fact that his physical make-up excludes him from certain types of vocational and social activities, although some concealed resentment may remain as an appreciable ingredient of his total personality. Conversely, special physical fitnesses make certain other types of adjustment particularly congenial.

2. GROUP MEMBERSHIP DETERMINANTS

The members of any organized enduring group tend to manifest certain personality traits more frequently than do members of other groups. How large or how small are the groupings one compares depends on the problem at hand. By and large, the motivational structures and action patterns of Western Europeans seem similar when contrasted to those of Mohammedans of the Near East or to Eastern Asiatics. Most white citizens of the United States, in spite of regional, ethnic, and class differences, have features of personality which distinguish them from Englishmen, Australians, or New Zealanders. In distinguishing group-membership determinants, one must usually take account of a concentric order of social groups to which the individual belongs, ranging from large national or international groups down to small local units. One must also know the hierarchical class, political or social, to which he belongs within each of these groups. How inclusive a unit one considers in speaking of group-membership determinants is purely a function of the level of abstraction at which one is operating at a given time.

Some of the personality traits which tend to distinguish the members of a given group from humanity as a whole derive from a distinctive biological heritage. Persons who live together are more likely to have the same genes than are persons who live far apart. If the physical vitality is typically low for one group as contrasted with other groups, or if certain types of endocrine imbalance are unusually frequent, the personalities of the members of that group will probably have distinctive qualities.

In the greater number of cases, however, the similarities of character within a group are traceable less to constitutional factors than to formative influences of the environment to which all members of the group have been subjected. Of these group-membership determinants, culture is with little doubt the most significant. To say that "culture determines" is, of course, a highly abstract way of speaking. What one actually observes is the interaction of people. One never sees "culture" any more than one sees "gravity." But "culture" is a very convenient construct which helps in understanding certain regularities in human events, just as "gravity" represents one type of regularity in physical events. Those who have been trained in childhood along traditional lines, and even those who have as adults adopted some new design for living, will be apt to behave predictably in many contexts because of a prevailing tendency to conform to group standards. As Edward Sapir has said:

> All cultural behavior is patterned. This is merely a way of saying that many things that an individual does and thinks and feels may be looked upon not merely from the standpoint of the forms of behavior that are proper to himself as a biological organism but from the standpoint of a generalized mode of conduct that is imputed to society rather than to the individual, though the personal genesis of conduct is of precisely the same nature, whether we choose to call the conduct "individual" or "social." It is impossible to say what an individual is doing unless we have tacitly accepted the essentially arbitrary modes of interpretation that social tradition is constantly suggesting to us from the very moment of our birth.

Not only the action patterns but also the motivational systems of individuals are influenced by culture. Certain needs are biologically given, but many others are not. All human beings get hungry, but no gene in any chromosome predisposes a person to work for a radio or a new car or a shell necklace or "success." Sometimes biologically-given drives, such as sex, are for longer or shorter periods subordinated to culturally acquired drives, such as the pursuit of money or religious asceticism. And the means by which needs are satisfied are ordinarily defined by cultural habits and fashions. Most Americans would go hungry rather than eat a snake, but this is not true of tribes that consider snake meat a delicacy.

Those aspects of the personality that are not inherited but learned all have—at least in their more superficial and peripheral aspects—a cultural tinge. The skills that are acquired, the factual knowledge, the basic assumptions, the values, and the tastes, are largely determined by culture. Culture likewise structures the conditions under which each kind of learning takes place: whether transmitted by parents or parental substitutes, or by brothers and sisters, or by the learner's own age mates; whether gradually or quickly; whether renunciations are harshly imposed or reassuringly rewarded.

Of course we are speaking here of general tendencies rather than invariable facts. If there were no variations in the conceptions and applications of cultural standards, personalities formed in a given society would be more nearly alike than they actually are. Culture determines only what an individual learns as a member of a group—not so much what he learns as a private individual and as a member of a particular family. Because of these special experiences and particular constitutional endowments, each person's selection from and reaction to cultural teachings have an individual quality. What is learned is almost never symmetrical and coherent, and only occasionally is it fully integrated. Deviation from cultural norms is inevitable and endless, for variability appears to be a property of all biological organisms. But variation is also perpetuated because those who have learned later become teachers. Even the most conventional teachers will give culture a certain personal flavor in accord with their constitution and peculiar life-experiences. The culture may prescribe that the training of the child shall be gradual and gentle, but there will always be some abrupt and severe personalities who are temperamentally disposed to act otherwise. Nor is it in the concrete just a matter of individuality in the strict sense. There are family patterns resultant upon the habitual ways in which a number of individuals have come to adjust to each other.

Some types of variation, however, are more predictable. For example, certain differences in the personalities of Americans are referable to the fact that they have grown up in various sub-cultures. Jones is not only an American; he is also a member of the middle class, an Easterner, and has lived all his life in a small Vermont community. This kind of variation falls within the framework of the group determinants.

The values imbedded in a culture have special weight among the group membership determinants. A value is a conception, explicit or implicit, distinctive of an individual or characteristic of a group, of the desirable which influences the selection from available modes, means, and ends of action. It is thus not just a preference, a desire, but a formulation of the *desirable*, the "ought" and "should" standards which influence action.

The component elements of a culture must, up to a point, be either logically consistent or meaningfully congruous. Otherwise the culture carriers feel uncomfortably adrift in a capricious, chaotic world. In a personality system, behavior must be reasonably regular or predictable, or the individual will not get expectable and needed responses from others because they will feel that they cannot "depend" on him. In other words, a social life and living in a social world both require standards "within" the individual and standards roughly agreed upon by individuals who live and work together in a group. There can be no personal security and no stability of social organization unless random carelessness, irresponsibility, and purely impulsive behavior are restrained in terms of private and group codes. If one asks the question, "Why are there values?" the reply must be: "Because social life would be impossible without them; the functioning of the social system could not continue to achieve group goals; individuals could not get what they want and need from other individuals in personal and emotional terms, nor could they feel within themselves a requisite measure of order and unified purpose." Above all, values add an element of predictability to social life.[1]

Culture is not the only influence that bears with approximate constancy upon all the members of a relatively stable, organized group. But we know almost nothing of the effects upon personality of the continued press of the impersonal environment. Does living in a constantly rainy climate tend to make people glum and passive, living in a sunny, arid country tend to make them cheerful and lively? What are the differential effects of dwelling in a walled-in mountain valley, on a flat plain, or upon a high plateau studded with wide-sculptured red buttes? Thus far we can only speculate, for we lack adequate data. The effects of climate and even of scenery and topography may be greater than is generally supposed.

Membership in a group also carries with it exposure to a social environment. Although the social and cultural are inextricably intermingled in an individual's observable behavior, there is a social dimension to group membership that is not culturally defined. The individual must adjust to the presence or absence of other human beings in specified numbers and of specified age and sex. The density of population affects the actual or potential number of face-to-face relationships available to the individual. Patterns for human adjustment which would be suitable to a group of five hundred would not work equally well in a group of five thousand, and vice versa. The size of a society, the density of its population, its age and sex ratio are not entirely culturally prescribed, although often conditioned by the interaction between the technological level of the culture and the exigencies of the physical environment. The quality and type of social interaction that is determined by this social dimension of group membership has, likewise, its consequences for personality formation.

Before leaving the group-membership determinants, we must remind the reader once more that this conception is merely a useful abstraction. In the concrete, the individual personality is never directly affected by the group as a physical totality. Rather, his personality is molded by the

---

[1] Fuller treatment of the concept of values will be found in C. Kluckhohn, "Values and Value-Orientations in the Theory of Action: An Exploration in Definition and Classification," in T. Parsons and E. Shills, (eds.), *Toward a General Theory of Action* (Cambridge; Harvard University Press, 1951).

particular members of the group with whom he has personal contact and by his conceptions of the group as a whole. Some traits of group members are predictable—in a statistical sense—from knowledge of the biological, social, and cultural properties of the group. But no single person is ever completely representative of all the characteristics imputed to the group as a whole. Concretely, not the group but group agents with their own peculiar traits determine personality formation. Of these group agents, the most important are the parents and other members of the individual's family. They, we repeat, act as individuals, as members of a group, and as members of a sub-group with special characteristics (the family itself).

3 . ROLE DETERMINANTS

The culture defines how the different functions, or roles, necessary to group life are to be performed—such roles, for example, as those assigned on the basis of sex and age, or on the basis of membership in a caste, class, or occupational group. In a sense, the role determinants of personality are a special class of group-membership determinants; they apply to strata that cross-cut most kinds of group membership. The long-continued playing of a distinctive role, however, appears to be so potent in differentiating personalities within a group that it is useful to treat these determinants separately.

Moreover, if one is aware of the role determinants, one will less often be misled in interpreting various manifestations of personality. In this connection it is worth recalling that, in early Latin, *persona* means "a mask"—*dramatis personae* are the masks which actors wear in a play, that is, the characters that are represented. Etymologically and historically, then, the personality is the character that is manifested in public. In modern psychology and sociology this corresponds rather closely to the role behavior of a differentiated person. From one point of view, this constitutes a disguise. Just as the outer body shields the viscera from view, and clothing the genitals, so the public personality shields the private personality from the curious and censorious world. It also operates to conceal underlying motivations from the individual's own consciousness. The person who has painfully achieved some sort of integration, and who knows what is expected of him in a particular social situation, will usually produce the appropriate responses with only a little personal coloring. This explains, in part, why the attitudes and action patterns produced by the group-membership and role determinants constitute a screen which, in the case of normal individuals, can be penetrated only by the intensive, lengthy, and oblique procedures of depth psychology.

The disposition to accept a person's behavior in a given situation as representative of his total personality is almost universal. Very often he is merely conforming, very acceptably, to the cultural definition of his role. One visits a doctor in his office, and his behavior fits the stereotype of the physician so perfectly that one says, often mistakenly, "There indeed is a well-adjusted person." But a scientist must train himself to get behind a man's cultivated surface, because he will not be able to understand much if he limits his data to the action patterns perfected through the repeated performance of the roles as physician, as middle-aged man, as physician dealing with an older male patient, etc.

4. SITUATIONAL DETERMINANTS

Besides the constitutional determinants and the forces which will more or less inevitably confront individuals who live in the same physical environment, who are members of a society of a certain size and of a certain culture, and who play the same roles, there are things which "just happen" to people. Even casual contacts of brief duration ("accidental"— i.e., not foreordained by the cultural patterns for social interrelations) are often crucial, it seems, in determining whether a person's life will proceed along one or another of various possible paths. A student, say, who is undecided as to his career, or who is about equally drawn to several different vocations, happens to sit down in a railroad car next to a journalist who is an engaging and persuasive advocate of his profession. This event does not, of course, immediately and directly change the young man's personality, but it may set in motion a chain of events which put him into situations that are decisive in molding his personality.

The situational determinants include things that happen a thousand times as well as those that happen only once—provided they are not standard for a whole group. For example, it is generally agreed that the family constellation in which a person grows up is a primary source of personality styling. These domestic influences are conditioned by the cultural prescriptions for the roles of parents and children. But a divorce, a father who is much older than the mother, a father whose occupation keeps him away from home much of the time, the fact of being an only child or the eldest or youngest in a series—these are situational determinants.

Contact with a group involves determinants which are classified as group-membership or situational, depending on the individual's sense of belongingness or commitment to the group. The congeries of persons among whom a man accidentally finds himself one or more times may affect his personality development but not in the same manner as those social units with which the individual feels himself allied as a result of shared experiences or of imaginative identification.

5. INTERDEPENDENCE OF THE DETERMINANTS

"Culture and personality" is one of the fashionable slogans of contemporary social science and, by present usage, denotes a range of problems on the borderline between anthropology and sociology, on the one hand, and psychology and psychiatry, on the other. However, the phrase has unfortunate implications. A dualism is implied, whereas "culture *in* personality" and "personality *in* culture" would suggest conceptual models more in accord with the facts. Moreover, the slogan favors a dangerous simplification of the problems of personality formation. Recognition of culture as one of the determinants of personality is a great gain, but there are some indications that this theoretical advance has tended to obscure the significance of other types of determinants. "Culture and personality" is as lopsided as "biology and personality." To avoid perpetuation of an over-emphasis upon culture, the writers have treated cultural forces as but one variety of the press to which personalities are subjected as a consequence of their membership in an organized group.

A balanced consideration of "personality in nature, society, and culture" must be carried on within the framework of a complex conceptual scheme which explicitly recognizes, instead of tacitly excluding, a number of types of determinants. But it must also not be forgotten that any classification of personality determinants is, at best, a convenient abstraction.

A few illustrations of the intricate linkage of the determinants will clarify this point. For example, we may instance a network of cultural, role, and constitutional determinants. In every society the child is differently socialized according to sex. Also, in every society different behavior is expected of individuals in different age groups, although each culture makes its own prescriptions as to where these lines are drawn and what behavioral variations are to be anticipated. Thus, the personalities of men and women, of the old and the young, are differentiated, in part, by the experience of playing these various roles in conformity with cultural standards. But, since age and sex are biological facts, they also operate throughout life as constitutional determinants of personality. A woman's motivations and action patterns are modified by the facts of her physique as a woman.

Some factors that one is likely to pigeonhole all too complacently as biological often turn out, on careful examination, to be the product of complicated interactions. Illness may result from group as well as from individual constitutional factors. And illness, in turn, may be considered a situational determinant. The illness—with all of its effects upon personality formation—is an "accident" in that one could predict only that the betting odds were relatively high that this individual would fall victim to this illness. However, when the person does become a patient, one can see that both a constitutional predisposition and membership in a caste or class group where sanitation and medical care were substandard are causative factors in this "accidental" event. Similarly, a constitutional tendency towards corpulence certainly has implications for personality when it is characteristic of a group as well as when it distinguishes an individual within a group. But the resources of the physical environment as exploited by the culturally-transmitted technology are major determinants in the production and utilization of nutritional substances of various sorts and these have patent consequences for corpulence, stature, and energy potential. Tuberculosis or pellagra may be endemic. If hookworm is endemic in a population, one will hardly expect vigor to be a striking feature of the majority of people. Yet hookworm is not an unavoidable "given," either constitutionally or environmentally: the prevalence and effects of hookworm are dependent upon culturally enjoined types of sanitary control.

Complicated interrelations of the same sort may be noted between the environmental and cultural forces which constitute the group membership determinants. On the one hand, the physical environment imposes certain limitations upon the cultural forms which man creates, or it constrains toward change and readjustment in the culture he brings into an ecological area. There is always a large portion of the impersonal environment to which men can adjust but not control; there is another portion which is man-made and cultural. Most cultures provide technologies which permit some alterations in the physical world (for example, methods of cutting irrigation ditches or of terracing hillsides). There are also those artifacts (houses, furniture, tools, vehicles) which serve as instruments for the gratification of needs, and, not infrequently, for their incitement and frustration. Most important of all, perhaps, culture directs and often distorts man's perceptions of the external world. What effects social suggestion may have in setting frames of reference for perception has been shown experimentally. Culture acts as a set of blinders, or series of lenses, through which men view their environments.

Among group-membership determinants, the social and cultural factors are interdependent, yet analytically distinct. Man, of course, is only one of many social animals, but the ways in which social, as opposed to solitary, life modifies his behavior are especially numerous and varied. The fact that human beings are mammals and reproduce bi-sexually creates a basic predisposition toward at least the rudiments of social living. And the prolonged helplessness of human infants conduces to the formation of a family group. Also, certain universal social processes (such as conflict, competition, and accommodation) are given distinct forms through cultural transmission. Thus, while the physically strong tend to dominate the weak, this tendency may be checked and even to some extent reversed by a tradition which rewards chivalry, compassion, and humility. Attitudes towards women, towards infants, towards the old, towards the weak will be affected by the age and sex ratios and the birth and death rates prevalent at a particular time.

The social and cultural press likewise interlock with the situational determinants. There are many forces involved in social interaction which influence personality formation and yet are in no sense culturally prescribed. All children (unless multiple births) are born at different points in their parents' careers, which means that they have, psychologically speaking, somewhat different parents. Likewise, whether a child is wanted or unwanted and whether it is of the desired sex will make a difference in the ways in which it will be treated, even though the culture says that all children are wanted and defines the two sexes as of equal value.

A final example will link the constitutional with both the group-membership and situational determinants. Even though identical twins may differ remarkably little from a biological standpoint, and participate in group activities which are apparently similar, a situational factor may intrude as a result of which their experiences in social interaction will be quite different. If, for instance, one twin is injured in an automobile accident and the other is not, and if the injured twin has to spend a year in bed, as the special object of his mother's solicitations, noticeable personality differences will probably develop. The extent to which these differences endure will depend surely upon many other factors, but it is unlikely that they will be entirely counteracted. The variations in treatment which a bed-ridden child receives is partly determined by culture (the extent to which the ideal patterns permit a sick child to be petted, etc.), and partly by extra-cultural factors (the mother's need for nurturance, the father's idiomatic performance of his culturally patterned role in these circumstances, etc.).

6. SIMILARITIES AND DIFFERENCES IN PERSONALITY

In conclusion, let us return for a moment to the observed fact that every man is "like all other men, like some other men, like no other man." In the beginning there is (1) the organism and (2) the environment. Using this division as the starting point in thinking about personality formation, one might say that the *differences* observed in the personalities of human beings are due to variations in their biological equipment and in the total environment to which they must adjust, while the *similarities* are ascribable to biological and environmental regularities. Although the organism and the environment have a kind of wholeness in the concrete behavioral world which the student loses sight of at his peril, this generalization is substantially correct. However, the formulation can be put more neatly in terms of field. There is (1) the organism moving through a field which is (2) structured both by culture and by the physical and social world in a relatively uniform manner, but which is (3) subject to endless variation within the general patterning due to the organism's constitutionally-determined peculiarities of reaction and to the occurrence of special situations.

In certain circumstances, one reacts to men and women, not as unique organizations of experience, but as representatives of a group. In other circumstances, one reacts to men and women primarily as fulfilling certain roles. If one is unfamiliar with the Chinese, one is likely to react to them first as Chinese rather than as individuals. When one meets new people at a social gathering, one is often able to predict correctly: "That man is a doctor." "That man certainly isn't a businessman, he acts like a professor." "That fellow over there looks like a government official, surely not an artist, a writer, or an actor." Similarities in personality created by the role and group-membership determinants are genuine enough. A man is likely to resemble other men from his home town, other members of his vocation, other members of his class, as well as the majority of his countrymen as contrasted to foreigners.

But the variations are equally common. Smith is stubborn in his office as well as at home and on the golf course. Probably he would have been stubborn in all social contexts if he had been taken to England from America at an early age and his socialization had been completed there. The playing of roles is always tinged by the uniqueness of the personality. Such differences may be distinguished by saying, "Yes, Brown and Jones are both forty-five-year-old Americans, both small-businessmen with about the same responsibilities, family ties, and prestige—but somehow they are different." Such dissimilarities may be traced to the interactions of the constitutional and situational determinants, which have been different for each man, with the common group-membership and role determinants to which both have been subjected.

Another type of resemblance between personalities cuts across the boundaries of groups and roles but is equally understandable within this framework of thinking about personality formation. In general, one observes quite different personality manifestations in Hopi Indians and in white Americans—save for those common to all humanity. But occasionally one meets a Hopi whose behavior, as a whole or in part, reminds one very strongly of a certain type of white man. Such parallels can arise from similar constitutional or situational determinants or a combination of these. A Hopi and a white man might both have an unusual endocrine condition. Or both Hopi and white might have had several long childhood illnesses which brought them an exceptional amount of maternal care. While an over-abundance of motherly devotion would have had somewhat different effects upon the two personalities, a striking segmental resemblance might have been produced which persisted throughout life.

In most cases the observed similarities, as well as the differences, between groups of people are largely attributable to fairly uniform social and cultural processes. When one says, "Smith reminds me of Brown," a biologically inherited determinant may be completely responsible for the observed resemblance. But when one notes that American businessmen, for example, have certain typical characteristics which identify them as a group and distinguish them from American farmers and teachers it can hardly be a question of genetic constitution. Likewise, the similarities of personality between Americans in general as contrasted with Germans in general must be traced primarily to common press which produces resemblances in spite of wide variations in individual constitutions.

To summarize the content of this chapter in other terms: The personality of an individual is the product of inherited dispositions and environmental experiences. These experiences occur within the field of his physical, biological, and social environment, all of which are modified by the culture of his group. Similarities of life experiences and heredity will tend to produce similar personality characteristics in different individuals, whether in the same society or in different societies.

Although the distinction will not always be perfectly clear-cut, the readings which follow will be organized according to this scheme of constitutional, group-membership, role, and situational determinants, and the interactions between them. It is believed that this will assist the reader in keeping steadily in mind the variety of forces operative in personality formation and the firm but subtle nexus that links them.

Part Two

SECTION II

CONSTITUTIONAL DETERMINANTS

INTRODUCTION

WHILE ONE cannot drive a fence through an individual to indicate which traits are biologically determined and which produced by the environment, it is nevertheless important to keep the biological basis of human nature constantly in mind. All personalities must adjust to such physiological tensions as those of food and sex hunger.

People not only have bodies. They have bodies that differ in accord with the age and sex of the individual. For example, the adolescent's personality problems are often a function of the incongruity between varying phases of his growth.[1] The sheer anatomical and physiological differences between men and women have consequences for the characteristic personality manifestations of the two sexes. These have been treated in a number of illuminating papers, such as: Gregory Zilboorg, "Masculine and Feminine: Some Biological and Cultural Aspects";[2] Willard R. Cooke, "The Differential Psychology of the American Woman";[3] Erich Fromm, "Sex and Character";[4] William Egleston Galt, "The Male-Female Dichotomy in Human Behavior";[5] and Karen Horney, "The Problem of Feminine Masochism."[6]

But age and sex differences are not the only biological variations of importance for personality. Every individual inherits through the germ plasm certain distinctive potentialities and limitations. No amount of training can make a musical genius or a champion athlete out of the majority of men. Every human organism has inherent predispositions towards special types of behavior, which tend to manifest themselves more or less regardless of the particular life experience to which the organism is subjected. Pediatricians agree that newborn infants of the same sex vary as to

---

[1] See Lawrence K. Frank, "Adolescence and Public Health," *American Journal of Public Health*, Vol. 31 (1941), pp. 1143–50.

[2] *Psychiatry*, Vol. 7 (1944), pp. 257–69.

[3] *American Journal of Obstetrics and Gynecology*, Vol. 49 (1945), pp. 457–72.

[4] *Psychiatry*, Vol. 6 (1943), pp. 21–31.

[5] *Ibid.*, pp. 1–14.

[6] *Psychoanalytic Review*, Vol. 22 (1935), pp. 241–57.

activity rates[7] and energy potentials. However, though ordinary experience and scientific study both offer presumptive evidence for the significance of the person's inborn nature, clear-cut proof is scanty because of the difficulty of isolating the purely hereditary factors. The most satisfactory data consist in the demonstration of fairly constant correlations between characteristic personality manifestations and patterns of physical traits,[8] peculiarities of the autonomic nervous system, regular occurence in certain family lines of certain abnormalities of personality (even where the environment of family members has varied widely), and types of body build. Present knowledge does not make possible the disentangling of the precise role of endocrine glands, features of brain structure, special properties of other nervous tissue, and other factors basic to such associations. It can only be said on a statistical basis that, for example, an individual of a given body-type is more likely (whatever the underlying reasons) to show certain personality traits than an individual of another particular body type.

[7] Chapple has maintained that this is true for adults also. See Eliot Chapple, " 'Personality' Differences As Described by Invariant Properties of Individuals in Interaction," *Proceedings of the National Academy of Sciences*, Vol. 26 (1940), pp. 10–16.

[8] See David M. Levy, "Aggressive-Submissive Behavior and the Froehlich Syndrome," *Archives of Neurology and Psychiatry*, Vol. 36 (1936), pp. 991–1020.

CHAPTER

3

Hudson Jost AND
Lester Warren Sontag

* THE GENETIC FACTOR IN AUTONOMIC NERVOUS-SYSTEM FUNCTION

The first two papers in this section are somewhat technical. Indeed, it will be noted that all the readings in this group employ mathematical methods. The complexities involved are such that only by treating a sizable number of cases and comparing the observed frequencies of association with those that might have occurred from chance alone can investigators hope to reach trustworthy conclusions. The study of twins also affords an invaluable opportunity for progress in this difficult field. Twins formed from a single fertilized egg (monozygotic twins) have identical heredity; twins formed from different eggs (dizygotic twins) are ordinarily reared in a highly similar environment, but their biological heredity is merely of the order of that of ordinary brothers and sisters.[1]

The article by Drs. Jost and Sontag presents valuable evidence, from the study of twins, siblings, and unrelated children, that constitution as expressed in the functioning of the autonomic nervous system may be, at

NOTE: Reprinted from *Psychosomatic Medicine*, Vol. 6 (1944), pp. 308–10, by permission of the authors and the American Society for Research in Psychosomatic Problems. (Copyright, 1944, by The Williams and Wilkins Company.)

[1] The second paper in this section also uses the twin method. For another type of approach involving the twin situation, see Douglass W. Orr, "A Psychoanalytic Study of a Fraternal Twin," *Psychoanalytic Quarterly*, Vol. 10 (1941), pp. 284–96.

[73

*least in part, inherited at birth. Their paper is oriented to the genetic pre-
disposition towards such manifestations of personality as persistent high
blood pressure, some forms of heart disease, and other conditions regarded as "psychosomatic" (i.e., involving both "mind" and "body").
However, the implications of their findings for personality as such are
obvious and important.*

1. INTRODUCTION

This paper will present the results of an attempt to demonstrate a genetic
factor in the patterns of autonomic nervous-system function in children.
The existence of such a genetic factor in autonomic activity or response
might go far toward explaining the familiar predisposition toward such
psychosomatic conditions as benign essential hypertension, coronary disease, mucous colitis, Raynaud's disease, etc.—a predisposition which
many workers believe they have observed.

The malfunction of an organ as the result of emotional stresses implies at once the greater-than-average physiological component of the
individual's emotional responses to stimulus, a lowered general threshold
of stimulation, or a lowered specific organ resistance predisposing it to
malfunction. By measuring the activity of certain organs which are dependent for their function on autonomic stimulus, it is possible to acquire
some picture of the state of activity of the autonomic nervous system itself, and then to compare the degrees of that activity and also the selective
expression of it in two or more individuals. Measurement of skin resistance, systolic and diastolic blood pressure, heart rate, salivation, respiration and vaso-motor persistence time form a pattern which expresses, to
some degree at least, the function and state of activity of the autonomic
nervous system. Inspection and statistical treatment of such measures on
a large group of children make it possible to determine whether monozygotic twins closely approximate each other in autonomic pattern, whether
siblings are less alike than the monozygotic twins, and finally whether
random selected unrelated children, matched for age, resemble each other
in autonomic pattern less closely than do the siblings and monozygotes.

2. DATA

The data for this analysis were gathered during the past three years at the
Fels Research Institute. This is part of the longitudinal study being carried
on by this group. Longitudinal scores of physical and psychological
growth are taken as well as those of physiological reactivity described in
this paper. The method and measures have been described by Wenger and
Ellington,[2] but a brief review of the process will be given here.

---

[2] M. A. Wenger, and M. Ellington, "The Measurement of Autonomic Balance in
Children: Method and Normative Data," *Psychosomatic Medicine*, Vol. 5 (1943),
p. 241.

Measures of vaso-motor persistence, salivary output, heart period, standing palmar conductance, reclining volar conductance, respiration period, and pulse pressure were obtained under the following standard conditions. All the measurements were made during the hours of 9–12 A.M., at least an hour after breakfast, in a quiet room in which the temperature and humidity were controlled at 74°–76° and 40% respectively. The subjects were all from the Fels study and were between six and twelve years of age. There were 62 subjects in the 1940 study, 74 in the 1941 study, and 81 in the 1942 study.

Vaso-motor persistence was elicited by means of a firm slow stroke applied over the biceps of the left arm by a stimulator calibrated to deliver approximately 250 grams of pressure. Two strokes were made in the form of an X, the stop watch was started at the incidence of the first stroke, and the time was recorded to the nearest minute at which the location of the "crossing" was no longer a matter of certainty. The persistence time ranged from three to thirty minutes. *Salivary output* was measured for a period of five minutes. The saliva was collected in a graduated centrifuge tube by means of a simple suction pump. The subject was instructed to bring the saliva to the front of the mouth and it was taken out by the suction. *Heart period* was obtained from minute samples taken over a twenty minute period. The heart was recorded electrocardiographically and the period consisted of the time in seconds between the QRS peaks. *Standing palmer conductance* was obtained through the use of Darrow's circuit. The electrodes were placed in the palms of each hand and the subjects were instructed to stand erect with their weight supported equally on both feet. *Volar skin conductance* was made in the same manner as the measure above, except that the electrodes were placed on the forearm just above the wrist and the subjects were in a reclining position. *Respiration period*: the respiration was recorded on a polygraph moving at a speed of three centimeters per second. The period was time between peaks on the respiratory curve. *Pulse pressure* was obtained from six measures of systolic and diastolic blood pressure taken in the twenty-minute period.

The measure of *"autonomic balance"* is a composite of the seven variables above. This measure has been described by Wenger. It is obtained by weighting the values of the various measures, the weights having been derived from a factor analysis. The measure expresses the degree of sympathetic or parasympathetic preponderance.

3. METHOD

Two methods of analysis were used. The first consisted of correlating the paired individuals in the three groups on the various measures. The scores of each twin were correlated with those of his partner, those of each child with those of his siblings, and a large sample of the scores of unrelated children were correlated with each other. The age factor was eliminated by the use of standard scores computed for each age level. The number of pairs for the years 1940, 1941, and 1942 were 5, 5 and 6 for the twins; 10, 19, and 25 for the siblings; and for the unrelated group 361, 324, and 324 respectively. This unrelated group represents a random comparison of cases and is sufficiently large to be stable.

The second method of analysis consisted of deriving difference scores between the paired individuals. For example, a pair of twins might have standard scores on a particular measure of 45 and 48; in this case their difference score would be 3. Two siblings might have scores of 48 and 55; in this case the difference score would be 7. Two unrelated children

TABLE I

*Correlations of the Twin, Sibling, and Unrelated Groups,
for the Years 1940, 1941, and 1942*

YEAR	CORRELATION	N
1940		
Twins	.434	5
Siblings	.255	10
Unrelated	.164	361
1941		
Twins	.470	5
Siblings	.406	19
Unrelated	.017	364
1942		
Twins	.489	6
Siblings	.288	25
Unrelated	.080	300

might have scores of 45 and 60; here the difference score would be 15. In all instances the lowest scores indicate that the individuals are little different in the particular measurement.

4. RESULTS

TABLE I presents the correlations of the seven variables for the twins, siblings, and unrelated groups. In all cases, the correlations are in the direction predicted by the hypothesis that there is a genetic factor in autonomic nervous-system function. The twins are more highly correlated than are the siblings and the siblings show higher correlations than the unrelated group. The measures are also consistent over the three-year period. The correlations are not significantly different, but are suggestive in that they all point in the same direction; that is, toward a genetic factor. For the years 1940, 1941, and 1942, the twin correlations are .434, .470, and .489 respectively; the sibling correlations are .255, .406, and .288 respectively; and the unrelated group correlate .164, .017, and .080 for the three years.

TABLE II presents the data obtained from the differential analysis. The means, number of pairs, critical ratios, and probabilities for all the groups are given here. The more striking differences may be enumerated.

The measure of *vaso-motor persistence* is significant at the .01 level for all the three years between the twins and unrelated groups, and at the .01 level between the sibling and unrelated groups for the last two years. The difference between the twins and siblings in this measure is not significant, but is in the direction indicated by the hypothesis. The measure of *standing palmar skin resistance* is reliable at the .05 level between the three groups except for the 1941 measures between the twins and siblings and the siblings and unrelated groups; it is significant at the .05 level between the twin and unrelated groups for that year. *Reclining pulse pressure* is significantly different between the twins and siblings and the twins and unrelated groups for two of the three years. These are at the .05 and .01 levels for the 1940 and 1942 measures respectively. The differences between the siblings and the unrelated groups is significant at the .05 level for the 1942 measure only. All other differences are in the expected direction, but are not significant. *Total salivation* is significantly different between the twins and unrelated, and the siblings and unrelated groups in two of the three years, but there is no significant difference between the twins and siblings; in fact, there is a reversal between these groups for the 1941 measure. The measure of *mean respiration period* is significant only in the 1941 measures and then only between the twins and unrelated, and the sibling and unrelated groups. The differences for the other years, although they are not significant, are in the expected direction. *Mean heart period* is significant for the same year and groups as is the respiration period. The critical ratios for all measures are low, but in the expected direction. *Reclining volar conductance* shows significant differences between the twins and siblings, and the twins and unrelated groups in the 1940 study, but no significant differences appear in the other years.

The measure "*autonomic balance*," which is a composite score of the above variables, shows some significance between the groups. The difference between the twins and siblings in the 1940 study is at the .05 level, as is the difference between the twins and unrelated groups for the same year. All the other measures are in the expected direction except for a reversal in the 1941 study between the twins and siblings. This is due to the fact that the highly-weighted measure of *total salivation* is reversed for this year.

5. SUMMARY

Measurements of vaso-motor persistence time, palmar standing skin resistance, reclining pulse pressure, salivation, respiration rate, reclining volar conductance, "heart period" and "autonomic balance" have been made during three different years upon a group of children between six

TABLE II

Means of the Differences in Physiological Measures of Monozygotic Twins, Compared with Those of Siblings and Unrelated Children, Plus a Similar Comparison between Siblings and Unrelated Children *

MEASURE	YEAR	TWIN MEAN	SIB'S MEAN	UN. R. MEAN	TWIN-SIB C. R.	TWIN-SIB P.	TWIN-UN. R. C. R.	TWIN-UN. R. P.	SIB-UN. R. C. R.	SIB-UN. R. P.
Vaso-motor Persistence	'40	2.60	6.70	11.80	1.35	—	6.75	.01	1.80	—
	'41	5.60	7.00	12.37	.62	—	3.98	.01	3.22	.01
	'42	1.17	4.45	11.26	2.06	—	6.25	.01	5.25	.01
Palmar Standing Skin Resistance	'40	3.60	8.83	11.34	2.64	.05	4.07	.01	2.44	.05
	'41	6.40	11.42	12.35	1.56	—	2.20	.05	.54	—
	'42	4.67	8.64	13.40	2.29	.05	6.10	.01	3.88	.01
Reclining Pulse Pressure	'40	3.00	9.00	11.08	2.61	.05	7.60	.01	.97	—
	'41	6.20	8.68	8.93	1.00	—	1.23	—	.16	—
	'42	4.50	8.76	12.98	2.46	.05	5.45	.01	3.36	.05
Total Salivation	'40	5.40	6.50	12.04	.55	—	5.22	.01	3.29	.01
	'41	7.80	6.00	12.44	-.92	—	3.04	.05	5.20	.01
	'42	8.67	9.24	9.97	.05	—	.41	—	.43	—
Mean Respiration Period	'40	7.20	9.40	12.10	.56	—	1.93	—	1.56	—
	'41	4.80	5.30	11.54	.25	—	4.24	.01	4.48	.01
	'42	5.80	10.10	11.48	1.26	—	1.96	—	.65	—
Mean Heart Period	'40	8.00	10.00	12.49	.62	—	1.60	—	1.51	—
	'41	7.20	7.60	12.81	.17	—	2.72	.05	4.26	.01
	'42	8.80	10.60	11.22	.45	—	.65	—	.34	—
Reclining Volar Conductance	'40	3.60	11.60	10.54	2.78	.05	6.07	.01	-.40	—
	'41	6.80	8.70	10.06	.52	—	1.00	—	.80	—
	'42	5.50	8.70	7.01	1.28	—	.74	—	-1.09	—
Autonomic Balance	'40	4.80	10.50	11.71	2.27	.05	3.01	.05	.55	—
	'41	11.50	8.28	13.91	-.21	—	.87	—	1.61	—
	'42	6.17	11.10	13.78	1.97	—	3.25	.01	1.24	—

* For reasons of space three columns in the original table have been eliminated and the data in them are summarized below:
The number of pairs of twins for the years 1940, '41, and '42, respectively, were 5, 5, and 6.
The number of pairs of siblings for the years 1940, '41, and '42, respectively, were 10, 19, and 25.
The number of pairs of unrelated children for the years 1940, '41, and '42, respectively, were 361, 324, and 324, except that for the measures of "mean heart period" and "reclining volar conductance" in 1940 the number of pairs was 310, and for the measure of "autonomic balance" the number of pairs in the three years were, respectively, 144, 129, and 307.

and twelve years of age. The group contained many pairs of siblings and six pairs of monozygotic twins. Comparison of the measures of siblings with those of twins shows them to be less alike than those of the twins. Similar comparisons show the unrelated less alike than the siblings and the unrelated less alike than the twins. Most of the differences are statistically significant. Correlations of the scores of pairs of twins, pairs of siblings and matched unrelated children show the correlations for the twins highest, siblings next, and unrelated least. These results suggest that what one might call "autonomic constitution" may be at least partially an inherited characteristic. These findings should help to explain the genetic predisposition to many of the psychosomatic diseases.

CHAPTER

4

Franz J. Kallmann

* THE GENETIC THEORY
OF SCHIZOPHRENIA

[An Analysis of 691 Schizophrenic Twin Index Families]

It has long been established that certain comparatively rare mental disorders (such as Huntington's chorea and juvenile amaurotic idiocy) are inherited in a statistically-predictable manner. An hereditary factor in manic-depressive psychosis has also been claimed.[1] Dr. Kallman presents the most impressive data thus far available that a quite common disorder of the personality is influenced by a specific genetic factor. He does not claim that all persons who have a certain heredity will become schizophrenic. He does show that those who have this predisposition are much more likely to respond to environmental pressures with these reactions. He rightly challenges the extreme environmentalists to show that an increase in the incidence of schizophrenia occurs under specified conditions among persons who have no blood relationship to other schizophrenes.

DESPITE notable changes in the attitude of contemporary psychiatry toward the constitutional problems of psychosomatic medicine, there is

NOTE: Reprinted from the *American Journal of Psychiatry*, Vol. 103 (1946), pp. 309–22, by permission of the author and the American Psychiatric Association. (Copyright, 1946, by the American Psychiatric Association.) The author has kindly permitted the omission of footnotes and of some diagrams with slight revision of the text.

[1] Abraham Myerson and Rosalie D. Boyle, "The Incidence of Manic-Depressive Psychosis in Certain Prominent Families," *American Journal of Psychiatry*, Vol. 98 (1941), pp. 11–21.

still a tendency to perpetuate the genetic theory of schizophrenia as a controversial issue.

Some arguments thrive largely on dialectic grounds and, from a scientific standpoint, are more apparent than real. Others are based on preconceptions which are kept alive by an ambiguous terminology and the pardonable tendency either to oversimplify a complex causality or to mistake it for obscurity. A main source of misunderstanding is the erroneous belief that acceptance of causation by heredity would be incompatible with general psychological theories of a descriptive or analytical nature, or that it might lead to a depreciation of present educational and therapeutic standards. Evidently, there is no point in presenting evidence of the inheritance of schizophrenia, if in subsequent statements the etiology of schizophrenic psychoses is likely to be listed as unknown, or if reservations are made regarding a similar psychotic syndrome labeled "dementia praecox," or if the given genetic mechanism is finally dismissed as unessential or non-Mendelian.

From a genetic point of view, the main question to be clarified is whether or not the capacity for developing a true schizophrenic psychosis is somehow controlled by inherited, predispositional elements. In order to settle this problem beyond any reasonable doubt, only three types of investigative procedure are available. They are:

1. The pedigree or family-history method,
2. The contingency method of statistical prediction, and
3. The twin-study method.

The investigation of individual family histories is the oldest, simplest, and most popular method of recording familial occurrence of an apparently hereditary trait. Such a pedigree [2] is often impressive to behold and sometimes is suggestive of the operation of heredity. If, for example, the mating of two psychotic parents is found, under certain circumstances, to be capable of giving rise to seven definite cases of schizophrenia among the offspring—that is, in all the children of this union who reached the age of maturity—it would seem inadequate to disregard the possible significance of the biological factor prerequisite for inheritance, namely, consanguinity. On the basis of this single observation, however, the genetic hypothesis would be no more conclusive than either the assumption of *folie à neuf* due to "psychic contagion" or the supposition that the psychosis of the father of this remarkable sibship was not "inherited" because, as was true in the example chosen, his parents had apparently been ordinary first cousins without schizophrenia.

Obviously, the general usefulness of the pedigree method is limited to the study of relatively rare unit characters which are easily traced and fairly constant in their clinical appearance. In more common traits and especially in irregularly expressed anomalies such as schizophrenia, it is

---

[2] The original paper contains a diagram of such a pedigree which had to be eliminated because of technical difficulties in publication.

necessary to employ statistical methods which demonstrate more clearly the effect of blood relationship.

This objective is accomplished by the *contingency method*, which compares the morbidity rates for representative samples of consanguineous and non-consanguineous groups. The results of such a procedure will indicate whether or not a given anomaly occurs more frequently in blood relatives of unselected index cases than is to be expected in the light of the normal average distribution of the trait in the general population. The available morbidity figures for schizophrenia, obtained with the contingency method, are summarized in TABLE III. The rates refer to different population and consanguinity groups and may have been compiled with different degrees of statistical accuracy. The samples differ in size as well as in uniformity, and many of them seem to have lived under socio-economic conditions which cannot be compared directly.

However, one essential point has been confirmed by all of these studies—namely, that the incidence of schizophrenia tends to be higher in blood relatives of schizophrenic index cases than it is in the general population. Concerning the offspring of schizophrenic index cases it has been shown that their morbidity rates range from 16.4 to 68.1 per cent, that is, from nineteen to about eighty times average expectancy, according to whether one or both of their parents are schizophrenic (TABLE III).[3] It is to be verified, therefore, that the chance of developing schizophrenia in comparable environments increases in direct proportion to the degree of blood relationship to a schizophrenic index case. If such evidence can be supplied, intransigent supporters of purely environmental theories should be expected to demonstrate with equally precise methods that a consistent increase in morbidity is found associated with particular environmental circumstances *in the absence* of consanguinity.

In order to establish the hereditary nature of a psychosis beyond the possibility of random contingency and in relation to the interaction of predispositional genetic elements and various precipitating or perpetuating influences acting from without, the best available procedure is the *twin-study method* in conjunction with an ordinary sibling study. Such a combination method has been adopted in our long-range studies of specific behavior disorders and has been called by us *Twin-Family Method*.[4] This approach provides six distinct categories of sibship groups reared under comparable environmental conditions; namely, monozygotic twins, dizygotic twins of the same sex, dizygotic twins of opposite sex, full siblings, half-siblings, and step-siblings. If the assumed genetic factor is negligible, the statistical expectation will be that the morbidity rates for

---

[3] A diagram showing graphically the expectation of schizophrenia and schizoid personality in descendants of schizophrenics has been omitted.

[4] A diagram showing the six categories of sibship groups has been omitted. A more detailed description of the method can be found in a previous report of F. J. Kallmann and D. Reisner, "The Twin Studies on Genetic Variations in Resistance to Tuberculosis," *Journal of Heredity*, Vol. 34, No. 9.

TABLE III

| | INCIDENCE OF SCHIZOPHRENIA IN CONSANGUINEOUS GROUPS RELATED TO: | | | | | | | | | | | |
| | ONE ORDINARY INDEX CASE OF SCHIZOPHRENIA | | | | | | | TWO ORDINARY INDEX CASES | | ONE TWIN INDEX CASE | |
	Incidence of Schizophrenia in General Population	Nephews and Nieces	First Cousins	Grand-Children	Half Siblings	Parents	Full Siblings	Children	Siblings	Children	Dizygotic Cotwins	Monozygotic Cotwins
Previous Morbidity Studies of Kallmann	.85	3.9	—	4.3	7.6	10.3	11.5	16.4	20.5	68.1	12.5	81.7
Range of Morbidity Rates of Other Investigators	.3-1.5	1.4-3.9	2.6	—	—	7.1-9.3	4.5-11.7	8.3-9.7	20.0	53.0	14.9	68.3

Schizophrenia Rates Obtained with the Contingency Method of Statistical Prediction full siblings and dizygotic twin partners should be about the same, but they should clearly differ from the rates for the other sibship groups.

One-egg twins are expected to show the highest concordance rate for a genetically determined disorder, even if brought up in different environments. Two-egg twins may be either of the same or of opposite sex, but genetically they are no more alike than any other pair of brothers and sisters who are born at different times. Half-siblings with only one parent in common should be about midway between the full siblings and the non-consanguineous step-siblings, if the given morbidity depends on the closeness of blood relationship rather than on the similarity in environment.

In order to obtain statistically representative material for the application of this method, our survey was organized on a statewide basis. The twin index cases (TABLE IV) were collected from the resident populations and new admissions of all mental hospitals under the supervision of the New York State Department of Mental Hygiene. The danger of bias on account of technical selective factors in the sampling of the material was avoided by referring the determination of the twin-index cases to the staffs of the hospitals co-operating in the survey. The only criteria for selection were that the reported cases be born by multiple birth and that they had been admitted with a diagnosis of mental disease.

The classifications of both schizophrenia and zygocity were made on the basis of personal investigation and extended observation. The twin diagnosis was based on findings obtained with the similarity method, since it is known now that monozygotic twins are not necessarily monochorial. The statistical analysis was limited to the families of 794 schizophrenic twin index cases whose cotwins were available for examination at the age of fifteen years. These index cases were reported within a period of nine years by twenty institutions, which in 1945 had a total resident population of 73,252 patients with 47,929 schizophrenics and 12,316 new admissions.

The random sampling of the 691 index pairs is indicated by the close correspondence between the statistically expected figure of 25.6 per cent for the proportion of monozygotic twin pairs in an unselected American twin group, and the actual percentage of 25.2 as obtained with the Weinberg Differential Method for the present study. It is in accordance with expectation that the main deficit is on the part of dizygotic twins of opposite sex. Altogether, there are 174 monozygotic and 517 dizygotic index pairs with schizophrenia in at least one member or, more precisely, 691 pairs constituted by 1,382 twins, of whom 794 were legitimate index cases. Of the dizygotic sets, 296 are same-sexed and 221 are opposite-sexed.

The excess of female over male index cases is almost 20 per cent. The ratio of white to non-white index cases is about 14:1. Approximately 70 per cent of the index cases are unmarried. The proportion of nuclear cases, characterized by hebephrenic or catatonic psychoses with the tendency to progression and deterioration, amounts to 68 per cent.

The various groups of relatives included in the analysis of these 691

TABLE IV

Racial and Diagnostic Distribution of the Twin Index Cases

| | ALL SCHIZOPHRENIC TWIN INDEX CASES REPORTED * | | | | | | | ALL COMPLETE INDEX PAIRS STUDIED † | | | | |
| | Marital status | | Racial distribution | | Diagnostic distribution | | | Mono-zygotic | Dizygotic | | Total number |
	Single	Married	White	Non-White	Nuclear	Periph-eral	Total number		Same sex	Opposite sex	
Male	292	70	337	25	253	109	362	75	132	221	317½
										2	
Female	266	166	405	27	290	142	432	99	164	221	373½
										2	
Total number	558	236	742	52	543	251	794	174	296	221	691

* Without index cases whose cotwins were unavailable at the age of 15 years

† The difference between 794 index cases and 691 index pairs is explained by the fact that 103 pairs both twin partners were reported as index cases and acceptable as such

twin index families are identified in TABLE V. There are 1,382 twins, 2,741 full siblings, 134 half-siblings, 74 step-siblings, 1,191 parents, and 254 marriage partners of twin patients, making a total of 5,776 persons who have been uniformly classified according to their mental, social, and genealogical conditions.

The collective schizophrenia rates for the different relationship groups are compared in TABLE VI. The variations in age distribution have been corrected by the use of the "Abridged Weinberg Method." The resulting morbidity rates are average expectancy figures valid for persons above the chief manifestation period, which in this study was assumed to extend from the age of fifteen to forty-four.

Regardless of whether the uncorrected or corrected rates are taken into account, they are in definite accordance with genetic expectation regarding both schizophrenia and schizoid personality. The corrected schizophrenia rate for full siblings amounts to 14.3 per cent, corresponding closely with the collective concordance rate for dizygotic twin pairs (14.7 per cent), although it clearly exceeds the rate for half-siblings (7.0 per cent). A comparison with our previous sibship figures reveals only minor variations which seem sufficiently explained by the different sampling procedures of sibship and descent studies. Our previous schizophrenia rates were 7.6 per cent for half-siblings, 11.5 per cent for full siblings, and 12.5 per cent for dizygotic cotwins.

The newly obtained morbidity figures for step-siblings and marriage partners of schizophrenic index cases are 1.8 and 2.1 per cent, respectively, showing a small excess over the general population rate of 0.85 per cent. So far as this excess is statistically significant, it is referable to the effect of mate selection rather than an expression of socially-induced insanity.

By contrast, the difference in concordance between two-egg and one-egg twin partners ranges from 14.7 to 85.8 per cent. An almost equally striking difference remains, if the comparison is limited to the groups of same-sexed dizygotic and separated monozygotic twin pairs (TABLE VIII). Their morbidity rates vary from 17.6 to 77.6 per cent, and this difference is still so pronounced that explanations on non-genetic grounds are very difficult to uphold. The total morbidity distribution as summarized in TABLE VIII is a rather clear indication that the chance of developing schizophrenia increases in proportion to the degree of consanguinity to a schizophrenic index case. The only other syndrome showing a significant increase in the index families is that of schizoid personality changes, whose genetically heterogeneous nature has been discussed in previous reports.

Concerning the total morbidity rate of 85.8 per cent for monozygotic cotwins, it should be borne in mind that the figure expresses the chance of developing schizophrenia in a comparable environment for any person that has survived the age of forty-four and is genetically identical with a schizophrenic index case, but is not distinguished by the fact of having been selected as the child of such an index case. The last point needs particular emphasis, since it apparently explains the difference between the

TABLE V

Number and Relationship of the Persons Included in the Survey

	Twins	Full Siblings	Half-siblings	Step-siblings	Parents	Husbands and Wives	TOTAL NUMBER
Living	1,198	1,682	84	47	618	221	3,850
Dead	184	1,059	50	27	573	33	1,926
Total Number	1,382	2,741	134	74	1,191	254	5,776

TABLE VI

Incidence of Schizophrenia and Schizoid Personality in the Twin Index Families

		RELATIONSHIP TO SCHIZOPHRENIC TWIN INDEX CASES					
	Parents	Husbands and Wives	Step-siblings	Half-siblings	Full Siblings	Dizygotic Cotwins	Monozygotic Cotwins
Statistically Uncorrected Rates	Number of Persons	1191	254	85	134	2741	517	174
	Cases of Schizophrenia	108	5	1	4	205	53	120
	Incidence of Schizophrenia *	9.1	2.0	1.4	4.5	10.2	10.3	69.0
Corrected Morbidity Rates	Schizophrenia †	9.2	2.1	1.8	7.0	14.3	14.7	85.8
	Schizoid Personality	34.8	3.1	2.7	12.5	31.5	23.0	20.7

* Related to all cases of schizophrenia and to all persons over age 15
† Related only to definite cases of schizophrenia and to half of the persons in the age group 15–44 (plus all persons over age 44)

morbidity rates of 68.1 and 85.8 per cent as found for the children of two schizophrenic parents and for the monozygotic cotwins of schizophrenic index cases, respectively. In fact, it is only by a comparison of these two figures that a satisfactory estimate can be obtained of the extent of biased sampling in a morbidity study dealing with children of schizophrenic index cases. In order to provide such a sample, schizophrenics must have had a chance of getting married and producing offspring.

According to our previous fertility studies (TABLE VII), the total re-

TABLE VII *

Marriage and Birth Rates of Schizophrenic Hospital Patients

	MARRIAGE RATE	BIRTH RATE PER MARRIAGE
General Population	71.0	3.3
Nuclear Group (hebephrenic and catatonic cases)	39.1	1.4
Peripheral Group (simple and paranoid cases without definite deterioration)	70.1	3.1
All Schizophrenics	50.3	1.9

* A diagram in the original article has been converted into this table.

productive rate of schizophrenic index cases is not more than about half that of a comparable general population. However, the decrease in fertility is much more pronounced in the nuclear group of schizophrenia, comprising the deteriorating types of hebephrenia and catatonia, than it is in the paranoid and simple cases. The consequence is that milder schizophrenic cases have a better chance of reproducing a schizophrenic child than have the more severe cases. If the children of one schizophrenic parent will often be the offspring of patients with lessened severity of their symptoms, the children of two schizophrenic parents may be expected to represent an even greater selection of potential schizophrenics in the direction of a highly resistant constitution. Obviously, such a process of natural selection does not operate in persons who have only the distinction of being the monozygotic cotwins of schizophrenic twin index cases.

Clinically it is very important that neither the offspring of two schizophrenic parents nor the monozygotic cotwins of schizophrenic index cases have a morbidity rate of 100 per cent as would be expected theoretically in regard to a strictly hereditary trait. This observation indicates a limited expressivity of the main genetic factor controlling schizophrenia,

TABLE VIII

Expectancy of Schizophrenia and Schizoid Personality in Blood Relatives of Schizo-phrenic Twin Index Cases

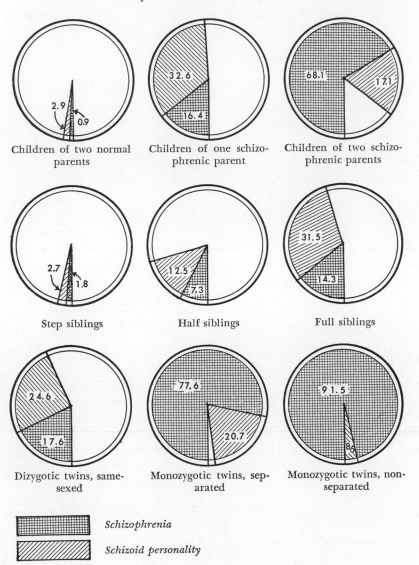

Children of two normal parents

Children of one schizo-phrenic parent

Children of two schizo-phrenic parents

Step siblings

Half siblings

Full siblings

Dizygotic twins, same-sexed

Monozygotic twins, sep-arated

Monozygotic twins, non-separated

Schizophrenia

Schizoid personality

but it should not be misinterpreted in the sense that the extent of the deficit is an adequate measure of the part played by non-genetic agents in the production of a schizophrenic psychosis. From a biological standpoint, the finding classifies schizophrenia as both preventable and potentially curable. The implication is that the main schizophrenic genotype is not fully expressed either in the absence of any particular factor of a precipitating nature or in the presence of strong constitutional defense mechanisms which in turn are partially determined by heredity. The statistical difference between observed and expected morbidity rates for unquestionably homozygous carriers of the schizophrenic genotype does not mean, however, that heredity is effective in only 70 to 85 per cent of schizophrenic cases, or that it is essential merely to the extent of 70 to 85 per cent in any one case.

In order to exclude the possibility that the entire difference in morbidity between monozygotic and dizygotic cotwins might be sufficiently *explained by factors other than genetic*, it is necessary to analyze the morbidity rates for the various sibship groups in relation to any developmental or environmental circumstances peculiar to twins and siblings. In this evaluation of significant similarities and dissimilarities in the life conditions of various relationship groups in our index families, a credible explanation should be sought especially for a finding which is never accounted for by exponents of purely "cultural" theories of schizophrenia. This rather striking observation is that over 85 per cent of our groups of siblings and dizygotic cotwins did *not* develop schizophrenia, although about 10 per cent of them had a schizophrenic parent, all of them had a schizophrenic brother or sister, and a large proportion shared the same environment with these schizophrenics before and after birth.

For anatomical reasons, prematurity of birth, instrumental delivery and reversal in handedness are more common in twins than in single-born individuals. However, no one of these factors appears to have any bearing on the occurrence of schizophrenia in persons who happen to be twins. There is practically no difference between concordant and discordant twin pairs in the frequency of premature birth or instrumental delivery. In discordant index pairs, the vast majority are alike in regard to handedness. In the group of twin pairs showing reversal in handedness, left-handedness occurs about as often in the non-schizophrenic twin partners as it does in schizophrenic twin index cases.[5]

The collective morbidity rates for the cotwins are modified by a variety of secondary factors, genetic as well as non-genetic, but certainly not to an extent which would explain the marked difference between the two types of twins. Variation in relation to the sex factor cannot exist in monozygotic twin pairs and is of equally limited extent in the groups of siblings and dizygotic twins. The range of the former group is from 12.3

---

[5] A table showing similarity and dissimilarity of delivery and handedness has been omitted, and the text of this paragraph has been slightly condensed.

to 16.1 per cent, and that of the latter group from 10.3 to 17.6 per cent (TABLE IX). The difference in morbidity remains constant regardless of whether the siblings and cotwins are male or female. This sex variation is an indication that fraternals belonging to the same sex as a given index case have a greater chance of being alike in any particular circumstances which may favor the manifestation of the schizophrenic genotype. It is clear, however, that these sex variations are by no means extensive enough to permit a non-genetic explanation for the entire difference, or a major

TABLE IX

Variations in the Schizophrenia Rates of Siblings and Twin Partners According to Sex and Similarity or Dissimilarity in Environment

	SIBLINGS OF TWIN INDEX CASES			DIZYGOTIC COTWINS			MONOZYGOTIC COTWINS		
	Male	*Female*	*Total Number*	*Male*	*Female*	*Total Number*	*Separated*	*Non-separated*	*Total Number*
Same-sexed	15.9	16.3	16.1	17.4	17.7	17.6	77.6	91.5	85.8
Opposite-sexed	12.5	12.0	12.3	10.5	10.2	10.3	—	—	—
Total Number	14.0	14.5	14.3	14.3	14.9	14.7	77.6	91.5	85.8

part of the difference, between the concordance rates of monozygotic and dizygotic twin pairs.

The main variations in the morbidity rate of monozygotic cotwins are apparently associated with age at disease onset, type of psychosis, and a variety of extrinsic factors causing significant changes in the physical development and general health status of one twin partner. Most of these modifications in susceptibility or resistance do not lend themselves to statistical analysis, and it is impossible here to enter into a discussion of individual twin histories. It is essential, however, to stress the great variability of such contingent influences, because it is this point which makes the etiology of schizophrenic processes so complex and a carefully adapted program of constructive therapeutic measures so important. Many of our twin histories indicate that incidental factors such as pregnancy, intercurrent disease, or a reducing diet which may have been responsible for the crucial difference between health and psychosis in one twin pair, will not have the same vital effect in others.

The morbidity rate for monozygotic cotwins varies from 77.6 to 91.5 per cent for those twin partners who were or were not separated for over five years prior to disease onset in the index twin (TABLE X). It has already been emphasized that this statistical difference is no adequate expression of the relative effect of extraneous circumstances on the development of schizophrenia in genetically alike persons. Separation is no exact measure

TABLE X

Concordance as to Schizophrenia in Separated and Non-separated Pairs of Monozygotic Twins

	SEPA-RATED * PAIRS	NON-SEPARATED PAIRS	TOTAL NUMBER
Number of Cotwins	59	115	174
Corrected Morbidity Rate of Cotwins	77.6	91.5	85.8

* Separated for five years or more prior to the onset of schizophrenia in the index twin

TABLE XI

Variations in the Average Age at Disease Onset and First Admission of the Monozygotic Twin Index Pairs Concordant as to Schizophrenia

	AVERAGE AGE IN YEARS			PERCENTAGE OF TWIN PAIRS SHOWING DIFFERENCES IN AGE AT ONSET OF SCHIZOPHRENIA					
	First Twin	*Second Twin*	*Difference between Twin Partners*	*No Difference*	*0.1–4 Years*	*4.1–8 Years*	*8.1–12 Years*	*12.1–16 Years*	*16.1–20 Years*
Onset of Disease	22.1	25.6	3.5	17.6	52.9	18.6	10.8	—	—
First Admission	26.0	30.3	4.3	26.5	38.2	21.6	7.8	3.9	2.0

of dissimilarity in regard to environmental agents precipitating schizophrenia. There are numerous factors of potential etiological significance, which are practically universal. In fact, our group of separated one-egg pairs includes twins who developed schizophrenia at almost the same time, although their separation took place soon after birth and led to apparently very different life conditions.

Conversely, even with similar environment it cannot be expected that the time of onset of a schizophrenic psychosis in genetically identical persons will be exactly the same. It is shown in TABLE XI that simultaneous occurrence of schizophrenia is found in only 17.6 per cent of monozy-gotic twin pairs. In about one-half of the index pairs (52.9 per cent) there is a difference of one month to four years, and in over one-quarter the difference may be from four to twelve years. Psychobiologically it is of interest to note that significant dissimilarities in symptomatology are ob-served only in twin partners who show a definite variation in age of onset.

The age discrepancies between twin partners remain about the same

TABLE XII

Distribution of Concordance and Discordance in Twin Index Pairs in Relation to Disease Onset and Environment

	NUMBER OF INDEX PAIRS		AVERAGE DURATION IN YEARS		DISCORDANT PAIRS IN PER CENT	
	Con-cordant	*Dis-cordant*	*Separation in Concordant Pairs*	*Discordance in Discordant Pairs*	*With Similar Environ-ment*	*With Dis-similar Environ-ment*
Monozygotic	120	54	11.1	8.5	61.1	38.9
Dizygotic Same-sexed	34	262	12.9	12.5	57.3	42.7
Dizygotic Op-posite-sexed	13	208	13.8	11.1	50.5	49.5
All Twin In-dex Pairs	167	524	11.8	11.5	55.0	45.0

if the comparison is based on the dates of first admission. The average age at disease onset is 22.1 years for the index twins, and 25.6 years for the cotwins.

There are also certain differences in the period of time during which either the twin partners were under observation before they were classi-fied as concordant or discordant as to schizophrenia, or during which the concordant pairs had been separated before the index twins developed their psychosis. A glance at TABLE XII will reveal, however, that these dif-ferences are entirely insufficient to explain the variations in morbidity between one-egg and two-egg types of twins. The separated concordant twins had lived apart for an average of 11.8 years before disease onset in the first twin, and the discordant index pairs had reached a total average age of thirty-three years at the time of their examination for this survey. All categories of cotwins had at that time been discordant for over eight years since the development of schizophrenia in the index cases.

It is more significant that similarity and dissimilarity of environment

TABLE XIII

Relationship between Similarity or Dissimilarity in Environment and Concordance or Discordance as to Schizophrenia in the Twin Index Pairs

| | COTWINS WITH SIMILAR ENVIRONMENT | | | COTWINS WITH DISSIMILAR ENVIRONMENT | | | ALL COTWINS | | |
| | Number of Cotwins | Rate of Cotwins in Per Cent | | Number of Cotwins | Rate of Cotwins in Per Cent | | Total Number | Rate of Cotwins in Per Cent | |
		Concordant	Discordant		Concordant	Discordant		Concordant with Dissimilar Environment	Discordant with Similar Environment
Monozygotic	114	71.1	28.9	60	65.0	35.0	174	22.4	19.0
Dizygotic	276	7.6	92.4	241	10.8	89.2	517	5.0	49.3
Total Number	390	26.2	73.8	301	21.6	78.4	691	9.4	41.7

are almost equally distributed among the discordant index pairs. The ratio for all discordant pairs is 5.5:4.5, and that for monozygotic pairs alone is 6:4.

Additional evidence against a simple correlation between closeness of blood relationship and increasing similarity in environment with correspondingly intensified pressure toward development of a psychosis is obtained by an investigation of the distribution of concordance and dis-

TABLE XIV

Distribution of Concordance in Relation to Similarity of Environment and Clinical Course of Schizophrenia

	CONCORDANT PAIRS IN PER CENT				
	Not Separated	*Separated— Similar Environment*	*Separated— Dissimilar Environment*	*Completely * Concordant*	*Incompletely * Concordant*
Monozygotic	50.8	16.7	32.5	67.5	32.5
Dizygotic	42.5	2.1	55.3	6.4	93.6
All Twin Index Pairs	48.5	12.6	38.9	50.3	49.7

* As related to the following four classifications:
Group I: No schizophrenia despite similar environment
Group II: Schizophrenia with little or no deterioration (recovery)
Group III: Schizophrenia with medium deterioration
Group IV: Schizophrenia with extreme deterioration

cordance in similar and dissimilar environments in both groups of index pairs (TABLE XIII). This analysis indicates that 22.4 per cent of all monozygotic pairs are concordant without similar environment, and that 49.3 per cent of all dizygotic twin partners remain discordant although they have been exposed to the same environment as an index case.

It may be of some interest that the concordance rate of monozygotic pairs varies from 65.0 to 71.1 per cent according to dissimilarity or similarity of environment, while there is no corresponding increase in the dizygotic group (10.8—7.6 per cent). There can be no doubt, however, that any such variation in relation to environment does not suffice to explain a ratio of 1:6 or 14.7:85.8 per cent, as has been obtained for the morbidity rates of dizygotic and monozygotic twin partners.

That heredity determines the individual capacity for development and control of a schizophrenic psychosis is demonstrated still more clearly, if the similarities in extent and outcome of the disease are taken as further criteria of comparison. This is the objective of the remaining tabulations (TABLES XIV–XVI) which compare the cotwin groups with completely and incompletely similar or dissimilar behavior to schizophrenia, instead of

TABLE XV

Variations in Resistance to Schizophrenia in the Twin Index Pairs

DEGREE OF RESISTANCE TO SCHIZO-PHRENIA	CLINICAL BEHAVIOR TO SCHIZOPHRENIA IN TWIN INDEX PAIRS				NUMBER OF TWIN PAIRS	
	First twin		*Second twin*		*Mono-zygotic*	*Di-zygotic*
	Sub-groups	*Clinical Classification*	*Sub-groups*	*Clinical Classification*		
Complete Dissimilarity	IV	Extremely De-teriorating Type of Schizophrenia	I	No Schizophrenia despite Similar Environment	0	91
	IV	Extremely De-teriorating Type of Schizophrenia	Ia	No Schizophrenia with Dissimilar Environment	0	62
Less Complete Dissimilarity	IV	Extremely De-teriorating type of Schizophrenia	II	Schizophrenia with Little or No Deterioration	9	21
	III	Schizophrenia with Medium Deterioration	I, Ia	No Schizophrenia (Regardless of Environment)	0	197
Complete Similarity	II	Schizophrenia with Little or No Deterioration (Recovery)	II	Schizophrenia with Little or No Deterioration	19	2
	III	Schizophrenia with Medium Deterioration	III	Schizophrenia with Medium Deterioration	33	0
	IV	Schizophrenia with Extreme Deterioration	IV	Schizophrenia with Extreme Deterioration	29	1
Less Complete Similarity	II	Schizophrenia with Little or No Deterioration	I, Ia	No Schizophrenia	54	120
	III	Schizophrenia with Medium Deterioration	II	Schizophrenia with Little or No Deterioration	20	14
	IV	Schizophrenia with Extreme Deterioration	III	Schizophrenia with Medium Deterioration	10	9
TOTAL NUMBER OF PAIRS	All Dissimilar Pairs				9	371
	All Similar Pairs				165	146
	Grand Total				174	517
RATIO	No Schizophrenia to Extremely Deteriorating Schizophrenia				0 : 174	1 : 3.5
	Dissimilar Resistance to Similar Resistance				3 : 55	3 : 1

comparing the twin groups with and without psychotic symptoms as was done by the use of morbidity rates.

Complete similarity has been assumed when both twins either recovered from a mild psychosis with little or no defect (Group II) or reached about the same degree of medium (Group III) or extreme deterioration (Group IV). On the basis of this classification, complete concordance is found in 67.5 per cent of the concordant one-egg twin pairs, but only in 6.4 per cent of the dizygotic pairs (TABLE XIV).

Complete dissimilarity means that the cotwins developed no psychosis despite similar environment (Group I), while the index twins showed an extremely deteriorating type of psychosis. Such a difference does not occur in the group of monozygotic twins, but it ensues in about every sixth dizygotic pair under similar environmental conditions (TABLE XV). This finding implies that the chance of a rapidly progressive psychosis (low resistance) is practically zero for a schizophrenic patient who is the monozygotic twin of, or genetically identical with, a person who remains free of schizophrenic manifestations under similar environmental circumstances. However, the chance of developing a very destructive type of psychosis is 1:3.5, if the person is merely the patient's sibling or dizygotic twin, which means that he is as likely to differ in the inherited elements for a satisfactory resistance as are two brothers or sisters.

In comparing the total groups with dissimilar and similar behavior to schizophrenia, incomplete similarity denotes a difference of only one step between two of the four sub-groups; and incomplete dissimilarity, a difference of two steps. This comparison yields a ratio of 3:55 for the monozygotic pairs, and a ratio of 3:1 for the dizygotic pairs. The difference in similarity of resistance between the two types of twins is expressed by a ratio of 1:55, which far exceeds the difference found in their original morbidity rates. In other words, similar behavior to schizophrenia is about eighteen times more frequent than dissimilar behavior in monozygotic twins, although dissimilarity predominates in dizygotic twin partners.

TABLE XVI expresses the same difference in resistance between one-egg and two-egg twins in rates rather than in ratios, identifying less complete and complete dissimilarity in behavior to schizophrenia with favorable and very favorable resistance, and similar behavior in the deteriorating sub-groups with insufficient resistance. In the monozygotic group, five out of one hundred cotwins of schizophrenic index cases show a tendency to favorable resistance and none shows very favorable resistance, if their twin partners are insufficiently resistant. In the dizygotic group, however, favorable resistance is seen in seventy-two out of one hundred cotwins of insufficiently resistant index cases, and very favorable resistance in about thirty.

The finding indicates that *constitutional resistance* to the main genotype of schizophrenia is determined by a genetic mechanism which is probably non-specific and certainly multifactorial. Taking into account the results of biometric investigations, there is reason to believe that this constitutional defense mechanism is a graded character and somehow correlated with the morphological development of mesodermal elements. For various reasons it does not seem likely, however, that the genetic mechanisms controlling susceptibility and lack of resistance to schizophrenia—that is, the ability to develop a schizophrenic psychosis and the inability to counteract the progression of the disease—are entirely identical with each other. If they are identifiable, it is possible without qualification to accept the recent suggestions of Penrose and Luxenburger that inheritance of schizophrenia may be "the result of many factors."

TABLE XVI *

Rates of Similar and Dissimilar Resistance to Schizophrenia

	DISSIMILAR BEHAVIOR TO SCHIZOPHRENIA (FAVORABLE RESISTANCE)		SIMILAR BEHAVIOR AND RESISTANCE	
	Complete	*Less Complete*	*Complete*	*Less Complete*
Monozygotic Twins	0	5.0	46.6	94.8
Dizygotic Twins	29.6	71.8	0.6	28.2

* A diagram in the original article has been converted into this table.

As far as the *specific predisposition* to schizophrenia is concerned, that is, the inherited capacity for responding to certain stimuli with a schizophrenic type of reaction, the findings of the present study are conclusively in favor of the genetic theory. Our conclusion is that this predisposition depends on the presence of a specific genetic factor which is believed by us to be recessive and autosomal.

The hypothesis of recessiveness is borne out by the taint distribution in the ancestry of our index cases and by an excess of consanguineous marriages among their parents. Of 211 twin index pairs without schizophrenia in their known ancestry, twelve sets (5.7 per cent) originated from consanguineous paternal matings. Of the remaining index pairs, 95 were found to have a schizophrenic parent; 283 had no schizophrenic parent, but schizophrenic cases in the collateral lines of ancestry; and in 102 pairs the available information about the ancestors was considered inadequate. This excess of consanguineous parental marriages in the present survey appears quite convincing, even if a part of it may be due to the fact that our index cases are twins.

Psychiatrically it should be evident that the *genetic theory of schizophrenia*, as it may be formulated on the basis of experiment-like observations with the twin family method, does not confute any psychological concepts of a descriptive or analytical nature, if these concepts are adequately defined and applied. There is no genetic reason why the manifestations of a schizophrenic psychosis should not be described in terms of narcissistic regression or of varying biological changes such as defective homeostasis or general immaturity in the metabolic responses to stimuli. Genetically it is also perfectly legitimate to interpret schizophrenic reactions as the expression either of faulty habit formations or of progressive maladaptation to disrupted family relations. The genetic theory explains only *why* these various phenomena occur in a particular member of a particular family at a particular time.

The general meaning of this genetic explanation is that a true schizophrenic psychosis is not developed under usual human life conditions unless a particular predisposition has been inherited by a person from both parents. Genetically it is also implied that resistance to a progressive psychosis does not break down without certain inherited deficiencies in constitutional defense mechanisms, the final outcome of the disease being the result of intricate interactions of varying genetic and environmental influences. Another genetic implication is that a schizophrenic psychosis can be both prevented and cured. The prerequisite is that the psychosomatic elements, which may act as predispositional, precipitating, or perpetuating agents in such a psychosis, are morphologically identified, and that the complex interplay of etiologic and compensatory mechanisms is fully understood. Pragmatic speculation will be no aid in reaching this goal.

SUMMARY

1. The methods available for genetic investigations in man are the pedigree or family history method, the contingency method of statistical prediction, and the twin study method.

2. A study of the relative effects of hereditary and environmental factors in the development and outcome of schizophrenia was undertaken by means of the "Twin Family Method." The study was organized with the co-operation of all mental hospitals under the supervision of the New York State Department of Mental Hygiene. The total number of schizophrenic twin index cases, whose cotwins were available for examination at the age of fifteen years, was 794.

3. In addition to 1,382 twins, the 691 twin index families used for statistical analysis include 2,741 full siblings, 134 half-siblings, 74 step-siblings, 1,191 parents, and 254 marriage partners of twin patients. The random sampling of these twin index pairs is indicated by the distribution of 174 monozygotic and 517 dizygotic pairs, yielding a ratio of about 1:3.

4. The morbidity rates obtained with the "Abridged Weinberg Method" are in line with the genetic theory of schizophrenia. They amount to 1.8 per cent for the step-siblings; 2.1 per cent for the marriage partners; 7.0 per cent for the half-siblings 9.2 per cent for the parents; 14.3 per cent for the full-siblings; 14.7 per cent for the dizygotic cotwins; and 85.8 per cent for the monozygotic cotwins. This morbidity distribu-

tion indicates that the chance of developing schizophrenia in comparable environments increases in proportion to the degree of blood relationship to a schizophrenic index case.

5. The differences in morbidity among the various sibship groups of the index families cannot be explained by a simple correlation between closeness of blood relationship and increasing similarity in environment. The morbidity rates for opposite-sexed and same-sexed two-egg twin partners vary only from 10.3 to 17.6 per cent, and those for non-separated and separated one-egg twin partners from 77.6 to 91.5 per cent. The difference in morbidity between dizygotic and monozygotic cotwins approximates the ratio of 1:6. An analysis of common environmental factors before and after birth excludes the possibility of explaining this difference on non-genetic grounds.

6. The difference between dizygotic and monozygotic cotwins increases to a ratio of 1:55, if the similarities in the course and outcome of schizophrenia are taken as additional criteria of comparison. This finding indicates that constitutional inability to resist the progression of a schizophrenic psychosis is determined by a genetic mechanism which seems to be non-specific and multifactorial.

7. The predisposition to schizophrenia—that is, the ability to respond to certain stimuli with a schizophrenic type of reaction—depends on the presence of a specific genetic factor which is probably recessive and autosomal.

8. The genetic theory of schizophrenia does not invalidate any psychological theories of a descriptive or analytical nature. It is equally compatible with the psychiatric concept that schizophrenia can be prevented as well as cured.

CHAPTER

5

R. Nevitt Sanford

---

* PHYSICAL AND PHYSIOLOGICAL

CORRELATES OF PERSONALITY

STRUCTURE

In the study of constitutional determinants of personality, the most convincing results have, up to this point, been achieved in two lines of investigation. The first, the genetic, has been illustrated in the two preceding papers. The second is the relationship between body build and personality. Dr. Sanford's is the first of three readings that report findings in this area. He also deals with observed correlations between personality syndromes and properties of the endocrine glands, the parasympathetic nervous system, and diet. The endocrine glands may well turn out to be the principal forces controlling body build. The terminology referring to "needs" (such as "Blamavoidance") is that developed by Murray and his associates in Explorations in Personality.

ALL the manifest personality syndromes in our group, with the exception of Willing Obedience, are correlated positively (.26 to .52) with Tall, Narrow Build. The highest correlations with this type of physique appear

NOTE: Reprinted from the chapter "Correlates of Manifest Personality Syndromes" in *Physique, Personality, and Scholarship*, by R. Nevitt Sanford, et al. (*Monographs of the Society for Research in Child Development*, Vol. 8, Ser. No. 34, No. 1, 1943), pp. 526–9. This extract (reprinted by permission of the author and of the Society for in the case of those syndromes—Counteractive Endocathection, Self-sufficiency, Guilt, and Remorse—which express, according to our hypothesis, pronounced inner life or inner emotional conflict. A consideration of individual subjects shows that our positive correlations are due not so much to the fact that the tall and thin subjects tend to be high as to the fact that the short and wide subjects tend to be low on the present syndromes. Of the eight highest subjects on Tall Narrow Build, three are high on Self-sufficiency; of the seven low subjects on the Tall and Narrow syndrome, five are low on either Self-sufficiency or Counteractive Endocathection. Subjects of this latter group are, with only one exception, high on either Social Feeling or Placid Immobility, patterns which stand in contrast to those under discussion and which will be discussed later. This fits in with the fact that Willing Obedience, a pattern in which there is a large social component, fails to correlate with Tall, Narrow Build. Thus, though we cannot say of our conscientious, endo-cathected subjects that they tend to be tall and thin, we can affirm with some definiteness that they tend not to be short and wide. To take it the other way round, we may say of our short and wide people that they tend not to be characterized by inner life or emotional conflict, but to be either socially responsive or placidly immobile.

Conscientious Effort and Orderly Production correlate significantly, .41 and .40, respectively, with Co-ordination. Good Co-ordination seems to fit in well enough with these two syndromes and to contribute to a picture of smooth or harmonious functioning. One might suppose that the same inner structure—a Superego that is well integrated with a strong Ego—that gives rise to such traits as Conjunctivity, Endurance, the need for Order, and the need for Construction likewise promotes good physical co-ordination. At the same time it is reasonable to suppose that good physical co-ordination, present from the beginning, might be a factor which favors the development of a strong Ego. That Co-ordination should fail to correlate with the other syndromes of our groups leads us to stress certain distinctions between them and the two patterns under discussion. In Guilt and Remorse more than in Conscientious Effort or Orderly Production there is inner conflict, conflict between strong instinctual impulses and a strong Superego as we may suppose. The need for Blamavoidance and the need for Abasement can be understood as efforts, sometimes successful, to hold the bad impulses in check. If an individual is afraid of what he might do and must therefore bring inhibition or over-control to bear upon his actions, it is not difficult to see how his physical co-ordination might suffer. Guilt and Remorse, it may be re-

Research in Child Development) is from a detailed report of research carried on by the Harvard Psychological Clinic and the Department of Child Hygiene, Harvard School of Public Health. Forty-eight children between the ages of five and fourteen years were studied intensively. The investigation included a great deal of biological study and emphasized the interrelationship between physiological, intellectual, personality, and environmental variables.

called, is correlated .15 with Anxiety. No subject who was high on Guilt and Remorse and on Anxiety was high on Co-ordination. (Anxiety correlates —.24 with Co-ordination. Of the twelve high Anxiety subjects, five are low and one is high on Co-ordination.) In Counteractive Endocathection and Self-sufficiency, more than in the patterns which are significantly correlated with Co-ordination, there is inner life. We might say that the subject prefers to devote himself to the things of the mind or to inner preoccupations and through lack of practice in physical activity fails to develop good co-ordination. There would appear to be, however, no real interference with co-ordination, for the correlations are around zero. Willing Obedience appears to be unrelated to Co-ordination.

Several of the personality syndromes we are considering, Conscientious Effort, Guilt and Remorse, Willing Obedience, and the fantasy syndrome Strong Character—and only these from all our personality syndromes—show small positive correlations with the Advanced Hormone, Osseous Development syndrome (.22 to .28). Most of the variables which comprise these syndromes increase with age, and we may therefore think of these patterns as characteristic of our more mature subjects. In spite of the fact that the correlations are low—statistically not significant—they afford some slight evidence of the co-variation of physiological and psychological maturity.

In this connection a correlation of .30 between the Anti-social fantasy syndrome and the Advanced Hormone pattern should be noted. A glance at the subjects who figure in this correlation shows four reversals. Judy is high on the Antisocial fantasy syndrome but low on Advanced Hormone. She is a seven year old child whose antisocial tendencies are expressed in her behavior as well as in her fantasies: her tendency to steal has been a problem. Three subjects, Barb, Ned, and Ted, were low on the Antisocial fantasy syndrome but high on Advanced Hormone. Two of them were high on the Attack syndrome. If we turn now to the subjects who are high on Advanced Hormone, we find that *all* of them are high on one or another of the manifest or covert personality syndromes under discussion. Of the subjects who are low on Advanced Hormone only four are high on any one of these personality syndromes. The point to this discussion is this: all the personality syndromes considered here can be understood as reflecting a conflict in the subject between strong instinctual impulses and strong inhibiting forces. That these syndromes seem to be related to Advanced Hormone might be explained by the following hypothesis: onset of puberty produced by increasing amounts of circulating sex hormones in the blood stimulates the individual toward expression of instinctual biologic urges; as these latter become stronger the forces for inhibiting and controlling them become stronger, hence, a positive relationship between Advanced Hormone and those syndromes in which a manifest need to avoid blame and covert antisocial needs—with fear of retaliation—are essential components.

Several of the present personality syndromes correlate positively with the Parasympathetic Response syndrome: Counteractive Endocathection .37, Strong Character .44, Conscientious Effort .22. These correlations would seem to give an indication that it is inner emotional life that is accompanied by this particular pattern of autonomic activity. One might venture the hypothesis that the subjects who exhibit these syndromes have emotions but strive to inhibit them, to keep them to themselves, so that the tension is discharged through flushing, sweating, and so forth, rather than through verbal or muscular action. It is noteworthy that Para-sympathetic Response correlates negatively with all those manifest personality syndromes—Social Feeling, Good Fellowship, Lively, Self-expression—in which social responsiveness or outgoingness is the central component. If the subject "lets himself go" emotionally, it seems he is less likely to exhibit autonomic reactivity. Parasympathetic Response correlates .33 with Tall, Narrow Build, the type of physique which seems to go with pronounced inner life rather than with social responsiveness. It is possible on this basis to compose a broader picture in inner life, tallness, thinness, and parasympathetic response and to contrast, this picture with one of social responsiveness, shortness, wideness, and absence of parasympathetic activity. The correlations among the several syndromes are not high, but it is possible to pick out several subjects who exemplify in detail each of these broader pictures.

There are significant correlations between good General Diet and several of the present personality syndromes: Conscientious Effort .41, Orderly Production .52, Counteractive Endocathection .52. These correlations are to be explained, it seems, not on the basis that good diet gives rise to physiological processes which in turn promote these kinds of psychological reactions, but rather on the basis that in certain cases the type of family situation which favors good diet also favors these personality syndromes. We may postpone further discussion of these relationships, therefore, until we have examined the family backgrounds of subjects high on the present syndromes.

It must be noted that General Diet correlates .50 with Male Sex. This fact does not, however, seem to bear upon the above correlations of General Diet with personality syndromes, for the latter appear to be more or less unrelated to Male Sex. That the boys should rate higher than the girls on General Diet might be explained on the basis that they are more active physically and have greater caloric requirements than the girls, and hence are less likely to criticize and refuse food put before them.

CHAPTER

6

David M. Levy

PSYCHOSOMATIC STUDIES OF SOME
ASPECTS OF MATERNAL BEHAVIOR

In the introductory paragraph to his article, Dr. Levy refers (on the basis of previously published research) to the interrelationship between constitutional determinants and such situational determinants as infidelity of a husband, illness or deformity in children. The present paper, however, is devoted to a strictly constitutional factor—evidence for a correlation between certain physiological traits of women and their varying manifestations of "maternal" behavior. It is clear that, at one remove, this constitutional determinant in mothers becomes a situational determinant for their children. That is, the child who happens to have a mother who is strongly "maternal" or "non-maternal" has his personality formed in a special kind of environment.

IN RECENT studies of overprotection [6], in which a group of consistently and highly maternal mothers was investigated, it was found that their maternal behavior was a characteristic feature of the personality in childhood as well as in later life. In a study of maternal rejection, in which an investigation was made of mothers distinctly non-maternal, a consist-

NOTE: Reprinted from *Psychosomatic Medicine*, Vol. 4 (1942), pp. 223-7, by permission of the author and the American Society for Research in Psychosomatic Problems. (Copyright, 1942, by The Williams and Wilkins Company.) The appendix giving a sample questionnaire on maternal attitudes has been omitted.

ency of this type of behavior was found also in childhood. Further, it was noted that the same kind of experience which tended to make the maternal woman more maternal had the opposite result, if any, in the non-maternal: experiences, for example, of infidelity of the husband, a long period of relative sterility, illnesses or deformity in the offspring.

Such considerations, among others, led to a search for constitutional factors in maternal behavior.

Ordinary observation is in conformity with the theory of the constitutionally maternal. Indeed, some women appear to be natural-born mothers, regardless of previous experience or training; while others, in spite of courses in child care and apparently all the good will in the world, are inept as mothers and dislike the maternal role. That such differences are not based entirely on particular individual experience is attested by certain experiments in animals. It may be argued that analogies of animal and human behavior in this regard are beside the point. However, it seems unnecessary to point out that in animals, certainly in quadrupeds, as in man, the basic pattern of maternal behavior, in the form of protecting and feeding the infant, is essentially the same. Problems of overprotection and rejection have been described in apes. Natural differences in the strength of the maternal drives in animals have been observed. Of the experimental studies, most pertinent appear to be those of Wiesner and Sheard [8] in which the maternal drive was studied among virginal laboratory rats, reared in isolated cages. The retrieving of baby rats was used as a measure of the drive. Differences were found not only of the all-or-none variety, but also of degree of skill in the method of retrieving. Such differences, in view of their completely isolated lives, were attributed, most plausibly, to innate differences in the animals studied, probably to quantitative differences in the hormones related to specific maternal function.

A first task in our study was to formulate an interview dealing with the subject of maternal behavior. This was done by inquiring into the maternal behavior of quite obviously highly maternal and non-maternal women, using a number of leads derived from the study of maternal over-protection. The classification of mothers into high, low, and middle groups, on the basis of office interviews, is relatively simple. Besides the interview, one has the advantage of direct study of the child, and other sources of information, e.g., father, governess, and others. Mothers of patients are, furthermore, quite receptive to such an investigation, since it is an integral part of the psychiatric study of the child. After some experience, it was found that the interview could readily be applied to women, married or unmarried, though with childless women, naturally, the test of maternal behavior with their own children is lacking. The interviews were made the basis of questionnaires and, so far, have been applied to about four hundred college students. This study, however, is concerned primarily with the results of the interviews of seventy-two women, all but six mothers of children treated for various behavior problems. The exceptions were personal acquaintances, who may be regarded as a control group.

Examples of interviews scored in the high, low, and middle group will illustrate the classifications.

As an example of "high maternal," consider case 14, a mother of four children. As a young child, her favorite game was taking care of dolls, dressing them, putting them to bed. She played with dolls until age fourteen or fifteen. She used to make visits among her mother's friends to take care of their babies. When she thought of being a mother, she hoped to have six children, and have them as soon as possible. When she saw a pretty baby on the street, she had a strong urge to take it in her arms and hug it. She was a "baby-carriage peeker" before, as after, marriage. In her relations with men she was always maternal; much more, she said, than they liked.

Actually, she had four children and is now pregnant with her fifth. She had a nurse for her first child and was miserable, she said, because she couldn't take full care of it. She hated the hospital rule of not having the baby in her room. She fed all her children at the breast, and with ease. She had a copious supply of milk.

Her husband stated that she really spoiled the children; that every so often she fought against this tendency and became severe, to protect them from her spoiling. But the children "see through it."

There is no difficulty in rating the case just cited in the high maternal group. Nor is there difficulty in classifying the following case in the low maternal group. It concerns the mother of two children, who never played any "maternal" games in childhood, nor played a maternal role to another child. She had very little interest in dolls and stopped playing with them when about age six. When she saw a pretty baby on the street, she was not at all interested. As an adolescent, she never indulged in the fantasy of being a mother and having children. She was ambitious to get married, but never thought about having children. As a mother she has felt quite incompetent. She took her children off the breast after two weeks, because she didn't like it; she felt like a cow, she said. She still hates the physical care of children, though she is a dutiful mother and rather affectionate. She never was maternal towards men. Her interests have always been feminine, and she has been quite popular with men.

The midgroup contains every variety of low and high. In this group, all mothers are included who appear to come somewhere between high and low or contain mixtures of both. This is in contrast with the large group rated by questionnaires in which each reply to a question is scored and the total score determines the classification.

An example of a woman rated in the midgroup is the mother of one child, case 52. She played with dolls probably to age five or six. She was not especially interested in maternal play in childhood. For a period of two years, age twelve and thirteen, she used to look after some neighbor's children because she liked to. She never anticipated the number of her children or had fantasies of being a mother. When she saw a pretty baby on the street, she was interested but had no urge to hold it, or have one of her own. After marriage, however, she became pregnant very soon and willingly. She voluntarily took sole care of her child and is evidently an affectionate mother. Her pregnancy was very difficult and she was warned by her physician against further impregnation.

In relation to maternal behavior, the following inquiries were found to be most pertinent: play with dolls (the roles most frequently selected and the age when interest in dolls was over); voluntarily mothering in childhood; the number of children anticipated and general maternal fantasies; general response to babies; anticipation of care and of breast feeding; vocation and avocation; self-rating on a scale of maternal feeling toward children.

Of the variations in maternal patterns that offered some difficulty in classification were women who were very fond of babies only after they could talk sentences and were "little personalities" (at age two years and older); also, occasionally women who had no interest in babies until after marriage, and women who showed no interest in babies other than their own.

After the classifications were established and their accuracy determined by having them checked by others, an attempt was made to match them against various body measurements and data of menstruation. In a group rated by means of questionnaires, data were assembled on body weight, height, shoulder width, hip width, gross size of breast, diameter of areola, menstruation, and general body configuration.

In this study of seventy-two women the three classifications were checked against the duration in days of menstrual flow, corrected for a twenty-eight-day cycle.

Since of the two variables checked against each other, the classification of maternal behavior has been described in detail, it will be necessary to consider the accuracy of the data on menstruation.

It is now well-established that data on the length of the menstrual cycle, as revealed in the medical literature, have been notoriously inaccurate. Studies in which the environment was sufficiently well-controlled so that adolescent girls could receive menstrual pads only from designated nurses who kept proper records, and the use of specially designated calendars, have demonstrated a high degree of irregularity in the length of the cycle [1, 2, 3, 4, 5, 7]. Comparing preliminary statements received in the usual interview and day to day records on forms over a period of six months, McCance, Luff, and Widdowson [7] found that women's statements about regularity are erroneous, but the days of duration of flow, accurate. Arey stated as his opinion that women give false replies about regularity of the cycle because they think that irregularity means abnormality.

In the interviews from which the data for this study were derived, special pains were taken to secure as nearly accurate information as memory would allow.

Questions were asked to determine the year of onset, the irregularities of the years following the menarche, and the estimated year at which the adult type of menstruation was established. Questions were asked on the assumption that irregularity is the rule. The widest range in the length of the cycle and the usual length were determined by questioning. The same method was used to determine the duration of flow. Thus in Case 11, the range in length of cycle is four to seven weeks, the usual interval six weeks. The duration of flow is almost always six days. Accepting six weeks as the usual period, and six days as the usual duration, the corrected duration on the basis of a four-week cycle is four-sixths of six, or four.

An example of marked early irregularity with gradual improvement is seen in Case 14. Menstruation started at 13 years. The duration of the flow was ten days to two weeks, exceptionally eight days and two and a half weeks, until marriage at 21 years. After marriage, duration of flow was usually ten days, never less, until about age 27; since then, the range is eight to ten. The length of the cycle was 20 to 28 days until about age 26; since then it has become more regular, ranging 25 to 29 days, mostly 27. The mother is now 37 years old.

Of the 72 women interviewed, irregularities in the cycle of one week or longer occurred in 14 (20 per cent). The majority had irregularities under one week, mostly one to three days. As in other studies the large majority had 28-day intervals. As in other studies, also, a progressive trend toward regularity following the menarche was found.

In 15 per cent, the duration of the period was three days; in 25 per cent, four; in 29 per cent, five; in 13 per cent, six. The remainder, 18 per cent, were less than three days or more than six days duration. The percentages for the three-, four-, and five-day periods are each within 2 per cent of those compiled by Barer and Fowler, whose studies of blood loss during menstruation made the most accurate figures available on duration of flow, in 100 women, since they collected the menstrual blood. Our figure for the six-day period is 6 per cent less than theirs. Since the most important figure in our study is represented by the number of days of flow, their study offers a valuable check on the accuracy of our figures.

A rather high and significant positive correlation was found between maternal behavior and duration of menstrual flow.

$$r = .579$$
$$P.E._r = .053$$

Expressed in another way, the majority of the subjects with four-day periods or less were classified in the group of low maternal; the majority with periods six days or longer, in the high maternal. Those with a five-day period showed nearly the same proportion in low, middle, and high. The percentages of those in the group of high maternal ascended as the number representing duration of flow ascended. Further, as the correlation study indicates, the finding cannot be attributed to chance.

Questionnaire studies may reveal other correlations of somatic and maternal behavior, possibly, of figures bearing on lactation. So far, at least, one psychosomatic ingredient of maternal behavior appears to be established.

The correlations of the various data and "maternal score," as determined by questionnaire studies, are given in the summary table that follows.

TABLE XVII

Correlations with Score of Maternal Behavior of:

Duration of Menstrual Flow
(Utah) $r = -.072 \pm .00027$
(Smith-Bklyn) $r = \quad .47 \pm .037$

Diameter Areola
(Utah) $r = \quad .232 \pm .0472$
(Utah-Bklyn) $r = \quad .33 \pm .0357$

Size Nipple
(Utah) $r = \quad .137 \pm .0523$
(Smith-Bklyn) $r = \quad .064 \pm .0506$

Weight in Relation to Height
(Utah) $r = \quad .135 \pm .0513$
(Smith-Bklyn) $r = -.074 \pm .0499$

Height
(Utah) $r = -.151 \pm .0513$
(Smith-Bklyn) $r = \quad .022 \pm .0492$

Hip Width
(Utah) $r = -.078 \pm .052$

Hip Width—Shoulder Width Ratio
(Utah) $r = -.104 \pm .0519$

Menarche
(Utah) $r = -.138 \pm .0508$
(Smith-Bklyn) $r = \quad .087 \pm .0485$

In each group there were 160 to 223 students from the colleges designated.

REFERENCES

1. Arey, L. B.: "The Degree of Normal Menstrual Irregularity," *Amer. J Obstet. Gynaec.*, Vol. 37 (1939), pp. 12–29.
2. Barer, A. P., and Fowler, W. M.: "The Blood Loss during Normal Menstruation," *Amer. J. Obstet. Gynaec.*, Vol. 31 (1936), pp. 979–86.
3. Breipohl, W.: "Untersuchungen über den Menstruationszyklus in der Men-

archezeit," *Zentralbl. f. Gynäk.*, Vol. 61 (1937), pp. 1335–42; "Untersuchungen über des Menarchealter auf Grund von Kontrollen durch Lehrerinnen," *Arch. f. Gynäk.*, Vol. 161 (1936), pp. 399–401.

4. Engle, E. T., and Shelesnyak, M. C.: "First Menstruation and Subsequent Menstrual Cycles of Pubertal Girls," *Human Biol.*, Vol. 6 (1934), pp. 431–53.

5. Gunn, D. L., Jenkins, P. M., and Gunn, A. L.: "Menstrual Periodicity: Statistical Observations on a Large Sample of Normal Cases," *J. Obstet. Gynaec.*, Brit. Emp., Vol. 44 (1937), pp. 839–79.

6. Levy, D. M.: "Maternal Overprotection," *Psychiatry*, Vol. 1 (1938), pp. 561–91; Vol. 2 (1939), pp. 99–128, 563–97; Vol. 4 (1941), pp. 393–438, 567–626.

7. McCance, R. A., Luff, M. C., and Widdowson, E. E.: "Physical and Emotional Periodicity in Women," *J. Hyg.*, Camb., Vol. 37 (1937), pp. 571–611.

8. Wiesner, P. B., and Sheard, N. M.: *Maternal Behavior in the Rat* (Oliver and Boyd, London, 1933).

Part Two

SECTION III

INTERRELATIONS BETWEEN CONSTI-
TUTIONAL AND GROUP-MEMBERSHIP
DETERMINANTS

INTRODUCTION

DR. JOHN GILLIN has shown very graphically how, within limits set by anatomy and physiology, human beings have evolved a variety of distinctive responses to the same biological needs.[1] Dr. Cora DuBois has also pointed out how the hunger drive is channeled by the cultural patterns for child training and how, in turn, the results of this interaction ramify into the whole structure of the culture and play a part in determining characteristic and preferred personality traits.[2]

Though every biological fact is given a social meaning, the stuff provided by heredity is not infinitely plastic. As Dr. Wayne Dennis has observed, we must not infer that the fitting of man into social patterns is strictly analogous to the pouring of wax into a mold. "The metaphorical wax, neurological, glandular, and otherwise, which comprises the human individual" has tendencies towards shapes of its own. The work of the agents of a society in shaping new organisms in the direction of preexisting patterns is more comparable to that of a skillful gardener who starts with a natural setting which he can never completely alter or control. But, with a prearranged form in mind, he can limit this growth, encourage that, destroy some weeds entirely for at least a season, accelerate a flowering, shape by pruning, train vines to grow in directions they would not have followed if only natural forces had had their way.

It is this extraordinarily complicated but extraordinarily interesting subject—the interrelations between biological, social, and cultural determinants—to which this section is devoted.

The problem of the interrelationship between constitutional determinants, on the one hand, and the group-membership, role, and situational determinants, on the other hand, is the old problem of "nature and nurture." In this section, we shall be dealing with only that aspect of "nurture" that is a consequence of membership in a nation, tribe, social class, or other enduring and recognized group.

---

[1] "Custom and Range of Human Response," *Character and Personality*, Vol. 13 (1944), pp. 101–34.

[2] "Attitudes toward Food and Hunger in Alor," in *Language, Culture, and Personality*, edited by Leslie Spier (1941), pp. 272–81.

One of the most central problems of this area is that of "instinct"—that is, of human behavior that occurs as a result of strictly biological processes alone, regardless of training in a particular socio-cultural environment. Dr. Wayne Dennis has reported upon an ingenious experiment, which shows that a number of responses in human infants develop even when social stimulation is minimal.[3] Dr. Dennis has presented other evidence that culture has little influence on behavior during the first year of life.[4] In these and other writings, Dr. Dennis has developed the contrast between "autogenous" and "sociogenous" behaviors, that is, between those manifestations of personality that appear almost automatically as a consequence of biological maturation[5] and those that are markedly affected by socio-cultural training. This theory, if it be borne out by other investigations, is obviously of profound significance for students of personality in culture. It means that personality has a substratum that is common to all human beings everywhere. This implies common limitations as well as potentialities. The effects of membership in a group, of role-playing, of accidents in life-experience are constrained within boundaries set during the first year of life by the autogenous patterns.

---

[3] "Infant Development under Conditions of Restricted Practice and Minimum Social Stimulation," *Genetic Psychology Monographs,* Vol. 23 (1941), pp. 143–89.

[4] "Does Culture Appreciably Affect Patterns of Infant Behavior?" in *Readings in Social Psychology* (T. M. Newcomb and E. L. Hartley, ed., 1947), pp. 40–6.

[5] Margaret Mead, "On the Implications for Anthropology of the Gesell-Ilg Approach to Maturation," *American Anthropologist,* Vol. 49 (1947), pp. 69–78.

CHAPTER

7

Margaret Mead

ANTHROPOLOGICAL DATA ON THE
PROBLEM OF INSTINCT

In the following paper, Dr. Mead states succinctly the working assumptions cultural anthropologists ordinarily make with respect to "instinct." It is to be noted that she accepts the reality of the autogenous determinants when she says that "we may expect ultimately to identify in human beings an original nature which has very definite form or structure. . . ."

THE social anthropologists' contribution to our knowledge of original nature, of which the instinct problem is one formulation, is based upon the utilization of existing cultures as our experimental material for the plasticity of original endowment. In the face of the extreme difficulty of experimenting with human beings in organized social settings, the accidents of history, especially where cultures have been preserved from contact with the main stream of history, are used as data on the possible variations in human behavior. In such an approach, we are working with two sets of variables, the biological equipment of human beings and the historical forms within which this equipment has been patterned. We there-

NOTE: Reprinted from *Psychosomatic Medicine*, Vol. 4 (1942), pp. 396–7, by permission of the author and the American Society for Research in Psychosomatic Problems. (Copyright, 1942, by The Williams and Wilkins Company.)

fore "hold human nature constant" and assume, for the present, that human nature may be regarded as similar for Bushmen, Hottentots, Americans of 1941, etc. Were we not to do so, it would be impossible to conduct such a series of experiments. We can, then, collect data on the plasticity of human beings, under different cultural conditions, data ranging all the way from slight details—such as gestures of assent and dissent, parts of the body associated with introspective reports of emotion, phonetic systems used in the language, variations in the perception of intra-organic and extra-organic data—through patternings of interpersonal relations, and systems of social organization governing the relationships of social groups to each other and to the physical environment. On the basis of such data, we can do two things: (1) criticize existing hypotheses concerning the plasticity of man's original nature, because they fail to provide an adequate base for an explanation of some type of recorded behavior in other cultures, and (2) develop hypotheses which will broaden the present theories involved.

It is important to note that it is not suggested that anthropological data is advanced as *proof* of the final validity of any theory. It is assumed that hypotheses about the instinctive nature of man will ultimately become so refined that they can be put to experimental tests of a quite different order from the tests provided by "experiments in nature." Our data is hypothesis-forming, not theory-validating.

In "holding nature constant," the anthropologist is faced with certain problems: Should he assume a *tabula rasa?* Should he assume some contemporary theory of human nature, such as the series of Freudian assumptions on instinct, or the Hullian approach to the problems of drives? Should he make explicit provision for the possibility of hereditary constitutional types with systematic differences in innate equipment, stated perhaps as systematically different patterns of maturation, or relative strength of a number of constant drives, etc.? What weight should be given to the possible complications introduced by differences between homogeneous and heterogeneous physical stocks, with the possibility that, if there are important individual differences in instinctive equipment, these differences may have become systematically characteristic of groups with a high proportion of common ancestry? What distinction should he make between the usefulness of existing formulations on the instinct problem as "tools of observation," and as "basic assumptions in the interpretation of data?"

We have here two main questions: (1) What sort of contribution can the anthropologist make to the theory of instincts? (2) What procedures involving instinct hypotheses does he find most helpful in making these contributions? It will readily be seen that these two problems are interdependent. If an existing hypothesis about the number, relative strength, degree of opposition or mutual reinforcement of different drives, be used to define and organize the observations in the field, are the results as valuable, in terms of a contribution to the instinct theory, as if no such special assumptions are made? To this question, I think the answer is an unequivocal negative, on two counts. Accepting any current instinct hypothesis narrows and limits the lines of research in such a way as to obscure data on plasticity—which is all that we can be expected to supply —and, furthermore, material collected from a limiting point of view can not be used for a series of rechecks on developing hypotheses. Conditions of anthropological field work are such that it is usually impossible to make a series of field studies of a primitive people to correspond with a series of shifts in psychological theory. Our greater contribution lies in collecting materials in such a way as to provide useful cross-checks on instinct hypotheses.

This does not mean, however, that the assumption of a *tabula rasa* type of human nature provides anthropologists with a background from which the most relevant types of data on plasticity can be collected. If the *tabula rasa* hypothesis is used, all similarities in behavior found in different and apparently historically discontinuous areas of the world have to be explained in terms of diffusion and borrowing. Unless we assume some series of regular cultural evolution, a series of super-organic evolutionary laws, which there are at present no valid grounds for assuming, anthropological data on widespread similarities becomes relatively valueless, psychologically. No matter how widely separated in space and time, the historical explanation becomes the only acceptable one in the discussion, for instance, of similar initiatory practices in New Guinea, Africa, Tierra del Fuego, and Australia. Each piece of cultural behavior must be referred to some historically antecedent piece of behavior, the origins of which we can never hope to examine, or test, or even profitably speculate about. The examination of primitive cultures in these terms would leave us psychologically exactly where we started.

The most useful assumption seems to be that we may expect ultimately to identify in human beings an original nature which has very definite form or structure, and possibly systematic individual differences which may be referred to constitutional type within that original nature, but we must make our investigations without particularization in regard to that original nature. By following this course, we can bring under consideration those widespread similarities in cultural behavior which occur in different parts of the world, at different levels of cultural development. Similarities in the personalities of natives of the Manus tribe of the Admiralty Islands and the inhabitants of rural New England can be used as data on the degree of structuralization which original drives contribute to the formation of similar personalities under widely divergent historical circumstances, and furthermore can be used to focus attention upon those aspects of the cultural system which make the most significant contribution to the modification of the original human nature along these parallel lines.

I should like to discriminate between the value of using specific instinct hypotheses *in the field,* as a theoretical frame for the *collection* of data, and the value of testing out existing hypotheses in subsequent *analysis* of the data. Such distinctions as Anna Freud draws in her study of the Ego and its Defense Mechanisms can most illuminatingly be applied to anthropological data, and it is most probable that, in the course of application, new insights may be developed which will contribute to the development of psychological theories of instinct.

CHAPTER

8

Lawrence K. Frank

CULTURAL CONTROL AND PHYSIO-
LOGICAL AUTONOMY

In the following selection, Mr. Frank shows very neatly how certain homely biological regularities that we tend to think of as purely biological are never encountered among normal adult humans except as culturally patterned. He goes on to compare the organization of a personality to the organization of a culture.

IF WE start with the newborn infant, we see an organism that has, in varying degree, well-organized physiological processes, of which, for our present discussion, we may distinguish three: (1) the physiological process concerned with metabolism that operates to reduce blood sugar, generate gastric contractions and the resultant tension and discomfort we call "hunger"; (2) the accumulation of urine and feces in hollow viscera where increasing pressure serves to release the sphincters and thereby permit evacuation; (3) the extraordinarily well-developed capacity for organic response to what we call "emotional stimuli" that evoke the physiological disturbances of anger or rage and fear.

NOTE: *American Journal of Orthopsychiatry*, Vol. 8 (1938), pp. 622–6. (Copyright, 1938, by the American Orthopsychiatric Association, Inc.) The author and the American Orthopsychiatric Association have kindly given permission to reprint this paper with slight abridgment.

The processes through which the child surrenders his physiological autonomy and accepts the cultural control is a very concrete series of operations performed by the cultural agent, mother or nurse, who persuades, cajoles, coerces, or otherwise treats the child so that the following modifications take place:

1. The infant learns to regularize the reduction of blood sugar so that he can tolerate progressively longer intervals between the intake of food and gradually learns to accept for nourishment those foodstuffs which are favored by his culture as appropriate and desirable. This process may be described as the transformation of hunger into appetite, whereby the child becomes progressively more dependent upon the external situation to arouse the desire for specific foodstuffs as contrasted with the more elementary development of hunger as a function of intra-organic processes. In order to simplify this discussion, no extended description will be given to the weaning process whereby the child is forced to surrender the intimacy of bodily contact with the mother and the satisfaction of the need of sucking, and learns to substitute auditory stimuli of approval and praise for the tactual stimuli through which he has been obtaining his basic security and emotional equilibrium.

2. The infant learns through the process of toilet training two important physiological adjustments: (a) to recognize the accumulation of pressures in the bladder and rectum as warning signals or cues for inhibition of the automatic sphincter release and for activity such as calling for assistance or going to the designated vessel or place where sphincter release is permitted; (b) to learn to evacuate urine or feces not merely when physiologically needed but when demanded, i.e., when an evacuation is required by the mother or nurse. Through these readjustments, the physiological processes of elimination are transformed from purely physiological events to cultural events in the sense that their occurrence now becomes dependent upon external situations and cultural requirements rather than intra-organic functional needs. This process illustrates the meaning of the statement that the infant surrenders his physiological autonomy to cultural control.

3. The infant learns under parental supervision what situations offer permissible opportunities for emotional reactions and to what extent so that the primitive, elemental physiological response that we call "emotion" becomes patterned in accordance with the approved forms of expression and repression favored by the culture.

The significance of these three readjustments may be described in terms of the processes whereby culture is literally built into the organism as a series of definite physiological adjustments in various primitive organic functionings. These physiological adjustments may be regarded as a process of transformation of behavior into conduct in the sense that the child ceases to behave reflexly and instinctively, as he learns the required conduct toward the culturally defined situations upon which he becomes progressively dependent for functional activity and direction to his life.

If time permitted, it would be desirable to examine also the processes whereby the child learns to respect the inviolabilities that we call "private property" and "sanctity of the person," through a process of having his naïve behavior response to things and people blocked or prohibited by the parents or cultural agent until the child learns to tolerate exposure to the biologically adequate stimulus of forbidden things and persons without responding thereto. Attention should also be given to the other transformations in behavior whereby the child learns to perform certain formal actions in designated situations (manners, etiquette, roles, rituals, etc.) as compelled responses to those situations, *as culturally defined*. The important point to note here is that there is nothing in the biological, physical, chemical, or otherwise objectively described situation that gives any clue to or reason for the conduct of avoidance of inviolabilities or the performance of compelled formal actions. These learned actions are developed in the child only because a cultural agent has defined situations in such a fashion and with such coercion that the child must learn to respond thereto in the prescribed cultural pattern of avoidance or performance. Again we may see how culture is built into the organism in the sense that the individual child learns to respond to the culturally defined situation in the prescribed patterns, and develops a differential sensitivity to the psychological potency of the culture situation. The definition of situations is made by a cultural agent, who thereby invests situations with authority which thereafter operates to direct the child's conduct.

What we see, therefore, is that culture is not some mysterious entity or mystic force floating around in the universe; culture approached operationally is the sum total of the ways people pattern their functions and behavior into conduct and more specifically transmit those patterns to their children. Culture, then, is a process, an activity, not an entity or a thing. It is to be studied as operating within human organisms where it has been established in their very organic structure, functioning, and behavior. This statement, of course, does not rule out or ignore the existence of an enormous body of tools and techniques and of words and symbols, such as written documents, etc., in and through which learned conduct is carried out and the continuity of culture is provided, because those tools and those symbols are used by the cultural agent in orienting his behavior and in enforcing cultural patterns upon the young.

While there may be a fairly high degree of uniformity in the officially approved cultural patterns that parents are supposed to transmit to their children, nevertheless the way in which these cultural lessons are taught to the child is exceedingly diverse, with varying degrees of emphasis, of coercion, and rigidity. Thus each individual child receives a highly idiomatic or idiosyncratic version of the official culture which, imposed upon his unique hereditary constitution and disposition, gives rise to what we call "personality." But this conception of personality is inadequate if it does not also recognize that the individual organism, while undergoing this cultural patterning, with all of its variations and peculiarities of the family situation, also develops, along with each of these cultural patterns, a persistent affective reaction to the world of people and things around him to which he must make these required responses. This brings us, then, to a realization that the individual is engaged not only in the overt conduct demanded by the culture, but he is also feeling, at each moment, acceptance or rejection of the cultural world in which he lives, acquiescence or resistance to its demands, and other affective disturbances as he is forced to accept deprivations, repressions, or finds indulgences or releases.

If we are, therefore, to clarify our thinking and focus our investigation on the problem of culture and personality we may perhaps gain both clarification and precision by recognizing that what we call "culture" is more or less an abstraction for the total aggregate of patterned conduct within a given culture area wherein we may distinguish certain persistent uniformities of pattern. Personality emerges, then, as the unique individual organic manifestation of that culture as seen in an individual who, as a member of that group, utilizes the prescribed cultural patterns, but always idiomatically and with affective reactions that are peculiarly his own.

From studies of personality and of cultures it appears that man *exists* as an organism in a common public world of animals, plants, structures, and other physical objects and processes, but each individual *lives* in his private world of meanings and feelings, derived from the impact of culture that takes place in the specific personal relations between cultural agents and the child. Different cultures tolerate varying degrees of deviation from the socially sanctioned patterns of conduct. In our culture when the individual's private world deviates too far from the official culture we speak of mental disorders; when the individual's overt conduct transgresses the culturally defined inviolabilities, or conflicts with the cultural requirements, we speak of delinquency and criminality.

It may be useful to think of each individual as living in his "life space" which is his individualized way of ordering experience, as contrasted with the more generalized ways of ordering experience called science or art or religion. We need to free ourselves from a great many obsolete conceptions and to realize that it is largely a question of the focus of our interest whether we are concerned with questions of uniformity, inter-relationships, and regularity such as we see in culture, or with questions of individual variations and diversification of the unique configurations of patterns that we find in the human personality.

CHAPTER

9

Jurgen Ruesch

SOCIAL TECHNIQUE, SOCIAL STATUS,
AND SOCIAL CHANGE IN ILLNESS

Dr. Ruesch examines another class of biological event (namely, illness) that is caught in a cultural web. Those cultural patterns that define social classes—and the values and behavior associated with classes—are shown in American society to influence the predispositions of members of the various class groups towards differentiated kinds of illnesses. This subject has also been treated by Dr. Arnold Green in another frame of reference.[1]

Two types of psychological theory have contributed most effectively to the development of a workable social psychology. The first of these is psychoanalysis. Psychoanalytic knowledge and concepts are utilized prominently in a number of readings in this volume, notably those by Erikson (p. 185), Dai (p. 545), Dollard (p. 532), Fromm (p. 505), Greenacre (p. 498), and Waelder (p. 671). The second significant system is that known as "learning theory." Dr. Ruesch's conceptual scheme of drive, cue, response, and reward derives from this theoretical system. It It likewise appears outstandingly in the papers by Davis and Havighurst (p. 308) and Davis (p. 507).

CHARACTER formation is the result of interaction between the individual and his environment, and the most important factors in the environment are human beings. In disease, human contact is not only responsible for

NOTE: Dr. Ruesch very kindly wrote especially for this volume this summary of some of the important researches being carried on by the Langley Porter Clinic and the Division of Psychiatry of the University of California Medical School.

[1] "The Middle-Class Male Child and Neurosis," *American Sociological Review*, Vol. 11 (1946), pp. 31-41.

the spreading of epidemics, but the emotions elicited during and after interaction between persons may influence exacerbation of and recovery from illness. Duodenal ulcer cases, for example, demonstrate clearly the relationship between interpersonal relations and disease. While observation and analysis of human interaction can be studied from the standpoint of the individual, another approach consists of understanding interpersonal relations in terms of the culture in which these people live. Such an attempt has been made in the following paragraphs which are devoted to the description of some social and cultural factors in disease. Concepts and findings outlined below were the result of several detailed studies [4, 5, 6, 7, 8] concerned with traumatic neuroses following head injury, with psychological invalidism in chronic disease, with the personality of male ulcer bearers, and with the personality of female patients who were subjected to partial operative removal of the thyroid gland because of goiter.

Interaction between human beings can be conceived as being the result of social techniques of the individuals involved. The term *social technique*, therefore, would include all the methods used by an individual in approaching, managing, and handling other persons. Among the various interpersonal approaches known to be used in adulthood, we can distinguish between the long term techniques on the one hand, and those used as interpersonal tactics or the short term techniques on the other. Under long term techniques, one might mention social climbing or prestige seeking, maintenance of superiority and dominance, nurturance (mothering or fathering), conformance, co-operation, competition, rivalry, dependence, social decline, self-abasive, avoidant, isolating, aggressive-destructive techniques, and acquisition or use of others. Among the short term techniques one might mention testing out, unthawing, startling, joking, teasing, flattering, offending, seducing, threatening, bribing, and pitying. Though space precludes in this article to describe techniques in any detail, there are a few common factors worthy of consideration.

The concept of social technique can be separated into four essential components. These are: drive, cue, response, and reward. A drive impels a person to respond. The response as such is action, at first non-systematized, and later learned and systematized. Cues determine when and where an individual will respond and which response he will make. Whether a given response or action will be repeated depends on whether or not it is going to be rewarded. While the fields of psychology, psychiatry, and psychoanalysis are primarily concerned with the problem of drive, response, and reward, the subsequent presentation will emphasize the problem of cues.

The observation, the nature, and the management of cues is largely a social and cultural function. The cue, so to speak, constitutes the link between culture and individual. What makes persons of one and the same culture alike is the awareness, observation, and response to the same cues. A cue can be defined as a symbol or signal perceived in a complex situation which is helpful in solving the problem for the individual. It serves the purpose to reinforce or attenuate drives. Thus any stimulus may be thought of as having a certain drive value depending upon its outside strength, and a certain cue value depending upon its distinctiveness. Drives such as hunger or thirst originate within the individual and are present wherever he goes. But whether action is instigated to satisfy this drive may depend upon the presence of water, for example. These drives are under ordinary circumstances not specific enough to elicit the correct response without the support of more distinct cue stimuli. Noticing a cue, then, can either be a response which is learned, and which is called "paying attention," or the cues may be contained in the stimulus itself, be it exteroceptive or proprioceptive. Presence of rewards, either in terms of mastery of the situation, or in terms of love and affection received from parents or educators, will induce the individual to repeat a response, remember a cue, and develop the response to a skill. The reward and punishment values associated with the observation of cues and responses later are introjected. In adult life, then, a cue when perceived does at the same time elicit in the individual sensation of punishment or reward. It is this process which is to a large extent cultural in nature. Children are, at least in our predominantly middle class culture, rewarded for similar or matched behavior, and punished for dissimilar or non-matched behavior. Observation of cues, therefore, is identified as rewarding, if others observe the same or similar cues. This process seems to be the base for group behavior and is an essential prerequisite for the social stratification of our society.

In the process of learning the child may take over cues either consciously or unconsciously, a process which has been illustrated by Miller and Dollard [3]. In addition, cues which have been taken over from a model may either remain conscious (that is, the individual remains aware of them), or they may be repressed (that is, the individual is unaware that these cues exist, and that they reinforce his drives). This latter phenomenon is of great importance in psychiatry. Cues and responses which at one time were conscious, and then were repressed, can be made conscious again in psychotherapeutic procedures. The culturally important cues usually are taken over unconsciously, and therefore are not surrounded by a wall of defenses. Making the patient aware of their existence is a relatively easy task. For example, a patient can be made aware of the fact that he is a social climber, and that he reacts to all cues which deal with prestige, and that he disregards other cues which for his adjustment may be of equal importance.

Adjustment in life largely depends upon the ability to generalize from one set of cues to another similar set. Not only does the individual need to be aware of a large number of cues in readiness for action in a great number of situations, but he also must understand the multiple meaning of single cues. One cue if combined with another one may mean something entirely different than if appearing alone. The simultaneous perception of several cues in a social situation is the foundation on which the evaluation of one's own and other people's roles is based. Matters are complicated further if one considers that in social situations each individual can assume simultaneously multiple roles. For example, one can be father, husband, and son at the same time. This perception and learning of cues regarding orientation of self to others occurs already in early childhood. One of the elementary experiences of the child, for example, is the successive shift of his role as intimate of mother. At first one unit with her, later united in breast-feeding, and still later definite detachment trains the child in the orientation of self to others along a horizontal co-ordinate. Another fundamental experience of the baby is the perception of the adult as being taller, bigger, more powerful, and superior to himself. He also learns that insects, pets, and younger siblings are smaller, less powerful, weaker, and therefore inferior. It follows that the individual learns very early to orient himself along a vertical co-ordinate, which at first means size and power, and later obviously prestige. In the process of learning, the young baby is trained to imitate and copy others. Reward is given for similar or matched behavior, punishment for dissimilar behavior. Therefore the individual learns to consider similarity or dissimilarity with others, making orientation as to identity a primary concern. Last, but not least, the infant is assigned from early childhood a role as a family member, in which he experiences successively such fundamental relations as being son or daughter, brother or sister, and later mate, father or mother, and member of the out-group. In adult life the individual tends to generalize from his childhood experiences, assigning himself and others roles which have a definite bearing upon the past.

These first and early social experiences of the child are later modified and perpetuated in a different way. Orientation along a vertical co-ordinate leads to the problem of social status and prestige; orientation along a horizontal co-ordinate to problems of intimacy and mating; orientation as to identity to problems of group membership; and orientation as to the family role leads to the more general function assumed in family and public life. These four fundamental aspects of modifications of childhood roles were studied in an adult patient population. The questions which needed answering can be formulated as follows: (1) What is the influence of the childhood role upon disease in adulthood? (2) What is the influence of the childhood role regarding the attitude towards the physician? (3) What relationship exists between problems of prestige, social status, and illness? (4) What is the influence of change of culture and roles upon the pathogenesis of disease?

In attempting to answer the first question, the interview material relating to the childhood of the patients was arranged in such a manner as to give the best possible information regarding the patient's past roles in illustration of his techniques used in the management of parents and siblings. Information was gathered essentially under three headings. Under the term "authority," those functions of the parent which related to reward and punishment were included. Under the term "ideal model," those functions were summarized which invited the child either to accept or reject cues from the parent. Under "affection" were listed those memories which had a bearing upon the parents physical or abstract manifestation of affection and love. In reconstructing, now, the childhood role regarding the patient's function of son or daughter, it was found that considerable deviation from the norm existed in various disease groups. In the normal boy, for example, in recent American culture, the principal source of affection is the mother, while father constitutes the main source of authority and ideals. In the girl, this relationship is reversed after the age of three. The essential feature in the normal boy or girl can be seen in the separation of the source of affection from the source of authority and ideals. This dichotomy implies that whenever punishment takes place it is not connected with the withdrawal of love, because the functions are distributed between the two parents. In abnormal patterns, on the other hand, we find a disturbance in the distribution of parental functions. In the ulcer population, for example, not a single case showed a normal childhood pattern (see TABLE XVIII, below). Instead, mother orientation was predominant; either the mother was the sole source of authority, ideals, and affection, thus rendering her so dominant that rebellion against her was impossible; or, the mother was affectionate, but so inconsistent and weak that she did not constitute a balance against the extremely punitive father. Considering these two examples, which constituted in over 50 per cent of the cases the predominant childhood pattern, we see that the child as son developed techniques of managing and handling either the

TABLE XVIII

Frequency of Childhood Patterns in Per Cent

	DUODENAL ULCERS (*men*)
Normal Boy	0
Dominant Mother	35
Dominant Father	19
Idealized Mother and Punitive Father	31
Equally Dominant Parents	13
Lack of Affection	2

	THYROIDECTOMIZED WOMEN (*females*)
Normal Girl	31
Dominant Mother	12
Anxiety Type (punitive mother and weak father)	19
Hysterical Personality (rejecting parents)	31
Male Identification	7

dominant, over-protective mother, or the punitive father. The techniques were essentially those of submission, dependence, and avoidance of expression of anger. As far as maturation was concerned, these boys either remained infantile, because maturation would have meant clash with the all powerful mother or the punitive father, or the boys began to strive for authority, but had to do it in an exaggerated fashion, in order to compensate for and overcome the authority of their parents. Both techniques, namely this overt dependence, or covert dependence with overt counteraction, proved to be so one-sided that the patients ultimately broke down under the stress and strain of the conflict with their environment and within themselves.

Similar results were obtained in the group of thyroidectomized patients. Only one third of all the female patients showed normal childhood constellations regarding the distribution of parental functions. An additional one third was characterized either by a punitive mother and a loving but weak father, or by rejection from both parents. The social techniques in this latter group obviously were characterized by ambivalent identification. Ideals could not be accepted from mother because she was punitive, nor from father because he was weak, inconsistent and submissive to mother. In adulthood, then, these patients tried to establish relationships with others by means of copying, imitating, and identifying. This technique, which in infancy was necessary for the process of learning, was carried over into adulthood usually with unwarranted results.

Another interesting aspect was found in the relationship to the siblings. Ulcer patients, for example, are the younger or the youngest born children in the family. Their techniques, therefore, were adjusted to handling a number of superior and older siblings, rather than younger brothers and sisters regarding whom they showed unavoidable rivalry. In addition, they were separated in terms of age from their nearest siblings. It is a well known fact that in a series of siblings the younger born children are spaced more apart in terms of years. An age difference of one to two years does not separate children as play-mates, but a difference of four or five years means isolation. The older siblings then have their own groups and do not admit the younger ones to their cliques and gangs. In ulcers, the average age spacing from patient to the nearest older siblings was between three and four years, and the age spacing towards the nearest younger siblings was two and a half in one and almost five years in another group. For practical purposes, therefore, these patients grew up as "only children." The relationship to the parents thus became more important than it would have been if they had had siblings nearer their own age. An additional disturbing factor was found in the sex distribution of the brothers and sisters of the ulcer patients. Boys prevailed in one group, girls in another. Again, one-sided techniques were developed to handle this situation.

A direct transfer of childhood roles upon adult social situations was observed in the study of patients' attitudes and opinions regarding physicians. The relations of ulcer patients to doctors was largely governed by features not related to medicine at all, but was determined by the patient's attitudes towards his parents, his desire to receive attention, and to climb socially. Patients who had an idealized mother tended to phantasize about doctors in idealized terms. Patients with an authoritarian mother expressed desire for medical authority. Overtly dependent patients wanted nurturance from doctors, while the counteractors, strivers, and compensators saw in the doctor primarily a medical authority. Difficulties encountered in managing a dominant mother found their repercussions in adult life as a tendency to be more aware of psychological problems than if the problems were centered around the father. Similar were the findings in the thyroid group, inasmuch as the patients, because of their need for identification, made the doctors into strict authorities.

TABLE XIX

Relation to Siblings

		PER CENT	
	Brother-Sister Ratio	Youngest Child	Oldest Child
Duodenal Ulcers	1.3	32	18
Thyroidectomized Patients	1.0	24	29

Summarizing the analysis of the childhood circumstances of different groups of patients, one can state that the majority of these individuals were characterized by faulty social techniques, in terms of fewness and unawareness of cues, lack of flexibility of responses, and in appropriateness of the techniques with regard to the cultural frame of reference. Inconsistent behavior of the parents, lack of continuity in contact with the parents, one-sided rewards or punishments, and unusual social situations in terms of family constellations were responsible for the development of these abnormal techniques. While the techniques of the patients were suitable in the home environment, generalization to include the wider social environment was obviously difficult, if not impossible. Adjustment in adult life, then, was attempted by perpetuating situations similar to those enjoyed in childhood. However, sooner or later the environment made its influence known either by forcefully separating the patient from his set-up, or by exerting social pressure in terms of cultural ideals, which stood in contrast to those of the individual. These conflicting tendencies then precipitated the breakdown.

Considering now the relation of social status and class membership with regard to disease, use was made of Warner's [10, 2] concept of the six social classes of the American society. For use with patients, the interview technique was modified as follows: All individuals were rated on the basis of nine criteria; the three economic criteria included spending, houses, and property; the three social criteria were concerned with family structure, occupation, and associations; and the three ideological criteria were related to education, attitude towards public, and interests. Although these criteria did not constitute a direct continuum from the lower lower to the upper upper class, it represented a scale suitable for rating. The advantage of this method was seen in the fact that the various factors contributing to the over-all rating could be kept separate. This procedure later furnished the evidence for the concept that not all the features of an individual's behavior are rated by society with the same prestige value. In an old and well stratified society all manifestations of behavior of one and the same person usually have a consistent prestige value. Spending, for example, corresponds in terms of social class values to the house, to the attitudes towards the public, to the education and the family structure. In a more fluid society like the American one, and especially in cases of patients, this does not seem to be the case. An immigrant, for example, can advance rapidly in terms of economic status, while his social status will lag considerably behind. Such a scattering of behavior throughout various prestige classes leads, of course, to social isolation.

If an individual, a gold-miner, for example, is upper middle class in terms of economic status, lower middle class in terms of social status, and upper lower class in terms of ideologies, he will find only few people with a similar personality structure. If he wants to associate with an upper middle class person he has not enough education; if he wants to associate with an upper lower class he is resented because of his wealth. The only solution to this dilemma is to find another person with a similar distribution of prestige factors in the personality. A second solution consists of equilibrating social and ideological status to the economic status. This latter procedure illustrates another general rule; individuals try to behave in such a manner that all their manifestations comply with a consistent prestige value throughout. However, this striving may be thwarted because it presupposes a change of social techniques which usually are not very modifiable during the mature years of a person's life. Striving without success constitutes of course a source of stress and strain.

Class membership as such seems to be related to disease. The lower middle class with its preponderance of psychosomatic conditions can be characterized as the culture of conformance and excessive repressive tendencies. Because of lack of expressive facilities, the only possible solution for unsolved psychological conflicts seems to remain in physical symptom formation. In the lower class, where expression of anger is permitted, and where non-conformance and rebellion are sanctioned by the class ideals, there exist other means of expressing conflicts. Therefore, we are not amazed to see that the diseases of the lower classes are connected with exposure to machines and expressions of hostility as found in the incidence of fractures, accidents, and traumatic disease. On the other hand, we are also not surprised to find that in upper class families, with their overbearing superego and cultural traditions, there is a relatively large incidence of psychosis and psychoneurosis. Therefore, we can state that the lower class culture favors conduct disorders and rebellion, the middle class culture physical symptom formations and psychosomatic reactions, and the upper class culture psychoneurosis and psychosis of the manic depressive type.

The American society with its flexible social stratification, invites the individual to change his social status. Increase in prestige is a popular concept in America where everybody expects to have unlimited possibilities. To make good is a culturally accepted ideal. We are therefore not surprised to find that among the bearers of psychosomatic conditions there is an unusually large number of people with strivings to increase their status. We can distinguish the climbers, that is people who succeeded in changing their status, from strainers, who attempted change and who hope for increase in prestige and dream about it, but who did not succeed. For these individuals, social climbing is a great effort and a source of stress and strain. The individual remains unaware that many of the social techniques that he learned in his childhood became obsolete when he changes from one class to another. He forgets and is unaware that the roles and the cues that he knows can be inappropriate in a different social environment. He forgets that climbing brings about social isolation, leaving behind the security enjoyed in the old group while trying to grow roots in a new and different class of people. Social decliners on the other hand are infrequent in psychosomatic conditions, but prevalent in alcoholism, for example. Here the patient usually declines economically first, idealogically second, and last only comes his social status and standing. But none the less decline also brings about isolation and change of techniques. However, the techniques are simpler and easier to learn.

What are now the reasons for changing one's social status? Decline and climbing are largely functions of the attitude towards one's parents. The dynamic rules which govern this relationship can best be formulated as follows: Individuals who are able to identify and accept the values of their parents are likely to continue their trend of social mobility. This would mean that, if the parents were static, they would remain in the same social class. If, on the other hand, the parents were climbers or decliners, the children will do likewise. Rebellion, on the other hand, and inability to accept the parental values, combined with ambivalent identification, usually leads to a reversal of the social mobility trend. Parents who are declining, then, have children who do better than they did, and parents who are climbers will have declining children. Reversal of the parents' decline is seen, for example, in an attempt to recoup the family fortune dissipated by the parents or the rejection of low family standings and strivings for economic achievement. Reversal of the upward trend of the parents ranges from the decline of the psychopath through simple antisocial behavior to such reactions as belonging to political or artistically radical groups and up to a mere rejection of whatever the parents have achieved. It is thus remarkable that many people are actually striving to change their own social position because unconsciously they disapprove of the parents and want to hurt their pride. Social trends may fluctuate through a lifetime, and social mobility trends may be reversed when early adolescence is compared to mature behavior. As guilt feelings arise or feelings of hostility are overcome at a more mature age, there is a tendency to recomply with the ideals and status of the parents.

Social mobility and social status have essentially a therapeutic and prognostic value. The decliners are the people who will offer the poorest prognosis. They are not suitable for psychotherapy. As long as they take it out on society and act out their problem in decline of status, they are unapproachable. The climbers, in turn, offer a better prognosis. Their terrific need for achievement often leads them into situations which they are unable to master. In psychotherapy, then, some of the patients learn to give up social climbing as a defense mechanism, limiting themselves to activities which bring gratification regardless of the prestige involved, thus gradually becoming more successful.

Transition from childhood to adulthood, from the family circle to the wider social environment, from civilian to military life, from one social class to another one, from one American region to another, from rural to urban living, or from ethnic to American, really constitutes a culture change. It involves adaptive behavior, a change in social techniques. The individual may have learned the necessary responses to master life, but the cues change their meaning from culture to culture. It follows that the more diversified and flexible the techniques of an individual are, the more likely he is to survive in the new environment. The learning of additional cues, the reinterpretation of existing cues, and the relearning of multiple cues and roles can be successful if the individual is able to fall back upon a backlog of diversified responses. If, however, cues and responses are neither diversified nor flexible, acculturation is difficult or impossible. In addition, the same response may be rewarded in one culture and punished in another. The individual's attitude, therefore, has to change, and introjected reward and punishment values of cues, responses, and techniques have to be modified. This latter procedure is the most difficult component of acculturation. The more strict and specific these reward and punishment values of the old culture were, the more difficult they are to change.

The process of acculturation from ethnic to American is most difficult. It is a common observation that immigrants from various countries show differences in speed and degree of assimilation to the American culture. This, of course, is not only a function of the immigrant's personality, but it is also related to the similarity or difference of his native culture to that of his adopted country. Though nationals of a given country vary in their culture because of differences in class and caste, one can observe national features which run through all classes. The American culture, for example, can be described as that culture which is represented by the lower middle class, composed of people of Anglo-Saxon descent and of Protestant religion. It is the core culture, because the first settlers came from English speaking countries, because they were middle-class and Protestant, and because they set the cultural standards for the new country. All later immigrants were compelled to adapt to these standards. Public opinion in America is largely an expression of this core culture. We find it in novels, on the radio, in newspapers, public speeches and in the opinion of the man on the street.

Acculturation to the American core culture is a function of the number of cues and responses which an individual possesses in common with Americans. This number in turn can be expressed in terms of closeness or distance to the American core culture. In adaptation of this principle for use with patients, a four-step scale was devised which classified a variety of personality features of individuals in terms of being identical, similar, different, and remote with reference to the American core culture. The classification was based upon reports in the literature and upon observations by American individuals about foreign features observed in ethnics. The essence of this procedure thus consisted of organizing the habits, customs, beliefs, and attitudes of the patient with regard to what is considered desirable in America by the vast majority of people.

Cultural ratings complement the ratings obtained from social class membership and social mobility. While the social class aspects give a vertical assessment of the individual's standing and striving in our system of values, the cultural assessment pertains to a horizontal analysis of the individual's behavior. In social mobility upwards or downwards we find a striving for increase or decrease of prestige; in acculturation from ethnic to American, on the contrary, the individual strives for substitution of cues of essentially similar value. In the course of the latter process, substitution is not always successful. It may be impossible, for example, to carry on an original occupation or profession because of lack of opportunity or different regulations. Under such circumstances, a lawyer might be compelled to become a salesman, which in terms of values would constitute a decline, and in terms of acculturation would mean that similar responses and cues could not be found in the new culture. Therefore, we see that in the process of acculturation the seeking of similar cues with similar values, which then are acceptable to the individual as well as to the new environment, may be obstructed by a variety of factors.

Whenever an individual differs from the American core culture, public opinion will put pressure upon him to adapt. These unacculturated personality areas represent sources of conflicts as well as of stress and strain, leading invariably to disruption of proper functioning of the individual. Areas where difficulties in acculturation are encountered usually deal with responses and cues which are characterized by strong reward or punishment values. Cues, for example, which have purely technical value, are taken over easily, as can be seen by the role of the foreign worker in America. Whether people speak the language or know the culture does not matter; they all can work. But as soon as we touch upon such problems as eating habits, religion, attitudes towards parents, success, or public, and in matters of child rearing, we find that the learning of new cues and responses is difficult; first, because the old ones are characterized by reward value, and, second, because the new ones may have punishment values for the individual, being in contrast to the teaching of his old culture. The problem gets even more complicated in the second generation. Whenever the values of parents and environment coincide, there is no conflict. But, if the new culture offers reward value,

TABLE XX

Nativity and Parentage in Per Cent

	SELF		PARENTS		
	Native Born	*Foreign Born*	*Native*	*Foreign*	*Mixed*
Population at large	91	9	79	14	7
Chronic Disease and Psychological Invalidism	93	7	54	35	10
Duodenal Ulcers	92	8	69	23	8
Thyroidectomized Patients	86	14	56	35	9

while the parent punishes for acceptance or rejection of a given cue, the result is ambivalence, confusion, and possibly total rejection or rebellion against acceptance of any values.

The practical purpose of assessing the personality areas which resist acculturation is to uncover features related to fear, shame, guilt, or anxiety, which are all symptoms of conflicting ideals and restrictions imposed by conscience. These in turn, when uncovered, are accessible to psychotherapy, and ultimately lend themselves for transformation and speed up the process of acculturation. The same holds true for all other types of culture change and social contact. Ciocco [1] once made the statement that the mortality in males is higher in conditions in which the so-called chronic degenerative pathological processes predominate, and from diseases which might be attributed to the social function such as working and fighting. Conditions in which the reaction to the social environment are of importance are: alcoholism, suicide, ulcers; skin cancer, silicosis, fractures, and intoxications as a result of occupation; neurosyphilis, meningitis, pneumonia, and influenza as a result of exposure to infection. The females, on the contrary, have a higher mortality in relation to conditions which bear upon the function of reproduction and the endocrine system. This statistical survey bears out the concept that disease is closely related to the social function, especially in males. Culture change is a phenomenon which aggravates and imperils social contacts rendering proper adjustment even more complicated.

The practical purpose of class ratings is seen in a better understanding of the culture and the environment in which the patient lives. And the way he lives largely determines the diseases he will contract as well as his attitude toward recovery. The assessment of social techniques represents the basis of all understanding of human interaction. Faulty techniques lead to maladjustment, and culture change or abnormal social situations merely expose these innate difficulties.

One last word about a controversial issue of major importance. The great American ideal that everybody is alike has advantages and disadvantages. The constructive aspect is that people strive for assimilation and common denominators in their social techniques; the destructive aspect is the inability to tolerate differences from others. Difference thus arouses anxiety. In contrast, we have other, especially western European, countries in which the social class system as well as the ethnic groupings are rather fixed and rigid. In England and in Switzerland, for example, people are trained to accept differences as a biological and social phenomenon of life. Hence difference does not arouse anxiety. The disadvantage in this system is the slow speed of assimilation of new-comers and the lack of flexibility and of possibilities for the individual; but on the other hand it gives the individual a secure foundation. He knows that he is and will remain different from others. Therefore he will not strive for goals which are out of his reach. He uses his energy in the most productive way, concerned with adaptation to reality rather than being worried about his similarity to or difference from others. Flexibility and social change are in America the principle sources of insecurity, while in Europe stratification and rigidity result in frustration.

REFERENCES

1. Ciocco, A.: "Sex Differences in Morbidity and Mortality," *Quarterly Review of Biology*, Vol. 15 (1940), pp. 59–73, 192–210.
2. Goldschmidt, W. R.: *Social Structure of the California Rural Community* (Ph.D. Thesis, Univ. of California, August 1942).
3. Miller, N. E., and Dollard, J.: *Social Learning and Imitation* (New Haven, Yale University Press, 1941).

4. Ruesch J., and Bowman, K. M.: "Prolonged Post-Traumatic Syndromes Following Head Injury," *Am. J. Psychiatry*, Vol. 102 (1945), pp. 145–63.
5. Ruesch, J., Christiansen, C., Dewees, S., Patterson, L. C., Jacobson, A., and Soley, M. H.: "Psychological Invalidism in Thyroidectomized Patients," *Psychosomatic Med.*, Vol. 9 (1947), pp. 77–91.
6. Ruesch, J., Harris, R. E., and Bowman, K. M.: "Pre- and Post-Traumatic Personality in Head Injuries," *Assoc. Res. Nerv. Ment. Dis.*, Vol. 24 (1945), pp. 507–44.
7. Ruesch, J., Harris, R. E., Loeb, M. B., Christiansen, C., Dewees, S., and Jacobson, A.: *Duodenal Ulcer—a Socio-Psychological Study of Naval Enlisted Personnel and Civilians* (Berkeley & Los Angeles, University of California Press, 1948).
8. Ruesch, J., with Loeb, M. B., Harris, R. E., Christiansen, C., Dewees, S., Heller, S. H., and Jacobson, A.: "Chronic Disease and Psychological Invalidism," *Psychosomatic Med.*, Mon. 9 (1946).
9. Ruesch, J., Loeb, M. B., and Jacobson, A.: *Acculturation and Disease* (Psychological Monographs: General & Applied, 1948).
10. Warner, L., and Lunt, P. S.: *The Social Life of a Modern Community* (New Haven, Yale University Press, 1941).

CHAPTER

10

John W. M. Whiting

* THE FRUSTRATION COMPLEX IN KWOMA SOCIETY

The response to frustration is always modified by learning. Dr. Whiting deals with the modification of the aggressive response, suggesting that the interaction between the innate response and the nature of group membership patterns the kinds of learning modifications imposed on the biological drive. Clearly there are limits to the plasticity of human nature; equally clearly the nature of group life limits the kinds of modifications which can be imposed on innate responses.

THE purpose of this paper is to present the results of the application of the frustration-aggression hypothesis [3] to data gathered at Kwoma. Before these results are stated, however, a brief summary of the way in which Kwoma individuals react to frustration from infancy through adulthood will be made. A more complete description of Kwoma reactions to frustration can be found in a recent publication by the present writer [11].

The data for this paper were gathered by S. W. Reed and myself during a field trip to New Guinea in 1936–37. The Kwoma are a small tribe situated in the mountains just north of the Sepik River and about 250 miles from its delta. They are an agricultural, patrilineal, polygynous

NOTE: Reprinted from *Man*, Vol. 44, No., *115*, (1944), by permission of the author and the Royal Anthropological Institute of Great Britain and Ireland.

group, in which political authority resides primarily in the kin group, and in which sorcery is both an explanation for sickness and an important means of social control. Their contact with the whites has been minimal, having been restricted to a few government officials and traders.

Kwoma infants are cared for almost exclusively by their mothers. For approximately the first three years of his life a Kwoma infant sits in the lap of his mother during the day and lies by her side at night. It is the Kwoma mother's duty to care for all the needs of her child during this period. When, despite this constant care, Kwoma infants suffer frustration, crying is the response which becomes most firmly fixed. A Kwoma mother, whenever her infant cries, does her best to comfort him. If he is hungry she feeds him; if he is cold she warms him; if he is sick or in pain she tries to soothe him. Thus by removing the source of frustration or pain the Kwoma mother rewards crying as a response to these conditions. Toward the end of infancy, when the child begins to talk, he responds to frustration or pain by asking for help, and his mother complies with his request whenever it is possible for her to do so. Thus during infancy a frustration-dependence sequence is established.

The weaning period is a transition from the lap of the mother to the play group, which takes from one to two years to accomplish. During this time the child stays around the house for the most part, trying to maintain or to regain the close relationship to his mother, which she is forcing him to relinquish. His initial response to the frustration which he receives at this time is, as might be expected, one of dependence. First he asks his mother to permit him to do what he wants. When she refuses, he cries. When this fails, he aggressively tries to get his way by attacking her. The mother extinguishes all these responses by non-reward or mild punishment when she or the father is the frustrator, but continues to reward them by helping him when the frustration or pain comes from another source. When the child finds that his attempts at dependence and aggression fail, he finally submits. He gives up his old habits and learns new ones to take their place. For this he is rewarded both because these new habits are in themselves successful and because his parents praise him for them. Thus a frustration-submission sequence is established under the condition that the parent is the frustrator; the frustration-dependence sequence is still maintained with respect to other frustrators.

At the age of five or six the Kwoma child enters the play group, which consists of other boys of the hamlet, his classificatory brothers. His reaction to his bigger brothers is similar to his reaction to his parents. Since aggression leads to retaliation by a person bigger and stronger than he, since avoidance makes him miss the fun, and since dependence reactions are ignored unless he is in real danger, submission is his most adaptive and usual response.

The first instances of aggression to be rewarded are those directed toward younger siblings. Since they are smaller than he is, he can bully them and beat them without fear of effective retaliation, and thus he achieves direct reward for this response. Unless he is so violent in his aggression toward them that he seriously injures them, his parents praise him for this behaviour rather than punish him for it. Thus, both by direct reward and by social approval, the frustration-aggression sequence is established with respect to children younger and smaller than he.

Late in childhood, after the frustration-aggression sequence has been well practised, he uses it even toward his older siblings and parents. Usually when it occurs in this context, either it is combined with avoidance —a hit-and-run response—or it takes the verbal form of insult from a safe distance, or grumbling and criticism behind their backs.

Although submission, dependence, avoidance, and aggression all occur during childhood, the response which is idealized is that of aggression. Two contrasting epithets express this ideal. *Karaganda yikafa*, which means "little child," is used to criticize a child when he is too submissive or dependent, whereas *harafa ma malaka*, "big older brother man," is used to praise him for being appropriately aggressive. The latter term is applied to the men in the society who have the highest prestige, achieved largely by being bold and standing up for their rights. It is interesting, in the light of the above analysis, that the same words are used to describe a big older brother and the aggressive ideal of the culture.

When a Kwoma reaches adolescence, he must learn to behave in a special way toward his sisters, when they frustrate or threaten him. During childhood he has either avoided them, or behaved toward them as he would toward his brothers. At adolescence he must take the responsibility of protecting them from assaults on their virtue, and he must inhibit his own sexual feelings toward them. This creates a new series of frustrations, and the culture provides a special type of response—a joking relationship. He is permitted to accuse a sister of breaking incest rules, or to make disparaging remarks about her personal habits and appearance, insults which, if directed toward other persons, would lead to serious physical combat, but which evoke laughter and retaliation in a similar vein when applied to a sister.

Toward frustrations imposed by the men who philander with his sisters, an adolescent is rewarded for reacting with aggression. Toward the frustrations of the male relatives of the girls he courts, he reacts with avoidance in the form of hiding his love-making from them. His behaviour toward other persons remains similar to his behaviour during childhood.

Upon his marriage and initiation into adulthood, a Kwoma man must share the responsibility for social control with other men of the hamlet. He must help maintain the rights of the hamlet against encroachments by non-relatives and foreigners, and he must educate his children. Both of these demand aggressive behaviour.

The frustrations which he or his relatives may endure from non-relatives may be divided into three classes: capital crimes, property and sex offences, and sorcery. Murder, rape, arson, and theft of sorcery material are capital crimes, the appropriate response to which is, if possible, to kill the criminal. Anyone who commits such a crime, however, normally flees from the tribe before the death penalty can be imposed.

A man responds to theft, trespass, or adulterous behaviour with a female relative by threats, insults, and other forms of vituperation. Challenges to duels are frequently issued, but actual duels never seem to occur. The most effective expression of his aggression is the breaking of all the relationships which he may have with relatives of the criminal until reparation is made. Since breaking relationship involves the cancellation of all mutual kinship obligations, and the risk of retaliating by sorcery, which is inhibited within any kin relationship, relatives are likely to bring pressure to bear on the criminal to pay the reparation.

If the facts are in dispute, a court meeting is held, at which each litigant presents his case before a group composed of the adult male relatives of both parties. In presenting his case, each litigant gets to his feet and stamps up and down before the court, jabbing the air with a bone dagger. The plaintiff angrily shouts his accusation, accompanying it with as many insults as he can think of. The defendant behaves in a similar manner, angrily denying the charge, and insulting his accuser. The pressure toward maintaining the reciprocal relations which exist between the members of the court is usually sufficient to bring about a compromise.

The Kwoma man, finally, has the responsibility for curing any serious illness which may befall him or his family. Since all serious sickness is believed to be caused by sorcery, it falls in the category of frustration by human agents. Whenever anyone becomes ill, the men of the hamlet repair immediately to the men's house and beat an aggressive message on the gongs: "Stop all sorcery. One of our relatives is sick, and anyone who continues to practice sorcery must answer to us, the men of this hamlet." If the sick person does not get well as a result, the men call a sorcery court to establish the identity of the sorcerer. The procedure at this meeting is the same as that in the court for theft or trespass, described above.

When a Kwoma man is frustrated by a non-relative, therefore, his primary response is aggression, ranging in form and intensity from insulting the frustrator to killing him. Failure to act aggressively in any of these situations is severely criticized, whereas boldness, and readiness to be aggressive, bring high praise and the epithet of "big older brother man." Since these aggressive reactions are not carried out by an individual alone, but by an organized group of co-operating relatives, the dependence reaction is also employed: A frustrated man asks his relatives to help him express aggression toward those who frustrate him.

A Kwoma adult must educate his children. He must prevent them from deviating from the customs of the tribe. When his children do not behave properly, he is frustrated, both because their acts directly annoy him, and because they bring criticism on him for being a poor parent. A parent responds to such frustration by spanking, scolding, and threatening the child. Here again the frustration-aggression sequence is specified as proper.

Frustration may be imposed by foreigners as well as by other Kwoma. The Kwoma dwell in a head-hunting area, and they live in constant fear of being murdered in raids carried out against them by neighbouring tribes. The men of other tribes, who have married Kwoma women, not being subject to the sanctions which apply to Kwoma men, may cause frustration by refusing to pay the required bride-price. Finally, a member of some other tribe may cause sickness by practicing sorcery. This is not an important factor, since a foreigner is rarely blamed for sorcery. Kwoma men react to such frustrations, either by carrying out a retaliatory raid, or by participating in an intertribal court meeting to demand payment of the bride-price, or wergild for a murdered relative. They wait for a year after a murder before calling a court meeting, to give time for their "bellies to cool," as they put it in pidgin English.

The Kwoma also carry out head-hunting raids against tribes who have not directly frustrated them. They frequently displace, to the out-group, aggression which was generated within the tribe. This is evidenced by the fact that the only two raids, about which I was able to question participants, began with in-group quarrels. One informant said that he had organized a raid because his wife had teased him for not showing himself a man by taking a head. When he started to beat her for taunting him, she accused him of being a woman-beater but afraid of men. With his "belly still hot with anger" he had then organized a raid, in which he had taken a head. Another raid, organized during the period of field work, also grew out of a quarrel between a husband and wife. Finally, in the memory of living informants, the Kwoma had once waged a war of extermination against another tribe over territorial rights.

These data, taken in conjunction with certain principles established in the fields of anthropology and psychology, suggest a series of propositions which I should like to advance in the hope that they may be tested by future field workers.

The *first proposition* is derived from an hypothesis advanced by Miller and Dollard [6, p. 64]. Although the innately dominant emotional reaction to frustration is an indeterminate visceral response involving the sympathetic division of the autonomic nervous system, these emotional responses, it is suggested, may be modified by learning, so that frustration may result in anger, in fear, in a combination of the two, or in neither of them. W. B. Cannon [2] indicates that there is no physiological difference between anger and fear except perhaps in the thalamic responses involved. It is possible that these two emotions can be distinguished by the overt reactions which are associated with them, the emotion associated with aggression being defined as anger, and the emotion associated with avoidance being defined as fear. Since the emotional reactions to frustration are covert, it is difficult to study them by the usual field techniques, and thereby to test this hypothesis conclusively from Kwoma data. Informants reported that they were angry, for example, when they were frustrated by anyone who was breaking a rule of the culture, and that to be angry was part of the idealized pattern of the "big older brother man"; they also reported that they were, for example, afraid of sorcerers, enemy head-hunters, and of the relatives of the girls with whom they philandered; they claimed, finally, that they were neither angry nor afraid when a relative frustrated them by asking for help, when it was in accordance with custom that he should do so.

The *second proposition* is derived from the general theory of culture, and from the observation of infants in our own society. Although the innately dominant overt response to frustration is violent thrashing muscular movements of the arms, legs, and torso, loud crying, and holding the breath [10, pp. 146 ff.], these responses, it is suggested, are modified by learning, so that culturally defined responses take their place. From casual and unsystematic observation, it seemed that Kwoma infants reacted to frustration with violent random thrashing behaviour and crying, and the evidence summarized above indicates that these reactions are modified into culturally determined responses.

The *third proposition* is that the culturally relevant overt reactions to frustration may be divided into four classes of social behaviour: aggression, submission, dependence, and avoidance. These four classes were found to be adequate to describe overt frustration-reactions in Kwoma culture, and it is proposed that they may be useful in the analysis of other cultures.

Various authors have used other systems of classifying reactions to frustration. Dollard *et al.* [3, pp. 8–9] and Sears [9] have distinguished between instrumental and goal behaviour. This distinction was made primarily on the psychological level whereas the present analysis is made on a cultural level, and does not contradict the above, but was found more useful for a cultural analysis. The nine-fold classification suggested by Allport [1] was developed to describe frustration under conditions of social change; the classification employed in the present paper is intended to describe the types of reactions to daily life frustrations. Finally, Rosenzweig's [8] classification—adequate-inadequate, direct-indirect, defensive-perseverative, and specific-nonspecific—is made on the person-

Context	Emotional Response	Overt Response
Frustration—{	Anger / Fear / Other	Aggression / Avoidance / Dependence / Submission

ality level, and thus is not directly pertinent to the problem undertaken in this paper.

Propositions one and three may be combined in the above diagram. Thus, frustration may result in any combination of emotional and overt responses, the combination being determined by the frustrating context.

The *fourth proposition* is a corollary of the general theory of culture. It is that the reaction to frustration depends upon the culturally defined context, and particularly upon the social relationship between the frustrated person and the frustrator. The overt reaction of a Kwoma to frustration can be predicted only if both his social position and that of his frustrator are known, as well as the cultural context in which both the frustration and the response to it occur.

The basic definition of frustration employed here is "any event which interferes with a habit." This definition is in accord with that employed in *Frustration and Aggression* [3, p. 11]. The analysis of Kwoma data shows that social frustrations are usually symbolic events such as commands, threats, warnings, and insults, which imply physical punishment or attack, the withdrawal of help or support, or the withholding of gifts. The implication of this proposition is, therefore, that the reaction to frustration depends both upon who issues the command or threat, and upon when and where he does it.

The fourth proposition, suggested by the work of Benedict and Mead, is that any culture may idealize, i.e., consistently reward, one or another of the types of reaction to frustration. Aggression is the type chosen by Kwoma. The aggressive implication of the ideal of *harafa ma malaka,* "big older brother man," is stressed in their culture from childhood on. Submission, dependence, or avoidance may be found to be idealized in other cultures.

The *fifth proposition* is derived from Malinowski's theory of the functional integration of culture, and Sumner's assumption of the strain toward consistency. It is that the idealized reaction to frustration is correlated and integrated with other aspects of the culture. Since the burden of social control in Kwoma society is placed on all adult males, rather than upon a chief, a council of elders, or some other specialized governing body, bold aggressive men are demanded to effect social control. Thus the idealization of this aggressive reaction is integrated with the governmental organization of the tribe.

The *sixth proposition* is derived from the Freudian theory of displacement. It is that when the anger-aggression sequence has been established, and then inhibited, either by social sanctions or by fear of retaliation, the aggression tends to occur toward some person other than the frustrator. This accords essentially with the theory of displacement proposed in *Frustration and Aggression* [3, pp. 41–4], and applied to social data by Dollard [4]. The differentiation of emotional and overt responses made in this paper, however, makes it possible to give a somewhat more adequate account of this phenomenon. As Miller has pointed out [6, pp. 61–4], emotional responses act as acquired drives. Thus anger may motivate a person, until a displaced aggression response reduces his tension. Displacement is suggested by the fact that a Kwoma child is more aggressive toward his younger siblings than the frustrations imposed by them seem to warrant—he sometimes even attacks them when they have not frustrated him at all. It is probably motivated by anger generated by parents and older siblings, but not expressed toward them for fear of punishment. Similarly, head-hunting raids against tribes with whom the Kwoma have had but little contact, suggests that the mechanism of displacement operates here too. Both of the head-hunting expeditions discussed above began with an in-group quarrel.

The *seventh proposition* is more novel than the previous ones, and is put forward tentatively. Owing to certain universal conditions of social life (it is suggested), aggression, submission, dependence, and avoidance will be likely to be specified as proper responses to frustration in given contexts in all societies. The analysis of Kwoma culture shows that the frustration dependence sequence occurs both during infancy, when the infant is too immature to care for himself, and later in life as a part of co-operative reactions to common frustrations; that the frustration-submission reaction occurs in childhood during the process of socialization; that the frustration-avoidance reactions occur in children when they are frustrated by persons bigger and stronger than they, and in adults when they are frustrated by persons whose supporting group is more powerful than their own; and, finally, that the frustration-aggression reaction occurs in education and social control. Since education, social control, co-operation, and variation in the power of the supporting group may be assumed to be universal cultural aspects, and helplessness of infants and disparities in the size and strength of individuals are universal biological facts, it may be assumed that these four reactions will be likely to occur in all cultures.

REFERENCES

1. Allport, G. W., Bruner, J. S., and Jandorf, E. M.: "Personality under Social Catastrophe," *Character and Personality*, Vol. X, No. 1 (Sept., 1941), pp. 1–22.
2. Cannon, W. B.: *Bodily Changes in Pain, Hunger, Fear, and Rage*, 2nd ed., (New York, D. Appleton and Co., 1929).
3. Dollard, John, Doob, L. W., Miller, N. E., Mowrer, O. H., and Sears, R. R.: *Frustration and Aggression* (New Haven, Yale University Press, 1939).
4. Dollard, John: "Hostility and Fear in Social life," *Social Forces*, Vol. XVII, No. 1 (Oct., 1938), pp. 15–26.
5. Miller, N. E.: "The Frustration-Aggression Hypothesis," *Psychological Review*, Vol. XLVIII, No. 4 (July, 1941), pp. 337–42.

6. Miller, N. E., and Dollard, John: *Social Learning and Imitation* (New Haven, Yale University Press, 1941).

7. Mowrer, O. H.: "Some Research Implications of the Frustration Concept as Related to Social and Educational Problems," *Character and Personality*, Vol. VII (Sept., 1938–June, 1939), pp. 129–35.

8. Rosenzweig, Saul: "A General Outline of Frustration," *Character and Personality*, Vol. VII (Sept., 1938–June, 1939), pp. 151–60.

9. Sears, R. R.: "Non-Aggressive Reactions to Frustration," *Psychological Review*, Vol. XLVIII, No. 4 (July, 1941), pp. 343–6.

10. Sherman, Mandel, and Sherman, I. C.: *The Process of Human Behaviour* (New York, Norton, 1929).

11. Whiting, J. W. M.: *Becoming a Kwoma* (New Haven, Yale University Press, 1941).

11

Frieda Goldman-Eisler

* BREASTFEEDING AND CHARACTER FORMATION

It is a truism that the exhaustive, patient empirical testing of hypotheses is basic to the development of a mature science. But as things stand today, many of the hypotheses in the social sciences have not yet been formulated in terms which permit empirical validation. Although there may be good reasons for this, as Dr. Gregory Bateson suggests,[1] psychoanalytic theory has frequently been criticized for the indefiniteness of its propositions. That this criticism is not entirely justified is clear from Sear's [2] work, as well as from the paper which follows. Here Dr. Goldman-Eisler has formulated a current psychoanalytic theory in definitive terms, tested it, and found that it is supported by her findings.

A point worth noting arises out of the last section of this paper: the operation of testing is not necessarily limited to mere validation or invalidation, but may well, in the hands of a skillful researcher, yield new hypotheses and further leads.

---

[1] "Sex and Culture," *Annals of the New York Academy of Sciences*, Vol. XLVII, pp. 647–60.

[2] "Survey of Objective Studies of Psychoanalytic Concepts," *Social Science Research Council Bulletin* 51, (1943), 156 pp.

NOTE: This paper is reprinted with slight abridgement from three papers first published in the *Journal of Personality*, Vol. 17, (1948), pp. 83–103; Vol. 19, (1950), pp. 189–96; and the *Journal of Mental Science*, Vol. XCVII, (1951), pp. 765–82. (Copyright 1948, 1950, 1951.)

INTRODUCTION

THE description of adult character in terms of childhood experience is one of the basic principles of psychoanalytic characterology, and indeed the ontogenetic approach to human personality is the essence of the theory and method of psychoanalysis. While this approach has for some time past exercised an effective influence on practical child guidance work as well as in the sphere of education, recent years have brought a challenge to "historical" theories of human motivation mainly under the influence of the Gestalt school, in particular of Kurt Lewin's topological and vector psychology. Lewin denied the explanatory value of genetic concepts and substituted for them the "Principle of Concreteness" [18], according to which "only existing facts can influence behavior" [19]. Allport [5], developing the theory of the "contemporary character of human motives," finds that motivational concepts such as those of "belongingness," "the field," or "closure" developed by Köhler, Koffka, Lewin, and others of the Gestalt persuasion give a satisfactory account of human motivation without the necessity of resorting to the conception of "instinctive reinforcement" and past experience [4]. The controversies that were conducted about this question in recent years [28] had, however, to remain mere disputations as it became clear that no conclusive evidence existed as yet either in favor or against the genetic view of the etiology of human behavior.

Nevertheless, suggestive results have already been obtained in different fields of behavior study.

The first line of evidence comes from observations of anthropologists relating customs of nursing, weaning, and toilet training among primitive tribes to the personal characteristics in their adult communities. The conclusions from this field support the genetic view showing that generous suckling and later weaning are related to generosity, optimism [9, 20, 21], and co-operative, peaceful behavior [20]; whereas ungenerous suckling and early weaning coincide with arrogance, aggression, and impatience [20], a tendency to collect and hoard food, competitiveness, and quarrelsomeness [9], and attitudes of acquisition and retention, love of property, hostility, suspicion, and a nostalgic sadness [9].

More corroborative evidence comes from experimental animal psychology, where the work of J. McV. Hunt established a direct relation between the period of early feeding frustration and hoarding behavior in the albino rat [15]. Hunt found that albino rats which had been frustrated at the age of 24 days hoarded, after an intervening period of five months' normal feeding, not only two and a half times as much as their litter mate controls (significant at between 5 and 1 per cent levels), but showed a significant difference in their hoarding behavior from the group which had been submitted to early frustration at a later age (32 days).

Margaret A. Ribble [24] approached the problem from a different angle by studying the overt behavior and the reactions of 600 small infants during nursing and weaning periods. As a result she came to the conclusion that there is an innate need in the infant for contact with the mother. She observed that in cases where babies had not been "mothered" or had been suddenly deprived of maternal care two reactions might develop: (1) a negativistic reaction manifesting itself as a refusal to suck, in a complete loss of appetite, a failure to assimilate food, and more or less hypertension or rigidity of body muscles, and (2) a depressive and regressive quiescence of even more sinister implications, stupor-like and bearing resemblance to the clinical picture of shock or "marasmus" (wasting away). The introduction of a "foster-mother" to give personal care and massage to these infants brought, she observed, dramatic results in restoring appetite, alertness, and reflex excitability. Concerning weaning, Ribble comes to the conclusion that favorable sucking during the first four months is an important condition for further positive development. Children who have been frustrated in their earlier suckling and in primary mothering seemed to give up sucking later and less spontaneously.

While these studies lend strong support to the principles of psychoanalytic characterology, further investigation is needed for the following reasons:

(A) The conclusions of anthropologists from the prevalent character type within a primitive tribe and its methods of child-rearing have two disadvantages: (1) They come to us as necessarily inexact observations in the field and not as quantitative data, and therefore cannot claim high reliability. (2) The anthropologists who have related these factors have only established their coincidence within one community; their sequence within the same individuals was left uncertain. Indeed, as their data stand they admit equally well of the interpretation that the character of the adults is the determinant, as that it is the result of the prevailing customs of child-rearing.

(B) Animal studies such as Hunt's experiments with albino rats, while highly suggestive in showing the genetic determination in the behavior of the adult, are not conclusive with respect to human character-formation.

(C) Ribble's work with infants shows the relationship between nursing experience and the psychological reactions of anxiety, withdrawal, and depression in the infant, but leaves unanswered the question of how far such reactions leave a permanent trace in the adult character make-up.

In the present investigation an attempt was made to find relevant evidence (a) through the quantitative and factorial study of adult character assessed by means of verbal rating scales, and (b) by relating the emerging syndromes to data concerning the length of breastfeeding obtained from the subjects' mothers. In studying adult character in relation to early feeding experience a considerable advantage was derived from the fact that psychoanalysis related certain character traits to definite phases in the libidinal development of the child, and grouped them with reference to their genetic roots into oral, anal, and genital characteristics. In this research we are concerned with oral character traits and their origin. Karl Abraham's and Edward Glover's studies of the phenomenon of orality in its characterological manifestations [3, 14] offer a hypothesis, the concrete elaboration of which gives direction and guidance to the quantitative study of this problem.

I. THE PSYCHOANALYTIC ORAL CHARACTER

The concept of "oral character" in psychoanalysis derives from three basic assumptions: those of infantile sexuality, of the existence of certain stages in its development, and of the influence of the vicissitudes of the libido on character-formation [1, 2, 11].

Oral character traits are assumed to originate from repressed or deflected oral impulses which are dominant during the nursing period and which have undergone transformation into certain permanent behavior patterns by the processes of reaction-formation, displacement, or sublimation [14]. Two main syndromes of bipolar significance seem to emerge from Abraham's and Glover's studies [3, 14], which the writers refer to two etiological factors. The basic conditions for the development and fixation in character of the one or the other syndrome are assumed to be the experiences of gratification or frustration attached to the oral stage of libido development. Determined by these would be the degree of fixation at, and consequent contribution to, the formation of character of either of the two subphases of the oral stage which Abraham [1] came to distinguish: the earlier "suckling" level with its freedom from all manifestations of ambivalence and the later "biting" or "oral-sadistic" level characterized by hostile and destructive feelings. Thus in a child who had suffered disappointment or deprivation in the early "sucking" period the need for oral gratification would seek compensatory satisfaction at the "biting" level. Persons whose sucking impulses had been left ungratified might also aim at satisfaction at the anal level and develop anal characteristics. Thus an admixture of anal character traits would have to be expected in the character make-up determined by oral dissatisfaction. Indeed, Abraham explicitly warns against expecting a "pure culture" of oral character traits [3], but conceives of orality rather as a predispositional fixation to be blurred, modified, or enriched by admixtures derived from later stages. Characteristics deriving from a satisfied sucking period may, he considers, "coincide in important respects with others derived from the final genital stage" [3, p. 404], on the basis of the common factor of freedom from manifestations of ambivalence (pre-ambivalence in the case of the sucking, and post-ambivalence in respect to the genital stage). Thus, if Abraham's argument is valid, we should expect an antithesis, not between oral and anal character traits, but an orally satisfied *cum* genital *versus* orally dissatisfied *cum* anal polarity.

Isolating the oral characteristics from later admixtures, however, Glover postulated, on the basis of the conditions of the individual's reactions attached to the oral stage, which he listed [14], the following derivative characteristics which might become manifest in a positive or negative form: omnipotence, ambivalence in object relations, sensitiveness in regard to getting and to the maternal function of environment, quick emotional discharge, alteration of mood, and rapid motor reactions. Actual analyses of oral subjects are reported to have yielded corresponding character traits [3, 14], which are grouped below according to their supposed genetic roots.

Traits of the orally gratified type. This type is described as being distinguished by an unperturbable optimism ("excess of optimism which is not lessened by reality experience" [14, p. 136]), by generosity, bright and sociable social conduct, accessibility to new ideas, and ambition accompanied by sanguine expectation.

Traits of the orally ungratified type. This type is characterized by a profoundly pessimistic outlook on life, sometimes accompanied by moods of depression and attitudes of withdrawal, a passive-receptive attitude, a feeling of insecurity, a need for the assurance of getting one's livelihood guaranteed, an ambition which combines an intense desire to climb with a feeling of unattainability, a grudging feeling of injustice, sensitiveness to competition, a dislike of sharing, and an impatient importunity. (Abraham, as distinct from Glover, seems not to link impatience with the dissatisfied type specifically, but with both oral character types.)

Fixation at the oral-sadistic level, understood to be compensatory for earlier disappointments, would show itself in ambivalence of feeling; traits of hostility, dislike, and envy; an aggressive use and omnipotent valuation of speech; covetousness; and maliciousness in social conduct.

Glover suggests that the archaic oral disposition does not result merely in an exaggeration of oral characteristics, but in an overstress of all the associated characteristics, omnipotence, narcissism, viewing, and that an exaggeration of the motor accompaniments of oral activity is of special significance.

Ernest Jones [16] describes under the term of "God Complex" a type which fulfils these conditions to a large extent. The basic feeling tone of this character he considers to be that of omnipotence; it might manifest itself either in the form of an excessive narcissism and exhibitionism or, by way of a reaction-formation, as excessive modesty and self-effacement. The most characteristic traits of the latter form Jones considers to be a tendency to aloofness (desire for inaccessibility), a stress on privacy, seclusion and secretiveness, unsociable and unsocial attitudes, and omnipotence phantasies often displaced to the sphere of knowledge. An omnipotent attitude to time, unpunctuality, and impatience in waiting would also usually be met with in such people.

Edmund Bergler [6], describing the syndrome of "oral pessimism" which he regards as a neurotic symptom, also emphasizes its strong narcissistic element, which he considers to be the result of a traumatic wound to the infantile delusion of omnipotence. Its outstanding trait would be a compulsion to repeat the experience of the original frustration (supposedly caused by the loss of mother's breast). By way of being fixated to frustration, the oral pessimist would derive pleasure from anticipating calamity and disappointment, and a masochistic satisfaction from being the victim. Outstanding characteristics of this type apart from pessimism and masochism would be a strong oral aggressiveness and, as a reaction to the mother's breast, a tendency either to seek what must ever remain unattainable or to develop by way of restitution an attitude of autonomy and independence.

The combined descriptive and etiological aspects in the psychoanalytic exposition of oral character traits raise two separate questions which determine the two sections of the present investigation: the first section aims at setting up from the data at hand character types corresponding to the syndromes of the oral character as described in psychoanalytical literature; the second investigates the way in which such types as well as individual character traits may be related to early experiences of frustration or gratification in feeding measured in terms of the length of breastfeeding.

THE EXPERIMENTAL POPULATION

The experimental population for this part of the investigation consisted of 115 adults, 47 men and 68 women between the ages of 18 and 35; 3 subjects were over 35. Forty-one were psychology students of University College, London, and Birkbeck College, London; 21 were students of the University Extension and W.E.A. classes in psychology (by occupation teachers, social workers, clerks, engineers, and housewives); 38 subjects were friends and friends of friends; and 15 were neurotic patients in psychoanalytic treatment. Socially the sample was a fairly homogeneous group belonging to what is commonly described as the "middle class."

THE RATING SCALES

The data analyzed in the present paper are derived from 19 rating scales designed to measure as many character traits: two types of scales were used: (a) short statements in aphoristic form which will be referred to as "Proverbs and Aphorisms," and (b) questionnaires. There were four scales in the former category constructed to measure the traits of Optimism, Pessimism, Passivity, and the Desire for the Unattainable (to be called in future simply Unattainability); questionnaires were used for the following traits: Autonomy, Ambition, Aggression, Change, Dependence, Deliberation, Endocathexis, Guilt, Nurturance, Aloofness, Conservatism, Sociability, Impulsion, and Displaced Oral Aggression.

Proverbs and aphorisms. These scales were composed of short statements giving expression to moods, sentiments, and emotional attitudes belonging to the respective traits. A list of 60 such statements was compiled, drawn largely from various books of proverbs, quotations, aphorisms. This list was at first administered individually to a trial group of 12 subjects who were interested in the experiment. These were asked to state their emotional reactions on a five-point scale ranging from $+2$ to -2. The subject's introspections and comments were then noted and his views about the form of the test as a whole as well as the meaning of each statement were discussed. Statements for which the subject had not offered an interpretation in line with the aim of the particular scale concerned or where he had not, at least, readily agreed to one suggested to him were eliminated. A few statements which were suggested by some of the subjects were, after critical deliberation and discussion with other subjects, included in the new scale. To disguise the continuity of the scales, the items belonging to the four traits tested in this list were given in random order. To achieve a wide enough interval between items of the same scale, "jokers" (i.e., undiagnostic items) were included. No quantitative validation of the items was undertaken at this stage.

Questionnaires. These were of the analytic type. The advantage of using the analytic scale technique lies in avoiding the danger of subjects overrating themselves in desirable traits: the items cover a range of presumed manifestations of the traits investigated dealing primarily with questions of the testee's behavior rather than his self-evaluation. Discussion with the subjects of the trial groups showed that the aim of disguising the symptomatic significance of the items was, to a fair extent, achieved. The random arrangement of the items was another factor making for the subjects' unawareness of the purpose of the experiment. All questionnaires except the one measuring Displaced Oral Aggression are modifications of the H. A. Murray scales [22] of the same description. Three scales appear under a different name in this investigation. Murray's Affiliation has been changed into Sociability, Rejection into Aloofness, and Sameness into Conservatism. The advantage of these scales derives from the psychological approach guiding their construction: the reference to psychoanalytic concepts and the attention given to abstracting basic "needs" or source traits from many different behavior manifestations. The scale measuring "Displaced Oral Aggresion" is largely based on Glover's [13] description of oral-sadistic speech-formation. The scoring, as in "Proverbs and Aphorisms," was on a five-point scale from $+2$ to -2.

Administration of the scales. The main consideration in deciding the method of administering the scales was that of anonymity. It was felt that an important condition for the truthfulnes and uninhibitedness of responses was that the subject should feel unobserved and unknown. A large number of the tests were distributed through other people and returned by mail. Students at the University were asked to deposit their completed tests in the Psychology Department and Evening Class students of the writer's own course were asked to return them all on one particular evening so that no individual identification was possible. With a group of 14 subjects, friends of the writer, anonymity could not be preserved. This disadvantage seemed, however, discounted by a greater spirit of co-operativeness shared by all but two, who, perhaps significantly, had offered to take part in the experiment although not asked. For the majority, however, the writer, checking against his own acquaintance with their personalities, was fairly well satisfied that their marks were genuine. The close acquaintanceship and tolerance of friendship seemed, in fact, to discourage false responses, at least where insight existed. Characteristic in this respect was the comment of one of these subjects: Marking +2 in answer to the question "Do you often let yourself go when you are angry?" she added: "Can't deceive *you* on this point." The unwillingness of the two uncooperative subjects in this group to divulge information about themselves revealed itself in their marking behavior. One of them marked predominantly nought (o) and the other answered the overwhelming majority of questions in the negative. Both were eliminated from the sample.

MEASURES OF INTERNAL CONSISTENCY AND RELIABILITY

(A) *Internal consistency*. Two measures of the internal consistency of the scales are given: (1) through the so-called test of internal consistency by which the diagnostic value of their items was determined against the criterion of the total raw scores and (2) through the intercorrelation table of traits, which by reflecting the trait relationships yields a certain measure of the consistency of the rating scales as a whole.

(1) *Index of item consistency*. The diagnostic value of test items was measured by the tetrachoric correlation coefficient using the upper and lower halves of the sample, and dividing them along the + and o borderline. In other words, the marks on each item were divided into two sections, one containing the +2 and +1 ratings, and the other the o's, −1's and −2's. Each of these two groups were again dichotomized with respect to the ratings on the total score and the cell frequencies for the fourfold table obtained. The coefficients were then read from Thurstone's graphs for tetrachoric correlation coefficients, Chi Squares were also computed from these fourfold tables, and items which did not show a highly significant difference (at the .01 level) between the reactions of the two criterion groups were discarded for scoring. Three items which did not satisfy this requirement but which showed good discriminative values between the upper and lower quarters were retained. It is of course understood that these figures are liable to shrinkage in a new group, but as their values can be regarded as rather high a considerable margin is left. Description of the traits together with their reliability coefficients is presented in the Appendix.

(2) In addition, a check on the above method is provided by the fact that most traits contained in the total battery of 19 scales are there in their opposites or near opposites. If their correlations and intercorrelation are to make "psychological sense" they must fulfil certain conditions. Thus Optimism and Pessimism, Exocathexis and Endocathexis, or Aloofness and

Sociability would be expected to show fairly high negative correlations. Coefficients not satisfying such an expectation would be serious pointers to these scales not measuring what they propose to measure. An examination of our actual intercorrelation table (TABLE XXI) shows that the above conditions are fulfilled to satisfaction. All traits of a clearly bipolar significance have high negative correlations. Their average value, which we may term the index of bipolarity, is −.50. This measure is based on the following seven trait pairs: Optimism-Pessimism (−.63), Exocathexis-Endocathexis (−.60), Sociability-Aloofness (−.50), Nurturance-Passivity (−.43), Change-Conservatism (−.50), Impulsion-Deliberation (−.75), Dependence-Autonomy (−.11).

(B) *Reliability*. The reliability of each scale was measured by the split-half method of correlating odd and even numbered items and corrected by the Spearman-Brown prophecy formula. The reliability coefficients for the 19 traits used in this research ranged between .90 and .55, and the average reliability coefficient was .76. This figure is, however, somewhat reduced by the two exceptionally low values for the tests of Dependence (.56) and Conservatism (.55). On the whole the reliability coefficients which are given with each test below are high considering that there was an average of no more than eight items in each scale. Doubling the length of the scales would, on the basis of the Spearman-Brown prophecy formula, increase the average coefficient for reliability for all 19 scales from .76 to .86.

TREATMENT OF DATA

There are various methods available for analyzing a set of psychological variables with a view to ascertaining their underlying basic components. With such data as in our research, where we are dealing with self-assessments of persons, Burt [8] has shown that by correlating persons equivalent information can be obtained, as by correlating tests or traits. The advantage of factorizing persons instead of traits when dealing with subjective data lies in the fact that in eliminating the general factor we eliminate the subjective bias which enters into any form of self-assessment. In researches concerned with type-psychology, as in the present, this factor corresponds to a kind of halo, probably involving also the person's attitude to the test as well as a general tendency to give high or low marks, and obscuring the outline of his character-profile.

For the purpose of dealing with large samples which exclude the application of the Q-technique of correlating persons, Burt has developed a simplified procedure which he proved to yield virtually the same results as the correlation of persons by the method of summation. By this procedure the standardized measurements of traits of each person (after the subtraction of the person's averages from the original scores) are correlated with a "key personality," the hypothetical or ideal type, instead of being correlated with those of all the other persons in the sample. The coefficient of

TABLE XXI

Intercorrelation of Traits

	Opt.	Exc.	Nur.	Soc.	Amb.	Change	Unatt.	Delib.	Imp.	Cons.	Dep.	Guilt	Aggr.	Auto.	Endo.	Oral	Aloof.	Pass.
Opt.																		
Exc.	.51																	
Nur.	.46	.35																
Soc.	.20	.31	.17															
Amb.	.27	.14	−.17	−.11														
Change	.16	−.05	.30	.21	.01													
Unatt.	−.34	−.22	.06	−.01	.22	.32												
Delib.	−.11	.05	.00	−.15	−.26	−.35	−.02											
Imp.	−.38	−.16	−.05	.08	.04	.44	.13	−.75										
Cons.	.21	−.19	−.11	−.50	−.15	−.50	−.14	.42	−.39									
Dep.	−.10	−.15	−.22	.39	−.32	.04	.11	.19	−.36	.06								
Guilt	−.15	.00	−.02	−.20	−.09	.12	.24	.23	−.17	.21	.47							
Aggr.	−.31	−.16	−.27	−.17	.00	.00	.28	−.38	.35	−.05	−.05	−.16						
Auto.	−.23	−.47	−.29	−.20	.07	.17	.07	−.12	.19	−.20	−.11	−.47	.10					
Endo.	−.29	−.60	−.16	−.40	.16	−.04	.32	.08	.04	.20	.21	.05	.08	.31				
Oral	−.36	−.07	−.19	−.19	−.23	−.02	−.07	−.42	.12	−.14	−.08	−.13	.33	.10	−.09			
Aloof.	−.37	−.55	−.65	−.50	−.09	−.29	−.02	.05	.00	.00	−.07	.05	.24	.41	.33	.12		
Pass.	−.48	−.34	−.43	−.01	−.40	−.01	−.25	−.16	.21	.02	.17	−.01	.32	.06	.03	.37	.44	
Pessim.	−.63	−.55	−.22	−.34	.05	−.14	−.02	.27	−.10	.16	−.02	.38	.11	.13	.20	.11	.37	−.03

the correlation of each person's measurements with those of the hypothetical type gives us a score equivalent to any particular person's standing with respect to the hypothetical type pattern set up.

Following this procedure standardized measurements were obtained. For every trait the mean is zero and the standard deviation unity. A bipolar clustering of the traits was noted towards the tail ends of the distribution with a gradual breaking up of the pattern towards the center; in other words, towards the extreme ends of the sample distribution there are a considerable number of persons whose trait measurements cluster in a clearly defined way forming syndromes or types of a certain description. It is from these measurements that the "key personality" or "standard type" is to be derived.

MEASURING THE "ORAL TYPE"

In setting up the key personality to serve as a standard of comparison it was necessary to decide what special set of measurements was to be taken for the standard. Burt [8] obtained the key measurements of his ideal extravert-introvert type by averaging all the figures for his group, first reversing, or reflecting, the signs for those belonging to one half of the group. In the present research the procedure followed was somewhat different: instead of the whole sample being used to arrive at the standard type, only 20 persons at each end of the distribution, 40 people in all, were selected; and the "key personality" was abstracted from the more extreme and marked characters, leaving out the intermediate group of mixed types.

In order, however, to guard against the element of arbitrariness and subjective choice which might enter into the selection of those 40 key persons and to allow for a self-selection of this group in close agreement with the existing trait relationships, the intercorrelation table of traits was made the guide and criterion according to which each subject was allotted to either one of the key or the intermediate groups. The majority of the 19 traits seemed to group themselves in two antithetical clusters: Optimism, Exocathexis, Nurturance, Sociability, and Ambition correlating positively at one end of the trait distribution; and Pessimism, Passivity, Aloofness, Endocathexis, Autonomy, and Oral Aggression constituting the opposite pole. The trait measurements of each person were then carefully examined; and those persons corresponding most closely with the above arrangement, by showing the two antithetical clusters irrespective of their signs, were selected. Those who showed positive measures for the first group of traits will be called "oral optimists," and those who showed positive measures for the second group of traits "oral pessimists." Twenty people at each end of the distribution were selected who qualified for these categories, the minimum requirement being that 8 out of the 11 characteristic traits should cluster at each end of the table in the expected antithetical way. The measurements for the 20 "oral pessimists" and the 20 "oral optimists" were totaled—the groups being kept separate—and averaged.

THE RESULTS

The measurements (in standard units) appear in columns I and II of TABLE XXII. In Chart XXIII we have presented them graphically with the standard measures expressing the saturations of the type with the traits given at the scale on the left. The relative symmetry of the two graphs or "profiles" seems to support the assumption that the traits form a bipolar syndrome. Thus a pooling of the two groups by totaling, after reversing the signs of the optimist group and averaging their measurements, was

TABLE XXII

Measurements of Traits

Traits	I Upper Group	II Lower Group	III IN STANDARD MEASURE Total	IV Average
Pessimism	+1.417	−0.894	+46.23	+1.155
Passivity	+0.966	−1.134	+42.01	+1.050
Aloofness	+0.826	−0.685	+30.23	+0.755
Oral Aggression	+0.862	−0.335	+23.94	+0.598
Endocathexis	+0.687	−0.485	+23.44	+0.586
Autonomy	+0.919	−0.234	+23.08	+0.577
Aggression	+0.445	−0.339	+15.69	+0.392
Guilt	−0.025	−0.587	+11.23	+0.280
Dependence	+0.364	−0.151	+10.31	+0.257
Conservatism	+0.135	−0.167	+6.06	+0.151
Unattainability	−0.052	+0.204	+5.13	+0.128
Impulsion	+0.194	+0.337	+2.86	+0.071
Deliberation	−0.092	−0.072	−0.40	−0.010
Change	−0.091	+0.395	−9.74	−0.243
Ambition	−0.309	+0.223	−10.66	−0.266
Sociability	−0.695	+0.536	−24.63	−0.615
Nurturance	−0.958	+0.611	+31.38	−0.784
Exocathexis	−0.952	+0.701	−33.07	−0.826
Optimism	−1.192	+1.086	−45.57	−1.139

considered justified. The simple totals appearing in column III and column IV show them reduced to ordinary standard measure. These measurements represent the "oral pessimist" type. A reversal of signs would give the "oral optimist" type.

This procedure may be criticized on the ground of what might be called its impurity of method. If the aim of the investigation was to test a psychoanalytic hypothesis, namely the relation of oral character types to weaning, it would have been more strictly logical to set up the ideal or hypothetical profile which was to serve as the criterion of orality, without reference to the actual intercorrelation table, but directly from psychoanalytic description. However, there were difficulties in the way of following this course. For although psychoanalytic writers agree with respect to the basic oral traits different writers put different emphasis on different traits or combinations of traits. Syndromes are sometimes suggested, but on other occasions traits are enumerated singly without a clear

CHART XXIII

Graphic Presentation of Measurements of Traits

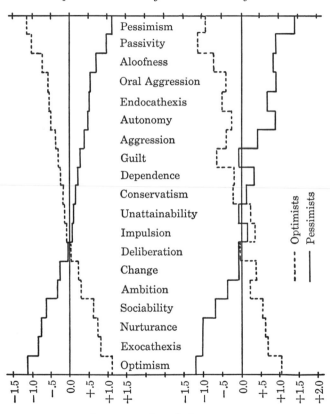

Average of pooled groups

idea being given of how they combine, or whether in fact they invariably do combine. Moreover, the psychoanalytic theory is phrased in a descriptive and qualitative form, so that it was still incumbent upon the investigator to find its quantitative expression. Thus, while psychoanalytic literature was responsible for the selection of the particular traits tested, their quantitative expressions have emerged from the intercorrelations of tests. As it happened these were very much, though not entirely, what one would have expected on the basis of the theory.

DISCUSSION

In referring to the newly found type factor as "oral" a presupposition was made as to its psychological nature which has yet to be justified. Two assumptions are implied in using the terms "oral pessimist" and "oral optimist." The first is that there is a close correspondence between the type obtained from our data and the oral character as described by psychoanalytic writers, and the second assumption is that this type has been etiologically determined by experiences at the oral level. It is with the first assumption that we are dealing in this paper: What is the relation of our standard type to the "oral character" of psychoanalysis?

In TABLE XXIV we have set out the main oral syndromes of psychoanalysis, checking each trait against the equivalent test results (as far as such have been obtained). The correspondence between the hypothesis and the experimental data is striking, and highly positive. All test measurements, except those for Impatience and Unattainability, are in the expected direction. Moreover, those traits which have received the greatest emphasis by most writers on the subject, i.e., omnipotence (optimism and pessimism), the passive-receptive attitude, narcissism, and generosity are also represented by the highest quantitative assessments; in other words, those traits which have impressed themselves on psychoanalysts most clearly as of oral connotation are precisely those whose contribution to the type in terms of test measurements is the most important.

From the figures in TABLE XXII and the "profiles" embodying them, the following personality types emerge:

I. *The Oral Optimist.* The measurements belonging to this type show a pronouncedly optimistic personality (Optimism +1.13, Pessimism −1.15) with a positive interest in external events (Exocathexis +0.82) as well as people (negative Aloofness −0.75), a generous inclination towards protecting the weak and needy (Nurturance +0.78), sociable (Sociability +0.61), unaggressive (Aggression −0.39, Oral Aggression −.59) and moderately ambitious (Ambition +0.26), with a tendency to welcome new things and experiences (Change +0.24) rather than clinging to the old (negative Conservatism −0.15). The correspondence with the description of the orally gratified type of psychoanalysis given previously in this paper is nearly complete.

II. *The Oral Pessimist.* This type shows a composition very similar to the orally ungratified type of psychoanalysis. As already mentioned, the main emphasis is on pessimism (pessimism +1.15, optimism −1.13) and the passive-receptive attitude (Passivity +1.05). The traits of aloofness (Aloofness +0.75 and negative Sociability −0.61), withdrawal (Endocathexis +0.58 and negative Exocathexis −0.82), a disinclination to be helpful and protective (negative Nurturance −.78) and narcissistic self-sufficiency (Autonomy +.57) bear a strong relation to Ernest Jones's "God Complex." What is particularly notable is the relative minor contri-

TABLE XXIV

Relation of Oral Phases to Scores on Trait Tests

(Figures represent standard measures; see TABLE XXII

Glover's basic characteristics of an 'archaic oral fixation'
{
Omnipotence
(tests: Optimism   1.13
        Pessimism   1.15)
Narcissism
(tests: Autonomy   + .57
        Aloofness   + .75)
Viewing (no test)
Exaggerated motor accompaniments
of oral activity
(test: Oral Aggression   + .59)
}

ORAL PHASE

SUCKING
(Pre-ambivalent)

Gratified (Omnipotence preserved)		Ungratified * (Omnipotence shaken)		BITING * (Oral-Sadistic)	
Traits		*Traits*		*Traits*	
Optimism		Pessimism		Hostility	
(tests: Optimism	+1.13	(tests: Pessimism	+1.15	(test: Aggression	+.39)
Pessimism	−1.15)	Optimism	−1.13)	Envy (no test)	
Generosity		Passive-receptive attitude		Covetousness (no test)	
(tests: Nurturance	+ .78	(tests: Passivity	+1.05	Aggressive use and omnipo-	
Passivity	−1.05)	Nurturance	− .78)	tent valuation of speech	
Social conduct: Bright, sociable		Attitude of withdrawal		(Oral aggression	+.59)
(tests: Sociability	+ .61	(tests: Endocathexis	+ .58	Impatience (Glover)	
Exocathexis	+ .82)	Exocathexis	− .82	(tests: Impulsion	+.07
Accessible new ideas		Sociability	− .61	Deliberation	−.01)
(tests: Change	+ .24)	Aloofness	+ .75)		
Conservatism	− .15)	Desire for the unattainable			
Ambition		(test: Unattainability + .13)			
(test: Ambition	+ .26)				

Impatience (Abraham)
(tests: Impulsion   +.26
      Deliberation   −.08)

God Complex (Ernest Jones)		Oral Pessimism (Edmund Bergler)	
Basic feeling tone = Omnipotence		Narcissism (tests as above)	
(tests: Optimism	1.13	Pessimism (tests as above)	
Pessimism	1.15)	Oral Aggressiveness	
Manifesting itself in:		(tests: Oral aggression	+.59
Narcissism		Aggression	+.39)
(tests: Autonomy	+.57		
Aloofness	+.75)	Autonomy and Independence	
Exhibitionism (no test)		(test: Autonomy	+.59)
or		Desire for the Unattainable	
Excessive modesty Self-effacement } (test: Guilt +.28)		(test: Unattainability	+.13)
Aloofness			
Stress on privacy Seclusion } (test: Aloofness +.75)			
Secretiveness (no test)			
Unsocial			
(test: Nurturance	−.78)		
Unsociable			
(tests: Sociability	−.61		
Aloofness	+.75)		
Curious (no test)			
Unpunctual (no test)			
Impatient			
(test: Impulsion	+.07)		

* Satisfaction sought at oral-sadistic level.

bution of Guilt (0.13) and Dependence (0.25) to the syndrome in comparison with those of the narcissistic and omnipotent traits. Narcissistic withdrawal of libido and self-sufficiency (Aloofness, Endocathexis, Autonomy) are far more characteristic of oral pessimism than self-reproach, superego anxiety, and feelings of unworthiness (Guilt), or anaclitic helplessness (Dependence). A comparison of the contribution to the type by the latter brand of passivity, which is one involving object attachment with that of the purely egocentric kind measured by the test of Passivity, shows the latter to be four times as large (Passivity −1.05, Dependence −0.25). The positive correlation of the tests of Dependence with Sociability (0.39) and Guilt (.47) seems to point to a common element in these traits, probably one of need for object relations, the comparative independence of which seems a marked characteristic of the oral pessimist syndrome as emerging in this research. This overstress of the narcissistic element, emphasized by Glover [11, p. 149], as indicative of an archaic oral disposition, seems to suggest that the "oral pessimist" syndrome is not only quantitatively less intense but also of a qualitatively different constitution from the typical manic-depressive psychosis through the study of which Abraham gained the first insight into oral character traits. The idea of a schizoid influence seems to suggest itself. Bergler [6] and Wisdom [27] have also brought into relief the characteristic importance for this type of a pervading narcissism setting up a defensive barrier against aggression turning upon the self. "What is peculiar," Bergler concludes in his study of oral pessimism, "is that oral pessimism is not identical with any of the known syndromes of psychopathology, that we find in this group schizoid, cycloid and apparently hysteric types. . . . Oral pessimism is constructed like a neurotic symptom and represents a narcissistic defense mechanism of the Ego, offering protection to the severely wounded illusions of omnipotence" [6]. Similarly Wisdom [27] interprets the attitude shown by Schopenhauer, in whom he finds most features of a typical manic-depressive psychosis represented. The fact that the latter, in spite of his profoundly pessimistic outlook on life, intellectually repudiated suicide and never attempted to take his own life, he considers to be indicative of narcissism accepted without guilt and of the continued existence of this even in the depressed state.

The contribution to the syndrome by the test of Displaced Oral Aggression seems to confirm Glover's views on oral-sadistic tendencies in speech and their significance as an aspect of the orally dissatisfied person. With a standard measure of 0.59 this trait takes fourth place among the 10 tests making a positive contribution to the oral pessimist type. Glover's emphasis or the "special significance" of exaggerated motor accompaniments of oral activity (such as grinding teeth, thrusting hand in mouth, etc.) as indicators of an "archaic oral disposition" also accords with the data of this research: Question 7 of the scale under discussion which inquires into such tendencies has the positive correlation of 0.52 with the test as a whole. We have here an indication of an oral-sadistic source-trait underlying motor as well as verbal aggression and contributing to the syndrome of oral pessimism. The element of oral impulsion in this type is thus clearly suggested.

The contributions to our type of Ambition, Change, and Conservatism are, though slight, in the direction suggested in the hypothesis. Measured against the importance attributed to it by Glover [13], the trait of Unattainability as diagnosed by our scale, however, seems to bear little relation (s.m. = +.128) to the orally dissatisfied type. Nor does the oral type found from our data seem to owe anything to the traits of Impulsion (0.07) and Deliberation (0.01). This fails to bear out Glover's version, which assumes Impatience (if Impulsion measures impatience) to be one of the outstanding characteristics of the orally dissatisfied type, but does not contradict Abraham's view, according to which impatience is linked with orality generally. With respect to the latter hypothesis our data are in the right direction: the values for Impulsion in the oral pessimist as well as the oral optimist clusters are positive, +.19 and +33 respectively (average +.26); equally the values for Deliberation are negative for both oral types (−.09 and −.07). Extreme oral characters at both ends of the distribution seem therefore to be inclined to impulsiveness, whereas deliberation seems rather to go with an absence of either strong oral optimism or pessimism.

SUMMARY

This investigation was undertaken with the aim of relating experimental data to psychoanalytic theory. The specific hypothesis tested was that of the "oral character." This involved two questions: (a) one concerned with the objective verification of the psychoanalytic description of oral character traits and syndromes, and (b) the other referring to their etiology. We have dealt thus far with the first part of the investigation.

1. Nineteen traits were tested by means of rating scales, using 115 adult subjects. The average number of items in one scale was 8.

2. The traits tested were: Optimism, Pessimism, Exocathexis, Endocathexis, Nurturance, Passivity, Sociability, Aloofness, Oral Aggression, Autonomy, Aggression, Guilt, Dependence, Ambition, Impulsion, Deliberation, Change, Conservatism, Unattainability.

3. Five of the scales have been newly constructed (Optimism, Pessimism, Passivity, Unattainability, Oral Aggression), and the rest are based on the H. A. Murray questionnaires in *Explorations in Personality*.

4. An item analysis was carried out with all scales, and reliability coefficients calculated. The average index of item consistency was .61. Reliability coefficients ranged from .90 to .56, the average being .76.

5. Burt's simplified type factor analysis correlating persons with a "standard type" was carried out. The "standard type" was abstracted from the standardized trait measurements of 40 persons, 20 at each end of the

distribution. These were selected in accordance with trait-relationships as shown by the intercorrelation table of traits.

6. A bipolar type factor was found, highly saturated with positive and negative manifestations of omnipotence and narcissism, and characterized mainly by the traits of Pessimism, Passivity, Aloofness, Oral Aggression (Verbal), Endocathexis, and Autonomy at one end; and Optimism, Exocathexis, Nurturance, and Sociability at the opposite end of the trait distribution.

7. The composition of this factor shows a striking correspondence to psychoanalytic oral character types as described by Abraham, Glover, Jones, and Bergler.

II. THE ETIOLOGY OF THE ORAL CHARACTER IN PSYCHOANALYTIC THEORY

As to the etiological assumptions which underlie the psychoanalytic classi-fication of oral character traits, they are in no way unequivocal and uni-versally agreed on among psychoanalysts. On the question of the type of experience most likely to induce an oral fixation Abraham considers early frustration or overindulgence of equal effect.

In either case the child would take leave of the sucking stage with dif-ficulty "and aim at compensation at the next stage" since its need for pleas-ure has either not been sufficiently gratified or has become too insistent [3]. Glover, while taking into account the question of oral disposition and of the *relative* length of an individual optimum period of "oral primacy," considers that "the shortening of this period is more likely to prove trau-matic" [14] than the lengthening.

Edmund Bergler [6] takes Abraham's point of view of the equivalence of frustration and overindulgence as traumatic factors and quotes Freud in support of the thesis that "the greed of the child after his first feed is alto-gether insatiable, that he never resigns himself to his loss," also referring to the latter's caustic remark that he would not be surprised if the analysis of a primitive who was given the nipple even when he could walk and talk, would reveal a grudge of having been deprived of the breast.

However, in his later writings Freud stresses the importance of the ex-ternal situation as a traumatic factor, viz., "frightening *instinctual* situa-tions can in the last resort be traced back to *external* situations of danger"; and he adds that it is not the objective injury that is feared but a state of traumatic excitation (previously experienced) "which cannot be dealt with in accordance with the norms of the pleasure principle" [10].

Sandor Rado [23], referring to Freud, points out that "the deepest fixa-tion-point in the melancholic (depressive) disposition is to be found in the 'situation of threatened loss of love,' more precisely, the hunger situation of the suckling baby." Thus according to this point of view it would, in the last resort, not be the endopsychic instinctual situations such as the greed and insatiability of the child but the external deprivation (hunger and loss of love), which, via an unpleasurable emotional excitation, would be the traumatic basis of depression.

Against this view Klein maintains that the actual experience of the "loss of the loved object" already meets with an intrapsychic situation of a successful introjection of the "good" object which decides whether the loss of the breast before and during weaning will result later on in a depressive state. "No doubt, the more the child can at this stage develop a happy relationship to its real mother, the more will it be able to overcome the depressive position. But *all* [italics mine] depends on how it is able to find its way out of the conflict between love and uncontrollable hatred and sadism" [17]. This depressive position characterized by cannibalistic desires directed at the mother and the guilt feelings arising therefrom Klein places between the third and fifth month [12].

THE PROBLEM

The implications of the etiological hypotheses of these writers seem to be fundamentally twofold:

(a) That the weaning trauma with its psychological manifestations of melancholia and oral pessimism is induced by an external situation which may be that of frustration manifested in early weaning rather than late, according to Glover, or that of frustration as well as of overindulgence (taking an overlong period of breastfeeding as a measure of the latter), according to Abraham and Bergler, and

(b) That the intrapsychic *instinctual* situation, the child's greed and his capacity to overcome his inner conflict between love and hatred of the object will in the last resort determine whether he will remain fixed to the oral level, or, as Klein terms it, in his depressive position.

The above survey throws the speculative character of psychoanalytic views concerning the etiology of oral depression clearly into focus. References are made by Abraham [3], Glover [14], Stärke [26], Bernfeld [7] —in full consciousness of the tentativeness of their views—to supporting material gained from the analyses of patients. It cannot, however, be assumed that these analyses have unearthed memories directly connected with the experiences of early frustration, particularly weaning, the first year of life being hardly accessible to psychoanalytic anamnesis. These authors rather seem to refer to later fantasies containing the motif of weaning the relation of which to the original trauma (of weaning) must be considered too uncertain to warrant its value as evidence. The analyses of children from which Klein derived her hypothesis again are based only on fantasy material and its interpretation, no steps having been taken to ascertain the objective data concerning the child's early feeding.

This is, of course, not surprising as it is not part of the orthodox psychoanalytic technique to ascertain data outside the therapeutic process by which the patient produces material. However, discussing Rank's theory of the "trauma of birth" Freud emphasized that with an etiological theory which "postulates a factor whose existence can be verified by observation, it is impossible to assess its value as long as no attempt at verification has been made" [10, p. 35].

Weaning, a frustration imposed on the child from the outer world and assumed to be a traumatic impact is an objectively verifiable factor. The above quoted etiological hypotheses concerning oral depression can thus be put to an objective test.

In this part of our study we tested two of the psychoanalytic hypotheses concerning the etiology of oral character:

(a) Glover's assumption that the shortening of the period of breast-feeding, i.e., early weaning, is more likely to prove traumatic than the lengthening, i.e., late weaning, and (b) Abraham's that an overlong as well as too short a period were traumatic.

The problem now arose how to define early and late weaning.

Referring to the literature we find that Ribble [24] with due consideration of the factors of individual differences in the rate of development and the effects of frustrating situations on this development has "found definitely that the sucking impulse spontaneously wanes as new oral activities become differentiated and are brought into momentum around the fourth month of life." Ribble feels that if sucking has been satisfactory up to that period weaning does not present problems, and in fact becomes spontaneous. On the basis of her observations she thought it justified to assume that, under normal conditions, weaning after a sucking period of four months would meet with a more favorable instinctual situation than weaning before that period. The same critical period in the instinctual development is also indicated by Klein, who places the depressive position between the third and fifth month.

In setting up our hypothesis that oral pessimism (depression) is a function of early weaning, it was therefore the fourth month which was chosen to separate the two groups to be tested. Early weaning we define as weaning at the ages of up to four months, and late weaning as weaning at five months of age and more. Caution may, however, be advised as to this definition with respect to the intermediate months in the age-scale of weaning, i.e., 5–8 months, since the possibility that *individual differences in rate of development* may blur the picture of the effect of weaning in these months on character-formation is to be considered. A clearer picture of this effect might be expected from comparing the early weaning group as defined above with the group of those subjects who were taken off the breast at not earlier than nine months of age, i.e., who were given what in the general medical opinion is the optimum period of sucking.

DATA

The data for this investigation were:

(a) Scores for oral pessimism obtained on the basis of the verbal rating scales for 19 character traits administered to 100 adult subjects and the type pattern derived therefrom, as described in the first part of this paper, and (b) information obtained from the mothers of these subjects as to whether the child had been breastfed or bottlefed from the start; and if breastfed, at what age it was taken off the breast.

An analysis of variance testing the scores of the two groups of early and late weaners (referring to both of the definitions of late weaning given above) for the significance of their differences was then computed.

TYPE SCORES

In order to measure each person's standing with respect to the type factor or pattern of oral pessimism or optimism, the individual measurements for each of the 19 character traits were correlated with the measurements of the standard pattern. The coefficients thus obtained express to what degree the person approximates to the oral pessimist type. The figures ranged from $+.97$ to $-.88$ and may actually be taken as the person's coefficients of saturation with the oral type factor. It should be noted that they do not depend on the *level* of response but that they are a measure of the approximation of the pattern of traits within each person to the standard pattern regardless of the size of the individual scores.

The question addressed to the mothers of the 100 subjects as to whether the child has been breastfed or bottlefed from the start; and, if breastfed, at what age it was taken off the breast, yielded the distribution of periods, shown in TABLE XXV, of weaning in the sample investigated:

TABLE XXV

Age at Weaning

	AGE	NO. OF SUBJECTS	GROUP
Bottlefed from the start		10	A
Breastfed	Two months and less	18	B
	Three months	9	C
	Four months	3	D
	Five months	4	E
	Six months	10	F
	Seven months	7	G
	Eight months	3	H
	Nine months	23	I
	Over nine months	13	J
		100	

This distribution shows the largest frequencies in the first two weaning groups (B and C) and the last two weaning groups (I and J). This may perhaps give some weight to the deliberation that certain general factors come into play determining the time of weaning, which in the intermediate months are complicated by other individual factors. What these factors determining the time of weaning might be is at present a matter of mere speculation. To the writer it seems that the attitude of the mother, either rejecting or accepting, might be one. It may be likely that a rejecting mother would have thrown off the burden of breastfeeding by the third or fourth month rather than drag on further. Mothers who carry on would probably be of the type who, discounting interference of other factors, would go through with breastfeeding to the conventionally ap-

TABLE XXVI

Analysis of Variance Testing the Difference between Early Weaners
(Weaned at Not Later than Four Months of Age) and Late Weaners
(Weaned at Five Months and More)

Source	Degrees of Fr.	Sum of Squares	Mean Square
Within weaning group............88		22.267	.253
Between early and late weaning group.......................... 1		2.033	2.033
	89	24.300	.2730

$$F = 8.04 \text{ (F for 1 \& 80 df at 1\% level} = 6.96)$$
$$= \Sigma^2.073 \quad \Sigma = .271$$

Analysis of Variance Testing the Difference between Early Weaners (as Above)
and Late Weaners (Weaned at Nine Months and More)

Source	Degree of F.	Sum of Squares	Mean Squares
Within weaning group............64		16.271	.254
Between early and late weaning groups...................... 1		1.904	1.904
	65	18.175	.280

$$F = 7.50 \text{ (F for 65 df at 1\% level} = 7.04)$$
$$= \Sigma^2.093 \quad \Sigma = .305$$

proved and medically advised age of nine months. This question calls for investigation, in order to throw more light on any relation between weaning and character which might be found. Other factors precipitating weaning might arise from the child's constitution and the intensity of the demands he makes on the mother, necessitating his being taken off the breast.

TREATMENT OF DATA AND RESULTS

An analysis of variance was then computed to test any possible difference in the scores on oral pessimism in the early and late weaning groups whereby both definitions of late weaning described above (five months and more, and nine months and more) were referred to successively. Below in TABLE XXVI are set out the corresponding tables, F-ratios and correlation coefficients.

We see that the difference in oral pessimist scores between early and late weaning whereby late weaning is defined in the two above described ways is significant, as tested through the F-ratio in both cases, beyond the 1 per cent level. The correlation between early weaning and oral pessimism, using the whole distribution of weaning groups and separating early and late weaning at the fourth month, is .271; if in the late weaning group only those taken off the breast at nine months and after are included, the correlation is slightly higher, viz., .305.

Testing Abraham's and Bergler's assumption that frustration as well as overindulgence determine oral pessimism, by comparing the early weaners with those subjects whose breastfeeding was continued after their ninth month of age we find the following results: the mean score for the former group in oral pessimism is +.138, for the latter group −.346. In other words, those who experience early frustration in breastfeeding tend in their adult character makeup to lean toward the oral pessimist end of our type pattern, while those who have enjoyed an overlong period of breastfeeding show a marked tendency to develop oral optimist character traits. The correlation between early weaning and oral pessimism on the one hand and overlong breastfeeding and oral optimism on the other is .368, significant beyond 1 per cent.

This result supports Glover's contention that the shortening of the weaning period must be regarded as traumatic factor rather than the lengthening, if by traumatic we mean inducing oral pessimism or depression. At the same time the above result also coincides with the observations of those anthropologists that a long period of breastfeeding and optimism, generosity, etc., were related.

DISCUSSION

Attempting to answer our question concerning the influence of breastfeeding on character-formation from our data, we can therefore say: that a significant (beyond the 1 per cent level) correlation has been found between early weaning and oral pessimism. Or, in other words, early weaning tends to be associated with oral pessimism, and late weaning tends to be associated with oral optimism to the extent of r = .27, respectively, .31.

These results seem therefore to indicate that the length of breastfeeding is a significant factor in the etiology of oral pessimism and probably depression; but the size of the correlation also shows that there are other factors involved in oral pessimism which account for it to a greater extent, and which still await investigation.

As to the effect of the weaning trauma itself, neither do we suggest that the duration of breastfeeding as such need be regarded as a last cause in the etiological chain; for it may equally be only a symptomatic manifestation of more fundamental factors, such as the attitude of the mother, the constitution of the child, and the interaction of both. Here too we need further investigation.

III. DATA REGROUPED BY FACTOR ANALYSIS

While the above results confirm the psychoanalytic theory of oral character types in most of its essential aspects, one important characteristic, the significance of which as an oral trait ranks extremely high in psychoanalytic literature [3, 14], forms an exception. This is the trait of impatience, measured in this investigation by the scales of impulsion, and, negatively,

CHART XXVII

Plot of the Factor Loadings of 19 Tests with Reference to Factor Axes I and II

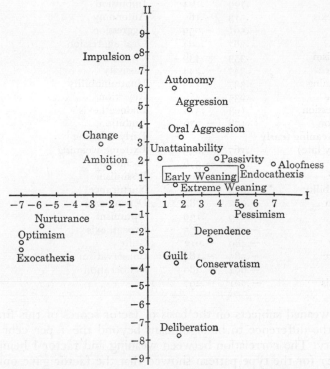

deliberation. These two scales make practically no contribution to the oral pessimist type profile, and the intercorrelation table of traits shows that it is outside the optimist or pessimist clusters. Its independence suggested a second bipolar factor, and the table was therefore submitted to a factor analysis (simple summation method). Two factors were extracted and their saturations with each of the 19 traits are given in TABLE XXVIII. The two columns of saturations are plotted in CHART XXVII.

The first factor corresponds to the type profile of oral pessimism and correlates with it .694. It differs from it in as far as factor I is most highly saturated with aloofness, which is only third in importance in the type; and passivity takes four place as against second place in the type. Moreover oral aggression and autonomy, which were fourth and sixth place respectively in the trait hierarchy of the type, move to tenth and eleventh place, thus contributing little to the factor. (We shall see below that these latter traits appear prominently in factor II.)

An analysis of variance discriminating between the two groups of early

CHART XXVIII

Saturation of Factors

	F_1.	F_2.		F_2.
Aloofness	.709	.164	Impulsion	.782
Endocathexis	.513	.160	Autonomy	.600
Pessimism	.501	−.059	Aggression	.476
Passivity	.387	.204	Oral aggression	.318
Conservatism	.353	−.438	Change	.275
Dependence	.347	−.244	Passivity	.204
Early weaning	.337	.129	Unattainability	.201
Aggression	.235	.476	Rejection	.164
Oral aggression	.193	.318	Endocathexis	.160
Deliberation	.178	−.776	Ambition	.139
Extreme weaning (early and very late)	.167	.059	Early weaning	.129
			Extreme weaning	.059
Guilt	.163	−.376	Sociability	−.012
Autonomy	.150	.600	Pessimism	−.059
Unattainability	.054	.201	Nurturance	−.184
Impulsion	−.052	.782	Dependence	−.244
Ambition	−.219	.139	Optimism	−.280
Change	−.253	.275	Exocathexis	−.297
Sociability	−.489	−.012	Guilt	−.376
Nurturance	−.585	−.184	Conservatism	−.438
Optimism	−.694	−.280	Deliberation	−.776
Exocathexis	−.701	−.297		

and late weaned subjects on the basis of factor scores of this first factor showed the difference to be significant beyond the 1 per cent level of probability. The correlation between weaning and factor I being .287 as against .27 for the type pattern showed that the factor gave only an insignificantly better prediction than the type.

FACTOR II OF IMPATIENCE-AGGRESSION-AUTONOMY AND ITS RELATION TO BREASTFEEDING

Our main interest in this paper, however, belongs to factor II. A person who would be highly saturated with this factor would be characterized by extreme impulsiveness and a disinclination to wait or refrain from action (impulsion, .761; deliberation, −.766), by independence or a disinclination to conform (autonomy, .560), by aggression, general and oral (aggression, .479; oral-aggression .303), and by a tendency to change, to seek the new, rather than to stick to the well-known and familiar (change, .267; conservatism, −.436). Aloofness (.234), passivity (.214), the desire for the unattainable (unattainability, .209), and withdrawal (endocathexis, .160; exocathexis, −.316) also contribute to the syndrome but to a much smaller extent. Pessimism plays an insignificant role in this factor. Recalling psychoanalytic writers on the subject, particularly Abraham and Glover, there is no doubt that a person like the one described above contains in his personality what these writers considered to be essential aspects of the oral character. None of the psychoanalytic writers have, however, discriminated between two independent factors, although there are implicit suggestions of an active reaction as compared with a passive submission to oral frustration (I).

The question remains to be answered what the relation of factor II to oral frustration is. The idea that "oral frustration, oral impatience, oral sadism, are inseparable" [25] is a firmly entrenched tenet of the psychoanalytic theory of orality. As factor II is highly loaded with impatience (impulsion) and aggression (sadism), its relation to weaning (if early weaning be taken as a measure of oral frustration) should demonstrate with what degree of confidence such a statement as the above may be made. Again factor scores were computed, showing the degree with which each individual was saturated with factor II, and the two weaning groups compared for differences. Although the differences were in the expected direction, those subjects who had been weaned early having a mean factor score of .0344, and those weaned late (after four months of breastfeeding), a mean factor score of −.0234, the analysis of variance showed that these differences were not significant; nor were the differences between the early weaned subjects and those who had been given breastfeeding for nine months and longer. In other words, our data do not confirm the psychoanalytic contention that oral frustration, impatience and oral aggression are inseparable or even related. Clearly early weaning seems to have not nearly as much bearing on the impatience-aggression syndrome as it had on the syndrome of pessimism, passivity, aloofness and endocathexis.

However, a hypothesis treated as a fact by so many careful and highly trained (although of course also highly indoctrinated) observers, and stated again and again with the utmost confidence, must still command attention. A further analysis of the data was therefore undertaken. This analysis was based on factor scores of oral pessimism (factor I). The sample was divided into four groups: impulsive and early weaned, non-impulsive and late weaned subjects. Those who had scores on factor II higher than .200 were apportioned to the impulsive group; those whose factor scores were below .200 to the non-impulsive. Late weaning was weaning after a period of breastfeeding lasting four months. The mean scores of oral pessimism for each of these groups are given in TABLE XXIX.

TABLE XXIX

Mean Scores

	Mean.	Numbers in group.
Early impulsive	516.7	8
Early non-impulsive	427.7	22
Late impulsive	413.6	11
Late non-impulsive	341.2	49

An analysis of variance was carried out comparing the differences between these groups. The F-ratio was 4.38, significant at the 1 per cent level of probability. T-tests were then calculated in order to compare the individual groups. TABLE XXX shows the results.

TABLE XXX

Analysis of Variance testing the difference between early weaned impulsive, early weaned non-impulsive, late weaned impulsive, and late weaned non-impulsive subjects with respect to oral pessimism (factor I).

Source	Degrees of freedom.	Sum of squares.	Mean square.
Between groups	3	281,657.5	93,885.8
Within groups	86	1,843,670.9	21,438.0
	89	2,125,328.4	

F = 4.38 (F for 3 and 80 df. at 1 per cent level = 4.04.)

T-tests.

	Early impulsive.	Early Non-impulsive.	Late impulsive.	Late Non-impulsive.
Early-impulsives	..	..	..	..
Early non-impulsives	1.47	..	..	..
Late impulsives	1.52	0.26	..	..
Late non-impulsives	3.14 *	2.30 †	1.48	..

* Significant at the 1 per cent level.
† Significant at the 5 per cent level.

The greatest difference, significant at the 1 per cent level, can be observed between the early-impulsive and the late non-impulsive group, and the difference between the early impulsive and late impulsive groups is significant at the 5 per cent level. These, one might argue, are the differences between early and late weaning, quite independent of any effect of the factor of impulsiveness. This, however, does not explain the following features in the above array: (a) that no significant difference exists between the early non-impulsive and the late impulsive group, and (b) the consistency of the decrease in oral pessimism as the scale moves from those who are impulsive as well as early weaned towards those who are neither.

For the purpose of further clarification, frequency distributions for the early and late weaning groups were plotted along the oral pessimism scale of factor I, saturations showing the location on this scale of the members of the impulsive groups of early as well as of late weaners (CHART XXXI). From these we can see that of the eight impulsive subjects in the

CHART XXXI

Frequency Distribution for Early and Late Weaning Plotted along the Oral Pessimism Scale of Factor I

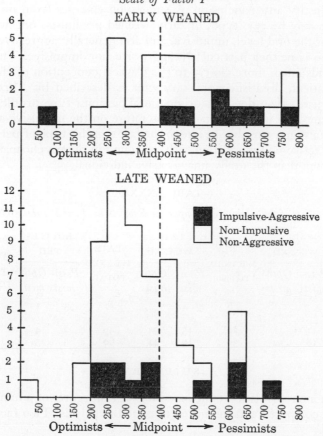

early weaning group, seven are above the mid-point of oral pessimism. (It may be noted that the one who is not is an extreme optimist.) The distribution of late weaners, on the other hand, shows the impulsive subjects to be distributed all along the scale from optimism to pessimism.

The proportions of impulsive and non-impulsive subjects in the two weaning groups and in the total sample are presented in TABLE XXXII. From this it may be seen that among those who had been weaned early, the proportion of pessimists, with an impulsive and aggressive personality, to optimists of the same disposition, was 7:1, while the ratio of pessimists to optimists among those with a placid disposition (non-impulsive, non-aggressive) was 11:11 or 1:1. No such contrast exists among the group of subjects who had enjoyed long breastfeeding, testifying to the fact that over the whole sample the two factors of oral pessimism and impulsion-aggression are independent.

Whether this relatively greater predominance of early weaned oral pessimists over optimists among the impulsive-aggressive individuals (7:1). as compared with the equal balance of the same types among placid individuals (11:11), is a matter of chance, or has any significance, is a question which must be left to a further investigation; the above numbers are too small to justify any conclusion. A suggestion emerges from our samples that there may be two types of early weaned pessimists: one who is aggressive at the oral level, impulsive, and also generally aggressive, and the other who is neither particularly aggressive nor impulsive. The former corresponds much more closely to the classical conception of the aggressive, impatient, dissatisfied oral character as described by Abraham and Glover [3, 14] than does our original oral pessimist type profile.

The emphasis in psychoanalytic literature on the impulsive-aggressive oral pessimist may possibly be due to the fact that this type of person being the more active will more frequently be found in the psychoanalytic consulting room than the orally unaggressive individual.

TABLE XXXII

Proportion of Impulsive Non-Impulsive Subjects in Two Weaning Groups

	EARLY WEANED Pessi-mists	Opti-mists	EARLY WEANED TOTAL	LATE WEANED Pessi-mists	Opti-mists	LATE WEANED TOTAL	BOTTLE-FED Pessi-mists	Opti-mists	BOT-TLE-FED TOTAL	TOTAL SAM-PLE
Impulsive	7	1	8	4	7	11	3	1	4	23
Non-im-pulsive	11	11	22	15	34	49	2	4	6	77
Sums	18	12	30	19	41	60	5	5	10	100

SUMMARY

1. An intercorrelation table based on the trait scores of a hundred subjects was computed and submitted to a factor analysis. Two factors were extracted.

2. Factor I corresponded in most aspects to the syndrome of oral pessimism as described by psychoanalysts, and correlated .694 with the previously derived type pattern. It was, however, poorly saturated with impulsion, aggression and autonomy, which psychoanalysts consider to be important traits of oral character. It was significantly related to early weaning.

3. Factor II was highly saturated with impulsion, aggression and autonomy. It was not significantly related to early weaning.

4. Individuals who had high scores on factor II and who had been weaned early had significantly higher scores on factor I (oral pessimism) than the rest of the sample.

5. A suggestion to be investigated emerges that the concurrence in an individual of early weaning and of the characteristics of impulsiveness-aggression-autonomy, as defined in factor II, may possibly promote the development of oral pessimism.

Appendix.

The Traits.

(Reliability coefficients are presented in parentheses following the description of each trait.) *

Optimism.—Optimism and pessimism are antithetical expressions in character-formation of an omnipotent and magical relation to reality. Optimism deriving from this unconscious source would be upheld by the individual against reasonable expectation. ($r = .90$.)

Pessimism: Inability to accept frustrating experiences as part of reality. Defence against disappointments through summary and advance resignation, anticipation of disappointment. ($r = .74$.)

Passivity: Passive-receptive attitude, hedonism, indolence, self-indulgence. ($r = .81$.)

Desire for the unattainable: Intense desire to climb combined with a feeling of unattainability, of difficulty in achievement, of the insuperable, grudging incapacity to get on. ($r = .78$.)

Displaced oral aggression (verbal): Aggressive use of speech, "omnipotent valuation of speech," "incisive speech." Question 7, although not referring to speech but to the oral expression of rage or the motor accompaniments of oral activity, was included in this scale.

Its correlation of $+.52$ with the total score of the scale seems to confirm that there is a common factor operating in primary oral aggression and aggressiveness in speech. ($r = .76$.)

Aggression: The desire to injure or inflict pain, to overcome opposition forcefully, to fight, to revenge an injury, to attack, to oppose forcefully ($r = .78$.)

Aloofness: Negative tropism for people, attitude of rejection. ($r = .73$.)

Ambition: Tendency to overcome obstacles, to exercise power, to strive to do something difficult as well and as quickly as possible, to attain a high standard, to excel one's self. ($r = .68$.)

Autonomy: Those who wish neither to lead nor to be led, those who want to go their own way, uninfluenced and uncoerced. Independence of attitude. ($r = .88$.)

---

* The reliability of each scale was measured by the split-half method of correlating odd and even numbered items and corrected by the Spearman-Brown prophecy formula.

Dependence: Tendency to cry, plead or ask for nourishment, love, protection or aid. Helplessness, insecurity. (r = .56.)

Guilt: "Conscience," inhibiting and punishing images, self-torture, self-abasement. (r = .90.)

Change: A tendency to move and wander, to have no fixed habitation, to seek new friends, to adopt new fashions, to change one's interests and vocation. Inconsistency and instability. (r = .71.)

Conservatism: Adherence to certain places, people, and modes of conduct. Fixation and limitation. Enduring sentiments and loyalties, persistence of purpose; consistency of conduct; rigidity of habits. (r = .56.)

Impulsion: The tendency to act quickly without reflection. Short reaction time, intuitive or emotional decisions. The inability to inhibit an impulse. (r = .82.)

Deliberation: Inhibition, hesitation and reflection before action. Slow reaction time, compulsive thinking. (r = .87.)

Exocathexis: The positive cathexis in practical action and co-operative undertakings. Occupation with outer events: economic, political and social occurrences. A strong inclination to participate in the contemporary world of affairs. (r = .72.)

Endocathexis: The cathexis of thought or emotion for its own sake. A preoccupation with inner activities, feelings, fantasies, generalizations, theoretical reflections, artistic conceptions, religious ideas; withdrawal from practical life. (r = .73.)

Nurturance: Tendency to nourish, aid, and protect a helpless object. To express sympathy, to mother a child, to assist in danger. (r = .79.)

Sociability: Tendency to form friendships and associations; to join, and live with, others. To co-operate and converse socially with others. To join groups. (r = .70.)

The rating scales for 19 character traits.

Optimism.

Rating scale.	Rel. coeff. r = .90.	$r_t.$*
1. Nothing is impossible to a willing heart		.74
2. In the end justice will prevail		.60
3. Every cloud has a silver lining!		.63
4. There is sure to be enough to go round		.92
5. Ill luck is good for something		.52
6. Atomic energy will bring progress and happiness to humanity		.54
7. Most people can be trusted		.47
8. The world is a wheel and it will all come round right		.73

Pessimism.

Rating scale.	Rel. coeff. r = .74.	$r_t.$
1. It is misery to be born, pain to live, grief to die		.76
2. Greed and lust are the most powerful motive forces of mankind		.44
3. Hardly anyone cares much what happens to you		.48

* Index of internal consistency measured by the tetrachoric correlation coefficient.

4. "Oh life! thou art a galling load
   Along a rough, a weary road" 64
5. There is sure to be a snag somewhere 41
6. Real friends are hard to find 52
7. Accept the present evil, lest a greater one befall you . .58
8. Hope only brings disappointment 58

Passivity.

Rating scale.   Rel. coeff. r = .81. r_t.

1. It is better to sit still than rise and fall 44
2. Oh, for a life of ease and luxury! . . . . .00 ($\emptyset$ = .41) *
3. "An idle life is life for me,
   Idleness spiced with philosophy" 60
4. Good luck is more help than hard work 65
5. Work has no place in paradise 58
6. It is better to do nothing than make a mistake . . .58
7. Comfort is indispensable for a contented life 50
8. Competition is not for me; it irritates rather
   than stimulates me 45

Desire for the Unattainable.

Rating scale.   Rel. coeff. r = .78. r_t.

1. I like best that which flies beyond my reach 55
2. I never get what I really want 80
3. Other people seem to get away with things—I can't . .64
4. How seldom the goal is ever reached in human affairs . .42
5. I often think that other people are more lucky than I am .38
6. The more you want something the less you get it . . .76
7. To attain what we desire means also to realize the vanity
   of our wishes 65
8. Life is a wild-goose chase 84
9. Dreams are the better part of life 67

Displaced Oral Aggression (Verbal).

Rating scale.   Rel. coeff. r = .76. r_t.

1. When your anger is aroused do you tend to express
   yourself in "strong" language? 62
2. Dou you like colourful invective? 71
3. Do you tend to make biting or sarcastic remarks when
   criticizing other people? 66
4. Do you find yourself frequently disagreeing with, and
   contradicting people? 73
5. Are you fond of arguing? 47
6. Do you like a debate to be hard-hitting? 67
7. When in a rage do you tend to give motor expression to
   your feelings, such as stamping your feet, grinding
   your teeth, pushing your fist in your mouth, or biting
   your fingernails or handkerchief or other objects, or
   breaking or tearing something? 52

---

* The $\emptyset$. coefficient is based on the correlation of the upper and lower quarters of the sample.

Aggression.

Rating scale. Rel. coeff. r = .78. r_t.

1. If somebody annoys you are you apt to tell him what you think of him? 76
2. Do you try to get your own way regardless of opposition? 57
3. If you come across a domineering person are you inclined to put him in his place? 69
4. Do you often blame other people when things go wrong? 66
5. Are you considered aggressive by some of your acquaintances? 80
6. Are you apt to express your irritation rather than restrain it? 78
7. Do you often let yourself go when you are angry? . .64
8. Are you apt to rebuke your friends when you disapprove of their behaviour? 59

Aloofness.

Rating scale. Rel. coeff. r = .73. r_{tt}.

1. Are you intolerant of people who bore you? 55
2. Do you maintain reserve when you meet strangers? . .49
3. Do you get annoyed when your time is taken up by people in whom you are not interested? 78
4. Do you avoid close intimacies with other people? . . .58
5. Do you usually keep yourself somewhat aloof and inaccessible? 60
6. Have you frequently been reproached for your scornful manner in argument, particularly with people whose ideas you considered inferior to your own? . . .70
7. Do you often tend to express your resentment against a person by having nothing more to do with him? . .55
8. Do you generally prefer the company of older, talented or generally superior people? 53
9. Do you find the company of uninteresting people insufferable in the highest degree? 71

Ambition.

Rating scale. Rel. coeff. r = .68. r_t.

1. Do you feel that life can offer hardly any substitute for great achievement 66
2. Are you rather set upon accomplishing some valuable piece of work some time in your life? 77
3. Does your admiration for certain great men inspire you to emulate them in one way or another? 78
4. Are you employing most of your energies in the pursuit of your career? 55
5. Do you feel that a far goal is deserving your efforts more than a daily duty? 72

6. Do you often dream about your future successes? . . .52
7. Would you say that ambition is a strong motive force in your life? 46

Autonomy.

Rating scale.   Rel. coeff. r = .88. r_t.

1. Do you feel that you cannot do your best work when you are in a subservient position? 52
2. Does it make you stubborn and resistant when others attempt to coerce you? 60
3. Do you often act contrary to custom or to the wishes of your parents? 70
4. Do you argue against people who attempt to assert their authority over you? 58
5. Do you try to avoid situations where you are expected to conform to conventional standards? 76
6. Do you go your own way regardless of the opinion of others? 62
7. Are you generally disinclined to adopt a course of action dictated by others? 55
8. Do you tend to disregard the rules and regulation that hamper your freedom? 74
9. Do you set independence and liberty above everything? .44
10. Are you apt to criticize whoever happens to be in authority? 54

Dependence.

Rating scale.   Rel. coeff. r = .56. r_t.

1. Do you usually tell your friends about your difficulties misfortunes? 76
2. Do you think of yourself sometimes as neglected and unloved? 72
3. Are you rather easily discouraged when things go wrong? 80
4. Do you sound the opinions of others before making a decision? 66
5. Do you feel lost and helpless when you are left by someone you love? 46
6. Are you apt to complain about your sufferings and hardships? 72

Guilt.

Rating scale.   Rel. coeff. r = .90. r_t.

1. Do you tend to be rather submissive and apologetic when you have done wrong? . . . . .24 ($\emptyset$ = .43)
2. Do you feel nervous and anxious in the presence of superiors? 53
3. Are you sometimes depressed by a feeling of your own unworthiness? 50

4. Do you often ask yourself: "Have I done right?" .   .   .73
5. Do you feel sometimes that people disapprove of you? .   .52
6. Are you frequently troubled by pangs of conscience? .   .60
7. Do you sometimes experience a vague feeling of anxiety
    as if you had done wrong and would be found out? .   .68

Change.

Rating scale.	Rel. coeff. r = .71.	r_t.

1. Do you like variety and contrast; always willing to wel-
   come change? 50
2. Do you frequently start new projects without waiting to
   finish what you have been doing? 50
3. Do you find that novel prospects—new places, new peo-
   ple, new ideas—appeal strongly to you? 65
4. Could you cut your moorings—leave your home, family
   and friends—without suffering great regret? .   .   .46
5. Do you find that your likes and dislikes change quite fre-
   quently? 47
6. Are you quick to discard the old and accept the new;
   new fashions, new methods, new ideas? 82
7. Are you fickle in your affections? 52
8. Have you often experienced rather marked swings of
   mood from elation to depression? 65

Conservatism.

Rating scale.	Rel. coeff. r = .56.	r_t.

1. Are you somewhat disturbed when your daily habits are
   disrupted by unforeseen events? 62
2. Do you find that you react with a certain resistance to
   untested innovations? 54
3. Do you find that a well-ordered mode of life with regu-
   lar hours and an established routine is congenial to
   your temperament? 82
4. Can you endure monotony without much fretting? .   .73
5. Do you prefer to associate with your old friends, even
   though by so doing you miss the opportunity of meet-
   ing more interesting people? 38
6. Are you usually consistent in your behaviour, going
   about your work in the same way, frequenting the
   same preferred places, following the same routes, etc.?   .64

Impulsion.

Rating scale.	Rel. coeff. r = .82.	r_t.

1. Do you often act on the spur of the moment? .   .   .67
2. Are you inclined to express your wishes without much
   hesitation? 49
3. Do you often act impulsively just to blow off steam? .   .91
4. Are you quick in commenting on most occasions? .   .48
5. Do you often do a thing in a hurry just to get it over? .   .46

6. When you have to act, are you usually quick to make up your mind? 42
7. Do you sometimes start talking without knowing exactly what you are going to say? 68
8. Are you easily carried away by an emotional impulse? . .83
9. Are you apt to say anything—though you may regret it later—rather than keep still? 78
10. Would you call yourself spontaneous in speech and action? 74

Deliberation.

Rating scale. Rel. coeff. r = .87. r_t.

1. Do you repress your emotions more often than you express them? 47
2. Do you think much and speak little? 74
3. Are you slow in deciding upon a course of action? . .90
4. Do you usually consider a matter from every standpoint before you form an opinion? 64
5. Are you slow to fall in love? 52
6. Do you usually make a plan before you start to do something? 45
7. When you are confronted by a crisis are you apt to become inhibited and do nothing? 43
8. Do you usually do things slowly and deliberately? . .74
9. Do you dislike making hurried decisions? . . .59
10. Are you good at repartee, quick retorts, snap judgements? (Neg.) 46

Exocathexis.

Rating scale. Rel. coeff. r = .72. r_t.

1. Do you find that you can deal with an actual situation better than cope with general ideas and theories? . .62
2. Do you like being in the thick of action? . . .49
3. Do you like to do things with your hands? . . .62
4. Would you call yourself a practical person? . . .60
5. Would you rather take an active part in contemporary events than read and think about them? . . .71
6. Do you like to have people about most of the time? . .66

Endocathexis.

Rating scale. Rel. coeff. r = .73. r_t.

1. Are you inclined to withdraw from any kind of action . .34 ($\phi = .50$)
2. Do you prefer to know rather than do? . . .57
3. Do you spend a lot of time philosophizing with yourself? .75
4. Do you think more about your private feelings or theories than you do about the practical demands of every-day existence? 78
5. Would you rather write a fine book than be an important public figure? 64

6. Do you like above all to discuss general questions—scientific and philosophical—with your friends? . .45

7. Are you more interested in aesthetic and moral values than in contemporary events? 63

8. Do you find at social gatherings that you don't seem to enjoy yourself as other people? 53

9. Are you apt to brood for a long time over a single idea? . .56

10. Do you dislike everything to do with money—paying, selling, bargaining? . . . . .34 ($\emptyset = .39$)

Nurturance.

Rating scale.   Rel. coeff. r = .79. r_t.

1. Are you drawn to people who are sick, unfortunate, or unhappy? 78

2. Do you feel great sympathy for any "underdog" and are you apt to do what you can for him? 69

3. Do you often go out of your way to feed, pet, or otherwise care for an animal? 43

4. Do you enjoy putting your own affairs aside to do someone a favour? 68

5. Are you considered, by some of your friends, as too good-natured, too easily taken in? 53

6. Do you enjoy playing with children? 80

7. Do you find it difficult to lend things (books, money, etc.) to other people? (neg.) 55

Sociability.

Rating scale.   Rel. coeff. r = .70. r_t.

1. Do you make as many friends as possible and are always on the lookout for more? 83

2. Do you accept social invitations rather than stay at home alone? 78

3. Do you make a point of keeping in close touch with the doings and interests of your friends? 50

4. Do you go out of your way to be with your friends . .52

5. Do you feel that friendship is more important than anything else? 75

6. Do you enjoy yourself very much at parties and social gatherings? 74

REFERENCES

1. Abraham, K.: "The First Pregenital Stage of the Libido," *Selected Papers* (London, Hogarth Press, 1916).
2. Abraham, K.: "A Short Study of the Development of the Libido, Viewed

in the Light of Mental Disorders," *Selected Papers* (London, Hogarth Press, 1924).

3. Abraham, K.: "The Influence of Oral Erotism on Character-Formation," *Selected Papers* (London, Hogarth Press, 1942).

4. Allport, G. W.: *Personality: A Psychological Interpretation* (New York, Holt, 1937), Chap. 7.

5. Allport, G. W.: "Personality, A Symposium. III: Geneticism versus Ego-Structure in Theories of Personality," *Brit. J. Educ. Psychol.*, Vol. 16 (1946), p. 57.

6. Bergler, E.: "Zur Problematik des 'oralen' Pessimisten," *Imago*, Vol. 20 (1934), pp. 330–76.

7. Bernfeld, S.: *The Psychology of the Infant* (London: Kegan Paul, 1929).

8. Burt, C.: *The Factors of the Mind* (London: University of London Press, 1940).

9. Erikson, Erik Homburger: "Childhood and Tradition in Two American Indian Tribes," *The Psychoanalytic Study of the Child* (London: Morrow, 1945). Vol. 1, pp. 319–50.

10. Freud, S.: *Inhibition, Symptom and Anxiety* (London: Hogarth Press, 1936).

11. Freud, S.: *Three Contributions to a Theory of Sex, The Basic Writings of Sigmund Freud* (New York: Modern Library, 1938).

12. Glover, E.: "Examination of the Klein System of Child Psychology," *The Psychoanalytic Study of the Child* (London, Imago, 1945), Vol. 1, pp. 75–118.

13. Glover, E.: "The Significance of the Mouth in Psychoanalysis," *Brit. J. Med. Psychol.*, Vol. 4 (1924), pp. 134–55.

14. Glover, E.: "Notes on Oral Character-Formation," *Int. J. Psycho-Anal.*, Vol. 6 (1925), pp. 131–53.

15. Hunt, J. McV.: "The Effects of Infant Feeding-Frustration upon Adult Hoarding in the Albino Rat," *J. Abnorm. Soc. Psychol.*, Vol. 36 (1941), pp. 338–60.

16. Jones, E.: "The God Complex," *Essays in Applied Psychoanalysis* (International Psychoanalytical Press, 1923).

17. Klein, Melanie: "The Psychogenesis of Manic-Depressive States," *Int. J. Psycho-Anal.*, Vol. 16 (1935), pp. 145–74.

18. Lewin, Kurt: *Principles of Topological Psychology* (New York: McGraw-Hill, 1936).

19. Lewin, Kurt: "Psychoanalysis and Topological Psychology," *Bull. Menninger Clinic*, Vol. 1 (1937), pp. 202–10.

20. Mead, Margaret: *Sex and Temperament* (New York: Morrow, 1935).

21. Money-Kyrle, R.: *Superstition and Society* (London: Hogarth Press, 1939).

22. Murray, H. A.: *Explorations in Personality* (London and New York: Oxford University Press, 1938).

23. Rado, Sandor: "The Problem of Melancholia," *Int. J. Psycho-Anal.*, Vol. 9 (1928), pp. 9–32.

24. Ribble, Margaret A.: "Infantile Experience in Relation to Personality Development," in J. McV. Hunt (ed.): *Personality and the Behavior Disorders* (New York: Ronald Press, 1944), Vol. 2, pp. 621–51.

25. Sharpe, Ella: "Hamlet," *Collected Papers on Psychoanalysis* (London: Hogarth Press, 1950).

26. Stärke, A.: "The Castration Complex," *Int. J. Psycho-Anal.*, Vol. 2 (1921), pp. 179–201.

27. Wisdom, J. C.: "The Unconscious Origin of Schopenhauer's Philosophy,"
    Int. J. Psycho-Anal., Vol. 26 (1945), pp. 44–52.
28. "Is the Doctrine of Instincts Dead? A Symposium," *Brit. J. Educ. Psychol.*;
    Vol. 11 (1941), pp. 155–72 (Cyril Burt); Vol. 12 (1942), pp. 1–9 (Philip E.
    Vernon) Vol. 13 (1943), pp. 1–15 (Cyril Burt).

CHAPTER

12

Eric Homburger Erikson

* GROWTH AND CRISES OF THE "HEALTHY PERSONALITY"

"Normal" and "abnormal," "healthy" and "unhealthy," are sometimes crudely used to signify approbation and disapprobation respectively. But stripping the value elements from these terms for scientific usage does not change matters for the parent, educator, or therapist who prefers normality and health to abnormality and illness.

The Midcentury White House Conference on Children and Youth selected for its main theme the "Healthy Personality." Faced with the task of contributing to this theme general insights derived from clinical and applied psychoanalysis, Mr. Erikson set out to formulate, instead of the pathological potentials of the infantile stages postulated in psychoanalysis, the possible personality gains that accrue from each stage. "In naming a series of basic balances," Mr. Erikson states, "on which the psycho-social health of a personality seems to depend, I found myself implying a latent universal value system which is based on the nature of human growth, the needs of the developing ego, and certain common elements in child training systems. This is a puzzling problem: I have myself contributed to studies demonstrating the dynamic relevance of cultural relativity. In postulating now, for example, a specific infantile stage during

NOTE: Thanks are due the author for reorganizing and shortening portions of this paper, originally written for the Josiah Macy, Jr. Foundation for the Fact-finding Committee of the Midcentury White House Conference, and published as Supplement II to the Transactions of the Fourth Conference on Infancy and Childhood (M. J. E. Senn, ed.), 1950.

The reader is urged to consult Dr. Erikson's recent book, *Childhood and Society*, where two papers published in the first edition of the present volume have been rewritten and expanded and, since they have become available in book form, omitted from this revised edition.

which something best circumscribed by the word 'initiative' is ready to emerge, to receive special attention in child training systems, and to undergo a decisive dynamic crisis, I must face the objection that 'initiative' gains prominence only in the American or Western scene. Yet, without some form of initiative, defined as it must be within the network of a culture's values, no ego could develop, no economy could exist, no value could make sense. Maybe we should not deny that there are trees just because we know now that we are apt to overlook the forest for the trees."

THE Fact-finding Committee of the White House Conference on Childhood and Youth has asked me to repeat here in greater detail a few ideas set forth in my recent book. There the matter of the healthy personality emerges, as if accidentally, from a variety of clinical and anthropological considerations. Here it is to become the central theme—an embarrassing fact when one is to cope with the matter as an "expert."

The psychoanalyst knows relatively more about the dynamics and cure of the disturbances which he treats daily than about the prevention of such disturbances. He must remember, too, that the prevention and even the cure of disease only foster the necessary condition for health. A person who is not sick is not necessarily healthy in every sense of the word. Health implies a state of surplus energy, vitality, and awareness. In regard to physical well-being, it is understood that the presence of health is more than the absence of disease, although widespread epidemiological prevention is necessary to keep people alive and free from disease. It is a strange fact in our civilization (and well worth discussion in the group for whom I think I am writing this) that the expert on mental disease is being forced, often not so very much against his will, into the role of the expert on mental health. But in the long run we want to do more than immunize our children against neurotic suffering. Prevention, based on what we know about disease, is only a necessary beginning. Strictly speaking, we cannot even be sure that we know what causes neurotic suffering until we have an idea of the nature of real health. This we have only begun to investigate.

An expert, in addition to some verifiable fact, consistent theory, and technical skill, must have some meaningful *insight*. This he must gain so that he may act effectively and decisively even while his knowledge, his theory, and his skills change and develop.

I consider it my task to approach the matter of the healthy personality from the genetic point of view: How does a healthy personality grow or, as it were, accrue from the successive stages of increasing capacity to master life's outer and inner tasks and dangers?

Within all of the limitations mentioned, then, I shall present a series of insights, trusting that the readers will hear experts from neighboring fields and that, in addition, they may be lucky enough to secure the advice of an expert on experts.

I

ON HEALTH AND GROWTH

Whenever we try to understand growth, it is well to remember the *epigenetic principle* which is derived from the growth of organisms *in utero*. Somewhat generalized, this principle states that anything that grows has a *ground plan*, and that out of this ground plan the *parts* arise, each part having its *time* of special ascendancy, until all parts have arisen to form a *functioning whole*. . . . At birth the baby leaves the chemical exchange of the womb for the social exchange system of his society, where his gradually increasing capacities meet the opportunities and limitations of his culture. How the maturing organism continues to unfold, not by developing new organs, but by a prescribed sequence of locomotor, sensory, and social capacities, is described in the child-development literature. Psychoanalysis has given us an understanding of the more idiosyncratic experiences, and especially the inner conflicts, which constitute the manner in which an individual becomes a distinct personality. But here, too, it is important to realize that in the sequence of his most personal experiences the healthy child, given a reasonable amount of guidance, can be trusted to obey inner laws of development, laws which create a *succession of potentialities for significant interaction* with those who tend him. While such interaction varies from culture to culture, it must remain within the *proper rate and the proper sequence* which govern the *growth of a personality* as well as that of an organism. . . . Personality can be said to develop according to steps predetermined in the human organism's readiness to be driven toward, to be aware of, and to interact with, a widening social radius, beginning with the dim image of a mother and ending with mankind, or at any rate that segment of mankind which "counts" in the particular individual's life.

It is for this reason that, in the presentation of stages in the development of the personality, we employ *an epigenetic diagram* analogous to one previously employed for an analysis of Freud's psychosexual stages.[1] It is, in fact, the purpose of this presentation to bridge the theory of infantile sexuality (without repeating it here in detail), and our knowledge of the child's physical and social growth within his family and the social structure. An epigenetic diagram looks like this:

The darkly shaded squares signify both a sequence of stages (I to III) and a gradual development of component parts; in other words the diagram formalizes a progression through time of a differentiation of parts. This, of course, makes little sense as an empty checkerboard without sig-

---

[1] See Part I of the author's *Childhood and Society* (New York, Norton, 1950).

PERSONALITY

	COMPONENT 1	COMPONENT 2	COMPONENT 3
Stage I	I₁	I₂	I₃
Stage II	II₁	II₂	II₃
Stage III	III₁	III₂	III₃

nificant items to fill the squares. I introduce the principle of the diagram here only in order to indicate (1) that each item of the healthy personality to be discussed is systematically related to all others, and that they all depend on the proper development at the proper time of each item; and (2) that each item exists in some form before "its" decisive and critical time normally arrives.

If I say, for example, that a *sense of basic trust* is the first component of mental health to develop in life, a *sense of autonomous will* the second, and a *sense of initiative* the third, the purpose of the diagram may become clearer.

	COMPONENT 1	COMPONENT 2	COMPONENT 3
First Stage (about first year)	BASIC TRUST	Earlier form of AUTONOMY	Earlier form of INITIATIVE
Second Stage (about second and third years)	Later form of BASIC TRUST	AUTONOMY	Earlier form of INITIATIVE
Third Stage (about fourth and fifth years)	Later form of BASIC TRUST	Later form of AUTONOMY	INITIATIVE

This diagrammatic statement, in turn, is meant to express a number of fundamental relations that exist among the three components, as well as a few fundamental facts for each.

Each comes to its ascendance, meets its crisis, and finds its lasting solution (in ways to be presently described in detail) toward the end of the stages mentioned. However, this does not mean that they begin there nor that they end there. All of them begin with the beginning, in some form, although we do not make a point of this fact, and we shall not confuse things by calling these components different names at earlier or later stages. A baby may show something like "autonomy" from the beginning, for example, in the particular way in which he angrily tries to wriggle his head free when tightly held. However, under normal conditions, it is not until the second year that he begins to experience the whole *critical alternative between being an autonomous creature and being a dependent one;* and it is not until then that he is ready for a *decisive encounter* with his environment, an environment which, in turn, feels called upon to convey to him its *particular ideas and concepts of autonomy and coercion* in ways decisively contributing to the character, the efficiency, and the health of his personality in his culture.

It is this *encounter,* together with the resulting crisis, which is to be described for each stage. Each stage becomes a *crisis* because incipient growth and awareness in a significant part-function goes together with a shift in instinctual energy and yet causes specific vulnerability in that part. One of the most difficult questions to decide, therefore, is whether or not a child at a given stage is weak or strong. Perhaps it would be best to say that he is always vulnerable in some respects and completely oblivious and insensitive in others, but that at the same time he is unbelievably persistent in the same respects in which he is vulnerable. It must be added that the smallest baby's weakness gives him power; out of his very dependence and weakness he makes signs to which his environment (if it is guided well by a responsiveness based both on instinctive and traditional patterns) is peculiarly sensitive. A baby's presence exerts a consistent and persistent domination over the outer and inner lives of every member of a household. Because these members must reorient themselves to accommodate his presence, they must also grow as individuals and as a group. It is as true to say that babies control and bring up their families as it is to say the converse. A family can bring up a baby only by being brought up by him. His growth consists of a series of challenges to them to serve his newly developing potentialities for social interaction.

Because of a radical *change in perspective,* each successive step is also a potential crisis. There is, at the beginning of life, the most radical change of all: from intrauterine to extrauterine life. But in postnatal existence, too, such radical adjustments of perspective as lying relaxed, sitting firmly, and running fast must all be accomplished in their own good time. In adult life they exist side by side; but adults "regress" along the familiar path: when tired by running (running after one thing or another) we sit down to rest and eat, and lie down to sleep. The interpersonal perspective, too, changes rapidly in early life, and often radically, as is testified by the proximity in time of such opposites as "not letting mother out of sight" and "wanting to be independent." As we proceed, it may seem to the reader as if we, too, were changing our perspective, as if even our conceptual approach changed from stage to stage; but this is to a large extent determined by the very different *capacities* which use different *opportunities* to become full-grown components of the ever new configuration that is the growing personality. In describing this growth and its crises as a development of a series of alternative basic attitudes, we take recourse to such terms as "a sense of autonomy," or "a sense of doubt." Like a "sense of health" or a "sense of not being well," such senses pervade surface and depth, consciousness and the unconscious. They are ways of experience, accessible to introspection; ways of behaving, observable by others; and objective inner states determinable by test and analysis. It is important to keep these three dimensions in mind, as we proceed.

II

BASIC TRUST VERSUS BASIC MISTRUST

I

For the first component of a healthy personality I nominate a sense of *basic trust*, which I think is an attitude toward oneself and the world derived from the experiences of the first year of life. By "trust" I mean what is commonly implied in reasonable trustfulness as far as others are concerned and a simple sense of trustworthiness as far as oneself is concerned. When I say "basic," I mean that neither this component nor any of those that follow are, either in childhood or in adulthood, especially conscious. In fact, all of these criteria, when developed in childhood and when integrated in adulthood, blend into the total personality. Their crises in childhood, however, and their impairment in adulthood are clearly circumscribed.

In *adults* the impairment of basic trust is expressed in a *basic mistrust* (which pervades much of what in psychiatry is called schizophrenia or paranoia). It characterizes individuals who withdraw into themselves in particular ways when at odds with themselves and with others. These ways, which often are not obvious, are more strikingly represented by individuals who regress into psychotic states in which they sometimes close up, refusing food and comfort and becoming oblivious to companionship. Insofar as we hope to assist them with psychotherapy, we must try to reach them again in specific ways in order to convince them that they can trust the world and that they can trust themselves.[2]

It is from the knowledge of such radical regressions and of the deepest and moot infantile layers in our not-so-sick patients that we have learned to regard basic trust as the cornerstone of a healthy personality. Let us see

---

[2] Frieda Fromm-Reichmann, *Principles of Intensive Psychotherapy* (Chicago, University of Chicago Press, 1950).

what justifies our placing the crisis and the ascendancy of this component at the beginning of life.

As the newborn is separated from his symbiosis with the mother's body, his inborn and more or less co-ordinated ability to take in by mouth meets the breasts' and the mother's and the society's more or less co-ordinated ability and intention to feed him and to welcome him. At this point he lives through, and loves with, his mouth; and the mother lives through, and loves with, her breasts.

For the mother this is a late and complicated accomplishment, highly dependent on her development as a woman; on her unconscious attitude toward the child; on the way she has lived through pregnancy and delivery; on her and her community's attitude toward the act of nursing—and on the response of the newborn. To him the mouth is the focus of a general first approach to life—the *incorporative* approach. In present-day psychiatry this stage is therefore usually referred to as the "oral" stage. Yet it is clear that, in addition to the overwhelming need for food, a baby is, or soon becomes, receptive in many other respects. As he is willing and able to suck on appropriate objects and to swallow whatever appropriate fluids they emit, he is soon also willing and able to "take in" with his eyes whatever enters his visual field. His tactual senses, too, seem to "take in" what feels good. In this sense, then, one could speak of an *"incorporative stage,"* one in which he is, relatively speaking, receptive to what he is being offered. Yet many babies are sensitive and vulnerable, too. In order to ensure that their first experience in this world may not only keep them alive but also help them to coordinate their sensitive breathing and their metabolic and circulatory rhythms, we must see to it that we deliver to their senses stimuli as well as food in the proper intensity and at the right time; otherwise their willingness to accept may change abruptly into diffuse defense.

Now, while it is quite clear what *must* happen to keep a baby alive (the minimum supply necessary) and what *must not* happen, lest he be physically damaged or chronically upset (the maximum early frustration tolerable), there is increasing leeway in regard to what *may* happen; and different cultures make extensive use of their prerogatives to decide what they consider workable and insist upon calling necessary. Some people think that a baby, lest he scratch his own eyes out, must necessarily be swaddled completely for the better part of the day and throughout the greater part of the first year; also, that he should be rocked or fed whenever he whimpers. Others think that he should feel the freedom of his kicking limbs as early as possible, but also that he, as a matter of course, be forced to cry "please" for his meals until he literally gets blue in the face. All of this depends on the culture's general aim and system. The writer has known some old American Indians who bitterly decried the way in which we often let our small babies cry because we believe that "it will make their lungs strong." No wonder (these Indians said) that the white man, after such an initial reception, seems to be in a hurry to get to the "next world." But the same Indians spoke proudly of the way their infants (breast fed into the second year) become blue in the face with fury when thumped on the head for "biting" the mother's nipples; here the Indians, in turn, believed that "it's going to make good hunters of them."

There seems to be some intrinsic wisdom, some unconscious planning, and much superstition in the seemingly arbitrary varieties of child training: what is "good for the child," what *may* happen to him, depends on what he is supposed to become and where. We shall come back to this.

The simplest and the earliest social modality is *"to get,"* not in the sense of "go and get," but in that of receiving and accepting what is given; and this sounds easier than it is. For the groping and unstable newborn's organism learns this modality only as he learns to regulate his readiness to get with the methods of a mother who, in turn, will permit him to co-ordinate his means of getting as she develops and co-ordinates her means of giving. The mutuality of relaxation thus developed is of prime importance for the first experience of friendly otherness: from psychoanalysis one receives the impression that in thus *getting what is given*, and in learning to *get somebody to do* for him what he wishes to have done, the baby also develops the necessary groundwork to *get to be* the giver, to "identify" with her.

Where this mutual regulation fails, the situation falls apart into a variety of attempts to control by duress rather than by reciprocity. The baby will try to get by random activity what he cannot get by central suction; he will activate himself into exhaustion or he will find his thumb and damn the world. The mother's reaction may be to try to control matters by nervously changing hours, formulas, and procedures. One cannot be sure what this does to a baby; but it certainly is our clinical impression that in some sensitive individuals (or in individuals whose early frustration was never compensated for) such a situation can be a model for a radical disturbance in their relationship to the "world," to "people," and especially to loved or otherwise significant people.

There are ways of maintaining reciprocity by giving to the baby what he can get through good artificial nipples and other forms of feeding and by making up for what is missed orally through the satiation of other than oral receptors: his pleasure in being held, warmed, smiled at, talked to, rocked, and so forth. Besides such "horizontal" compensation (compensation during the same stage of development) there are many "longitudinal" compensations in life: compensations emerging from later stages of the life cycle.

During the "second oral" stage the ability and the pleasure in a more active and more directed incorporative approach ripens. The teeth develop and with them the pleasure in biting *on* hard things, in biting *through* things, and in biting *off* things. This *active-incorporative* mode characterizes a variety of other activities (as did the first incorporative mode). The eyes, first part of a passive system of accepting impressions as they come along, have now learned to focus, to isolate, to "grasp" objects from the vaguer background and to follow them. The organs of hearing similarly have learned to discern significant sounds, to localize them, and to guide an appropriate change in position (lifting and turning the head, lifting and turning the upper body). The arms have learned to reach out determinedly and the hands to grasp firmly. As will be noted, we are more interested here in the overall *configuration* and *integration* of developing approaches to the world than in the *first appearance of specific abilities* which are so well described in the child-development literature.

With all of this a number of interpersonal patterns are established which center in the social modality of *taking* and *holding on to* things— things which are more or less freely offered and given, and things which have more or less a tendency to slip away. As the baby learns to change positions, to roll over, and very gradually to establish himself on the throne of his sedentary kingdom, he must perfect the mechanisms of grasping and appropriating, holding and chewing all that is within his reach.

The *crisis* of the oral stage (during the second part of the first year) is difficult to assess and more difficult to verify. It seems to consist of the coincidence in time of three developments: (1) a physiological one; namely, the general tension associated with a more violent drive to incorporate, appropriate, and observe more actively (a tension to which is added the discomfort of "teething" and other changes in the oral machinery); (2) a psychological one; namely, the infant's increasing awareness of himself as a distinct person; and (3) an environmental one; namely, the mother's apparent turning away from the baby toward pursuits which she had given up during late pregnancy and postnatal care. These pursuits include her full return to conjugal intimacy and may soon lead to a new pregnancy.

Where breast feeding lasts into the biting stage (and, generally speaking, this has been the rule) it is now necessary to learn how to continue sucking without biting, so that the mother may not withdraw the nipple in pain or anger. Our clinical work indicates that this point in the individual's early history provides him with some sense of basic loss, leaving the general impression that once upon a time one's unity with a maternal matrix was destroyed. Weaning, therefore, should not mean sudden loss of the breast and loss of the mother's reassuring presence too, unless, of course, other women can be depended upon to sound and feel much like the mother. A drastic loss of accustomed mother love without proper substitution at this time can lead (under otherwise aggravating conditions) to acute infantile depression [3] or to a mild but chronic state of mourning which may give a depressive undertone to the whole remainder of life. But even under more favorable circumstances, this stage seems to introduce into the psychic life a sense of division and a dim but universal nostalgia for a lost paradise.

[3] Rene Spitz, "Hospitalism," *The Psychoanalytic Study of the Child* (New York: Internat. Univ. Press, 1945).

It is against the combination of these impressions of having been deprived, of having been divided, and of having been abandoned, all of which leave a residue of basic mistrust, that basic trust must be established and maintained.

2

What we here call "trust" coincides with what Dr. Therese Benedek has called "confidence." If I prefer the word "trust," it is because there is more naïveté and more mutuality in it: an infant can be said to be trusting, but it would be assuming too much to say that he "has confidence." The general state of trust, furthermore, implies not only that one has learned to rely on the sameness and continuity of the outer providers but also that one may trust oneself and the capacity of one's own organs to cope with urges; that one is able to consider oneself trustworthy enough so that the providers will not need to be on guard or to leave.

In the psychiatric literature we find frequent references to an "oral character," which is a characterological deviation based on the unsolved conflicts of this stage and which expresses itself in more or less pathological and irrational ways of approaching the world (either too pessimistically or too optimistically) from the sense of being helpless and dependent (or from the sense) of being greedy and grasping, or worse. Wherever oral pessimism becomes dominant and exclusive, infantile fears, such as that of "being left empty," or simply of "being left," and also of being "starved of stimulation," can be discerned definitely in the depressive forms of "being empty" and of "being no good." Such fears, in turn, can give orality that particular avaricious quality which in psychiatry is called "oral sadism," that is, a cruel need to get and to take in ways harmful to others. But there is an optimistic oral character, too, one which has learned to make giving and receiving the most important thing in life; and there is "orality" as a normal substratum in all individuals, a lasting residuum, of this first period of dependency on powerful providers. It normally expresses itself in our dependencies and nostalgias, and in our all too hopeful and all too hopeless states.

The pathology and irrationality of these trends depend entirely on the degree to which they are integrated with the rest of the personality (unity of personality) and the degree to which they fit into the general cultural pattern and use approved interpersonal techniques for their expression. The integration of the oral stage with all the following ones results in a combination of faith and realism.

Here, as elsewhere, we must, therefore, consider as a topic for discussion the expression of *infantile urges* in *cultural patterns* which one may (or may not) consider a pathological deviation in the total economic or moral system of a culture or a nation. One could speak, for example, of the invigorating belief in "chance," that traditional prerogative of American trust in one's own resourcefulness and in Fate's store of good intentions. This belief, at times, can be seen to degenerate in large-scale gambling; or in "taking chances" in the form of an arbitrary and often suicidal provocation of Fate; or in the insistence that one has not only the right to an equal chance but also the privilege of being preferred over all others who have invested in the same general enterprise. In a similar way all the pleasant reassurances which can be derived (especially in good company) from old and new taste sensations, from inhaling and imbibing, from munching and swallowing and digesting, can turn into mass addictions neither expressive of, nor conducive to, the kind of basic trust which we have in mind.

Here we are obviously touching on phenomena the analysis of which would call for a comprehensive approach both to personality and to culture. This would be true also for an epidemiological approach to the problem of the oral character, the schizoid character, and the mental diseases seemingly expressive of an underlying weakness in oral reassurance and basic trust.

A related problem for discussion is the belief (reflected in much of contemporary obstetric and pediatric concern with the methods of child care) that the establishment of a basic sense of trust in earliest childhood makes adult individuals less dependent on mild or malignant forms of addiction, on self-delusion, and on the needs for avaricious appropriation. Of this, little is known; and the question remains whether healthy orality makes for a healthy culture or a healthy culture makes for healthy orality, or both.

At any rate the psychiatrists, obstetricians, pediatricians, and anthropologists, to whom the writer feels closest, today would agree that *the firm establishment of enduring patterns for the balance of basic trust over basic mistrust* is the first task of the budding personality and therefore first of all a task for maternal care. But it must be said that the *amount of trust* derived from earliest infantile experience does not seem to depend on absolute *quantities of food or demonstrations of love* but rather on the *quality* of the maternal relationship. Mothers create a sense of trust in their children by that kind of administration which in its quality combines sensitive care of the baby's individual needs and a firm sense of personal trustworthiness within the trusted framework of their community's life style. (This forms the basis in the child for a sense of identity which will later combine a sense of being "all right," of being oneself, and of becoming what other people trust one will become.) Parents must not only have certain ways of guiding by prohibition and permission; they must also be able to represent to the child a deep, an almost somatic conviction that there is a meaning to what they are doing. In this sense a traditional system of child care can be said to be a factor making for trust, even where certain items of that tradition, taken singly, may seem unnecessarily cruel. Here much depends on whether such items are inflicted on the child by the parent in the firm traditional belief that this is the only way to do things or whether the parent misuses his administration of the baby and the child in order to work off anger, alleviate fear, or win an argument, with the child or with somebody else (her mother, her husband, her doctor, or her priest). This latter kind of situation we shall refer to as the parental *exploitation of the inequality* between adult and child.

In times of change—and what other times are there, in our memory? —one generation differs so much from another that items of tradition often become disturbances. Conflicts between mother's ways and one's own self-made ways, conflicts between the expert's advice and mother's ways, and conflicts between the expert's authority and one's own self-willed ways may disturb a mother's trust in herself. Furthermore, all the mass transformations in American life (immigration, migration, and Americanization; industrialization, urbanization, mechanization, and others) are apt to disturb young mothers in those tasks which are so simple yet so far-reaching. No wonder, then, that the first section of the first chapter of Dr. Benjamin Spock's book[4] is entitled "Trust Yourself." But while it is true that the expert obstetrician and pediatrician can do much to replace the binding power of tradition by giving reassurance and guidance, he does not have the time to become the father-confessor for all the doubts and fears, angers and arguments, which can fill the minds of lonely young parents. Therefore, I would like to reiterate that a book like Dr. Spock's needs to be read in study groups where the true psychological spirit of the town meeting can be created; that is, where matters are considered to be agreed upon not because somebody said so, but because the free airing of opinions and emotions, of prejudices and of errors has led to a general area of relative consent and of tolerant good will.

This chapter has become unduly long. In regard to the matters discussed here, it is too bad that one must begin with the beginning. We know so little of the beginnings, of the deeper strata of the human mind. But since we have already embarked on general observations, a word must be said about one cultural and traditional institution which is deeply related to the matter of trust, namely, religion.

As a psychologist it is not my job to decide whether religion (or, for that matter, tradition) should or should not exist, or whether it should or should not be confessed and practiced in particular words and rituals. The psychological observer must ask whether or not in the area under observation religion and tradition are living psychological forces creating the kind of faith and conviction which permeates a parent's personality and thus reinforces the child's basic trust in the world's trustworthiness. The psychopathologist cannot avoid observing that there are millions of people who cannot really afford to be without religion, and whose pride in not having it is that much whistling in the dark. On the other hand, there are millions who derive faith from other than religious dogmas, that is, from sources of fellowship, of work completion, of social action, of scientific pursuit, and of artistic creation. And again, there are millions who profess faith, yet in practice mistrust both life and man. With all of these in mind, it seems worth while to speculate on the fact that religion through

---

[4] Benjamin Spock, *The Common Sense Book of Baby and Child Care* (New York, Duell, Sloan & Pearce, 1945).

the centuries has served to restore a sense of trust at regular intervals in the form of faith while giving tangible form to a sense of evil which it promises to ban. All religions have in common the periodical childlike surrender to a Provider or providers who dispense earthly fortune as well as spiritual health; the demonstration of one's smallness and dependence through the medium of reduced posture and humble gesture; the admission in prayer and song of misdeeds, of misthoughts, and of evil intentions; the admission of inner division and the consequent appeal for inner unification by divine guidance; the need for clearer self-delineation and self-restriction; and finally, the insight that individual trust must become a common faith, individual mistrust a commonly formulated evil, while the individual's need for restoration must become part of the ritual practice of many, and must become a sign of trustworthiness in the community.

Whosoever says he has religion must derive a faith from it which is transmitted to infants in the form of basic trust; whosoever claims that he does not need religion must derive such basic faith from elsewhere.

III

AUTONOMY VERSUS SHAME AND DOUBT

I

A survey of some of the items discussed in Dr. Spock's book under the major headings "The One-Year-Old" and "Managing Young Children" will enable those of us who, at this time, do not have such inquisitive creatures in our homes to remember our skirmishes, our victories, and our defeats:

196. Feeling his oats.
197. The passion to explore.
199. He gets more dependent and more independent at the same time.
203. Arranging the house for a wandering baby.
204. Avoiding accidents.
    Now's the time to put poisons out of reach.
206. How do you make him leave certain things alone?
207. Dropping and throwing things.
255. Children learn to control their own aggressive feelings.
256. Biting humans.
261. Keeping bedtime happy.
264. The small child who won't stay in bed at night.

My selection is intended to convey the inventory and range of problems described, though I cannot review here either the doctor's excellent advice or his good balance in depicting the remarkable ease and matter-of-factness with which the nursery may be governed at this as at any other stage. Nevertheless, there is an indication of the sinister forces which are leashed and unleashed, especially in the guerilla warfare of unequal wills; for the child is often unequal to his own violent drives, and parent and child unequal to each other.

The over-all significance of this stage lies in the maturation of the muscle system, the consequent ability (and doubly felt inability) to co-ordinate a number of highly conflicting action patterns such as "holding on" and "letting go," and the enormous value with which the still highly dependent child begins to endow his autonomous will.

Whereas psychoanalysis has enriched our vocabulary with the word "orality" to designate the importance of the mouth in the first year, it has added the term "anality" to include the particular pleasurableness and willfulness which attaches to the eliminative organs at one or another period at this stage. The whole procedure of evacuating the bowels and the bladder as completely as possible is, of course, enhanced from the beginning by a premium of "feeling good" which says in effect, "well done." This premium, at the beginning of life, must make up for quite frequent discomfort and tension suffered as the bowels learn to do their daily work. Two developments gradually give these "anal" experiences the necessary volume: the arrival of better formed stool and the general co-ordination of the muscle system which permits the development of voluntary release, of dropping and throwing away. This new dimension of approach to things, however, is not restricted to the sphincters. A general ability, indeed, a violent need, develops to drop and to throw away and to alternate withholding and expelling at will.

As far as anality proper is concerned, at this point everything depends on whether the cultural environment wants to make something of it. There are cultures where the parents ignore anal behavior and leave it to older children to lead the toddler out to the bushes so that his compliance in this matter may coincide with his wish to imitate the bigger ones. Our Western civilization, and especially certain classes within it, have chosen to take the matter more seriously. It is here where the machine age has added the ideal of a mechanically trained, faultlessly functioning, and always clean, punctual, and deodorized body. In addition it has been more or less consciously assumed that early and rigorous training is absolutely necessary for the kind of personality which will function efficiently in a mechanized world which says "time is money" and which calls for order-liness, punctuality, and thrift. Indications are that in this, we have gone too far; that we have assumed that a child is an animal which must be broken or a machine which must be set and tuned—while, in fact, human virtues can grow only by steps. At any rate our clinical work suggests that the neurotics of our time include the "over compulsive" type, who is stingy, retentive, and meticulous in matters of affection, time, and money, as well as in matters concerning his bowels. Also, bowel and bladder training has become the most obviously disturbing item of child training in wide circles of our society.

What, then, makes the anal problem potentially important and difficult?

The anal zone lends itself more than any other to the expression of stubborn self-insistence on contradictory impulses because, for one thing, it is the model zone for two conflicting modes which must become alternating; namely, *retention* and *elimination*. Furthermore, the sphincters are only part of the muscle system with its general ambiguity of rigidity and relaxation, of flexion and extension. This whole stage, then, becomes a battle for *autonomy*. For as he gets ready to stand on his feet more firmly, the infant delineates his world as "I" and "you," "me" and "mine." Every mother knows how astonishingly pliable a child may be at this stage, if and when he has made the decision that he *wants* to do what he is supposed to do. It is impossible, however, to find a reliable formula for making him want to do just that. Every mother knows how lovingly a child at this stage will snuggle and how ruthlessly he will suddenly try to push the adult away. At the same time the child is apt both to hoard things and to discard them, to cling to possessions and to throw them out of the windows of houses and vehicles. All of these seemingly contradictory tendencies, then, we include under the formula of the retentive-eliminative modes.

The matter of mutual regulation between adult and child now faces its severest test. If outer control by too rigid or too early training insists on robbing the child of his attempt *gradually* to control his bowels and other functions willingly and by his free choice, he will again be faced with a double rebellion and a double defeat. Powerless in his own body (sometimes afraid of his bowels) and powerless outside, he will again be forced to seek satisfaction and control either by regression or by fake progression. In other words, he will return to an earlier, oral control, that is, by sucking his thumb and becoming whiny and demanding; or he will become hostile and willful, often using his feces (and, later, dirty words) as ersatz ammunition; or he will pretend an autonomy and an ability to do without anybody to lean on which he has by no means really gained.

This stage, therefore, becomes decisive for the ratio between love and hate, for that between co-operation and willfulness, and for that between the freedom of self-expression and its suppression. From a sense of *self-control without loss of self-esteem* comes a lasting sense of autonomy and pride; from a sense of muscular and anal impotence, of loss of self-control, and of parental over-control comes a lasting sense of doubt and shame.

To develop autonomy, a firmly developed and a convincingly continued stage of early trust is necessary. The infant must come to feel that basic faith in himself and in the world (which is the lasting treasure saved from the conflicts of the oral stage) will not be jeopardized by this sudden violent wish to have a choice, to appropriate demandingly, and to eliminate stubbornly. *Firmness* must protect him against the potential anarchy of his as yet untrained sense of discrimination, his inability to hold on and to let go with circumspection. Yet his environment must back him up in his wish to "stand on his own feet" lest he be overcome by that sense of having exposed himself prematurely and foolishly which we call shame, or that secondary mistrust, that looking back after a double-take, which we call doubt.

Shame is an infantile emotion insufficiently studied. Shame supposes that one is completely exposed and conscious of being looked at—in a word, self-conscious. One is visible and not ready to be visible; that is why we dream of shame as a situation in which we are stared at in a condition of incomplete dress, in night attire, "with one's pants down." Shame is early expressed in an impulse to bury one's face, or to sink, right then and there, into the ground. This potentiality is abundantly utilized in the educational method of "shaming" used so exclusively by some primitive peoples, where it supplants the often more destructive sense of guilt to be discussed later. The destructiveness of shaming is balanced in some civilizations by devices for "saving face." Shaming exploits an increasing sense of being small, which paradoxically develops as the child stands up and as his awareness permits him to note the relative measures of size and power.

Too much shaming does not result in a sense of propriety but in a secret determination to try to get away with things when unseen, if, indeed, it does not result in deliberate shamelessness. There is an impressive American ballad in which a murderer to be hanged on the gallows before the eyes of the community, instead of feeling appropriately afraid or ashamed, begins to berate the onlookers, ending every salvo of defiance with the words, "God damn your eyes." Many a small child, when shamed beyond endurance, may be in a mood (although not in possession of either the courage or the words) to express defiance in similar terms. What I mean by this sinister reference is that there is a limit to a child's and an adult's individual endurance in the face of demands which force him to consider himself, his body, his needs, and his wishes as evil and dirty, and to believe, without bitter doubt, in the infallibility of those who pass such judgment. Occasionally he may be apt to turn things around, to become secretly oblivious to the opinion of others, and to consider as evil only the fact that they exist: his chance will come when they are gone, or when he can leave them.

Many a defiant child, many a young criminal, is of such makeup, and deserves at least an investigation into the conditions which caused him to become that way.

To repeat: muscular maturation sets the stage for experimentation with two simultaneous sets of social modalities—*holding on* and *letting go*. As is the case with all of these modalities, their basic conflicts can lead in the end either to hostile or to benign expectations and attitudes. Thus, "to hold" can become a destructive and cruel retaining or restraining, and it can become a pattern of care: "to have and to hold." To "let go," too, can turn into an inimical letting loose of destructive forces, or it can become a relaxed "to let pass" and "to let be." Culturally speaking, these modalities are neither good nor bad; their value depends on whether their hostile implications are turned against an enemy or a fellow man—or against the self.

The last-named danger is the one best known to psychiatry. Denied the gradual and well-guided experience of the autonomy of free choice, or weakened by an initial loss of trust, the sensitive child may turn against himself all his urge to discriminate and to manipulate. He will *overmanipulate* himself, he will develop a *precocious conscience*. Instead of taking possession of things in order to test them by repetitive play, he will become obsessed by his own repetitiveness; he will want to have everything "just so," and only in a given sequence and tempo. By such infantile obsessiveness, by dawdling, for example, or by becoming a stickler for certain rituals, or by stammering, the child then learns to gain power over his parents and nurses in areas where he could not find large-scale mutual regulation with them. Such hollow victory, then, is the infantile model for a compulsion neurosis. As for the consequences of this for adult character, they can be observed in the classical compulsive character which we have mentioned. We must add to this the character dominated by the wish to "get away with" things—yet unable to get away even with the wish. For while he learns evasion from others, his precocious conscience does not let him really get away with anything, and he goes through life habitually ashamed, apologetic, and afraid to be seen; or else, in a manner which we call "overcompensatory," he evinces a defiant kind of autonomy. The real inner autonomy, however, is not carried on the sleeve.

2

But it is time to return from these considerations of the abnormal to a study of the headings which transmit the practical and benevolent advice of the children's doctor. They all add up to this: be firm and tolerant with the child at this stage, and he will be firm and tolerant with himself. He will feel pride in being an autonomous person; he will grant autonomy to others; and now and again he will even let himself get away with something.

Why, then, if we know how, do we not tell parents in detail what to do to develop this intrinsic, this genuine autonomy? The answer is: because when it comes to human values, nobody knows how to fabricate or manage the fabrication of the genuine article. My own field, psychoanalysis, having studied particularly the excessive increase of guilt-feelings beyond any normal rhyme or reason, and the consequent excessive estrangement of the child from his own body, attempted at least to formulate what should *not* be done to children. These formulations, however, often aroused superstitious inhibitions in those who were inclined to make anxious rules out of vague warnings. Actually, we are learning only gradually what exactly *not* to do with *what kind* of children at *what age*.

People all over the world seem convinced that to make the right (meaning *their*) kind of human being, one must consistently introduce the senses of shame, doubt, guilt and fear into a child's life. Only the patterns vary. Some cultures begin to restrict early in life, some late, some abruptly, others more gradually. Until enough comparative observations are available, we are apt to add further superstitions, merely because of our wish to *avoid* certain pathological conditions, without even knowing definitely all the factors which are responsible for these conditions. So we say: Don't wean too early; don't train too early. But what is too early and what is too late seem to depend not only on the pathologies we wish to avoid but also on the values we wish to create, or, to put it more honestly, on the values we wish to live by. For no matter what we do in detail, the child will feel primarily what we live by, what makes us loving, co-operative, and firm beings, and what makes us hateful, anxious, and divided in ourselves.

There are, of course, a few matters of necessary avoidance which become clear from our basic epigenetic point of view. It will be remembered that every new development carries with it its own specific vulnerability. For example, at around eight months the child seems to be somehow more aware, as it were, of his self; this prepares him for the impending sense of autonomy. At the same time he becomes more cognizant of his mother's features and presence. Sudden or prolonged separation from his mother at that time apparently can cause a sensitive child to experience an aggravation of the experience of division and abandonment, arousing violent anxiety and withdrawal. Again, in the first quarter of the second year, if everything has gone well, the infant just begins to become aware of the autonomy discussed in this chapter. The introduction of bowel training at this time may cause him to resist with all his strength and determination, because he seems to feel that his budding will is being "broken." To avoid this feeling is certainly more important than to insist on his being trained just then because there is a time for the ascendancy of autonomy and there is a time for the sacrifice of autonomy; but obviously the time for a meaningful sacrifice is *after* one has acquired and reinforced a core of autonomy and has also acquired more insight. The more exact localization in time of the most critical growth periods of the personality is becoming established only now. In the meantime parents would do well to learn how and where to discuss things with others; to ascertain through discussion with other parents in what respects they can freely choose their methods of child training according to common sense and temperament and to enjoy their children without worry. They must also learn when and where to consult the experts who master the itinerary of possible damages and who master the methods of finding out what may have happened to a disturbed child. Often what happened could not have been avoided; and usually the unavoidable cause of trouble is not one thing but the coincidence in time of many changes which upset the child's orientation. He may have been involved in a special growth period when the family moved to a new place. Perhaps he was forced to begin learning words all over again when the grandmother who had taught him his first words suddenly died. This event may have necessitated a trip on the part of the mother, which exhausted her because she happened to be pregnant at the time, and so forth. Given the right spirit toward life and its vicissitudes, a parent can handle such matters with the help of the pediatrician or guidance expert. For the expert's job should be (to quote Dr. Frank Fremont-Smith) "to set the frame of reference within which choice is permissible and desirable."

Where such a frame of reference, or such information, does not meet with a wish on the mother's part to execute responsible choices and to enjoy her children in her and their way, she may have to ask another expert what is wrong with her; or, if she is not alone in her problem, she may well ask others who are interested in the same question: What is wrong with us? For in the last resort (as comparative studies in child training have convinced many of us) the kind and degree of a sense of autonomy which parents are able to grant their small children depends on the dignity and the sense of personal independence which they derive from their own lives. Again, just as the sense of trust is a reflection of the parents' sturdy and realistic faith, so is the sense of antonomy a reflection of the parents' dignity as individuals.

As was the case with "oral" personality, the compulsive personality (often referred to as "anal" in the psychiatric literature) has its normal aspects and its abnormal exaggerations. If well integrated with other compensatory traits, some compulsiveness is useful in the administration of matters in which order, punctuality, and cleanliness are essential. The question is always whether we remain the masters of the rules by which we want to make things more manageable (not more complicated) or whether the rules master the ruler. But it often happens, in the individual as well as in group life, that the letter of the rules kills the spirit which created them.

3

If I may seemingly digress here, it becomes apparent that the basic need of the individual for a delineation of his autonomy in the adult order of things is taken care of by the system of "law and order," be it the simple distribution of privileges and duties in a small tribe or in the complicated system of a modern state. Political organization assigns with the power of government certain privileges of leadership and certain obligations of conduct; while it assigns to the ruled certain obligations of compliance and certain privileges of remaining autonomous and self-determining. Where this whole matter becomes blurred, however, the matter of individual autonomy becomes an issue of mental health, as well as one of political reorientation. Where large numbers of people have been prepared in childhood to expect from life a high degree of personal autonomy, pride, and rich opportunities for those who are ready to grasp them, and then in later life find himself ruled by superhuman organizations and machinery too intricate to understand, the result may be deep chronic disappointment not conducive to healthy personalities. . . . Such considerations may serve as a basis for the kind of discussion which I wish to recommend. For the sense of autonomy in the child (a sense richly fostered in American childhood in general) must be backed up by the preservation in economic and political life of a high sense of autonomy and of initiative, both of which in some classes and regions are challenged by the complication and mechanization of modern life. All great nations (and all the small ones) are being enveloped in the problems of the organization of larger units, larger spheres, and larger interdependencies which by necessity redefine the role of the individual. It is important for the spirit of this country, as it is for that of the world, that the American sense of individual autonomy and initiative be preserved in this time of transition and that new forms of communication between groups be developed, so that a consciousness of equality and individuality may grow out of the necessity for divided function within the increasing complexity of organization. For wherever typical past experience or the over-all social development in the present upset the sense of autonomy, a number of fears are aroused which find expression in anxiety on a large scale, often individually slight and hardly conscious, but nevertheless strangely upsetting to people who seemingly, on the surface, have what they want or what they seem to have a right to expect. Besides irrational fears of losing one's autonomy—"don't fence me in"—there are fears of being sabotaged in one's free will by inner enemies; of being restricted and constricted in one's autonomous initiative; and, paradoxically enough, at the same time of not being completely controlled enough, of not being told what to do. While many such fears are, of course, based on the realistic appraisal of dangers inherent in complex social organizations and in the struggle for power, safety, and security, they seem to contribute to psychoneurotic and psychosomatic disturbances, as well as to the easy acceptance of slogans which seem to promise alleviation of conditions either by excessive and irrational regimentation.

Here one may disagree as to what details need and ought to be discussed. I presented this problem as only one example of a larger problem; namely, that the sense of autonomy, which arises, or should arise, in the second stage of childhood, is fostered by a handling of the small individual which expresses a sense of dignity and independence on the part of the parents and a confident expectation that the kind of autonomy fostered earlier will not be frustrated later. This, in turn, necessitates a relationship of parent to parent, of parent to employer, and of parent to government which reaffirms the parent's essential dignity regardless of his social position. It is important to dwell on this point because much of the shame and doubt, much of the indignity and uncertainty which is aroused in children is an expression of the parents' frustrations in marriage, in work, and in citizenship.

IV

INITIATIVE VERSUS GUILT

I

Having found a firm solution of his problem of autonomy, the child of four and five is faced with the next step—and with the next crisis. Being firmly convinced that he *is* a person, the child must now find out *what kind* of a person he is going to be. And here he hitches his wagon to nothing less than a star: he wants to be like his parents, who to him appear very powerful and very beautiful, although quite unreasonably dangerous. He "identifies with them," that is, he plays with the idea of how it would be to be them. Three strong developments help at this stage, yet also serve to bring the child closer to his crisis: (1) he learns to move around more freely and more violently and therefore establishes a wider and, so it seems to him, an unlimited radius of goals; (2) his sense of language becomes perfected to the point where he understands and can ask about many things just enough to misunderstand them thoroughly; and (3) both language and locomotion permit him to expand his imagination over so many things that he cannot avoid frightening himself with what he himself has dreamed and thought up. Nevertheless, out of all this he must emerge with a sense of unbroken initiative as a basis for a high and yet realistic sense of ambition and independence.

One may ask here—one may, indeed—what are the criteria for such an unbroken sense of initiative? The criteria for all the senses discussed here are the same: a crisis, beset with fears, or at least a general anxiousness or tension, seems to be resolved, in that the child suddenly seems to "grow together" both psychologically and physically. He seems to be "more himself," more loving and relaxed and brighter in his judgment (such as it is at this stage). Most of all, he seems to be, as it were, self-activated; he is in the free possession of a certain surplus of energy which permits him to forget failures quickly and to approach what seems desirable (even if it also seems dangerous) with undiminished and better aimed effort. In this way the child and his parents face the next crisis much better prepared.

We are now approaching the end of the third year, when walking is getting to be a thing of ease, or vigor. The books tell us that a child "can walk" much before this; but from the point of view of personality development he cannot really walk as long as he is only able to accomplish the feat more or less well, with more or fewer props, for short spans of time. He has made walking and running an item in his sphere of mastery when gravity is felt to be *within*, when he can forget that he is doing the walking and instead can find out what he can do *with it*. Only then do his legs become an unconscious part of him instead of being an external and still unreliable ambulatory appendix. Only then will he find out with advantage what he now *may* do, along with what he *can* do.

To look back: the first way-station was prone relaxation. The trust based on the experience that the basic mechanisms of breathing, digesting, sleeping, and so forth have a consistent and familiar relation to the foods and comforts offered gives zest to the developing ability to raise oneself to a sitting and then to a standing position. The second way-station (accomplished only toward the end of the second year) is that of being able to sit not only securely but, as it were, untiringly, a feat which permits the muscle system gradually to be used for finer discrimination and for more autonomous ways of selecting and discarding, of piling things up—and of throwing them away with a bang.

The third way-station finds the child able to move independently and vigorously. He is ready to visualize himself as being as big as the perambulating grownups. He begins to make comparisons and is apt to develop untiring curiosity about differences in sizes in general (and later of sexual parts in particular). He tries to comprehend possible future roles, or at any rate to understand what roles are worth imitating. More immediately, he can now associate with those of his own age. Under the guidance of older children or special women guardians, he gradually enters into the infantile politics of nursery school, street corner, and barnyard. His learning now is eminently intrusive and vigorous: it leads away from his own limitations and into future possibilities.

The *intrusive mode*, dominating much of the behavior of this stage, characterizes a variety of configurationally "similar" activities and fantasies. These include the intrusion into other bodies by physical attack; the intrusion into other people's ears and minds by aggressive talking; the intrusion into space by vigorous locomotion; the intrusion into the unknown by consuming curiosity.

This is also the stage of infantile sexual curiosity, genital excitability, and occasional preoccupation and overconcern with sexual matters. This "genitality" is, of course, rudimentary, a mere promise of things to come; often it is hardly noticeable as such. If not specifically provoked into precocious manifestation by especially strict and pointed prohibitions ("If you touch it, the doctor will cut it off") or special customs (such as sex play in groups), it is apt to lead to no more than a series of fascinating experiences which soon become frightening and pointless enough to be repressed during the ascendancy of that human specialty which Freud called the "latency" period, that is, the long physiological delay separating infantile sexuality (which in animals is followed by maturity) and physical sexual maturation.

The sexual orientation of the boy is focused on the phallus and its sensations, purposes, and meanings. While erections undoubtedly occur earlier (either reflexively or in response to things and people who make the child feel intensively), a focused interest may now develop in the genitalia of both sexes, as well as an urge to perform playful sex acts, or at least acts of sexual investigation. The increased locomotor mastery and the pride in being big now and *almost* as good as father and mother receives its severest setback in the clear fact that in the genital sphere one is vastly inferior; furthermore it receives an additional setback in the fact that not even in the distant future is one ever going to be father in sexual relationship to mother, or mother in sexual relationship to father. The very deep emotional consequences of this insight make up what Freud has called the Oedipus complex.

Psychoanalysis verifies the simple conclusion that boys attach their first genital affection to the maternal adults who have otherwise given comfort to their bodies and that they develop their first sexual rivalry against the persons who are the sexual owners of those maternal persons. The little girl, in turn, becomes attached to her father and other important men and jealous of her mother, a development which may cause her much anxiety, for it seems to block her retreat to that self-same mother, while it makes the mother's disapproval ever so much more magically dangerous because unconsciously "deserved."

Girls often have a difficult time at this stage, because they observe sooner or later that, although their locomotor, mental, and social intrusiveness is increased equally with, and is as adequate as, that of the boys, thus permitting them to become perfect tomboys, they lack one item: the penis. While the boy has this visible, erectable, and comprehensible organ to which he can attach dreams of adult bigness, the girl's clitoris only poorly sustains dreams of sexual equality. She does not even have breasts as analogously tangible tokens of her future; her maternal drives are relegated to play fantasy or baby tending. On the other hand, where mothers dominate households, the boy, in turn, can develop a sense of inadequacy because he learns at this stage that while a boy can do well in play and work, he will never boss the house, the mother, and the older sisters. His mother and sisters, in fact, might get even with him for vast doubts in themselves by making him feel that a boy (with his snails and puppy-dog tails) is really an inferior if not a repulsive creature. Both the girl and the boy are now extraordinarily appreciative of any convincing promise of the fact that someday they will be as good as father or mother—perhaps better; and they are grateful for sexual enlightenment, a little at a time, and patiently repeated later. Where the necessities of economic life and the simplicity of its social plan make the male and female roles and their specific powers and rewards comprehensible, the early misgivings about sexual differences are, of course, more easily integrated in the culture's design for the differentiation of sexual roles.

This stage, then, adds to the inventory of basic social modalities in both sexes that of "making" in the older and today slangier sense of "being on the make." There is no simpler, stronger word to match the social modalities previously enumerated. The word suggests enjoyment of competition, insistence on goal, pleasure of conquest. In the boy the emphasis remains on "making" by head-on attack; in the girl it sooner or later changes to "making" by making herself attractive and endearing. The child thus develops the prerequisites for *masculine* and *feminine initiative*, that is, for the selection of social goals and perseverance in approaching them. Thus the stage is all set for entrance into life, except that life must first be school life. Fortunately, or unfortunately, the child here represses many of his fondest hopes and most energetic wishes. His exuberant imagination is tamed and he learns the necessary self-restraint and the necessary interest in impersonal things—even the three R's. This often demands a change of personality that is sometimes too drastic for the good of the child. This change is not only a result of education but also of an inner reorientation, and it is based on a biological fact (the delay of sexual maturation) and a psychological one (the repression of childhood wishes). For those sinister Oedipus wishes (so simply and so trustingly expressed in the boy's assurance that he will marry mother and make her proud of him, and in the girl's that she will marry father and take much better care of him), in consequence of vastly increased imagination and, as it were, the intoxication of increased locomotor powers, seem to lead to secret fantasies of terrifying proportions. The consequence is a deep sense of *guilt*—a strange sense, for it forever seems to imply that the individual has committed crimes and deeds which, after all, were not only not committed but also would have been biologically quite impossible.

While the struggle for autonomy at its worst concentrated on keeping rivals out, and was therefore more an expression of *jealous rage* most often directed against encroachments by *younger* siblings, initiative brings with it *anticipatory rivalry* with those who were there first and who may, therefore, occupy with their superior equipment the field toward which one's initiative is directed. Jealousy and rivalry, those often embittered and yet essentially futile attempts at demarcating a sphere of unquestioned privilege, now come to a climax in a final contest for a favored position with one of the parents; the inevitable and necessary failure leads to guilt and anxiety. The child indulges in fantasies of being a giant and a tiger, but in his dreams he runs in terror for dear life. This, then, is the stage of fear for life and limb, including the fear of losing (or on the part of the girl the conviction that she may have lost) the male genital as punishment for the fantasies attached to infantile genital excitement.

All of this may seem strange to readers who have only seen the sunnier side of childhood and have not recognized the potential powerhouse of destructive drives which can be aroused and temporarily buried at this stage, only to contribute later to the inner arsenal of destructiveness which is ever ready to be used when opportunity provokes it. By using the words "potential," "provoke" and "opportunity," I mean to emphasize that there is little in these inner developments which cannot be harnessed to constructive and peaceful initiative if only we learn to understand the conflicts and anxieties of childhood and the importance of childhood for mankind. But if we should choose to overlook or belittle the phenomena of childhood, or to regard them as "cute" (even as the individual forgets the best and the worst dreams of his childhood), we shall forever overlook one of the eternal sources of human anxiety and strife.

2

It is at this stage of initiative that the great governor of initiative, namely, *conscience*, becomes firmly established. The child now feels not only ashamed when found out but also afraid of being found out. He now hears, as it were, God's voice without seeing God. Moreover, he begins automatically to feel guilty even for mere thoughts and for deeds which nobody has watched. This is the cornerstone of morality in the individual sense. But from the point of view of mental health, we must point out that if this great achievement is overburdened by all too eager adults, it can be bad for the spirit and for morality itself. For the conscience of the child *can* be primitive, cruel, and uncompromising, as may be observed in instances where children learn to constrict themselves to the point of overall inhibition; where they develop an obedience more literal than the one the parent wishes to exact; or where they develop deep regressions and lasting resentments because the parents themselves do not seem to live up to the new conscience which they have fostered in the child. One of the deepest conflicts in life is the hate for a parent who served as the model and the executor of the conscience but who (in some form) was found trying to "get away with" the very transgressions which the child can no longer tolerate in himself. These transgressions often are the natural outcome of the existing inequality between parent and child. Often, however, they represent a thoughtless exploitation of such inequality; with the result that the child comes to feel that the whole matter is not one of universal goodness but of arbitrary power. The suspiciousness and evasiveness which is thus mixed in with the all-or-nothing quality of the supergo, that organ of tradition, makes moralistic man a great potential danger to himself and to his fellow men. It is as if morality, to him, became synonymous with vindictiveness and with the suppression of others.

It is necessary to point to the source of such moralism (not to be mistaken for morality) in the child of this age because infantile moralism is a stage to be lived through and worked through. The consequences of the guilt aroused at this stage (guilt expressed in a deep-seated conviction that the child as such, or drive as such, is essentially bad) often do not show until much later, when conflicts over initiative may find expression in a self-restriction which keeps an individual from living up to his inner capacities or to the powers of his imagination and feeling if not in sexual impotence or frigidity. All of this, of course, may in turn be "overcompensated" in a great show of tireless initiative, in a quality of "go-at-itiveness" at all cost. This many adults overdo to a point where they can never relax when they feel that their worth as people consists entirely in *what they are doing*, or rather in *what they are going to do next*, and not in what they are, as individuals. The strain consequently developed in their bodies, which are always "on the go," with the engine racing, even at moments of rest, is a powerful contribution to the much-discussed psychosomatic diseases of our time.

Here I must admit again to psychopathological bias, and say again that we must learn to look beyond pathology. Pathology is only the sign that valuable human resources are being neglected, that they have been neglected first of all in childhood. Well-meaning, well-guided, and well-trained parents in a functioning cultural and economic system can certainly help their children through this and any other stage with much energy to spare.

The problem is again one of mutual regulation. Where the child, now so ready to overrestrict himself, can gradually develop a sense of paternal responsibility, where he can gain some insight into the institutions, functions, and roles which will permit his responsible participation as an adult, he will soon find pleasurable accomplishment in wielding tools and weapons, in manipulating meaningful toys, and in taking responsibility for himself and for younger children.

For such is the wisdom of the ground plan that at no time is the individual more ready to learn quickly and avidly, to become big in the sense of sharing obligation and performance rather than power, in the sense of *making things, instead of "making" people*, than during this period of his development. He is also eager and able to combine with other children for the purpose of constructing and planning, instead of trying to boss and coerce them; and he is able and willing to profit fully by the association with teachers and ideal prototypes.

Parents often do not realize why some children suddenly seem to think less of them and seem to attach themselves to teachers, to the parents of other children, or to people representing occupations which the child can grasp: firemen and policemen, gardeners and plumbers. The point is that children do not wish to be reminded of the principal inequality with the parent of the same sex. They remain identified with this same parent; but for the present they look for opportunities where superficial identification seems to promise a field of initiative without too much conflict or guilt.

Often, however (and this seems more typical of the American home than of any other in the world), the child can be guided by the parent himself into a second, a more realistic identification based on the spirit of equality experienced in doing things together. In connection with comprehensible technical tasks, a companionship may develop between father and son, an experience of essential *equality in worth*, in spite of the *inequality in time schedules*. Such companionship is a lasting treasure not only for parent and child but for mankind, which so sorely needs an alleviation of all those hidden hatreds which stem from the exploitation of weakness because of mere size or schedule.

Only a combination of early prevention and alleviation of hatred and guilt in the growing being, and the consequent handling of hatred in the free collaboration of people who feel *equal in worth although different in kind or function or age*, permits a peaceful cultivation of initiative, a truly free sense of enterprise. And the word "enterprise" was deliberately chosen. For a comparative view of child training suggests that it is the prevalent economic ideal, or some of its modifications, which is transmitted to the child at the time when, in identification with his parent, he applies the dreams of early childhood to the as yet dim goals of an active adult life.

v

INDUSTRY VERSUS INFERIORITY

I

One might say that personality at the first stage crystallizes around the conviction "I am what I am given," and that of the second, "I am what I will." The third can be characterized by "I am what I can imagine I will be." We must now approach the fourth: "I am what I learn." The child now wants to be shown how to get busy with something and how to be busy with others.

This trend, too, starts much earlier, especially in some children. They want to watch how things are done and to try doing them. If they are lucky they live near barnyards or on streets around busy people and around many other children of all ages, so that they can watch and try, observe and participate as their capacities and their initiative grow in tentative spurts. But now it is time to "go to school." In all cultures, at this stage, children receive some systematic instruction, although it is by no means always in the kind of school which literate people must organize around teachers who have learned how to teach literacy. In preliterate people much is learned from adults who become teachers by acclamation rather than by appointment; and very much is learned from older children. What is learned in more primitive surroundings is related to the basic skills of *technology* which are developed as the child gets ready to handle the utensils, the tools, and the weapons used by the big people: he enters the technology of his tribe very gradually but also very directly. More literate people, with more specialized careers, must prepare the child by teaching him things which first of all make him literate. He is then given the widest possible basic education for the greatest number of possible careers. The greater the specialization, the more indistinct the goal of initiative becomes; and the more complicated the social reality is, the vaguer the father's and mother's role in it appears to be. Between childhood and adulthood, then, our children go to school; and school seems to be a world all by itself, with its own goals and limitations, its achievements and disappointments.

Grammar school education has swung back and forth between the extreme of making early school life an extension of grim adulthood by emphasizing self-restraint and a strict sense of duty in doing what one is *told* to do, and the other extreme of making it an extension of the natural tendency in childhood to find out by playing, to learn what one must do by doing steps which one *likes* to do. Both methods work for some children at times but not for all children at all times. The first trend, if carried to the extreme, exploits a tendency on the part of the preschool and grammar-school child to become entirely dependent on prescribed duties. He thus learns much that is absolutely necessary and he develops an unshakable sense of duty; but he may never unlearn again an unnecessary and costly self-restraint with which he may later make his own life and other people's lives miserable, and in fact spoil his own children's natural desire to learn and to work. The second trend, when carried to an extreme, leads not only to the well-known popular objection that children do not learn anything any more but also to such feelings in children as are expressed in the by now famous remark of a metropolitan child who apprehensively asked one morning: "Teacher, *must* we do today what we *want* to do?" Nothing could better express the fact that children at this age *do* like to be mildly coerced into the adventure of finding out that one can learn to accomplish things which one would never have thought of by oneself, things which owe their attractiveness to the very fact that they are *not* the product of play and fantasy but the product of reality, practicality, and logic; things which thus provide a token sense of participation in the world of adults. In discussions of this kind it is common to say that one must steer a middle course between play and work, between childhood and adulthood, between old-fashioned and progressive education. It is always easy (and it seems entirely satisfactory to one's critics) to say that one plans to steer a middle course, but in practice it often leads to a course charted by avoidances rather than by zestful goals. Instead of pursuing, then, a course which merely avoids the extremes of easy play or hard work, it may be worth while to consider what play is and what work is, and then learn to dose and alternate each in such a way that play is play and work is work. Let us review briefly what play may mean at various stages of childhood and adulthood.

The adult plays for purposes of recreation. He steps out of his reality into imaginary realities for which he has made up arbitrary but none the less binding rules. But an adult must not be a playboy. Only he who works shall play—if, indeed, he can relax his competitiveness.

The playing child, then, poses a problem: whoever does not work shall not play. Therefore, to be tolerant of the child's play the adult must invent theories which show either that childhood play is really the child's work or that it does not count. The most popular theory, and the easiest on the observer, is that the child is nobody yet and that the nonsense of his play reflects it. According to Spencer, play uses up surplus energy in the young of a number of mammalians who do not need to feed or protect themselves because their parents do it for them. Others say that play is either preparation for the future or a method of working off past emotion, a means of finding imaginary relief for past frustrations.

It is true that the content of individual play often proves to be the infantile way of thinking over difficult experiences and of restoring a sense of mastery, comparable to the way in which we repeat, in thought and endless talk, experiences that have been too much for us. This is the rationale for play observation, play diagnosis, and play therapy. In watching a child play, the trained observer can get an impression of what it is the child is "thinking over," and what faulty logic, what emotional dead end he may be caught in. As a diagnostic tool such observation has become indispensable.

The small world of manageable toys is a harbor which the child establishes, returning to it when he needs to overhaul his ego. But the thing-world has its own laws: it may resist rearrangement or it may simply break to pieces; it may prove to belong to somebody else and be subject to confiscation by superiors. Often the microsphere seduces the child into an unguarded expression of dangerous themes and attitudes which arouse anxiety and lead to sudden *play-disruption*. This is the counterpart, in waking life, of the anxiety dream; it can keep children from trying to play just as the fear of night terror can keep them from going to sleep. If thus frightened or disappointed, the child may regress into daydreaming, thumb-sucking, masturbating. On the other hand, if the first use of the thing-world is successful and guided properly, the *pleasure of mastering toy things* becomes associated with the *mastery of the conflicts* which were projected on them and with the *prestige* gained through such mastery.

Finally, at nursery-school age playfulness reaches into the world shared with others. At first these others are treated as things; they are inspected, run into, or forced to "be horsie." Learning is necessary in order to discover what potential play content can be admitted only to fantasy or only to play by and with oneself; what content can be successfully represented only in the world of toys and small things; and what content can be shared with others and even forced upon them.

What is infantile play, then? We saw that it is not the equivalent of adult play, that it is not recreation. The playing adult steps sideward into another, an artificial reality; the playing child advances forward to new stages of real mastery. This new mastery is not restricted to the technical mastery of toys and *things;* it also includes an infantile way of mastering *experience* by meditating, experimenting, planning, and sharing.

2

While all children at times need to be left alone in solitary play (or later in the company of books and radio, motion pictures and video, all of which, like the fairy tales of old, at least *sometimes* seem to convey what fits the needs of the infantile mind), and while all children need their hours and days of make-believe in games, they all, sooner or later, become dissatisfied and disgruntled without a sense of being useful, without a sense of being able to make things and make them well and even perfectly: this is what I call the *sense of industry*. Without this, the best entertained child soon acts exploited. It is as if he knows and his society knows that now that he is psychologically already a rudimentary parent, he must begin to be somewhat of a worker and potential provider before becoming a biological parent. With the oncoming latency period, then, the normally advanced child forgets, or rather "sublimates" (that is, applies to more useful pursuits and approved goals) the necessity of "making" people by direct attack or the desire to become papa and mamma in a hurry: he now learns to win recognition by producing things. He develops industry, that is, he adjusts himself to the inorganic laws of the tool world. He can become an eager and absorbed unit of a productive situation. To bring a productive situation to completion is an aim which gradually supersedes the whims and wishes of his idiosyncratic drives and personal disappointments. As he once untiringly strove to walk well, and to throw things away well, he now wants to make things well. He develops the pleasure of work completion by steady attention and persevering diligence.

The danger at this stage is the development of a sense of *inadequacy and inferiority*. This may be caused by an insufficient solution of the preceding conflict: he may still want his mummy more than knowledge; he may still rather be the baby at home than the big child in school; he still compares himself with his father, and the comparison arouses a sense of guilt as well as a sense of anatomical inferiority. Family life (small family) may not have prepared him for school life, or school life may fail to sustain the promises of earlier stages in that nothing that he has learned to do well already seems to count one bit with the teacher. And then, again, he may be potentially able to excel in ways which are dormant and which, if not evoked now, may develop late or never.

Good teachers, healthy teachers, relaxed teachers, teachers who feel trusted and respected by the community, understand all this and can guide it. They know how to alternate play and work, games and study. They know how to recognize special efforts, how to encourage special gifts. They also know how to give a child time, and how to handle those children to whom school, for a while, is not important and rather a matter to endure than to enjoy; or the child to whom other children are much more important than the teacher and who shows it.

Good parents, healthy parents, relaxed parents, feel a need to make their children trust their teachers, and therefore to have teachers who can be trusted. It is not my job here to discuss teacher selection, teacher training, and the status and payment of teachers in their communities—all of which is of direct importance for the development and the maintenance in children of a *sense of industry* and of a positive identification with those who *know* things and know how to *do* things. Again and again I have observed in the lives of especially gifted people that one teacher, somewhere, was able to kindle the flame of hidden talent.

The fact that the majority of teachers in the elementary schools are women must be considered here in passing, because it often leads to a conflict with the "ordinary" boy's masculine identification, as if knowledge were feminine, action masculine. Both boys and girls are apt to agree with Bernard Shaw's statement that those who can, do, while those who cannot,

teach. The selection and training of teachers, then, is vital for the avoidance of the dangers which can befall the individual at this stage. There is,
first, the above-mentioned sense of inferiority, the feeling that one will
never be any good—a problem which calls for the type of teacher who
knows how to emphasize what a child *can* do, and who knows a psychiatric
problem when she sees one. Second, there is the danger of the child's
identifying too strenuously with a too virtuous teacher or becoming the
teacher's pet. What we shall presently refer to as his sense of identity can
remain prematurely fixed on being nothing but a good little worker or a
good little helper, which may not be all he *could* be. Third, there is the
danger (probably the most common one) that throughout the long years
of going to school he will never acquire the enjoyment of work and the
pride of doing at least one kind of thing well. This is particularly of concern in relation to that part of the nation who do not complete what
schooling is at their disposal. It is always easy to say that they are born
that way; that there must be less educated people as background for the
superior ones; that the market needs and even fosters such people for its
many simple and unskilled tasks. But from the point of view of the healthy
personality (which, as we proceed, must now include the aspect of playing
a constructive role in a healthy society), we must consider those who have
had just enough schooling to appreciate what more fortunate people are
learning to do but who, for one reason or another, have lacked inner or
outer support of their stick-to-itiveness.

It will have been noted that, regarding the period of a developing sense
of industry, I have referred to outer hindrances but not to any crisis (except a deferred inferiority crisis) coming from the inventory of basic
human drives. This stage differs from the others in that it does not consist
of a swing from a violent inner upheaval to a new mastery. The reason
why Freud called it the latency stage is that violent drives are normally
dormant at that time. But it is only a lull before the storm of puberty.

On the other hand, this is socially a most decisive stage: since industry
involves doing things beside and with others, a first sense of division of
labor and of equality of opportunity develops at this time. When a child
begins to feel that it is the color of his skin, the background of his parents,
or the cost of his clothes rather than his wish and his will to learn which
will decide his social worth, lasting harm may ensue for the *sense of
identity*, to which we must now turn.

VI

IDENTITY VERSUS SELF-DIFFUSION

I

With the establishment of a good relationship to the world of skills and
to those who teach and share the new skills, childhood proper comes to an

end. Youth begins. But in puberty and adolescence all sameness and continuities relied on earlier are questioned again because of a rapidity of body growth which equals that of early childhood and because of the entirely new addition of physical genital maturity. The growing and developing youths, faced with this physiological revolution within them, are now primarily concerned with attempts at consolidating their social roles. They are sometimes morbidly, often curiously, preoccupied with what they appear to be in the eyes of others as compared with what they feel they are and with the question of how to connect the earlier cultivated roles and skills with the ideal prototypes of the day. In their search for a new sense of continuity and sameness, some adolescents have to refight many of the crises of earlier years, and they are never ready to install lasting idols and ideals as guardians of a final identity.

The integration now taking place in the form of the ego identity is more than the sum of the childhood identifications. It is the inner capital accrued from all those experiences of each successive stage, when successful identifications led to a successful alignment of the individual's *basic drives* with his *endowment* and his *opportunities*. In psychoanalysis we ascribe such successful alignments to "ego synthesis"; this writer has tried to demonstrate that the ego values accrued in childhood culminate in what he has called *a sense of ego identity*. The sense of ego identity, then, is the accrued confidence that one's ability to maintain inner sameness and continuity (one's ego in the psychological sense) is matched by the sameness and continuity of one's meaning for others.

To go back into early childhood once more: a child who has just found himself able to walk, more or less coaxed or ignored by those around him, seems driven to repeat the act for the pure enjoyment of functioning and out of the need to master and perfect a newly initiated function. But he also acts under the immediate awareness of the new status and stature of "one who can walk," although different peoples and different people may express this according to a great variety of expectations: "one who will go far," "one who will be able to stand on his own feet," "one who will be upright," "one who must be watched because he might go too far," or sometimes "one who will surely fall." At any rate, to become "one who can walk" is one of the many steps in child development which suddenly give an experience of physical mastery and of cultural meaning, of pleasure in activity and of social prestige; it thus is one building stone of self-esteem. This self-esteem, confirmed at the end of the major crises, as outlined here, grows to be a conviction that one is learning effective steps toward a tangible future, that one is developing a defined personality within a social reality which one understands. The growing child must, at every step, derive a vitalizing sense of reality from the awareness that his individual way of mastering experience is a successful variant of the way other people around him master experience and recognize such mastery.

In this, children cannot be fooled by empty praise and condescending encouragement. They may have to accept artificial bolstering of their self-esteem in lieu of something better, but what I call their accruing ego identity gains real strength only from wholehearted and consistent recognition of real accomplishment, that is, achievement that has meaning in their culture.

A child has quite a number of opportunities to identify himself, more or less experimentally, with habits, traits, occupations, and ideas of real or fictitious people of either sex. Certain crises force him to make radical selections. However, this historical era in which he lives offers only a limited number of socially meaningful models for workable combinations of identification fragments. Their usefulness depends on the way in which they simultaneously meet the requirements of his maturational stage and his habits of adjustment. But should a child feel that the environment tries to deprive him too radically of all the forms of expression which permit him to develop and to integrate the next step in his ego identity, he will defend it with the astonishing strength encountered in animals who are suddenly forced to defend their lives. Indeed, in the social jungle of human existence there is no feeling of being alive without a sense of ego identity. To understand this would be to understand the trouble of adolescents better, especially the trouble of all those who cannot just be "nice" boys and girls, but are desperately seeking for a satisfactory sense of belonging, be it in cliques and gangs here in our country or in inspiring mass movements in others.

The ego identity develops out of a gradual integration of all identifications, but here, if anywhere, the whole has a different quality than the sum of its parts. It is the integration of the whole, not the quality or strength of the parts, which makes the difference. Under favorable circumstances children have the nucleus of a separate identity in early life; often they defend it even against the necessity of overidentifying with one of their parents. This is difficult to learn from patients, because the neurotic ego has, by definition, fallen prey to overidentification and to faulty identifications with disturbed parents, a circumstance which isolated the small individual both from his budding identity and from his milieu. But we can study it profitably in the children of minority-group Americans who, having successfully graduated from a marked and well-guided stage of autonomy, enter the most decisive stage of American childhood: that of initiative and industry.

Minority groups of a lesser degree of Americanization (Negroes, Indians, Mexicans, and certain European groups) often are privileged in the enjoyment of a more sensual early childhood. Their crises come when their parents and teachers, losing trust in themselves and using sudden correctives in order to approach the vague but pervasive Anglo-Saxon ideal, create violent discontinuities; or where, indeed, the children themselves learn to disavow their sensual and overprotective mothers as temptations and a hindrance to the formation of a more American personality.

On the whole, it can be said that American schools successfully meet the challenge of training children of play-school age and of the elementary grades in a spirit of self-reliance and enterprise. Children of these ages seem remarkably free of prejudice and apprehension, preoccupied as they still are with growing and learning and with the new pleasures of association outside their families. This, to forestall the sense of individual inferiority, must lead to a hope for "industrial association," for equality with all those who apply themselves wholeheartedly to the same skills and adventures in learning. Many individual successes, on the other hand, only expose the now overly encouraged children of mixed backgrounds and somewhat deviant endowments to the shock of American adolescence: the standardization of individuality and the intolerance of "differences."

The emerging ego identity, then, bridges the early childhood stages, when the body and the parent images were given their specific meanings, and the later stages, when a variety of social roles become available and increasingly coercive. A lasting ego identity cannot begin to exist without the trust of the first oral stage; it cannot be completed without a promise of fulfillment which from the dominant image of adulthood reaches down into the baby's beginnings and which creates at every step an accruing sense of ego strength.

2

The danger of this stage is *role diffusion;* as Biff puts it in Arthur Miller's *The Death of a Salesman,* "I just can't take hold, Mom, I can't take hold of some kind of a life." Where such a dilemma is based on a strong previous doubt of one's ethnic and sexual identity, delinquent and outright psychotic incidents are not uncommon. Youth after youth, bewildered by some assumed role, a role forced on him by the inexorable standardization of American adolescence, runs away in one form or another: leaving schools and jobs, staying out all night, or withdrawing into bizarre and inaccessible moods. Once "delinquent," his greatest need, and often his only salvation, is the refusal on the part of older youths, advisers, and judiciary personnel to type him further by pat diagnoses and social judgments which ignore the special dynamic conditions of adolescence. For if diagnosed and treated correctly, seemingly psychotic and criminal incidents do not in adolescence have the same fatal significance which they have at other ages. Yet many a youth, finding that the authorities expect him to be "a bum" or "a queer," or "off the beam," perversely obliges by becoming just that.

In general it is primarily the inability to settle on an occupational identity which disturbs young people. To keep themselves together they temporarily overidentify, to the point of apparent complete loss of identity, with the heroes of cliques and crowds. On the other hand, they become remarkably clannish, intolerant, and cruel in their exclusion of others who are "different," in skin color or cultural background, in tastes and gifts, and often in entirely petty aspects of dress and gesture arbitrarily selected as *the* signs of an in-grouper or out-grouper. It is important to understand (which does not mean condone or participate in) such intolerance as the necessary *defense against a sense of self-diffusion*, which is unavoidable at a time of life when the body changes its proportions radically, when genital maturity floods body and imagination with all manner of drives, when intimacy with the other sex approaches and is, on occasion, forced on the youngster, and when life lies before one with a variety of conflicting possibilities and choices. Adolescents help one another temporarily through such discomfort by forming cliques and by stereotyping themselves, their ideals, and their enemies.

It is important to understand this because it makes clear the appeal which simple and cruel totalitarian doctrines have on the minds of the youth of such countries and classes as have lost or are losing their group identities (feudal, agrarian, national, and so forth) in these times of world-wide industrialization, emancipation, and wider inter-communication. The dynamic quality of the tempestuous adolescences lived through in patriarchal and agrarian countries (countries which face the most radical changes in political structure and in economy) explains the fact that their youths find convincing and satisfactory identities in the simple totalitarian doctrines of race, class, or nation. Even though we may be forced to win wars against their leaders, we still are faced with the job of winning the peace with these grim youths by convincingly demonstrating to them (by living it) a democratic identity which can be strong and yet tolerant, judicious and still determined.

But it is equally important to understand this in order to treat the intolerances of our adolescents at home with understanding and guidance rather than with verbal stereotypes or prohibitions. It is difficult to be tolerant if deep down you are not quite sure that you are a man (or a woman), that you will ever grow together again and be attractive, that you will be able to master your drives, that you really know who you are,[5] that you know what you want to be, that you know what you look like to others, and that you will know how to make the right decisions without, once for all, committing yourself to the wrong friend, girl, or career.

Religions help the integration of such identity with "confirmations" of a clearly defined way of life. In many countries, nationalism supports a sense of identity. In primitive tribes puberty rites helped to standardize the new identity, often with horrifyingly impressive rituals.

Democracy in a country like America poses special problems in that it insists on self-made identities ready to grasp many chances and ready to adjust to changing necessities of booms and busts, of peace and war, of migration and determined sedentary life. Our democracy, furthermore, must present the adolescent with ideals which can be shared by youths of many backgrounds and which emphasize autonomy in the form of independence and initiative in the form of enterprise. These promises, in turn,

---

[5] On the wall of a cowboys' bar in the wide-open West hangs a saying: "I ain't what I ought to be, I ain't what I'm going to be, but I ain't what I was."

are not easy to fulfill in increasingly complex and centralized systems of economic and political organization, systems which, if geared to war, must automatically neglect the "self-made" identities of millions of individuals and put them where they are most needed. This is hard on young Americans, not because (or perhaps not only because) they are precious and willful, but because their whole upbringing, and therefore the development of a healthy personality, depends on a certain degree of choice, a certain hope for an individual chance, and a certain conviction in freedom of self-determination.

We are speaking here not only of high privileges and lofty ideals but also of psychological necessities. Psychologically speaking, a gradually accruing ego identity is the only safeguard against the *anarchy of drives* as well as the *autocracy of conscience*, that is, the cruel overconscientiousness which is the inner residue in the adult of his past inequality in regard to his parent. Any loss of a sense of identity exposes the individual to his own childhood conflicts—as could be observed, for example, in the neuroses of World War II among men and women who could not stand the general dislocation of their careers or a variety of other special pressures of war. Our adversaries, it seems, understand this. Their psychological warfare consists in the determined continuation of general conditions which permit them to indoctrinate mankind within their orbit with the simple and yet for them undoubtedly effective identities of class warfare and nationalism, while they know that the psychology, as well as the economy, of free enterprise and of self-determination is stretched to the breaking point under the conditions of long-drawn-out cold and lukewarm war. It is clear, therefore, that we must bend every effort to present our young men and women, again faced with dislocation and the horror of war, with the tangible and trustworthy promise of opportunities for a rededication to the life for which the country's history, as well as their own childhood, has prepared them. Among the tasks of national defense, this one must not be forgotten.

In suggesting selected themes for discussion, and as examples for other themes, I have tentatively referred to the relationship of the problem of trust to matters of adult faith; to that of the problem of autonomy to matters of adult independence in work and citizenship. I have pointed to the connection between a sense of initiative and the kind of enterprise sanctioned in the economic system, and between the sense of industry and a culture's technology. In searching for the counterpart and development of identity in social life, one confronts the problem of ideology and aristocracy. Here we use the word in its widest possible sense connoting the conviction that the best people rule and that that rule as defined in one's society develops the best in people. In order not to become cynically or apathetically lost, youths in search of an identity must somewhere be able to convince themselves that those who succeed acquire not only the conviction that they have proven to be better than others, but also the obligation of being the best, that is, of personifying the nation's ideals. In this country, as in any other, we have those successful types who become the cynical representatives of the "inside track," the "bosses" of impersonal machinery, and who thus misguide the adolescent search for an ideal of healthy functioning; we also have examples of humble and human success, ranging all the way from Babe Ruth to the family Eisenhower. In a culture once pervaded with the value of the self-made man, a special danger ensues from the idea of a synthetic personality: as if you are what you can appear to be, or as if you are what you can buy; and here the special influence of the entertainment industry on the nation's youths must be considered. This can be counteracted only by a system of education that transmits values and goals which determinedly aspire beyond mere "functioning" and "making the grade."

VII

THREE STAGES OF ADULTHOOD

I

Intimacy and Distantiation vs. Self-Absorption

Here childhood and youth come to an end; life, so the saying goes, begins: by which we mean work or study for a specified career, sociability with the other sex, and in time, marriage and a family of one's own. But it is only after a reasonable sense of identity has been established that real *intimacy* with the other sex (or, for that matter, with any other person or even with oneself) is possible. Sexual intimacy is only part of what I have in mind, for it is obvious that sexual intimacies do not always wait for the ability to develop a true and mutual psychological intimacy with another person. What I have in mind is that late-adolescent need for a kind of fusion with the essence of other people. The youth who is not sure of his identity shies away from interpersonal intimacy; but the surer he becomes of himself, the more he seeks it in the forms of friendship, combat, leadership, love, and inspiration. There is a kind of adolescent attachment between boy and girl which is often mistaken either for mere sexual attraction or for love. Except where the mores demand heterosexual behavior, such attachment is often devoted to an attempt at arriving at a definition of one's identity by talking things over endlessly, by confessing what one feels like and what the other seems like, and by discussing plans, wishes, and expectations. Where a youth does not accomplish such intimate relation with others—and, I would add, with his own inner resources—in late adolescence or early adulthood, he may either isolate himself and find, at best, highly stereotyped and formal interpersonal relations (formal in the sense of lacking in spontaneity, warmth, and real exchange of fellowship), or he must seek them in repeated attempts and repeated failures. Unfortunately, many young people marry under such circumstances, hoping to find themselves in finding one another; but alas, the early obligation to act in a defined way, as mates and as parents, disturbs them in the completion of this work on themselves. Obviously, a change of mate is rarely the answer, but rather some wisely guided insight into the fact that the condition of a true twoness is that one must first become oneself.

The counterpart of intimacy is *distantiation:* the readiness to repudiate, to isolate, and, if necessary, to destroy those forces and people whose essence seems dangerous to one's own. This more mature and more efficient repudiation (it is utilized and exploited in politics and in war) is an outgrowth of the blinder prejudices which during the struggle for an identity differentiate sharply and cruelly between the familiar and the foreign. At first, intimate, competitive and combative relations are experienced with and against the selfsame people. Gradually, a polarization occurs along the lines of the competitive encounter, the sexual embrace, and various forms of incisive combat.

Freud was once asked what he thought a normal person should be able to do well. The questioner probably expected a complicated, a "deep" answer. But Freud is reported to have said, *"Lieben und arbeiten"* (to love and to work). It pays to ponder on this simple formula; it gets deeper as you think about it. For when Freud said "love," he meant the expansiveness of generosity as well as genital love; when he said love *and* work, he meant a general work productiveness which would not preoccupy the individual to the extent that his right or capacity to be a sexual and a loving being would be lost. Thus we may ponder but we cannot improve on the formula, which includes the doctor's prescription for human dignity—and for democratic living.

Psychiatry, in recent years, has emphasized *genitality* as one of the chief signs of a healthy personality. Genitality is the potential capacity to develop orgastic potency in relation to a loved partner of the opposite sex. Orgastic potency here means not the discharge of sex products in the sense of Kinsey's "outlets" but heterosexual mutuality, with full genital sensitivity and with an over-all discharge of tension from the whole body. This is a rather concrete way of saying something about a process which we really do not understand. But the idea clearly is that the experience of the climactic mutuality of orgasm provides a supreme example of the mutual regulation of complicated patterns and in some way appeases the potential rages caused by the daily evidence of the oppositeness of male and female, of fact and fancy, of love and hate, of work and play. Satisfactory sex relations make sex less obsessive and sadistic control superfluous. But here the prescription of psychiatry faces overwhelming prejudices and limitations in parts of the population whose sense of identity is based on the complete subordination of sexuality and, indeed, sensuality to a life of toil, duty, and worship. Here, too, only gradual frank discussion can clarify the respective dangers of traditional rigidity and abrupt or merely superficial change.

2

Generativity vs. Stagnation

The problem of genitality is intimately related to the seventh criterion of mental health, which concerns parenthood. Sexual mates who find, or are on the way to finding, true genitality in their relations will soon wish (if, indeed, developments wait for the express wish) to combine their personalities and energies in the care of common offspring. The pervasive development underlying this wish I have termed *generativity*, because it concerns the establishment (by way of genitality and genes) of the next generation. No other fashionable term, such as creativity or productivity, seems to me to convey the necessary idea. Generativity is primarily the interest in establishing and guiding the next generation, although there are people who, from misfortune or because of special and genuine gifts in other directions, do not apply this drive to offspring but to other forms of altruistic concern and of creativity, which may absorb their kind of parental responsibility. The principal thing is to realize that this is a stage of the growth of the healthy personality and that where such enrichment fails altogether, regression from generativity to an obsessive need for pseudo-intimacy takes place, often with a pervading sense of stagnation and interpersonal impoverishment. Individuals who do not develop generativity often begin to indulge themselves as if they were their own one and only child. The mere fact of having or even wanting children does not, of course, involve generativity; in fact the majority of young parents seen in child-guidance work suffer, it seems, from the retardation of or inability to develop this stage. The reasons are often to be found in early childhood impressions; in faulty identifications with parents; in excessive self-love based on a too strenuously self-made personality; and finally (and here we return to the beginnings) in the lack of some faith, some "belief in the species," which would make a child appear to be a welcome trust of the community.

3

Integrity vs. Despair and Disgust

Only he who in some way has taken care of things and people and has adapted himself to the triumphs and disappointments adherent to being, by necessity, the originator of others and the generator of things and ideas —only he may gradually grow the fruit of the seven stages. I know no better word for it than integrity. Lacking a clear definition, I shall point to a few attributes of this state of mind. It is the acceptance of one's own and only life cycle and of the people who have become significant to it as something that had to be and that, by necessity, permitted of no substitutions. It thus means a new, a different love of one's parents, free of the wish that they should have been different, and an acceptance of the fact that one's life is one's own responsibility. It is a sense of comradeship with men and women of distant times and of different pursuits, who have created orders and objects and sayings conveying human dignity and love. Although aware of the relativity of all the various life styles which have given meaning to human striving, the possessor of integrity is ready to defend the dignity of his own life style against all physical and economic threats. For he knows that an individual life is the accidental coincidence of but one life cycle with but one segment of history; and that for him all human integrity stands and falls with the one style of integrity of which he partakes.

This, then, is a beginning for a formulation of integrity based on my experience in clinical and anthropological fields: it is here, above all else, where each reader, each discussant, each group, must continue to develop in his or its own terms what I have gropingly begun in mine. But I can add, clinically, that the lack or loss of this accrued ego integration is signified by despair and an often unconscious fear of death: the one and only life cycle is not accepted as the ultimate of life. Despair expresses the feeling that the time is short, too short for the attempt to start another life and to try out alternate roads to integrity. Such a despair is often hidden behind a show of disgust, a misanthropy, or a chronic contemptuous displeasure with particular institutions and particular people—a disgust and a displeasure which (where not allied with constructive ideas and a life of cooperation) only signify the individual's contempt of himself.

In order to approach or experience integrity, the individual must know how to be a follower of image bearers in religion and in politics, in the economic order and in technology, in aristocratic living, and in the arts and sciences. Ego integrity, therefore, implies an emotional integration which permits participation by followership as well as acceptance of the responsibility of leadership: both must be learned and practiced.

VIII

CONCLUSION

At this point, then, I have come close to overstepping the limits (some will say I have long and repeatedly overstepped them) that separate psychology from ethical philosophy. But in suggesting that parents, teachers, and doctors must learn to discuss matters of human relations and of community life if they wish to discuss their children's needs and problems, I am only insisting on a few basic psychological insights, which I shall try to formulate briefly in conclusion.

While we have, in the last few decades, learned more about the development and growth of the individual and about his motivations (especially unconscious motivations) than in the whole of human history before us (excepting, of course, the implicit wisdom expressed in the Bible or Shake-speare), increasing numbers of us come to the conclusion that a child and even a baby—perhaps even the fetus—sensitively reflect the quality of the milieu in which they grow up. Children feel the tensions, insecurities, and rages of their parents even if they do not know their causes or witness their most overt manifestations. Therefore, you cannot fool children. To develop a child with a healthy personality, a parent must be a genuine person in a genuine milieu. This, today, is difficult because rapid changes in the milieu often make it hard to know whether one must be genuine *against* a changing milieu or whether one may hope for a chance to do one's bit in the way of bettering or stabilizing conditions. It is difficult, also, because in a changing world we are trying out—we must try out—new ways. To bring up children in personal and tolerant ways, based on information and education rather than on tradition, is a very new way: it exposes parents to many additional insecurities, which are temporarily increased by psychiatry (and by such products of psychiatric thinking as the present paper). Psychiatric thinking sees the world so full of dangers that it is hard to relax one's caution at every step. I, too, have pointed to more dangers than to constructive avenues of action. Perhaps we can hope that this is only an indication that we are progressing through one stage of learning. When a man learns how to drive, he must become conscious of all the things that *might* happen; and he must learn to hear and see and read all the danger signals on his dashboard and along the road. Yet he may hope that some day, when he has outgrown this stage of learning, he will be able to glide with the greatest of ease through the landscape, enjoying the view with the confident knowledge that he will react to signs of mechanical trouble or road obstruction with automatic and effective speed.

We are now working toward, and fighting for, a world in which the harvest of democracy may be reaped. But if we want to make the world safe for democracy, we must first make democracy safe for the healthy child. In order to ban autocracy, exploitation, and inequality in the world, we must first realize that the first inequality in life is that of child and adult. Human childhood is long, so that parents and schools may have time to accept the child's personality in trust and to help it to be human in the best sense known to us. This long childhood exposes the child to grave anxieties and to a lasting sense of insecurity which, if unduly and senselessly intensified, persists in the adult in the form of vague anxiety—anxiety which, in turn, contributes specifically to the tension of personal, political, and even international life. This long childhood exposes adults to the temptation of thoughtlessly and often cruelly exploiting the child's dependence by making him pay for the psychological debts owed to us by others, by making him the victim of tensions which we will not, or dare not, correct in ourselves or in our surroundings. We have learned not to stunt a child's growing body with child labor; we must now learn not to break his growing spirit by making him the victim of our anxieties.

If we will only learn to let live, the plan for growth is all there.

13

Alfred M. Tozzer

BIOGRAPHY AND BIOLOGY

There is another significant dimension to the interrelationship between the biological and the cultural. Biological events and processes are imbedded in a socio-cultural matrix which may be described scientifically. But interpretations of human life are influenced by extra-scientific factors—for example, the popular intellectual fashions of a period. One of the characteristics of Western culture during the past hundred years has been the role of biological thinking in the prevailing intellectual climate. Darwin, Pasteur, Lister, and other biologists and physicians so caught the popular imagination that there has been a tendency to seek in biology a final explanation for human personality. Professor Tozzer shows how compulsive has been this bent among biographers and autobiographers.

CARLYLE once said: "A well-written life is almost as rare as a well-spent one."

The usual biographer has a fairly constant pattern in setting out on his research. The various qualities of his subject he usually assigns, without question, to that subject's forebears with a definite assumption of a biological connection. That we inherit many or most of our physical characteristics is a scientific fact. That we acquire, in the same way, character-

NOTE: A paper by this title was first published in the *American Anthropologist*, Vol. 35 (1933), pp. 418–32. The present selection, however, represents an hitherto unpublished rewriting (1945), which Dr. Tozzer kindly allows the editors to print with some abridgment. Thanks are also due to the editor of the *American Anthropologist* for permission to republish this paper.

istics of speech, of personality, of potentialities, is less easy to prove. At the moment, I am not so much concerned with the truth or falsity of inheritance of mental endowment and capability as I am in trying to show that the biographer, with little or no hesitation, specifies the ancestral origin of definite qualities, many of which have no remote chance of having been handed down through the genetic process. In this search of life histories, I shall use no order as to time or place, skipping, with no embarrassment, back and forth from Alexander to Mary Baker Eddy. I venture a beginning in the New Testament. A French Abbé once said of Jesus Christ: "God was His Father, but He also came of a very respectable family on His Mother's side."

William Graham Robinson, in his book *Life Was Worth Living*, expresses the idea of a biological model by writing: "In a collection of Memoirs, the author always makes a start by gracefully recalling his ancestry, and I, anxious that all shall be set forth properly and in order, must bow to convention, but I could wish that my ancestors had been more thoughtful in leaving interesting traces of their existence."

In 1827, Scott published a biography of Napoleon, naming himself simply as "the author of Waverley." He mentions that at Napoleon's birthplace in Corsica, "an ominous plaything" of his childhood is still preserved and exhibited. Scott's son-in-law, John Gibson Lockhart, in his life of the Corsican, also speaks of this "small brass cannon" as a "chosen plaything." Scott writes: "We leave it to philosophers to inquire whether the future love of war was suggested by the accidental possession of such a toy; or whether the tendency of the mind dictated the selection of it; or, lastly, whether the nature of the pastime, corresponding with the taste which chose it, may not have had each their action and reaction, and contributed between them to the formation of a character so warlike."

Napoleon's battles are history, but those waged between Environment and Heredity are still with us. Sir Osbert Sitwell brings out the two opposing sides in his charming autobiography with the significant title *Left Hand, Right Hand*, recalling the belief of the palmist that the lines of the left hand are inherited and those of the right hand "are modified by actions and environment."

One side or the other is usually taken in this discussion of nature versus nurture. Little standing has the pacifist who proposes to watch the struggle from a safe distance. I am simply concerned at present with the discussion of the misleading statements, mainly on the side of Heredity, which are so commonly presented in biographical studies. I shall not attempt to prove that the genealogically inclined biographer is wrong. I hope to show, however, that the hunt for famous ancestors has no closed season and that the drive often has to be carried into distant fields.

In eight out of ten of all biographical or autobiographical works, from Plutarch almost to Arthur Schlesinger, Jr., the author, in his first chapter, feels impelled to find something in his subject's parentage on which to hang characteristics and aptitudes. Formerly, these were all favorable to the person whose portrait was being painted. But, for almost twenty-five years, beginning in 1921 with Strachey's *Queen Victoria,* unfavorable characteristics are brought forward, a deplorable trait is now seldom overlooked, and the pathological approach is present in this new approach, the so-called psychological biography.

Unlike Scott with his fairness, most authors find no dilemmas—the cannon was there because one of Napoleon's ancestors was of sanguinary disposition and it had been put there, seemingly, by some sort of mysterious blood-transfusion.

Some of the greatest figures of early history were so far above the common herd that it was not human blood, but celestial that ran in their veins. Plutarch "allowed as certain" that Alexander was a descendant of Hercules, although it took thirty-eight generations to produce Alexander the Great from this ancestral god. One ingenious genealogist has traced the lineage of our first president back to Thorfinn and finally to Odin which, as someone once wrote, "is sufficiently remote, dignified, and lofty to satisfy the most exacting Welshman that ever lived."

When the subject of a memoir has had a distinguished ancestry, the author has an easy task. There is nothing more obvious than to have greatness thrust upon one by his forebears. "Fairly to follow the growth of a mind," writes one author in his own memoir, "one must begin by inspecting the soil wherefrom its roots drew nourishment. From what heritage did I set out?" Less elegantly, perhaps, a biographer of Washington observes: "We have always selected our race horses according to the doctrines of evolution, and we now study the character of a great man by examining first the history of his forefathers." The genealogists were kept busy for years supplying Washington with a proper English pedigree. They did not all agree, but the search went merrily on. In 1932, there appeared three massive and sumptuous volumes, the results of a long search in the colonial archives of the United States, the British Isles, and of Holland.

Most biographers fail to take note or to disagree with Lord Rosebery, who writes in his life of Chatham: "There is one initial part of a biography which is skipped by every judicious reader; that in which the pedigree of the hero is set forth, often with warm fancy, and sometimes at intolerable length."

Consider the vain searches that have been made to provide Shakespeare with some ancestral gleam foreshadowing his genius. One writer makes a determined effort when he says: "Shakespeare's father was at least extraordinarily *fond* of dramatic entertainments, if we may infer anything certain from the brief records of his office as Mayor of Stratford; for he appears to have given the players the kind of welcome that Hamlet admonished Polonius to bestow upon them."

The biographers of Franklin, of Kant, of Keats, of Lincoln, and of numerous others are hard pressed to find some eugenically qualifying ancestors. Franklin is fortunately provided with an uncle who seems to have borrowed books from a country squire. Franklin himself calls him "something of a lawyer." Kant was, after all, the son of a saddler, and Keats the son of the head ostler of the Swan Hoop.

The biographer is not always the one who can be blamed for the attempt to fabricate an eminent genealogical background. We read in a modern biography of Isaac Newton,

> There is nothing distinguished in his ancestry. A line of yeomen and farmers in whom the bad and the good were about equally mixed, held out no promise of eugenically flowering into the finest mathematical genius that the world has ever known. His lack of genealogical distinction irked Newton all his life. As if he were an upstart who had amassed a fortune by lucky speculations, he longed to be counted among the gentry and went about as far as a man of his truth-loving disposition and intellectual honesty could go in claiming a questionable kinship with a branch of the Scottish country nobility.

Samuel Johnson tells us in his *Lives of the Poets*, "Alexander Pope was born in London, May 22, 1688, of parents whose rank or station was never ascertained. We are informed," continues Johnson characteristically, "they were of gentle blood. . . ."

While we find Newton and Pope secretly partial to a patrician lineage, Napoleon, according to Lockhart, scorns any attempt to provide himself with an illustrious family tree. We read:

> In the after-days when Napoleon had climbed to sovereign power, many flatterers were willing to give him a lofty pedigree. To the Emperor of Austria, who would fain have traced his unwelcome son-in-law to some petty princess of Treviso, he replied, "I am the Rodolph of my race." And he silenced, on a similar occasion, a professional genealogist with, "Friend, my patent rates from Monte Notte" (his first battle).

It is unusual to find so honest a statement as appears in a life of a Governor of Massachusetts Bay Colony. The author writes: "When the whole career of Sir Henry Vane shall have been traced, and his principles, sentiments, and views fully presented, the reader will be inclined to regard with astonishment the fact that such a history was commenced, and such a character formed, under circumstances so very unlikely to lead to them." "Astonishment," then, is the word when neither heredity nor environment can be made to play a part.

The quest for the origins of genius in the ancestral strain sometimes reaches extreme lengths. The biographer of Tchaikovsky writes:

> Tracing back his pedigree, we do not find a single name connected with music. There is not one instance of a professional musician and only three can be considered amateurs. . . . All the rest of the family . . . not only lacked musical talent, but were indifferent to the art. Thus, it is impossible to ascertain from whom Peter Tchaikovsky inherited his genius, if, indeed, there can be any question of heredity. His one certain

inheritance seems to have been an abnormally neurotic tendency which probably came to him through his grandfather, Assier, who suffered from epilepsy. If it is true, as a modern scientist asserts, that "genius" is merely an abnormal physical condition, then it is possible that Tchaikovsky may have inherited his musical gı't, at the same time as his "nerves," from the Assier family.

One biographer confesses he is baffled in writing the life of Izaak Walton and states he had "a pedigree such as would puzzle Old Nick."

Sometimes it is not the quality of the ancestral blood, but the color that seems to count. In speaking of the anti-slavery movement in New England, Barrett Wendell, characteristically, finds some excuse for two of the four leaders. Theodore Parker was "born of country folk," and William Lloyd Garrison was "born of the poorer classes at Newburyport," and they "sprang from the lower classes of New England which never intimately understood its social superiors. A self-made man, however admirable, can rarely outgrow all the limitations of his origin." But Barrett goes on, "No such excuses may be pleaded for the two other anti-slavery orators." Wendell Phillips was born "of the oldest New England gentry," and Charles Sumner came from "a respectable family in Boston." It seems that a blue-blooded ancestry is not accountable in an abolitionist, but lack of blue corpuscles may constitute an extenuating circumstance.

When one parent has ties of blood with the gentility and the other with the soil, the successful offspring is easily explained as in Beveridge's *John Marshall*. "When the gentle Randolph-Isham blood mingled with the sturdier currents of the common people, the result was a human product stronger, steadier, and abler than either. So, when Jane Randolph became the wife of Peter Jefferson, a man from the grass roots, the result was Thomas Jefferson. The union of a daughter of Mary Randolph with Thomas Marshall, a man of the soil and forests, produced John Marshall."

In a few minutes spent in turning over the pages of the first volume of the *Dictionary of American Biography*, I found the following under the "life" of the painter, Edward A. Abbey. "To the father, little of the son's ability may be traced. The paternal grandfather was a pioneer in the application of electrotyping printing and in the invention of the internal exploding engine. Here we may suspect the origins of Edward A. Abbey's executive and mechanical capacity."

A modern writer finds it necessary, in treating of great national leaders with no famous heritage, to call his book *Gods and Little Fishes*. Incidentally, one of the "Gods" was Mussolini.

This search for potential chromosomes takes many forms; they may be minutely catalogued through several and sometimes distant generations, from the immediate maternal or paternal ancestor, from the race or races of the parents, from their nationalities, from the father's occupation, from the parents' religious faith, or from some other "carrier."

It is perhaps worthy of note that in this genealogical scrutiny the father's line of descent assumes first importance. A kind of primogeniture of traits is sought. Often the maternal line comes into the discussion only as a court of last resort.

Sir Osbert Sitwell I quote again. "When my father talked of 'the family,' he was indicating, of course, his own and not his mother's or my mother's; for the English tradition regards every child born in wedlock as being solely his father's, descended from his father and father's father." He adds parenthetically, "just as the English tradition considers a bastard as having no father, only a mother, and therefore, no ancestors." He continues, "The mother's family do not enter in, bear no responsibility, and derive no credit. My father, however, frequently noticed and never failed at the time to mention, with distaste, traits, physical and otherwise, occurring in me which he had observed in members of my mother's family, directing attention to them, too, with a sour look, as though I had in some way broken all the rules of the game of heredity and as though there were evidence, also, of original sin."

Another autobiographer writes: "I must really try to put some little interest into my mother's side of the family. I know next to nothing about it, but with a little ingenuity, much may be accomplished."

Predestination comes into every study of Byron. "These two unhappy children," writes one of his biographers, "half-brother and sister, represent the sorrows of their line to some purpose. They overshadowed them, black-winged, and when we understand better the problems of life and death, we shall know what the Greeks spoke of, the inexorable dooming of the Three Sisters. They meant in one word, Heredity." Poe, in biographical literature, shares with Byron the presence of the "black wings" of an evil inheritance.

A difficulty is encountered when there are several children and only one is of outstanding position. A second wife sometimes comes in to help. A biographer of Daniel Webster easily settled the dilemma by writing, "The father, old Ebenezer, has, as we shall see, his points, but it was Abigail Eastman who put into the Webster blood something that made it yield power. The first wife's five children, and the first three of the second wife's offspring (so strong, apparently, was the Websterian influence) amounted to nothing. Then . . . came Daniel Webster. Old Ebenezer finished off his second family with a daughter, who, fashioned after the Webster pattern, was just 'one of the crowd.' " The author admits that if the first wife had not died, there might have been a Daniel, but he "would have lacked half the ingredients that made up the man who looms so large in American history."

Still another obstacle is met when a great man, like the wicked and slothful servant, hides his talent in the earth rather than putting it out at interest in his offspring. A modern biographer of the Hohenzollerns writes: "It is told that when Napoleon met various nephews and successors of the great Frederick on a hunting expedition, and, in particular, had enjoyed a very thorough dose of the company of that monosyllabic simpleton, Frederick William III, he remarked (to Murat), 'How a genius ever found his way into that Hohenzollern family is, and remains, an eternally impenetrable mystery of creation!'" Silence usually covers the mediocre progeny of the great, as in the Psalm: "For when he dieth he shall carry nothing away; his glory shall not descend after him."

Other than the physical characteristics of race, such as skin color, the structure of the hair, and the shape of the head, we know next to nothing scientifically of racial inheritance; and yet it is at the basis of many world-wide prejudices.

We read in one place, "The numerous amorous affairs of the elder Dumas are interpreted as natural outcroppings of his African blood," although, with inexplicable contradiction, the author at the beginning of the book explains the fidelity of General Dumas to his wife as "consistent with the nature of an Uncle Tom who is loyal to his plantation Chloe." The same author, I might point out, has biographies of Balzac and George Sand, but, unfortunately, cannot in these cases use Negro blood to excuse sexual non-conformity.

What the biographer usually refers to as "racial strains" have, in many cases, nothing to do strictly with race. For an obvious example, there is no such thing as a French race. The three main divisions of the so-called Caucasian race are all found within the boundaries of France.

Each nationality, according to the pattern, has its potential virtues and its striking characteristics. These differ from authority to authority and from century to century, but few hesitate to trace them to the racial blood stream. It was the great commander and pioneer in New France, Cadillac, who "had so many Spanish characteristics, the laughing irony, the quenchless ardor, the chivalry . . . that we suspect he must have had Spanish blood of the Pyrenees from his mother's side of the house." This *post hoc propter hoc* reasoning comes in again and again in biography. Our hero seems French in some of his ways. Therefore, embryologically connected with France he must be.

James Parton ventures thus concerning Franklin: "A conjecture has been made respecting the remoter origin of the family which deserves mention only because it derives probability from Benjamin Franklin's peculiar cast of character." He goes on to say that the name *may be* of French origin, hence the "French traits of character" in Franklin are satisfactorily accounted for. He continues: "That sprightliness of mind, that mixture of gayety and prudence, of fancy and good sense, has frequently resulted from the union of the two races. Our Franklin, then, may have inherited with his solid English traits an infusion in vivifying Celtic blood."

A passage in John Morley's *Gladstone* is only one of many that mentions qualities supposedly characteristic of Scotch hormones. Morley writes: "An illustrious opponent once described him by way of hitting his singular duality of disposition, as an ardent Italian in the custody of a Scotsman. It is easy to make too much of race, but when we are puzzled by Mr. Gladstone's seeming contrarieties of temperament, his union of impulse with caution, of passion with circumspection, of pride and fire with self-control . . . we may perhaps find a sort of explanation in thinking of him as a highlander in the custody of a lowlander."

Continuing for a moment on the same so-called "racial" strain, Woodrow Wilson's Scotch ancestry "made him," we read, "magnificently confident in the conclusions to which his reasons led him . . . He, himself, was inclined to attribute the imaginative side of his nature to an Irish element in his origin." Few of Wilson's many biographers fail to mention the Scotch characteristics.

Racial ingredients are often treated as if they were a sort of chemical compound concocted in a test-tube, each race bringing in its own element, and one race, usually the Nordic or, more specifically, the English, forming the quintessence of quality. In a life of Walt Whitman, the author, in speaking of the poet's mother, Louisa Van Velsor, writes: "The Van Velsors were pure Dutch, but 'the Major' (her father) had married a young woman of Welsh descent and Quaker sympathies named Amy (Naomi) Williams. . . . It will thus be seen that Louisa Van Velsor was of mingled Dutch and Welsh blood, with an English strain for tempering." In spite of the English strain as a tempering element, the author adds, "she was almost illiterate," so that, like Carlyle's mother, she probably "learned painfully to write."

An author may go so far as to conjecture the childhood appearance of his hero by allowing his imagination to rest on the nationality of an ancestor. Whitman's mother, if you remember, was partly Dutch so the same biographer writes: "The little Walt must have looked like a sturdy, jolly Dutch baby, with singularly fair skin, hair 'black as tar' . . . and blue eyes that early caught the trick of gazing steadily."

A female autobiographer modestly writes: "From the *German* stock (her mother's grandparents), I probably inherited an industrious nature and love of neatness, law, and order. . . . From the French (her father's grandparents), no doubt I derived a lively imagination, the taste for what is artistic and choice, and the sense of propriety and economy with which my first teacher has always said I was naturally endowed."

These "racial" cytoplasms seem sometimes to turn savagely upon one another. Carleton Beals, in his life of Porfirio Diaz, says the moral collapse of Diaz "had its source in his mestizo origin," in the *internal* conflict between the Spanish and Indian strains.

A final example of racial genes going back three generations is taken from the life of William Dean Howells in the *Dictionary of American Biography*: "His ancestry was mixed, a Welsh ingredient predominating strongly on his father's side, and a Pennsylvania German on his mother's. An English great-grandmother sobered the Welsh ferment; an Irish grandfather (mother's father) aerated the Teutonic phlegm."

It will be remembered that a scandal wove itself around Horace Walpole's ancestry. Being born eleven years after his two brothers and being unlike them in tastes and appearance, a biographer made the suggestion that Horace was not the son of Sir Robert, but "of one of his mother's admirers," Lord Hervey, whom he resembled physically and mentally. Austin Dobson, who denies this vehemently, states that, "No suspicion ever crossed the mind of Horace." One might quote from Macbeth: "There's no art to find the mind's construction in the face."

The biographer is certainly on delicate ground when he attempts to erect a solid structure based on resemblances either physical or mental in which to house proof of extra-marital indiscretions.

Not long ago, a controversy occurred in England between Dean Inge and Lytton Strachey. The Dean cast doubt upon the paternity of Queen Elizabeth. Edward VI and his half-sister, Queen Mary, resemble their father, Henry VIII, but Elizabeth, according to the Dean, has no drop of Tudor blood; she "favors" her mother, thus she is a daughter of Anne, but not of Henry. This is new evidence for Anne's conviction and consequent decapitation. Strachey naturally objects to the Dean's thesis and points out that the question of facial difference was banished by the resemblance in character of Elizabeth and of her father.

The testimony of Green, in his *Short History of the English People*, again refutes the Dean's contention as he painstakingly does what today is called "screening" the various dispositions of Elizabeth, thus assigning some to one parent and some to the other. We read:

> Her moral temper recalled in its strange contrasts the mixed blood within her veins. She was at once the daughter of Henry and of Anne Boleyn. From her father she inherited her frank and hearty address, her love of popularity and or free intercourse with the people, her dauntless courage, and her amazing self-confidence. Her harsh, man-like voice, her impetuous will, her pride, her furious outbursts of anger, came to her with her Tudor blood. . . . But strangely in contrast with the violent outlines of her Tudor temper stood the sensuous, self-indulgent nature she derived from Anne Boleyn.

Religion often seems to show itself as a virulent factor in the hereditary matrix which may run on for generations. Thomas Bowman, in the *Dictionary of American Biography*, acquires a Scotch-Presbyterian gene from his mother and a Methodist gene from his father.

A biographer of Wilson stresses "the hereditary taint" of teachers and ministers in his ancestry which made him want "to manage men as well as to instruct them."

This religious thorn in the flesh may lose its pointedness, as in the case of Theodore Roosevelt. His "grandmother," according to a biographer, "on his father's side, was a Pennsylvania Quaker; but the religious strain in the stock was already thin, and by the time Roosevelt came in for his quota, it had almost lost its potency."

The endocrines of Quakerism seem especially vigorous, as Thomas Mott Osborne had "reformer and rebellion in his blood on his mother's side, for she came of the Quaker family of militant thought that produced Lucretia Mott, stanch abolitionist and leader of the movement for women's rights."

We hear something of the conflagration produced in Thomas Mann's somatic atoms when we learn: "From his Brazilian mother he inherited the malleable steel that, striking against the flint of his paternal Protestantism, gave him the spark of creative ability."

In Mary Baker Eddy we have, according to one writer, "the opposed elements of her inheritance . . . the ultimate sources of the contradictory features in the religion which she founded. . . . The Abigail Ambrose in Mrs. Eddy gave birth to the radiant hopes of Christian Science, and the dark Earth-bound spirit of Mark Baker supplied the doctrine of malicious animal magnetism."

Even occupations seem to carry with them germ plasms which are passed on from father to offspring. Mrs. Ruth Hanna McCormick inherited a "flair for politics." Richard Croker, born in Ireland, was the son of a veterinary; he acquired with his blood and race a passion for horses. President Eliot once said: "My father afterwards became interested in all public plantations, commons, etc. His grandson turned out a landscape architect by a distinct line of inheritance which, unfortunately, skipped me."

The noted Egyptologist, Flinders Petrie, modestly claims little for himself, deriving his entire archaeological equipment from a most assorted collection of ancestral protoplasms from his mother and father, to say nothing of two grandfathers, and a great-grandfather. "Looking back," he writes, "I can now see how much I owe to my forebears; partly carried from my grandfather Petrie's handling of men and material, and his love of drawing; from my great-grandfather Mitton's business ways and banking; from three generations of Flinders surgeons' love of patching up bodies; from my mother's love of history and knowledge of minerals."

The present William H. Vanderbilt, Jr., a former state senator from Rhode Island, inherits "from his father's family an interest in transportation which he has modernized in the development of a motor bus system in Rhode Island and in participating in the directorship of several aviation corporations. From his mother, Mrs. Elsie French Fitzsimmons, the President of the National Republican Woman's Committee of Rhode Island, he seems to receive his political talents as she is a descendant of Amos Tuck of New Hampshire, one of the original organizers of the party in the ante-Civil War period."

Another Republican hormone seems to have been responsible for Associate Justice Edward Terry Sanford's inheriting "from his Yankee father a stout Republican faith."

You will remember that the second horn of Scott's dilemma was the possibility that the famous canon might have been an "accidental possession." Thus, living with this death-dealing weapon as a toy produced

Napoleon's war-like proclivities. Environment is seldom mentioned in our biographer's program. As we have seen, it is utilized only when no vestige of a properly renowned protoplasm can be found.

Trevelyan, in his *John Bright*, is outspoken in his questioning the attribution of qualities to lineage: "There is no exact science of heredity," he writes, "and nothing is more conjectural than the derivation of a great man's qualities of mind and heart. But it is the tradition of the Bright family that John inherited much from his mother." He adds, significantly and truthfully, the other possibility, environment, when he says: "It is certain that he owed her much for the manner of his upbringing."

James Truslow Adams, in his study of *The Adams Family*, raises about the same questions as those previously noted by Scott. "Was it due to some mysterious result from the combination of Adams and Boylston blood far beyond the ken of science even today; or to some unfathomable synchronism between the peculiar qualities of the Adamses and the whole social atmosphere of the next few generations, a subtle inter-play of unknown forces; or to mere chance in a universe in which atoms rush and collide chaotically?"

Let us continue, for a moment, on the Adams family and recognize the subtle interplay of heredity and environment, neither of which can be separated or resolved. Bliss Perry, after hearing parts of this paper as given before the Massachusetts Historical Society, writes in his *Richard Henry Dana*: "Twentieth century biologists have taught us to receive with caution the old doctrine of the inheritance of personal characteristics through the dominant qualities of the male line. Chromosomes now seem strangely indifferent towards the conventions of biography, which have often made fathers more significant than mothers and invented arbitrary patterns of ancestral traits to which descendants have—conveniently for biographers—obligingly conformed. It is no longer possible to write the history of American families during three hundred years in the simple faith that 'Amurath to Amurath succeeds,' and that one Winthrop or Lee or Lincoln must inevitably resemble all other Winthrops, Lees, and Lincolns."

Professor Palmer, writing of his wife, draws largely upon environment to furnish him with data on her character. "All the child's early years were passed in a tract of smiling country, where hills, woods, fertile fields, and the winding stream of the Susquehanna expressed the beauty and friendliness of nature with nothing of its savagery. These gracious influences became a rich endowment. Nature did for her what it did for Wordsworth's Lucy, imparted to her its mystery, its poise, its solitude, its rhythmic change, its freedom from haste and affectation."

In the life of Leonardo da Vinci, the author has this to say: "The melancholy pervading the ancestral marish park, with its copses, its pools, and its sea view, could not but affect his childhood." This might all be true if we knew that there were copses and pools in his garden.

Perhaps the most famous example in literature of the alleged working of environment is in the chapter in Ruskin's *Modern Painters*, entitled, "The Two Boyhoods of Giorgione and of Turner," where he contrasts, on one hand, "the mountains and the sea," the "world of night life," the "marble city" of Giorgione's youth and, on the other, the "close-set block of houses" near Covent Garden where Turner was born. The "deep furrowed cabbage-leaves at the greengrocer's, magnificence of oranges in wheelbarrows round the corner." He continues, "Turner's foregrounds had always a succulent cluster or two of greengrocery at the corners. . . . He not only could endure, but enjoyed and looked for *litter* like Covent Garden wreck after the market. His pictures are often full of it from side to side."

In my casual search in the literature of biography, I have found only two claims of pre-natal influence on the "somatick atoms." A writer on the life of Raymond Foulche-Delboso, a leading Hispanist of the end of the last century, remarks, "The study of Spanish came after he had left school. This language was self-taught. His mother claims a pre-natal influence for his later vocation. She herself was studying Spanish during the time of her seclusion."

Before Frank Lloyd Wright, the architect, was born, he notes in his autobiography that his mother had said he would follow this profession as she was enamoured of buildings and she had taken the pains to hang on the walls of the room, which was to be his, pictures of the great English cathedrals.

The most "scientific" biography ever written is, I suppose, that of Francis Galton by Karl Pearson. Galton was a grandson of Erasmus Darwin, the poet-naturalist, and a half-cousin of Charles Darwin. His most famous work was *Heredity-Genius*, brilliant but questionable in its hypotheses. It was he who gave us the much abused term "eugenics."

With the greatest care, the author analyzes the chief physical characteristics of his fellow scientist, not omitting his "good looks," his ailments, and adds, to round out the picture, his "good digestion." These he sums up. "Thus we realize that, in most of his physical characters, Francis Galton was not a Darwin. . . . It is to the mental characters we must turn for likeness." The quest for Darwinian cytoplasm, not finding satisfactory results in the bodily features, falls back on the far more illusive mental qualities. Under letters running from *a* to *m*, he lists thirteen characteristics, among them being an even temper, great sympathy, ascetic rather than a sensuous nature, strong mechanical bent, power of observation, love of adventure, steadfastness of purpose, and mechanical ingenuity. At the end he adds: "We believe that several of these features are markedly Darwin, but others just as certainly come from different strains." This catalogue is accompanied by elaborate charts showing the genealogy on both sides of Galton's family. The author then proceeds to select a definite ancestor, some going back several generations, for each of the qualities. "Thus, as most men, Francis Galton was physically and mentally a blend of many ancestral traits." "What we do realize is that they (the traits) were not the product of his environment, whether of home, or school or college." Kings College, London, and Trinity College have indeed sins of omission if this statement be true. The home, the school, and the college rather than heredity may have been at work on Francis Galton's brothers, as they were *not* famous. The "kaleidoscope of heredity" is, the author is willing to admit, puzzling.

We have all heard of the remarkable list of the descendants of Jonathan Edwards: "12 college presidents, 265 college graduates, 65 college professors, 60 physicians, 100 clergymen, 75 army officers, 60 prominent authors, 100 lawyers, 30 judges, 80 public officers—state governors, city mayors, and state officials, 3 congressmen, 2 United States senators, and 1 vice-president of the United States."

Jonathan's grandfather, Richard Edwards, had two wives, Elizabeth Tuthill (or Tuttle) and Mary Talcott. In speaking of the line leading from the first wife, one of our authors writes,

> And now let us see how this legacy of God's wealth in blood or germ cells may enrich a creation with infinite usury. . . . All over America I have met her descendants. Many of them wear a gold badge known as the Tuthill emblem. It is a badge of honor. Everywhere they hold places of distinction. . . . "The Blood of Greatness" is the only phrase that would properly describe such a pedigree. . . . Elizabeth Tuthill was a marvellous girl, nearly three hundred years ago at Hartford, Connecticut. She married Richard Edwards, a great lawyer. They had one son and four daughters. They have all left their mark upon American blood. . . . Later in life Richard Edwards married Mary Talcott. She was an ordinary, every-day, commonplace woman. She had ordinary, every-day, commonplace children. The splendid heredity of Richard Edwards was swamped by the mating.

It is, then, this "marvellous girl," Elizabeth, whom we have to thank for all the college professors. The author is ready to admit that the factor of heredity may not have been "completely separated from the factor of environment. There can be little doubt," he adds, "that some members of moderate ability attained distinguished positions through family influence." He rather spoils this probable truth by adding, "Finally, then, we see actually and literally that from dogs to kings, from rats to college presidents, blood always tells."

But we are forgetting "the marvellous girl," Elizabeth. She is well worth our detailed attention. Parkes, in his life of *Jonathan Edwards*, writing of the grandfather, says: "There was, however, a skeleton in the family cupboard. Richard Edwards, at the age of twenty, had been inveigled into marriage by a rich and attractive New Haven girl, Elizabeth Tuttle (or Tuthill), who, within a few months, gave birth to a child whose paternity her husband disowned; aften twenty years of marriage she seems to have become insane, for her husband secured a divorce; one of her sisters committed infanticide; and one of her brothers was hanged for murdering another sister. Timothy Edwards was her second child.

One of Jonathan's sisters was distinctly queer, and two of his nieces became confirmed opium-eaters; his own youngest son, born too late to be educated by him, was a clever and erratic Don Juan; and one of his grandsons was Aaron Burr," whom we have met before. Jonathan's father was expelled from Harvard College, according to the quarter-bill book, and was assessed the heaviest fine recorded in this place. Infanticide, sororicide, sex delinquencies, insanity, the confirmed use of drugs, and Aaron Burr must be laid at Elizabeth's door quite as much as all the United States Senators and the Professors.

Chesterton writes: "Vanity, I find, is undoubtedly hereditary. Here is Mrs. Jones, who was very sensitive about her sonnets being criticized, and I found her little daughter in a new frock looking in the glass. The experiment is conclusive, the demonstration is complete; there, in the first generation, is the artistic temperament—that is vanity; and there in the second generation is dress—and that is vanity. We should answer, 'My friend, all is vanity, vanity, and vexation of spirit.' " He continues: "These are the three first facts of heredity. That it exists; that it is subtle and made of a million elements; that it is simple and cannot be made into those elements. To summarize; you know there is wine in the soup. You do not know how many wines there are in the world. . . . It is obvious that you can no more be certain of a good offspring than you can be certain of a good tune if you play two fine airs at once on the same piano."

From the days of Alexander the Great, who Plutarch "allowed as certain" was a descendant of Hercules, to the most modern of writers, the biographical pattern has remained fixed. The biologist has done little to alter this model, and an anthropologist ventures to step in and suggest a more careful scrutiny of scientific facts in writing of the lives of the great and the near-great.

Part Two

SECTION IV

GROUP-MEMBERSHIP DETERMINANTS

INTRODUCTION

ONE CAN'T get very far in studying organisms without studying their environment—and vice versa. For the organism and the environment constitute a bi-polar field. The individual can be explained only in terms of his relatedness to others in his group. Patterns of interpersonal relations are influenced by the physical and biological environment of the group, by the number of persons in the group and the age and sex distribution of these, and by culture. Of these various dimensions the cultural has been the most intensively studied. This is partly because there are some clear-cut resemblances between the structure of a personality and the structure of a culture.[1]

But the attempt to merge the field of culture and personality in a kind of binocular vision presents many difficulties. One difficulty arises from the student's submission to the cultural compulsives of his own time, from his unconsciously investing the philosophy of liberalism, for example, with an absolute quality.[2]

Some anthropologists have tended too much to view personality simply as the product of specific cultural conditions, particularly the patterns for the training of children. Some psychiatrists have, on the other hand, presented over-simple formulations which derived culture patterns almost wholly from the projection, sublimation, or symbolization of various unconscious dynamisms in individuals. Today there is increasing realization that both of these viewpoints must be modified and merged with other factors in wider abstractions. Child training procedures are not in any simple sense the cause of a given type of character structure. Rather, they are means of providing for the continuity and stability of social life by passing on to the coming generation a generalized feeling tone toward life and specific values that accord with the functional imperatives of that particular social system. The energy for implementing

---

[1] J. W. Woodward, "The Relation of Personality Structure to the Structure of Culture," *American Sociological Review*, Vol. 3 (1938), pp. 637–51.

[2] Gustav Ichheiser, "Misinterpretations of Personality in Everyday Life and the Psychologist's Frame of Reference," *Character and Personality*, Vol. 12 (1943), pp. 145–60.

the special objectives a given group has set for itself can be maintained only if roughly the same view of the world is built into children by comparable methods of child training which varyingly respect the child's natural maturation or interfere with it at particular times and in particular manners.

Complications arise from the existence of sub-cultures and of culture change. The very young child feels the impact of his sub-culture as a unit because only his immediate family act as the psychological agency of the society at this point. Only when the middle-class child goes to a public school and moves out into the wider society generally does the existence of Negro, lower-class, and a Southern regional sub-culture begin to have direct consequences for his personality. Perhaps the only usual exception occurs when the children of an aristocratic sub-group are reared largely by nurses or servants. Here the child presumably gains some sort of awareness of difference between the mores of class sub-cultures, while at the same time he may recognize the uniformity of a regional sub-culture equally shared by parents and servants.

Many studies in the field of culture and personality have been cast in too static a framework. In the contemporary world there tend to be generational sub-cultures in most societies. The presence of different and conflicting systems of conventional understandings makes for confusion and for tremendous problems in the integration of personality. As Gardner Murphy has recently remarked, "In a rapidly changing culture rigid patterns become dynamite instead of mortar."

In view of this dynamic aspect, great caution is needed in evaluating investigations that deal solely with central tendencies. If a public opinion study reveals that seventy per cent of German youngsters have an attitude toward the father's authority that is manifested by only thirty per cent of American youngsters, this is interesting and important. But for purposes of prediction in an unstable culture the remaining thirty per cent in each case is also significant. Also, it must be remarked that expressions of opinion may simply be verbalizations of a dominant ideal pattern of the culture and have a limited value unless checked with behavior.[3] In particular, they need to be checked with the behavior of the specific individual who uttered the opinion. Pictures of a culture obtained by sampling anthropological informants or by public opinion studies may be meaningless since the same individual appears at markedly different places on various curves. So far as prediction is concerned, an opinion is worth something insofar as it is genuinely revealing of the total emotional matrix in which it is rooted. An understanding of the culture-personality field that is significant for action must rest upon a valid description of the dynamic structure of the emotional forces operative in the group.

---

[3] Gwynne Nettler and Elizabeth H. Golding, "The Measurement of Attitudes Toward the Japanese in America," *American Journal of Sociology*, Vol. 52 (1946), pp. 31–9.

In the readings that follow it was not possible to illustrate all types of investigation of group-membership determinants. Additional approaches range from the study of dreams,[4] time perception [5] and other structurings of perception, behavior problems,[6] and case histories of individuals,[7] to general descriptions of culturally determined personality expression [8] and over-all accounts of typical personality in a culture.[9]

Differences of opinion as to whether there is a modal or typical personality structure in a group rest largely on conceptions of personality that are at variance. If personality is taken to include content such as general expectations of other people and of life in general, basic orientations to space and time, then clearly there are central tendencies in the personalities of one society as opposed to another. If personality is restricted to the special organization of emotional traits (timidity, aggressiveness, tempo of action, and the like) in the individual, then the clinician can see only the uniqueness of each person.[10]

---

[4] Katherine Luomala, "Dreams and Dream Interpretation of the Diegueno Indians of Southern California," *Psychoanalytic Quarterly*, Vol. 5 (1936), pp. 195–225.

[5] Edmund Bergler and Geza Roheim, "Psychology of Time Perception," *Psychoanalytic Quarterly*, Vol. 15 (1946), pp. 190–206.

[6] Marion R. Meyers and Hazel M. Cushing, "Types and Incidence of Behavior Problems in Relation to Cultural Background," *American Journal of Orthopsychiatry*, Vol. 6 (1936), pp. 110–17.

[7] A. I. Hallowell, "Shabwan: A Dissocial Indian Girl," *American Journal of Orthopsychiatry*, Vol. 8 (1938), pp. 329–40; and Melvin Tumin, "Some Fragments from the Life History of a Marginal Man," *Character and Personality*, Vol. 8 (1945), pp. 261–96.

[8] John J. Honigmann, "Morale in a Primitive Society," *Character and Personality*, Vol. 12 (1944), pp. 228–36.

[9] Ruth Landis, "The Personality of the Ojibwa," *Character and Personality*, Vol. 6 (1937), pp. 51–60; Jules Henry, "Personality of the Kaingang Indians," *Character and Personality*, Vol. 5 (1936), pp. 113–23; Dinko Tomasic, "Personality Development in the Zadruga Society," *Psychiatry*, Vol. 5 (1942), pp. 229–61; and "Personality Development of the Dinaric Warriors," *Psychiatry*, Vol. 8 (1945), pp. 449–93.

[10] Cf. Melvin Seeman, "An Evaluation of Current Approaches to Personality Differences in Folk and Urban Societies," *Social Forces*, Vol. 25 (1946), pp. 160–65.

14

Geoffrey Gorer

THE CONCEPT OF NATIONAL

CHARACTER

The members of the society, in accord with the cultural patterns that they themselves have internalized, show the new member of the group what the rewards in the game of life are and convince him that he wants these rewards. By early childhood a selective awareness has been, as it were, built into the new members of the group by the agents of the culture, especially in the family. Socialization is necessary to the continuity of group life, for it orders and channels not only what the members of the group do but also their motives for doing so. It is this similarity in motives in terms of group standards which Mr. Gorer stresses.

VIEWED from one aspect, the concept of national character is as old as history, and probably as old as human society. From Herodotus and the early Chinese travellers onwards, every writer who described societies foreign to his own described the typical characteristics of their members, most generally with a series of adjectives or attributes: brave, loyal, treacherous, avaricious, artistic, volatile, and so on. Differences between the typical characteristics of members of different societies were immediately recognized and labelled; as a generalization it might be said that the smaller the observing society, the more attention was paid to the differ-

NOTE: Reprinted from *Science News*, No. 18, pp. 105–22, 1950 (Penguin Books), by permission of the author and Penguin Books (copyright 1950).

enccs among the observed. A society as large as classical China could lump almost all outsiders together as "foreign devils"; societies as small as the city-states of classical Greece were very alert to the different emphases of Sparta, Boeotia, Athens, and so on.

These differences were noted as a fact of immediate apprehension; in those societies which tried to give logical, as opposed to theological, explanations of observable phenomena, these differences were explained in terms of whatever psychology was prevalent as an explanation of individual differences. Thus where the Hippocratic concept of humours was favoured, differences in national characteristics were ascribed to differences in climate; in aristocratic or pseudo-aristocratic societies explanations were found in heredity, often with ultimate recourse to validating myths (e.g. "the descendants of Ham"); environmentalists emphasized the landscape, materialists the technological development, and so on.

Differences in national character were as immediately apprehended as differences in individual character, or very nearly so. But just as the apprehension, the most subtle artistic description, of individual character preceded by centuries, indeed millennia, even the most fumbling attempts to give logical and dynamic explanations for these observed individual differences, so has there been a slightly longer lag between the observation of shared differences and attempts to give logical and dynamic explanations for them. Different people will obviously choose different landmarks which determine the birth of a new science. I would choose 1895, the year of the publication of Freud and Breuer's *Studien über Hysterie* [5] as the year in which the scientific study of individual psychology was born, and 1934, the year of the publication of Ruth Benedict's *Patterns of Culture* [3] as the birth year of the scientific study of national character.

I should perhaps make a couple of remarks about the choice of these books as landmarks. Both of them drew, to a greater or lesser extent, on the work and experiments which had been performed previously, but which the authors reinterpreted; and both authors advanced, in these first essays, hypotheses and guesses which they subsequently discarded. Thus Ruth Benedict's employment of Nietzchean typology was soon found to be quite inadequate; but her main, and at that time novel, emphasis on the psychological coherence of the varied institutions which make up a society laid the groundwork for the reinterpretation of earlier fieldwork, and set new and more rigorous criteria for fieldwork in the future.

The concept of national character is a concept deriving from cultural anthropology. It employs many of the hypotheses developed by Occidental psychologists in their study of members of their own cultures, but the emphasis is very different: it is the study of personality in culture, people in their social setting, not of the individual envisaged either as isolated or in a tacitly known society.

Cultural anthropology is the study of shared habits, of habits which are common either to all the members of a society, or at least to significant or relevant portions thereof. The basic assumption that underlies the description of a culture by the observations of a few months or years is that the relatively few observations (in comparison with the total life of a society) that the field-worker is physically able to make are representative of an infinitely larger series of identical or similar items of behaviour which members of the observed society will continue to perform (*ceteris paribus*) whenever the appropriate situation arises. In other words he is studying shared habits. This underlying assumption is perhaps not very often put into words, but it is absolutely basic to anthropologists' claims to scientific status. If we report patriliny, or totemism, or cross-cousin marriage, or decorated pottery, or brand-tillage, we assume that any or all of these customs were in existence before we made our field-trip and, without outside interference, will continue in existence after our departure. Without this assumption we are making anecdotal observations on unique events, so inadequately sampled that no sort of generalization would be possible.

Now habits have been studied very thoroughly over the last half-century, particularly by Pavlov [16] and the different schools of behaviourists; and the existence of a habit has at least two implications. The existence of a habit implies either instinct or *learning* (or the combination of the two); and the more elaborate the experiments have become, the smaller has grown the role allotted to instincts in the habitual behavior of mammals. It has been demonstrated fairly convincingly that even such typical behaviours as that of a cat playing with a mouse is learned, typically taught by the mother cat to her kittens, and may not be manifested by an isolated kitten reared artificially. As far as human beings are concerned, the only elaborate piece of instinctual behavior which is generally recognized is the sucking reflex of the newborn. Instinctual drives or needs are recognized, but the behaviour which will result in their gratification or alleviation has to be learned. As mothers and nurses know, small babies will learn very quickly whether screaming is an adequate technique for having their needs seen to; and it is quite practicable to extinguish this type of behaviour almost completely in a few weeks if quiet is considered more important than possible injury to the infant.

For habits to exist they have to be learned; for habits to be manifested there has to be some drive or *motive* for the behaviour, stronger than the fatigue involved in the performance. The systematic employment of the concept of drive or motive in the study of habit learning is due to Professor Clark L. Hull [10]; but though it was omitted from Pavlov's theory, it was implicit in his practice. Dogs or rats can be trained to perform very elaborate and complicated actions to get food; but if they are sated just before the experiment takes place they are more likely to go to sleep or comb their whiskers, than to run mazes, operate food boxes, ring bells, sit up and beg, or perform whatever habit they have learned.

If a habit is manifested, there has been learning in the past and there is motive in the present; this is universally true, both for mice and men. But if habits are shared, then there is at least a strong presumption that learning has been shared, and that the same or similar motives are activating the people who manifest these habits. These are presumptions, not axioms: various techniques of learning may produce the same habit (for example a rat may learn to run a maze correctly to get food at the goal, or to avoid an electric shock if it takes the wrong path), and the same habit may gratify different motives for the different performers (in the case of the rats, reduction of hunger in one case, avoidance of pain in the other). Such variations need to be consistently taken into account; but in the case of rats the various devices of learning and motives for performing seem relatively limited; the range is larger for human beings, but we have no reason to suppose it unmanageably large.

Shared habits imply the probability of shared learnings and shared motives. These implications have not always been followed in the anthropological study of shared habits. Until recent years learning has usually only been described (if at all) in the specialized and limited field of technical apprenticeship. There have been theories of motivation in abundance; but, with the partial exception of materialist functionalism (to which I shall revert), these motives have been derived from theoretical considerations and applied to all the cultural behaviour considered significant by the anthropologist. They have not been derived from the study of the end-results of the habits described.

The concept of national character, or basic character structure—both rather unfortunate terms which have grown up by accident—is an attempt to isolate and describe the motives shared by the members of a society who manifest the same shared habits or culture. Motives are never under any normal circumstances directly observable; they have to be deduced from the study and analysis of the actual results of given sequences of behaviour. To increase the probability that motives are accurately inferred, the types of learning and teaching employed in the society being investigated are studied as minutely and fully as possible. To the extent that congruence is found between the attitudes derived from the various learning situations, and the goals sought by the shared habits of the adults, to that extent it seems probable that the motives have been accurately described.

Insofar as he is a mammal, man shares the same basic physiological motives with all other mammals. These include the tensions produced by hunger, thirst, sexual desire, bladder and colon pressure; the avoidance of asphyxiation, excessive heat or cold, fatigue or other pain; and fear and anger in response to unrelieved tensions or situations threatening pain or displeasure. As a mammal, man cannot survive unless these needs are gratified; and every society known or imagined must develop institutions to take care of these needs for all its members.

The insistence on the physiological needs of human beings was Malinowski's [12] great contribution to social anthropology; the emphasis on the fact that even primitive people had to eat, and did not spend their lives demonstrating the solidarity of the tribe by ritual, or indulging in pre-logical thinking, or pursuing marriage arrangements which would render it almost impossible to bear or rear children, was a very considerable gain in realism.

But at first the baby was emptied out with the bath water. Instead of pre-logical thinking being imputed to primitives, completely logical thinking was ascribed to them; instead of all their behaviour being irrational, all their behaviour was relentlessly rational; they engaged in no actions which did not (taking into account their technological and scientific stage of development) lead more or less directly to concrete physiological satisfactions.

This construct had two major disadvantages. No human beings as completely rational as these primitives could be found anywhere outside the monographs of anthropologists; and it was impossible, by these hypotheses, to account for the very marked variations in practice and attitudes which characterized tribes in very similar stages of technological development.

All his life Malinowski was deeply hostile to psychoanalysis and, despite his contribution to the discussion concerning the universality of the Oedipus complex [11], he remained basically ignorant both of its evidence and its deductions. When further analysis showed that the completely rational primitive was an untenable figment, he had no construct by which to account for irrational behaviour except the undefined catch-all "culture." Thus, in the paper "Man's Culture and Man's Behaviour" [13] written in almost the last year of his life, he says:

Satisfaction . . . appears invariably as a cultural appetite rather than as the satisfaction of pure physiological drive. Breathing, as carried on by certain European communities within the non-ventilated and heavily modified atmospheres of enclosed rooms, would not satisfy an Englishman accustomed to a superabundance of fresh air. The satisfaction of appetite by food discovered to be unclean ritually, magically, or in terms of what is repugnant to a culture, does not lead to a normal state of satiety, but to a violent reaction, including sickness. The satisfaction of the sex impulses in an illicit or socially dangerous manner produces detumescence, but also conflicts which may lead in the long run to functional disease.

Thus culture determines the situation, the place and the time for physiological needs. It delimits it by general conditions as to what is licit or illicit, attractive or repulsive, decent or opprobrious. Although the act itself, as defined in terms of anatomy, physiology and interaction with the environment is constant, its prerequisites, as well as its consequences, change profoundly.

No one, I think, would question the facts to which Malinowski is alluding, but his explanations seem to me, at least, woefully inadequate. What is this reified "culture" who finds certain foods repugnant, decides that acts are licit or illicit, decent or opprobrious? Or even if this red herring be left aside, is "culture" a sufficient or adequate explanation for the "violent reactions" or "functional diseases" produced by physiologically neutral acts?

I think that the behaviour which Malinowski is delineating in this passage can be described as patterned or shared emotional responses to certain acts or situations. The vomiting, the functional disease, and so on, cannot (at least in our present stage of knowledge) be accounted for by the cultural modifications of physiological drives; they are emotional responses, and responses of a very violent nature. But these responses are not ascribed to specific individuals with idiosyncratic histories or disbalances; they are likely to be manifested by any member of the society under certain conditions. And these responses are not consciously provoked nor under voluntary control; they happen "automatically" in certain situations which vary more or less arbitrarily from culture to culture. Disgust and revulsion are learned habits, culturally modified emotions.

It seems neither consistent nor logical to assume that only negative reactions demonstrate the cultural modification of the emotions. The emotions felt when the cultural rules are followed and gratifications achieved in a legitimate manner are surely as much examples of the cultural modification of the emotions as the disgust or functional disease which follows their violations.

Malinowski and his followers have demonstrated in very great detail the fact that physiological urges are modified and transformed into cultural appetites. It is my basic argument that in an analogous fashion the expression and organization of the emotions common to members of a society (or a portion thereof) are also modified and transformed. The expression and organization of love and hate, anxiety and security, guilt and shame, self esteem and self-depreciation, and the like are modified and transformed in a fashion analogous to the physiological urges, but probably to a greater degree. The minimum and maximum amounts of food, for example, that an individual can survive with or dispose of, and also its composition, are physiologically regulated within fairly narrow limits; according to our present knowledge, the limits seems to be considerably wider concerning the maximum and minimum of hate or fear or love which human beings can tolerate, or without which they cannot survive. The observations on the basis of which the presence and extent of, say, love or hate are inferred, are likely to be more complicated than the observations on the basis of which, say, hunger or avarice are inferred, because a greater number of variables are involved; but this seems to me to be a difference in degree, not a difference in kind.

The material gathered by an anthropologist studying national character differs from that gathered by an anthropologist without this interest only in its greater completeness. As least that is the ideal. There is no aspect of social life which does not potentially provide data on the shared motives of the members of the society. Every activity has its emotional as well as its functional components. Differences of tension and relaxation, of friendliness or quarrelling, of formality or carelessness are all revealing. The jokes and insults, the proverbs, the causes for quarrels and the methods of reconciliation, the most often repeated myths and stories, the varying or rigid composition of work or play groups, the metaphors for technical processes, even habitual physical postures, can all provide clues. As these clues—what I have called elsewhere [9] the psychological surplus value to every activity—accumulate, they will tend to fall into certain patterns. These patterns can be tentatively identified, and then further evidence can be sought to confirm or disprove the prevalence of shared motives so identified; earlier notes can be inspected for material which will provide confirmation or disproof; exceptions must be noted, and explanations sought for them; and gradually, by a systematic process of trial and error, a formulation of the main operative motives shared by a majority of the members of the society being investigated will become possible.

The test of hypotheses about national character is the same as the test for any other scientific hypothesis; it is accurate prediction. If the hypotheses about the main shared motives are correct, it should be possible to predict how a group—never, it must be emphasized, specific individuals—will respond to a new situation. The one major advantage of developing hypotheses about the main shared motives of large and complex societies, which somewhat compensates for the enormous technical disadvantages, is that news reporting does give the possibility of testing hypotheses against behaviour in a systematic way which is impossible for more primitive societies.

Hypotheses such as I have outlined can be deduced entirely from the behaviour of adults. It will, I think, always be found that the number of relevant hypotheses is relatively small. This I would suggest is due to the principle of avoiding fatigue. For the purposes of analysis we may describe social activity in terms of a number of discrete institutions; but with rare exceptions people are constantly passing from one institution to another—often merely by passing from one room to another—and it would add considerably to the strain of daily living if each institution demanded a new type of predisposition and a new set of habits and expectations. This is not to suggest that there is no change in behaviour or attitudes as roles or institutions are changed; but these changes are likely to be coherently related to each other.

Field trips have only a short duration, as compared with the life of an individual, not to mention the life of society. If only the current motives of the adult are studied it is at least theoretically possible that there has been some major distortion operative during the whole period of observation—a threat of war for example.

Living societies do not, however, consist exclusively of adults. Even in the smallest communities there are a few representatives of each stage of development from the new-born infant, through children and adolescents, to the old, the senile and the dying. Now it is a basic assumption of all social science, though I do not think it has ever been stated, that societies will maintain their culture through an indefinitely long period of time (at least longer than a single life-time) unless they are disrupted by war or famine, by the introduction of more advanced technological processes, by epidemic disease, or by some other drastic interference which can be roughly described as external. This assumption implies that the present generation of adults will be replaced in due course by the present generation of children who, as adults, will have habits very similar to those of their parents; and it also renders it highly probable that the childhood learning of the contemporary adults was at least very similar to the learning which contemporary children are undergoing. So in a society which is not in the process of drastic change, the observable adults foreshadow the future development of observable children, observable children recapitulate the past experience of observable adults.

Newborn children differ from one another in inherent constitutions and almost certainly in temperament. It is possible that they also differ in intellectual and emotional potentialities; but despite a very great deal of work which has been done on this subject, this assumption is still unproved. So although each child (with the exception of identical twins) is born with its unique genetic constitutions, it seems reasonable to assume that in any large group the different potentialities cancel one another out, as it were, and that newborn infants represent a random sample of constitutions and temperaments. When human genetics have developed further, it may be necessary to modify or abandon this hypothesis; but in the meantime it seems more scientific to ignore genetic determinants when considering the development of a group of infants.

Habits are learned. The existence of a habit indicates prior learning. Therefore the existence of shared habits among the adults of a society postulates the probability of shared or similar learning some time earlier in their lives. But the fact that societies continue, though their personnel changes, implies that the learning of today's children is likely to be very similar to the earlier learning of today's adults; and consequently the study of the learning today's children are undergoing can be expected to throw light on the formative experiences of today's adults, and today's adults to foreshadow the final development of today's children. Consequently the study of child-training can give a dynamic and, as it were, three-dimensional picture of the shared wishes of adults.

The chief matters to be observed in investigating child-training are when, how, for what and by whom a child is rewarded or punished. Rewards and punishments may be concrete (food or slaps) or symbolical (praise or blame, the giving or withdrawing of love or attention, promises or threats). The inhibiting or postponing of an immediately desired physical gratification—food, excretion, movement—can be considered as punishment.

Whenever such observations have been made in a society which is not disrupted or in a state of violent transformation it has been found that the customs of child-training are strictly and remarkably patterned, and that the range of variation in the meting out of rewards and punishments to children by adults belonging to the same society (or, in larger societies, to the same sub-group) is relatively very small, particularly as compared with the very great variations between one society and another. This is a matter of empirical observation, and not of theory; as far as I know, it was not predicted by anyone before repeated observations had been made.

Although each individual experience is unique in some respects, the great majority of the members of any stable society have undergone a great deal of common learning; and it seems logical to assume that this common learning is antecedent to the common habits and common emotions manifested by the adults.

In a way this is a hen-egg situation. Owing to their prior experience the adults, the parents, have certain predispositions, ideas, and wishes which will tend to make them select specific types of behaviour to reward or punish in their children; and these rewards and punishments will lay the foundation for adult characters with predispositions, ideas, and wishes similar to those of the parents. Society is continuum, and one can start one's investigations anywhere.

The interpretation of the observations made on child-rearing is (like the interpretation of any other anthropological observations) a question of the theories which appear most probable to the interpreter. Ideally, in every aspect of anthropology, sufficient and sufficiently precise data should be presented so that alternative theoretical interpretations could be made. This ideal has been more or less realized in most of the studies of primitive national character with which I am acquainted.

Speaking for myself, the chief interpretive tools that I employ are the theory of learning developed by Clark L. Hull and his associates and followers, combined with the theory of individual psychological development developed by Sigmund Freud and his associates and followers. Hull's theory was developed from the study of experimental animals (dogs and rats) and is applicable to nearly all mammals; except for one technical point—it seems to me more useful to assume that generalization is primary and discrimination learned rather than the reverse—his postulates do not appear to need modification, though of course they cannot be applied with very great rigour, owing to the absence of laboratory conditions, and the far greater number of variables. Freud's constructs, on the other hand, are based on culture-limited observations, and the application of his theories to non-Occidental cultures has demanded considerable modifications; but the concept of the dynamic unconscious, the psychological derivatives of the processes of maturation, the various hypotheses concerning symbolism and symbolical behaviour, the theory of dreams, and the hypothesis that the nuclear family situation is the prototype of social behaviour all seem to have cross-cultural validity. Many other aspects of Freudian theory are being modified by data from non-Occidental societies.

The observation of childhood-training has a few other advantages besides those I have mentioned. Compared with adult life, the possible variables in the treatment of children up to about the age of six, say, are relatively limited, owing to the children's biological weakness and needs, and significant variations can be more readily isolated. Secondly, the techniques of child observation vary very little from culture to culture, and this is one of the comparatively rare situations where the observations of anthropologists on primitive cultures and the observations of specialists in our own society are strictly comparable [14]. One needs a relatively very sophisticated economist to discuss constructively data on primitive economics; but nearly all specialists in child psychology and development can discuss data on primitive child-rearing; and consequently the two branches of the discipline can be cross-fertilized with relative ease and mutual benefit. These are, I think the chief reasons why students of national character have tended to concentrate on the observation of children, though theoretically this is not necessary.

This discussion so far has been concerned with the analysis of national character based on prolonged observation and interviewing in the traditional fashion of social anthropology, which is to my mind the only really satisfactory technique. However, under the exigencies of war, I, together with some colleagues, attempted to make tentative analyses of the national character of a people while we were outside their territory [6, 7, 8]; since some of these analyses have had fairly wise diffusion, it would appear worth while discussing shortly the principles and techniques involved in such work.

The basic hypotheses are exactly the same as those employed in the analysis of primitive national character (except that one has to expect regional or class difference) but the data on the basis of which the deductions are made are of a somewhat different, and on the whole of a much less satisfactory nature. The interview is still the main source of data, but the material derived from interviews cannot on the whole be checked by observations, except of a very limited nature—the behaviour of the informants towards the interviewer and towards one another. Also the informants do not on the whole form an inter-acting community; and a great many of their activities are different because they are living outside their native country.

These drawbacks are so great that I think it would be unjustifiable to use this type of analysis in any situation in which it would be possible to use direct observation of living communities; and I myself have only employed such indirect methods for the analysis of the national character of peoples whose territory could not be visited, either because of the war or because of their internal policies. Working under such disadvantages the results can only be very approximate.

There are however some technical compensations for working on literate and technically advanced communities. In nearly all cases there is a considerable amount of sociological, historical, and theological work already performed by competent experts; and this material can be drawn on for the specific purposes of the investigation. Secondly, there is a great deal of symbolical material available for analysis: books and pictures, cartoons and jokes, works of philosophy and speculation, plays and films. Films are perhaps the most useful of all. Nearly all films are works of collaboration involving a considerable number of people, so that the unique personality of the artist is less likely to be operative than in other works of art or fiction. Secondly, the popular film must correspond to some predispositions of the audience. Thirdly, whatever the fiction, the background and the minor characters are likely to be fairly representative. And finally the film can be played through over and over again, which makes possible the most minute study of every detail, which is of course not the case in any other art form involving living people. There are two first-class studies of individual films—Gregory Bateson on *Hitlerjunge Quex* [2] and Erik Homburger Erikson on *The Childhood of Maxim Gorki* [4]; and Martha Wolfenstein and Nathan Leites's *Movies: A Psychological Study* [11] analyzes the plots of a considerable number of American, French, and English films to bring out the regularities and point the contrasts in the preferred fantasies of these three nationalities.

Finally, in the analysis of complex contemporary cultures there is the inestimable advantage that there are expert informants, individuals whose knowledge of the society under investigation, or some aspect of it, is both great and sophisticated. My own reliance on the expert informant is extremely great. These experts may or may not be members of the society being investigated.

For one reason or another, it is relatively seldom that I can quote these expert informants verbatim. Some have to refuse permission because of their official position, or because they want to hide their associations with foreigners, and I naturally respect such requests. Sometimes the names would be meaningless. For example, one of the most valued informants on the childhood of Great Russian peasants was a midwife, who had been working continuously in the Russian Countryside from 1911 until her capture by the Germans. It would be difficult to imagine a better authority on the matters within her competence; had she written and published what she said, I could have cited her as an unquestioned authority. But the spoken word is so deprecated today that using her material without the mediation of print is stigmatized as "hearsay."

This reliance on the spoken word seems to disconcert historians particularly. The use of a dozen documents is "scientific history"; the use of four hundred interviews "brilliant guesswork," even though the documents may be as private and informal as letters or diaries. This superstitious reverence for the written or printed word would not seem to be susceptible to argument.

A more reasonable ground for criticism is the relative absence of quantification, of numbers or percentages in the studies made so far. As far as this relates to unitary and observable practices, the criticism is justified. I should much prefer to be able to write that in a Great Russian community of X thousand, "A" per cent of the infants were swaddled tightly, "B" per cent were in "envelops" and "C" per cent had their limbs free, rather than to state that the traditional Great Russian custom, which is reported to be still widely practiced, is tight swaddling; that the "envelops" were an upper-class innovation introduced about 1900, probably from Poland and the

Baltic countries; and that the Soviets made a great deal of propaganda for Western-type infantile care, and that the available photographs suggest that in the State crêches of which the government was very proud, infants were not swaddled and that it seems probable that the present elite employed these "modern" methods in rearing their own children. To be able to make quantified statements would be a distinct scientific advance and would give greater validity to the hypotheses based on these observations. We can hope that quantification of this sort will be achieved one day, though the technical problems of sampling large populations for such variables have not yet been solved. No critic could be more conscious than I am myself of the ultimate desirability of such numerical statements.

The total concept of national character, however, is no more susceptible to quantification than is the total concept of a kinship system. The data from which a kinship system is abstracted can be quantified—one can say in a given community how many people have fathers with married elder sisters, and note with what terms each of them addresses the paternal aunt and her husband—but the diagrammed and classified kinship system, which, incidentally, is almost certainly unknown to all the people who employ it, is not a numerical concept, and the attempt to introduce quantifications would be what I consider pseudo-science. Kinship systems are based on the unique or idiosyncratic arrangement of a small number of basic relationships—G. P. Murdock [15] employs eight—and it is the structure and organization, not the components, which the kinship system describes. The analogy with the concept of national character is very close.

The study of national character resembles, or tends to resemble, those branches of scientific knowledge which have to do with the functioning of living organisms, such as the various aspects of physiology or ecology. Because of the striking advances of mechanics in the last century, there is a wide-spread belief—perhaps prejudice would be the more exact word—that all science should approximate to chemistry or engineering, where the results can be stated in numerical formulae. But there is a science of structure as well as a science of components; from astronomy to the structure of the atom, from physiology to psychoanalysis, scientific statements are statements of inter-relationship, not statements, at least in the first place, of quantity.

The concept of national character refers to the structuring and combination of traits or motives. As far as our present knowledge goes, there is no trait or motive which is uniquely confined to a single society; but still according to our present knowledge, the structure, the combination of motives, the national character of a society, is always unique, though in societies from the same culture area the differences may be comparatively slight. Because the concept of national character refers to structure or pattern, the employment of psychological tests as a tool of investigation has been relatively unrewarding, for most such tests are designed to demonstrate the presence or absence of one or a small group of traits, in nearly all cases among the members of a society presumed to be known. The chief exceptions to this generalization have been called the so-called "projective tests"; to the extent that the testing material is not culture-limited (as are for example the cards in Murray's T.A.T. or the Szondi test), significant results can, it would seem, be obtained by administering and interpreting these tests, though I doubt whether more information is got by these means than would be got through skilled interviewing in the same time. A considerable number of workers have applied the Rorschach test to members of primitive or non-Occidental societies [1]. According to my reading of the evidence, very little material on national character can be deduced from the scored profiles, which emphasize individual differences rather than shared traits; but if the actual protocols are used as though they were material from dreams or mythology, significant and suggestive regularities can be deduced from a series of tests.

The attempt to describe the national character of a people when away from their territory is, I admit, something of the nature of a *tour de force*, and the results are never more than probabilities, hypotheses to be tested against further material and fuller investigation. I, at least, do not expect these first formulations to be complete, accurate or final descriptions of the national characters of the people studied. Science always advances by a series of approximations.

I do, however, believe that national character is capable of being studied and analyzed, and that, to the extent that it is understood, it will be of some utility in avoiding those unnecessary misunderstandings and misinterpretations which may lead to war. Were the political situation not so desperate, it would probably be more desirable to wait another fifty years, while the techniques of anthropology, sociology, and psychology were further refined and elaborated, before attempting so complex and difficult a task with such relatively inadequate tools. But we do not have the choice; we are not even certain that we have another fifty years. It is a choice between doing what we can with the tools available, or doing nothing. I think the attempt is worth the inherent risks.

REFERENCES

1. Abel, T. M. and Hsu, F. L. K.: 1949. "Some Aspects of Personality of Chinese as Revealed by the Rorschach Test," *Journal of Projective Techniques*, Vol. XIII (1949), pp. 285–301.
2. Bateson, G.: "Cultural and Thematic Analysis of Fictional Films," *Annals of the New York Academy of Sciences*, Series II, Vol. IV (1943).
   Bateson, G.: An analysis of the Nazi film "Hitlerjunge Quex" (Mimeographed, 1944).
3. Benedict, R.: *Patterns of Culture* (Boston, Houghton Mifflin, 1934).
4. Erikson, E. H.: *Childhood and Society* (New York, Norton, 1950).

5. Freud, S. and Breuer, J.: *Studien über Hysterie* (Vienna, 1894).
6. Gorer, G.: "Themes in Japanese Culture," *Transactions of the New York Academy of Sciences*, Series II, Vol. 5 (1943), 106–124.
7. Gorer, G.: Burmese Personality (Mimeographed, 1943).
8. Gorer, G.: The Greek Community and the Greek Child (Mimeographed, 1943).
9. Gorer, G. and Rickman, J.: *People of Great Russia* (London, Cresset Press, 1949).
10. Hull, C. L.: *Principles of Behaviour* (New York, D. Appleton-Century, 1943).
11. Malinowski, B.: "Mutterrechtliche Famile und Oedipus-Komplex," *Imago*, Vol. X (1924), pp. 228–76.
12. Malinowski, B.: articles in the *Encyclopaedia of Social Sciences*.
13. Malinowski, B.: "Man's Culture and Man's Behaviour," *Sigma Xi Quarterly*, Vol. 29, Nos. 3, 4 (1941).
14. Mead, M.: "Social Anthropology and its Relation to Psychiatry," in F. Alexander and H. H. Ross (eds.): *Dynamic Psychiatry* (Chicago, University of Chicago Press, 1950).
15. Murdock, G. P.: *Social Structure* (New York, Macmillan, 1949).
16. Pavlov, I. P.: *Conditioned Reflexes*, trans. by G. V. Anrep (London, Oxford University Press, 1927).
17. Wolfenstein, M. and Leites, N.: *Movies: A Psychological Study* (Glencoe, Ill., Free Press, 1950).

CHAPTER

15

A. Irving Hallowell

AGGRESSION IN SAULTEAUX SOCIETY

With Dr. Hallowell's paper we arrive at a still more concrete case study on a yet smaller canvas. The materials come from a single Indian tribe, and only a single generalized psychological manifestation is treated. Dr. Hallowell demonstrates how the forming personality learns to control and express aggression by conformity to group habits. The culture channels manifestations of personality into socially acceptable forms. The individual in Saulteaux society is taught to anticipate in fantasy retaliation for his own actual and fantasied hostile acts towards others. Beliefs with regard to the supernatural, and especially sorcery, tend to keep overt aggression in face-to-face relations at a minimum but provide means for projecting, displacing, and rationalizing aggression.

THIS paper might have been given the paradoxical title "aggression among a patient, placid, peace-loving, non-literate people." For among the aborigines of North America, the Algonkians of the northern woodlands, in contrast with Indians like the Iroquois or the Sioux, have long enjoyed the reputation of being as mild-mannered and overtly *unaggressive* a people as could be found anywhere among the native tribes of this continent. Indeed, the statement made by Speck that among the Montagnais-Naskapi of the Labrador peninsula, "strife is scarcely present; violence strenu-

NOTE: Reprinted from *Psychiatry*, Vol. 3 (1940), pp. 395–407, with abridgment of footnotes, by permission of the author and the William Alanson White Psychiatric Foundation. (Copyright, 1940, by the William Alanson White Psychiatric Foundation, Inc.)

ously avoided; competition even courteously distained" [1] is just as outwardly true of that other branch of the Algonkians, the Saulteaux, whom
I shall discuss.

Yet Father Le Jeune, a Jesuit missionary who lived among the Montagnais-Naskapi in the early part of the seventeenth century qualifies this
idyllic picture of outward harmony by an astute observation. "It is
strange," he wrote, "to see how these people agree so well outwardly,
and how they hate each other within. They do not often get angry and
fight with one another, but in the depths of their hearts they intend a
great deal of harm. I do not understand how this can be consistent with
the kindness and assistance that they offer one another" [2].

What interested me when I stumbled across this characterization not
long ago was the fact that it could be applied with equal validity to the
Saulteaux. In my opinion, it goes to the very heart of the characterological
problem presented by them. What I propose to do in this paper is to offer
an explanation of what to Le Jeune was a psychological paradox, by examining the conditions under which ambivalent traits similar to those
noted by him are molded. While my explanation applies primarily to the
Berens River Saulteaux, there are so many analogies to my observations in
Le Jeune's account of seventeenth century Montagnais society that I
imagine a similar explanation applies to them also and perhaps even to
other Algonkian peoples.

The Berens River Saulteaux are an offshoot of the Ojibwa-speaking
peoples whose center of dispersion was in the neighborhood of Lake Superior. The Berens River empties into the eastern side of Lake Winnipeg
and the native population today, still hunters and fishermen by occupation, comprises between nine hundred and one thousand individuals. Officially, there are three bands of these Indians, one located at the mouth of
the river, another centered at Lake Pekangikum about two hundred and
fifty miles inland, and a third between these two. The only white inhabitants are traders, trappers, a handful of missionaries, and, in summer,
prospectors. Some Indians of the inland bands still remain unchristianized,
and acculturation processes have affected the three bands to a varying
degree, many aspects of aboriginal life still flourishing in considerable
strength.

To the casual observer, co-operation, laughter, harmony, patience, and
self control appear to be the key notes of Saulteaux interpersonal relations. These people have never engaged in war with the whites nor with
other Indian tribes. There are no official records of murder; suicide is
unknown; and theft is extremely rare. Open expressions of anger or quarrels ending in physical assault seldom occur. A spirit of mutual helpfulness
is manifest in the sharing of economic goods, and there is every evidence
of co-operation in all sorts of economically productive tasks. No one, in
fact, is much better off than his neighbor. Dependence upon hunting, fish-
ing, and trapping for a living is precarious at best, and it is impossible to
accumulate food for the inevitable rainy day. If I have more than I need,

I share it with you today because I know that you, in turn, will share your surplus with me tomorrow.

This is the superficial picture that the Saulteaux present to the observer, and it could be reinforced by a detailed description of the functioning of their economic and social institutions. But it is not a complete or fully realistic portrait. It does not expose the deeper psychological realities of Saulteaux life. While at first glance there seems to be no manifest evidence of aggression and while I myself did not at first sense the undercurrents of hostility that actually exist, on more intimate acquaintance I was forced to correct my earlier impressions. Furthermore, I came to the conclusion that in all probability the undercurrents of aggression that I shall describe were even stronger in the past when native beliefs flourished in full strength and acculturative influences had not set in.

I shall attempt to show how, under the cultural conditions imposed by Saulteaux society, aggressive impulses are both stimulated and channelled; what the range of behavior is that has aggressive meaning for them and the behavior considered appropriate to combat it; and what are the effective reinforcements of these patterns of behavior from other sources. I believe that the cultural moulding of such impulses provides a definite clue to the psychological source of certain personality traits which are typical of these Indians. These traits are, in part, a function of dealing with people in terms of certain culturally imposed modes of behavior. Saulteaux patterns of social behavior create a fundamental ambivalence in the interpersonal relations of individuals. My hypothesis is that the outwardly mild and placid traits of character which they exhibit and the patience and self restraint they exercise in overt personal relations are a socially constituted façade that often masks the hostile feelings of the individual. If some of the basic beliefs and concepts of these people were changed, their aggressive impulses would be reconstellated and the personality traits referred to would no longer assume their characteristic forms.

This absence of overt aggression in face-to-face situations is an outstanding feature of interpersonal relations in Saulteaux society. While in other non-literate societies, unformalized or even formalized modes of open aggression frequently are found, among the Saulteaux behavior of this kind is not sanctioned, except in the mildest form. The joking relationship between cross-cousins of opposite sex and the exchange of belittling remarks between men of different sibs are the only formalized patterns of such behavior that I can think of. Aside from these, hostility in face-to-face situations seldom occurs. Consequently, when expressions of hostility do break through in the form of verbal threats, gestures, or physical assault, they are taken more seriously than is the case in societies where such modes of behavior are commoner because they are socially sanctioned. I have never observed boys or girls in an exchange of blows. At the mouth of the river I know of a case where a man gave his brother a severe beating because he believed that the latter had attempted to seduce his wife. One couple is said to quarrel frequently and even to beat each other occasionally. Parents, I may add, seldom inflict corporal punishment upon their children.

Insulting remarks, direct rebuke or expressions of disapproval, open contradiction, sneering or calling a person names are avoided by the Saulteaux. Swearing, in the sense of blasphemy, is not a problem because their language is inadequate in this respect. Gestures that suggest an approach to physical assault like shaking one's fist in a person's face, are a deadly insult. It immediately suggests extreme hostility. Even pointing one's finger at a person, accompanied by an angry facial expression or insulting words, may be interpreted as a serious threat against one's life. Actual physical assault is only legitimate in self-defense. In the past, individuals who, according to native belief, were judged to be cannibals—*windīgowak* —were killed. But since such individuals were thought to be a menace to the lives of others, public opinion sanctioned their execution.

It must be recognized that Saulteaux society functions in terms of primary group relations, and in certain respects the kinship structure buttresses the discouragement of open aggression, just as the struggle to make a living may be partly responsible for food sharing and outward co-operation in economic tasks. The Saulteaux kinship system is centripetal in tendency in the sense that everyone with whom one comes in social contact not only falls within the category of a relative, but a blood relative, through the extension in usage of a few primary terms. There are no distinct terms for persons that in English we classify as relatives-in-law. This means that from the standpoint of a man, for example, all other men will fall into the following categories: grandfather, father, step-father (father's brother), mother's brother (father-in-law), brother, cross-cousin (brother-in-law), son, step-son (brother's son), sister's son (son-in-law), grandson [3]. Thus open quarrels would be tantamount, in terms of the formal kinship structure, to the expression of hostility between relatives. But since there are many other societies in which the same situation exists, I do not believe that the kinship structure as such is the basic factor in the discouragement of open aggression. There is no intrinsic reason, moreover, why relatives should not quarrel.

However, certain specific attitudes and behavior patterns that characterize certain classes of relationship among the Saulteaux tend to discourage *open* hostility and thereby fortify self-restraint. There is considerable emphasis laid, for instance, upon the solidarity of brothers and, in fact, of all relatives in the male line. This means that quarrels between brothers or other closely related males are more shocking than those between cross-cousins. And since the relation between a man and his mother's brother (= father-in-law) is a highly formalized one, requiring at all times a display of respect and even continence of speech, this type of habitual attitude and conduct in itself inhibits open aggression.

At any rate, in Saulteaux society the inhibition of openly expressed aggression in face-to-face situations is not only a socially sanctioned ideal, in actual fact it seldom breaks down. What Duncan Cameron said of the

Lake Nipigon Saulteaux more than a century ago is true of the Berens River people. He remarked that ". . . Indians, in fact, seldom quarrel when sober, even if they happen to hate each other" [4]. It will be noticed that Cameron's last phrase echoes the note of aggression struck by Le Jeune. The Indians whom both these men knew were by no means devoid of aggressive impulses, but they were observed to exercise restraint in the open expression of them.

The same is true of the Berens River Saulteaux. The problem, therefore, is to discover how individuals in this society dispose of their aggressive impulses. For it must not be supposed, in situations where any open display of hostility has been suppressed, that this is the end of the matter. In fact, complete inhibition of such hostile emotions is neither expected nor demanded.

Broadly speaking, the culturally sanctioned channels for hostility are of two kinds. The first is typified by all the unformalized ways and means that may be utilized in any human society for the *indirect* discharge of aggression. Gossip is as rife among the Saulteaux as among any other people, and many unpleasant and even scandalous things are said behind a person's back that no one would utter in a face-to-face situation. Even serious threats are made that may never be carried out.

The chief of one of the Berens River bands has been my interpreter and mentor. He has also been in office for many years. Again and again some of the Indians of his band have used devious means to remove him from his office, principally by accusations in anonymous letters sent to Ottawa or circulated locally by word of mouth. But when the band has met formally, no one has dared to accuse him openly of the things said behind his back. All sorts of accusations were made against a missionary up the river a few years ago, and from what I could gather a great many of them were true. But when a meeting was held to air the whole matter, it came to nothing because the Indians would not state their grievances to his face. This is typical. An example of another kind of covert unformalized aggression is the following:

Some years ago, several Berens River Indians who were out hunting came upon the traps of an Indian of the Sandy Lake Band. There is no love lost between the people of the Berens River and the Sandy Lake Indians; in fact, no marriages occur between the two groups. One of the hunters, egged on by his companions, defecated on one of the traps. Then he sprung the trap so that a piece of feces was left sticking out. Such an act was an insult to the owner of the trap and a deterrent to any animal that might approach it. It was a doubly aggressive act because, in addition to insulting the owner, it interfered with the purpose for which the trap was set and consequently menaced his making a living. That the Indian who did it recognized the nature of his act is proved by the fact that he later dreamed that a conjurer of the Sandy Lake Band tried to kill him by sorcery.

This leads to the second channel for the expression of aggression—

sorcery and magic. This society does not actually say: "Thou shalt love thy neighbor." What it does say is: "If I hate my neighbor, it is better that I should not openly quarrel with him." On the one hand, that is to say, the open display of aggression is not socially sanctioned, while on the other, unformalized types of *covert* aggression are tolerated and, in *addition*, institutionalized means of covert aggression are provided. The use of these not only enables me to injure my neighbor, but they permit the accomplishment of my ends without his knowledge. He can only suspect, but never be quite certain, that I am using sorcery and magic against him. It is this obverse side of the discouragement of open aggression that gives a potentially ambivalent aspect to all forms of co-operative effort and harmonious interpersonal relations among these people. For the availability of sorcery and magic in this society may just as truly be said to foster *covert* aggression as the cultural norms requiring self-control in face-to-face situations may be said to discourage overt aggression. With sorcery and magic at my disposal, in fact, I can vent my anger with greater effectiveness than would be possible by verbal insult or even a physical assault, short of murder. I can make a person suffer a lingering illness, interfere with his economically productive activities, and thus menace his living; I can also make his children ill or lure his wife away by love magic; I can even kill him if I wish. But when I meet him face to face, I will give no evidence of my hostility by gesture, word, or deed. I may even act with perfect suavity and kindness toward him and share the products of my hunt with him. Of course, he may suspect that I hate him or am angry, but, even if he goes so far as to accuse me openly of using sorcery and magic, I will not admit it. To do so would be tantamount to an open threat. And since he suspects me, it would do no good to deny hostile intentions because he probably would not believe me anyway. So I may not answer him at all, but just turn and walk away.

It can now be understood that, while there are no *official* records of murder among the Saulteaux, this does not correspond with the psychological realities of this society. Within their behavioral world, murder has occurred again and again but always as the result of sorcery, not of overt physical aggression. These Indians will not only name individuals who have met their death by sorcery, they will also name their reputed murderers. And they will go on to mention an even larger number of cases in which illness or failure in hunting have been caused by the malevolent action of human beings.

But the use of sorcery and magic as instruments of covert aggression is two-edged: whether an individual himself uses them or not, the threat that other individuals may be using them against him constantly lurks in the background of his daily life. In addition to the personal friction and unconscious hostilities that arise in all human relations, these people are dogged by anxieties and fears arising from the belief that human beings may do them positive harm in this covert way.

Such fears have very deep and wide ramifications in view of the fact that human viscissitudes of all sorts may be interpreted as due to magic or sorcery. Such emotions are precipitated under circumstances where the cultural definition of a situation weights the explanation of events in favor of sorcery or magic. Whereas we might attribute such inevitable hazards of life as illness, misfortune, or death to chance or impersonal causes of one sort or another, the individual in Saulteaux society is bound to consider magic or sorcery as a possible cause; in fact he is forced to do so because of the character and limits of the explanation of phenomena his culture offers him as alternatives. Consequently, it is inevitable that this explanation of events will be applied under appropriate circumstances, and that these cases, in turn, will later be adduced as evidence of the reality of sorcery. This vicious circle cannot be broken without radical cultural changes. The type of knowledge accessible to the Saulteaux is primarily a function of their culturally constituted attitudes towards the nature of the phenomenal world. It is impossible for them to think in terms of our category of natural causation, without utilizing the premises which have been built up in occidental culture as the result of the accumulation of scientifically acquired knowledge.

At the present day, as a result of acculturation processes, the explanation of events in terms of sorcery and magic appears to have been somewhat relaxed and diminished in range, especially in the band of Indians at the mouth of the river. But it has by no means died out. Individuals, for example, still suffer from *anicinabewapine*, Indian sickness. This is a modern term which shows that the contemporary Saulteaux clearly recognize a distinction between the diseases known to the whites and those which are peculiar to them. Indian sickness occupies an autonomous etiological category because it is "sent by someone," as they put it. More specifically, it is based upon the idea that some material object can be projected by sorcery into a person's body in order to cause illness. White doctors cannot cure this kind of disease, so native doctors still flourish who are prepared to treat it by removing the object. Illness or even death also may be caused by an attempt on the part of a conjurer to steal one's soul during sleep.

An examination of cases of illness attributed to both these types of sorcery indicates how this explanation extends the range of personal hostilities upon a thoroughly irrational and non-realistic plane. For, in such instances, suspicion usually falls upon some particular person. In fact, if an aggressor is not immediately identified, a conjurer may be employed to discover him by clairvoyant means. The reason for this is that countermeasures may have to be instituted in order to save the patient. The chief point of psychological interest here is what determines the selection of an aggressor. In many cases I have found that the determining factor is the belief held by the victim that he has offended the suspected aggressor in some way. In other words, the sick person is apprehensive because he feels that he has aroused another person's hostility by some act of his and that the offended person has retaliated by making him sick. This measure-for-measure philosophy is very deep seated among the Saulteaux, pervading many aspects of their social and economic life, so that it is not strange to find it operating in connection with aggressive patterns of behavior. Le Jeune, referring to the Montagnais, gives an amusing application of it which is psychologically diagnostic as well. He says: "They eat the lice they find upon themselves, not that they like the taste of them, but because they want to bite those that bite them." Once we understand that the Saulteaux always think and act in terms of this same principle, it is not difficult to see why an individual who thinks he may have offended another is quite ready to believe that he has been bewitched by the person he has offended. One case already mentioned illustrates this. The Indian who defecated on a trap thought that the owner of the trap had identified him because he dreamed that a conjurer of the Sandy Lake Band was attempting to steal his soul.

I have another account of a dream experience which was interpreted by the dreamer in the same way. In this instance, the dreamer was a boy who had made fun of a hunchback by imitating the way the cripple walked. The boy believed that the father of this cripple, a conjurer, attempted to retaliate by trying to steal his soul [5]. The next day he was ill. He soon recovered, but he considered it a close call. In other words, both these men were threatened with death because of what they had done. In another case a young man suffered from "Indian disease." It was thought that the man who had sent it was angry because the victim had beaten him in a dog team race. In this instance the young fellow died, despite the efforts of a native doctor who tried to "suck out" the malignant object. From a native point of view this was a clear case of murder.

In all these cases we would consider the reputed retaliation—attempted murder—entirely disproportionate to the offenses committed, if we considered them to be serious at all. But within the context of Saulteaux society, it is obvious that their evaluation is quite different. They were considered sufficient grounds for the counter-aggressive actions that menaced the lives of their perpetrators. In this respect, they are excellent psychological clues to the kind of behavioral world in which these people live and act. If individuals believe that such actions of theirs are sufficient to arouse such extreme forms of retaliation, we can only conclude that they, in turn, might be offended by acts of a similar nature. It seems reasonable to infer, therefore, that these Indians are extremely sensitive to the slightest trace of aggression. There is also the concomitant inference that individuals must be constantly on their guard lest they give offense to others. Consequently, the ground is well prepared in this society for the development of paranoid or pseudo-paranoid trends in individuals. And the genuine paranoid would, no doubt, find it relatively easy to build up a plausible structure of delusions. One man who suffers from deep-seated anxieties and exhibits some specific phobias once showed me several objects, a shell (*migis*), a couple of porcupine quills, a round stone and a rectangular stone, that had been magically projected against him but failed to do him harm because of his own powers. Even in the ordinary affairs of life, a mistrust of others is often exhibited that is unconsciously projected and rationalized as positive aggression, even without the involvement of sorcery and magic [6].

An informant once told me that his father had cautioned him thus: "Don't laugh at old people. Don't say to anyone that he is ugly. If that person is *mandauwizi* (has magic power), he will make you so." It is obvious from the phrasing of this statement that no abstract ethical ideal was involved. Avoidance of offense is clearly a matter of self-defense.

The cases cited also illustrate another point of psychological interest. It is apparent that it is the prevalent belief in sorcery that is of prime importance, not its actual practice. I do not mean to imply by this that no one really uses magic or sorcery to gain his ends. Quite the contrary is true. But from the standpoint of the individual who interprets his illness or misfortune as due to sorcery, it makes little difference whether in actual fact someone has gone through certain procedures or not. Such procedures in any case are secret and no one can be sure. The point is that the psychological effects are the same. This is why sorcery and magic are real to the Saulteaux. Everyone acts as if they were, and they thus become effective constituents of thought and feeling. From an objective point of view, the belief system of the Saulteaux thus fosters fantasy situations in which aggressive impulses become easily entangled with interpersonal relations in ways that may engender deep and irrational consequences. I may be easily led to believe that you are hostile to me and have bewitched me when this is not true at all. Yet there is no way of coming to grips with the situation in a realistic fashion. The following story illustrates this point, and, since I happen to know one of the participants well, I can speak with some assurance about his feelings in the matter.

About fifteen years ago at Treaty time, representatives of a number of the Berens River bands and some men from Deer Lake were camped at the mouth of the river. Joe, a young man of the Berens River Band, and Wabadjesi, a man from Deer Lake, were wrestling. They were having lots of fun, and a crowd gathered about them. But Joe was getting the better of Wabadjesi. The latter became irritated at this. Since there was such a crowd of men present, evidently his pride was hurt. "If you throw me to the ground," he said to Joe, "you will see sparks of fire." The narrator commented that he did this to scare Joe, who knew Wabadjesi had the reputation of being a powerful shaman.

"There won't be any sparks from the ground," Joe retorted, "but when you hit the ground you'll see plenty," and with that he let Wabadjesi go. But Wabadjesi was really mad by this time and said, "Watch out for me the first part of the winter. Then you'll know something about a Crane" (the generic term for some of the Indians at Deer Lake and Sandy Lake, derived from their totemic affiliations). To this Joe replied, "And you'll know something about a Cree." (Parenthetically it is stated that Joe was joking. He was not a Cree, but his mother was.) People began to talk about this exchange of open threats between the two men. Some of them even felt sorry for Joe, because Wabadjesi had such a big reputation as a man who was known to use sorcery and magic. Joe, on the other hand, had been to boarding school, was a Christian, and was entirely innocent of any malevolent intent.

The following winter the man who told me this story went trapping with Joe. They left the settlement in October. The first night out, just at dusk, they made camp in the shelter of a tall white spruce. They had hardly gotten settled when they heard what appeared to be the voice of a bird coming from the tree. They could not catch sight of it; neither could they identify its cry which kept sounding almost every ten minutes until midnight. They thought it strange but were not disturbed. The next morning they started off, and, continuing from that day throughout the winter, no matter where they went, a Pygmy Owl seemed to be accompanying them. This was an untoward sign, because sorcerers sometimes use these birds to accomplish evil ends. But neither man was ill.

In the spring after returning to the mouth of the river, they received news that Wabadjesi was very sick, in fact, that he was not expected to live. This made some people think that perhaps Joe really "was something." One man who had been present at the exchange of threats the previous summer even went so far as to say to him, half jokingly, "You almost killed Wabadjesi." "I did not intend to kill him," Joe replied in the same mood; "I had pity on his two boys."

This anecdote shows how, under certain circumstances, it is possible for an individual to be accused of using sorcery when in fact he has not done so. In this case I know that Joe did not entertain hostile feelings of any sort against Wabadjesi. Yet if Wabadjesi had died as a result of his illness, I think there is no doubt that people would have believed that Joe killed him. It certainly would have been the interpretation of the Deer Lake Indians, who are much less acculturated than those of the Berens River Band. For it must be noted, among other things, that Joe *did* return Wabadjesi's threat. At the same time it is interesting to observe that Joe on his part was a bit apprehensive during the winter. Had he been a pagan, however, he would no doubt have suffered more severely. Furthermore, I am convinced that if Joe *had* been taken ill he would have blamed Wabadjesi, and, if he had then followed the native pattern of behavior, he would have been fully justified in retaliating by some covert means. As it was, there is no doubt that Wabadjesi, on his side, believed that Joe had taken some aggressive action against him. He could hardly think otherwise, since he openly threatened Joe in the first place.

Another explanation offered me was that Wabadjesi undoubtedly tried to injure Joe but that for some reason or other, as sometimes happens, his magic turned upon him and caused the same sickness in him that he tried to project toward his victim.

One has only to multiply such episodes imaginatively to gain an idea of the extent to which the fantasies arising out of such situations can insinuate themselves into the fabric of Saulteaux society, engendering antagonisms that affect the interpersonal relations of individuals in various ways.

It is possible to cite other cases in which it can be shown how repressed —unconscious—aggressions of individuals are caught up and put into social circulation, as it were, through mechanisms of projection, displacement, and rationalization. Some frustrating experience of mine may release repressed aggression toward an associate or some other person who is in no way the cause of my frustration or even hostile to me. But the prevalent belief in magic and sorcery enables me to rationalize my own projected aggression as hostility directed toward myself emanating from the person I have unconsciously chosen. These are familiar dynamisms, but among the Saulteaux we can discern them operating in a cultural milieu that facilitates rather than nullifies or offsets the canalizing of unconscious hostilities through them.

In one instance that I know of, G accused two of his brothers-in-law (cross-cousins), with whom he was hunting, of using magic to attract animals to their traps. This is an aggressive act because magic should only be used when hunting alone. It is so powerful that there is no chance for my hunting partner to catch any fur if I use it. Consequently, the use of hunting magic by one man of a group is tantamount to depriving the others of part of their living. G felt frustrated because for a month and a half no animals came to his traps. "Not even a weasel," he said. Yet often he saw tracks circling them. All during this time his brothers-in-law were quite successful in their catch. G told me about the whole affair himself, and in this case there was no mention of any offense he may have committed against them. He was simply infuriated and vowed that the next time he went hunting with them he would use plenty of medicine.

Why was G so ready to accuse his brothers-in-law of hostility? I happen to know G quite well, and, without going into detail, I will merely state that I believe there is plenty of evidence in his personal history and behavior in general to suggest that he has constant difficulty in disposing of unconscious aggression. (The Rorschach record of this man supports many of the following deductions in a most convincing manner.) G is often at swords points with his father, and he is the man I mentioned previously who gave his brother a severe beating. This brother, moreover, is his mother's favorite among the sons, so that the source of G's aggression probably is very old and deep. Among other things, I strongly suspect that G has a considerable amount of repressed hostility toward his wife. The hunting episode I believe to be thoroughly symptomatic of G, and my guess is that it involves the displacement of some of his aggression toward his wife upon her relatives. He has given additional evidence of this, and it is facilitated by the fact that, whereas G is a militant Protestant, his wife and her people are strong Catholics.

On the hunting trip referred to, G had not only accused his brothers-in-law of using magic, he had gone on to insult them as Catholics by hinting that the local priests had sexual intercourse with the local nuns. I do not believe that it is accidental, either, that he launched into a tirade against X, one of his brothers-in-law, and his father-in-law as well. G's sister is married to X, and G told me the way he got her was by love magic—that is, unfairly. Then he went on to say that he believed that his own wife had only had one child because her father had given her medicine to prevent her impregnation. G's wife is an attractive girl and had many suitors before her marriage, among them G's brother. It was a forced marriage, following the girl's pregnancy. For some time G has suffered from the suspicion that his wife has been unfaithful. While he has discovered no proof, I happen to know that his suspicions are by no means groundless and that G's brother is involved. G feels disadvantaged in many other ways and seems unable to adjust himself. It is not surprising, therefore, to find G utilizing the beliefs of his culture to screen his own repressed hostilities. This permits a plausible defense against their emergence into his conscious life.

In still another case, Kiwetin, an old man, told me about a dream he had had in which one of his spiritual helpers informed him that a neighbor of his, a middle-aged woman, was using sorcery against him. Kiwetin had been sick, and this was the explanation he finally adopted when the medicine he took did no good. The woman was married; in fact she was notorious for her many husbands, her dynamic personality, and her reputed knowledge of sorcery. One of the interesting points about this case is the fact that Kiwetin emphasized the woman's outward display of kindness and amiability toward him. He was a widower and lived alone, and he said she often invited him into her house to have something to eat. But he always refused. He also said that she smiled pleasantly at him, but he knew this was put on, that "it was only on her face, not in her mind." In other words, Kiwetin's suspicions made any genuine friendliness with this neighbor impossible. I suspect there were unconscious involvements here, but I know nothing about this old man's personal history in any detail.

In this case, however, there was another psychological factor involved which I have not yet touched upon. It requires emphasis in order to round out the analysis of the strains and stresses to which a belief in magic and sorcery subjects individuals in Saulteaux society. Kiwetin was a man who, in his earlier days, was one of the leaders of the Midewiwin, an institution which offered curative services for payment. The leaders of the Midewiwin also possessed a great deal of information about magic of various kinds which they dispensed at a price. So in Kiwetin's account of the woman who he thought had bewitched him, he made it quite clear that he felt secure against any measures she might take. In his dream he even turned down the offer of one of his spiritual helpers to injure her. Kiwetin was fully confident that he could protect himself. His sense of security sprang from his belief in the power of his *own* supernatural helpers. In this respect he was in a different situation than G. The latter was a much younger man, who had been brought up during a period of fairly rapid acculturation. He did not claim to have any spiritual helpers. Consequently, he lacked the inner security of members of the older generation who believed that they had powerful guardian spirits to help them. Hence, G's belief that magic was being used against him made him feel the full impact of the hostility of others because he was exposed to it without protection.

It is important to recognize, then, that native belief does provide a means for protection against the covert hostility of others, even though from our point of view this is likewise on a fantasy level. Up until recent years every boy at the age of puberty or a little before was sent out to fast in the forest. It was then that he secured the blessings of supernaturals who were to be his guardian spirits through life. While there was no formalized fast of this kind for girls, they were not debarred in native theory from the acquisition of supernatural helpers. In Saulteaux society the belief in the power of one's guardian spirits provides the ultimate basis for a sense of security in the individual. Nevertheless, the supernatural protection thus afforded varies greatly from individual to individual. Some persons are thought to have a great deal of power derived from this source, others little. And, following the extremely individualistic patterns of this society, no one knows how strong an individual is until he exercises his power. A few individuals are known to have considerable power because they utilize it in connection with conjuring, curing, or in other professional ways. It is also possible to add to the power one acquires at puberty by purchasing a knowledge of magic from others. This is what the mide men did, thereby enhancing their prestige.

In the give and take of social life, therefore, the apprehension of aggression varied concomitantly with the reputed power of the aggressor. And the ambivalence I have referred to reached its apex in the attitude toward the mide men and conjurers whom everyone knew to be the most powerful men of all. They were greatly feared and highly respected at the same time. Outwardly they were treated with the utmost show of deference. Only an individual supremely confident of his own powers would run the risk of offending an individual of this class. But one could be less certain about others because there might be no reason for a man to exhibit or exercise his powers until he was in a tight place. It is also true that the mide men and conjurers often took advantage of the reputation they enjoyed. They were the ones who were sometimes openly aggressive in situations where other men would fear to be. In fact, the Indians say that formerly such individuals were constantly trying each other out.

An instance of this is the following story told to me by the son of the man called Owl. Pazagwīgabo, the one who provoked the duel by sorcery, was the arrogant leader of the Midewiwin [7]. He is reputed to have killed many persons by sorcery, and one informant described him as "savage looking in the eyes." The father of this informant warned him as a child never to play near Pazagwīgabo's camp. Owl, on the other hand, was a quiet-mannered man, small in stature, and said to be kind to everyone. He was not a mide, but he had the reputation of being an excellent doctor and he was a noted conjurer. These two men belonged to different local groups and seldom came into personal contact. But on the occasion to be described, Pazagwīgabo had come down the river to the Hudson's Bay Company post and found Owl sitting on the platform of the store with a lot of other Indians.

Pazagwīgabo went up to Owl and said: "You think you are a great man. But do you know that you are no good? When you want to save lives, you always bring that stone along. I don't believe it's good for anything." To this Owl replied: "Pazagwīgabo, leave me alone. I have never bothered you."

"You are not worth leaving alone," the mide said, and with this he grabbed Owl by the front part of his hair and threw him down. Owl simply got up and said, "Leave me alone." But Pazagwīgabo grabbed his hair a second time, and, when Owl made no resistance, he did it for a third time. Then Owl said: "Are you looking for trouble?"

"That's what I want," said Pazagwīgabo, "I want you to get mad. That's why I did this." So Owl replied: "All right. I know you have been looking for it for a long time. I know you think you are a great mide. You are nothing. If I point my finger at you you will be a dead man. But now I'll tell you something. Don't you do anything to my wife or my child. Do it to me, and I'll do the same."

"Ho! Ho! (Thanks, thanks)," said Pazagwīgabo, "Expect me at *kijegīzis* (kind moon). (The equivalent of February in our month count, so named because the winter is beginning to moderate.) I'll give you a chance to do what you like to me."

Sure enough, during *kijegīzis*, Owl was taken sick. He got the shell (i.e., a *mīgis*) all right, but of course he brought it out. (These shells are objects associated with the Midewiwin. Their projection into a person usually is sufficient to cause death. In the ceremony of the Midewiwin, their lethal effects are demonstrated as well as the power of the mide-men to revive victims. A mide would be expected to use such deadly shells in a sorcery duel. Owl's own power was sufficient to counteract the effect of Pazagwīgabo's magical weapon and even to eject the shell from his body.) But Owl was not sick long; he easily recovered. Shortly after this, news came down the river that Pazagwīgabo was sick. He was unable to walk. News spread about that he was getting worse and worse. Every once in a while Owl was heard to say, "Huh! Huh! I guess I nearly killed him. I did not mean to; I was only playing with him." When the *ni'kīgīzis* (goose moon) appeared (April) Pazagwīgabo only was barely able to walk about. At the beginning of the summer, Pazagwīgabo came down the river again with some other members of his band. They camped near Little Grand Rapids, where Owl lived. When Owl saw the old mide, he walked right up to him and said, "You know who Owl is now! I was only playing with you this time. I did not intend to kill you. But I never want to hear again what you said in this place. And I don't expect you to do again what you did here. Don't think you are such a great mide!"

Of course this story may be a biased version of the events that actually occurred. We do not know what Pazagwīgabo had to say. But as it stands it is a beautiful illustration of how overt aggression, even when it occurs, is almost immediately displaced to the level of sorcery (fantasy) for a showdown. It also exhibits in dramatic fashion the confidence of both men in their own powers. In a situation where their very lives were in imminent danger, this made each of them courageous. Although men of lesser power may have suffered greater apprehension in such a situation, yet every Saulteaux has a modicum of spiritual armor because of the blessings from the supernaturals that he has secured in his puberty fast. He is not completely at the mercy of hostile forces. A consciousness of this fact is the balancing factor in the total situation. One might even say that *some* confidence in one's ability to ward off possible aggression is in the nature of a psychological necessity. Otherwise, the apprehension and fears of individuals in this society might drift too easily into a paralyzing terror. In principle, confidence in one's ability to face the hostility of others, or to match hostility against hostility, might be said to give a realistic, healthy tone to situations where the warp and woof are so frequently constructed of fantasies.

However, apprehension and fear cannot be totally eliminated. This is not due entirely to the fact that a belief in sorcery and magic fosters such emotions. There is always the possibility that another man's power *may* be greater than my own. To provoke him to exercise them by offending him, is always a gamble. What may happen if I do suffer from delusions of grandeur is well illustrated by Pazagwīgabo. And still another aspect of the use of sorcery and magic has been referred to that may make me hesitate to use them at all: the possibility that if they fail to work for any reason, or my victim is stronger than I, they may be retroactive in effect. Instead of accomplishing my ends, I may fall ill or die by my own sorcery. But in such matters, it is not possible to be wise before the event. One has to risk defeat in attempting to achieve the ends desired. Thus both for the man who is confident of his powers, as well as for the common man, the best defense is to avoid offense if one seeks what the Saulteaux call *pīmädazīwin*—life in the fullest sense.

REFERENCES

1. Speck, Frank G.: "Ethical Attributes of the Labrador Indians," *Amer. Anthrop.*, Vol. 35 (1933), pp. 559–94.
2. Thwaites, Reuben Gold (editor): *The Jesuit Relations and Allied Documents*, 73 Vols. (1937); in particular, Vol. 12, No. 13.

3. For a summary description of the Saulteaux kinship system, see Hallowell, A. Irving: "Cross-Cousin Marriage in the Lake Winnipeg Area," in Davidson, Daniel Sutherland (editor): *Twenty-Fifth Anniversary Studies* (Philadelphia, Philadelphia Anthropological Society, 1937); in particular, pp. 95–110.

4. Masson, Louis F. R.: *Les Bourgeois de la Compagnie du Nord-Ouest* (Quebec, Cote et Cie, 1890), Vol. 2, p. 262; Cameron's observations are dated 1804.

5. The details of this dream as narrated to me are given in Hallowell, A. Irving: "Psychic Stresses and Culture Patterns," *Amer. J. Psychiatry*, Vol. 92 (1936), pp. 1291–1310.

6. See the account of J. D., in Hallowell, A. Irving: "Fear and Anxiety as Cultural and Individual Variables in a Primitive Society," *J. Sociol. Psychol.*, Vol. 9 (1938), pp. 25–47.

7. For further details about this man and the Midewiwin in this area, see Hallowell, A. Irving: "The Passing of the Midewiwin in the Lake Winnipeg Region," *Amer. Anthrop.*, Vol. 38 (1936), pp. 32–51.

CHAPTER

16

Dorothy Eggan

THE GENERAL PROBLEM OF
HOPI ADJUSTMENT

This paper deals with the problem of hostility in another American Indian society. Every culture must provide solutions to the problem: What to do about hate satisfaction and still preserve a well-knit fabric of interpersonal relations? In all societies there are alternative ways of manifesting aggression, although certain groups more than others structure life in such a way as to avoid tension-producing or competitive situations. Mrs. Eggan lays more emphasis upon economic factors than did Hallowell and does not restrict herself to the cultural aspect of the group-membership determinants, for she considers the effects of the physical environment in which the society makes its living. These matters have also been treated in another important paper on personality formation among the Pueblo Indians.[1]

This paper likewise stresses another influence of a situational character —namely, that of acculturation.

IN ANY prolonged contact with the Hopi Indians, an investigator who is interested in the psychic as well as in the more tangible phenomena of

NOTE: Reprinted, with some revisions by the author, from the *American Anthropologist*, Vol. 45 (1943), pp. 357–73. The author and the American Anthropological Association have kindly consented to slight abridgment and incorporation of some footnotes into the text.

[1] Esther S. Goldfrank, "Socialization, Personality, and the Structure of Pueblo Society," *American Anthropologist*, Vol. 47 (1945), pp. 516–39.

culture is struck with the notable mass "maladjustment" of these people, maladjustment being here defined, not in neurotic terms, but as a state in which friction predominates in personal relations, and in which the worst is anxiously and habitually anticipated. The comment has been made by numerous persons who have worked with them that one Hopi is a delightful friend; two are often a problem; and more than that number are frequently a headache. Discord is apparent in inter-tribal relations as well, as most Hopi will testify; the younger ones particularly express annoyance over the endless arguments which accompany any group attempt to reach a decision in tribal matters. Gossip is rampant throughout the villages; witchcraft is an ever present threat, one's relatives as well as others being suspect; and in some cases individuals fear that they may be witches without being aware of it. Even sisterly love in this strongly matrilineal society seldom runs smoothly, although sisters present a united front against the rest of their world.

The Hopis' mutual distrust and their fear of spirits are reflected in their dreams, of which a high percentage collected to date must be classed as "nightmares." Almost universal lip service is accorded a pleasing conception of life after death, which involves a simple form of ancestor worship, the ancestors being revered as *kachinas* and clouds who bring rain and all things necessary to life. Yet the Hopi betray a great fear of death and the dead and are reluctant to talk about these things; such fear is strongly apparent in their own interpretation of dreams, since to dream of a wide variety of objects "means death," although in actual use these symbols are associated with daily living rather than, or in addition to, death and burial.

Without doubt, growing confusion fathered by white contact—particularly since 1850—has augmented the seriousness of the situation. Here as elsewhere the conquerors and those who came after them met the sensitive pride and dignity of the Hopi with arrogance and impatient crudeness. That acculturation is not the sole cause of this confusion, however, is indicated by Hopi myths, by early accounts of the tribe, and by the frequent division and redivision of their villages. The disintegration of Oraibi into some five villages since 1906 is an extreme example of village dissension. It may be argued that responsibility for Oraibi's troubles rests squarely on white shoulders, since the original split was nominally over submission to white control, but Titiev's study of old Oraibi reveals many other important elements. And it must also be recognized that, had a pattern of agreement been more strongly woven into their habitual reactions, a more solid front either for or against *bahanas* (whites) might have been presented.

In seeming contradiction to the evidence cited, we find that the Hopi are a hospitable people who smile easily and give generously with a quiet dignity. Many exhibit a remarkable degree of reconciliation to white intrusion, some by ignoring white culture as far as possible, believing that theirs is the better though harder road, others by merging parts of it effectively with their own. In fact from surface indications one is inclined to envy Hopi peace, and it is entirely possible for a visitor who is not particularly interested in, nor sensitive to, psychological manifestations to be unaware of their inner turmoil; but a smiling face and a preference for one's own way of life—even an attitude of pity and contempt toward another way—do not necessarily mean that one's own society affords a maximum of psychic serenity. It is also obviously true that these exhibitions of unease are not exclusively Hopi; perhaps they are even inherent in compact community living, as experience in any small town or ingroup seems to indicate. But dismissing them thus evades the problem, nor does it suggest to what degree they may be minimized for greater group solidarity and individual comfort.

It seems reasonable to assume that man's reactions are shaped in part by pressures which derive from his total environment, despite the emphasis certain psychoanalysts place upon "instinctual" drives. It is difficult, perhaps impossible, to describe adequately the emotional gamut to which the Hopi were subjected by their physical milieu, but an understanding of it is prerequisite to an examination of their emotional reactions. Surveying these external conditions briefly, we find in northeastern Arizona a ragged plateau of sand and rock from which a series of arid mesas, averaging six thousand feet in altitude, descend to increasingly arid plains some thousand feet lower. It is a land of violent moods, of eternal thirst and sudden devastating rains, of searing heat and slow, aching cold. The flora and fauna have long since become discouraged and are sparse. Neither was it a promising environment for human habitation; yet for perhaps fifty generations, the two thousand or more constituents of the Hopi tribe have lived, entirely surrounded by their traditional enemies, on a "reservation" where several neighboring mesas disintegrate into the plains below.

Here the seasons marched by with scant regard for man's trampled ego. The Hopi could only sit and wait while sandstorms cut to ruin the young plants that meant "life for the people." Through long hot summer afternoons they daily scanned the sky, where thunderheads piled high in promise and were dispersed by "bad winds." When at last the rain fell, it often came in destructive torrents which gutted the fields and drained away into the arroyos. Of course these sedentary people, lacking even the relief through action with which their nomadic foes responded to these conditions, worked out a religion designed to control the unappeasable elements. But even this prop became a boomerang, since their beliefs held no promise of virtue through suffering; rather, *all* distress was equated with human failure, the consequence of "bad hearts." (This concept must not be confused with that of "two hearts," who are regarded as *powakas*—witches—and greatly feared. A "bad heart" is thought to be the result of quarrels, "bad thoughts," and other misbehavior.) Obviously when the Hopi danced for rains and the rains did not fall, they could not tolerate a weakened faith in the Gods, so they searched among their fellows for "someone whose heart was not right," and who thus "spoiled the ceremonies."

In earlier times, "the peaceful people" occupied villages at the foot of the mesas, where seepage from above, plus occasional freshets from the ordinarily dry arroyos, aided their dead, who returned as clouds to sprinkle the earth. But these sites were extremely vulnerable to attacking enemies, and fearing Spanish reprisals after the Pueblo Rebellion in 1680, as well as continual depredations from the surrounding nomadic tribes, they moved to the mesa tops which could be defended but to which all food, water, fuel, and building material had to be carried—except for the aid of the few burros left by the Spaniards—on overburdened backs.

When the boundaries of their "reservation" were finally defined by white law, the action had significance for the Hopi in several ways. First, an almost universal response to an articulate designation of restrictions appears to be one of antagonism, particularly when such prohibitions seem to give an advantage to competitors. In fact, such restrictions frequently suggest behavior that would otherwise never be attempted. This natural reaction among the Hopi was further intensified by the fact that they had long anticipated the return of an "elder brother" who instead of crystallizing existing limitations was to have dealt effectively with all their difficulties, thus permitting them to "inherit the earth." Second, government protection made organized physical aggression against their enemies largely unnecessary as well as unlawful. Among a peaceful people with strong sanctions against aggression, the necessity for occasional violence turned outward was a useful emotional vent, and the substituted verbal castigation of *bahana* dictators and Navajo intruders was not an adequate surrogate. Although in spite of scant moisture, ravaging sand storms, and determined rodents and worms, they had previously managed to coax sustenance from unwilling sand, and to stave off heat, cold, thirst, and marauders to at least an effective survival level, government protection came gradually to include a reluctant quality of dependence, therefore to lessen self-confidence and through this their security and pride. One Hopi has summarized the feelings of many by making it clear that in spite of wanting nothing to do with whites, the fear of them:

> . . . especially of what the United States Government could do, was one of the strongest powers that controlled us, and one of our greatest worries. . . . We might be better off if the whites had never come . . . but now we have learned to need them.

If we attempt to explain Hopi discord in the light of certain widely accepted psychological theories, we might expect to find some or all of the following: early and severe anal training; oral frustration in early childhood; strictly conceived and enforced taboos on early sexual experimentation and gratification; numerous restrictions on the child's manipulation of his early physical environment; strict parental control with resultant conflict; a constellation of circumstances which would permit the Oedipus complex to flourish; early and strict repression of aggression; and an inadequate foundation for dealing with fear which might accompany any one of these.

In view of the wide dissemination of today's psychological theories, it is not necessary to discuss the origins and consequences of such pressures as those listed above. Whether the drives involved in general human behavior are instinctual or not, the facts are that all children who have been observed, in whatever society, exhibit: a tendency for libidinal impulses to be fixated temporarily in oral and anal zones; a strong inclination to investigate their environment by the active expenditure of physical and mental energy; resistance to authority; emotional attitudes toward parents and caretakers; and a tendency toward aggression, if by aggression we mean the manipulation of one's environment to achieve one's own ends. Furthermore, although varied amounts of emphasis are placed upon one or another of these drives, students of personality are well agreed that drastic interference with any or all of them in childhood may produce a "low frustration threshold," or actual neuroses in later life.

The western-molded investigator of Hopi maladjustment, having grown up under a weight of interdictions in regard to most of these things, is biased by a psychology which is at once the result of experiencing them and experimentation with them. He therefore begins his "psychoanalytic" journey backward into Hopi childhood of forty or more years ago with confidence and a fellow feeling. But he finds to his surprise that few societies have provided a more uninhibited infancy. (It must be borne in mind throughout this discussion that, while frequent reference is made to observations and findings regarding Hopi childhood of the present decade, it is done only when evidence indicates that the same conditions, or intensifications of them, existed during comparable periods in the lives of adult Hopi of today, whose behavior is being examined.)

Rarely did an infant enter the scene unwanted—at least that was Hopi theory and more often than not the theory was fact. The first child was especially desired by his parents and by all of his relatives. Normally at the time of his birth the young couple was living with the wife's family, so that from the first day of his life the baby was surrounded by constant care and attention, not only from his mother but from his mother's mother and her siblings as well. This extended maternal family was joined by paternal aunts and relatives who acted in various ceremonial capacities and showered attentions upon the newcomer. Children subsequent to the first had not only their mother, their maternal and paternal relatives, but older siblings as well to indulge them. As soon as a girl child could bear the weight of a baby she was given the care of her next youngest sibling, or one of her mother's sister's children, and little boys were sometimes pressed into service as nursemaids, though not so frequently as girls, of course. Even in the relatively rare primary family—and this is true in the modern homes of today as well—the mother and child spent much time in the mother's mother's or mother's sister's households, where the infant still had many caretakers and a complete sense of security. At no time was he as *exclusively* dependent upon his "real" mother for comfort as is a child in our society; he was even given the breast of his mother's sister or other close relative in his mother's absence, if he cried for food. There was always, therefore, a convenient lap within reach when he was sleepy and a convenient breast when he was hungry. The writer has never known a Hopi child who was not approximately as content with his mother's mother or sisters (or if he had a sibling nurse, very often with her) as with his mother, but of course the caretaker seldom took him far from his source of food supply.

While the infant was small, he was bound to a cradle-board where his movements were restricted, except when he was removed for bathing, but Dennis was unable to discover any measurable effects of this on modern Hopi children. The child was released from the board as soon as he became restless in his bonds, and from then on had few restraints. A cry brought the breast or whatever attention seemed desired. There was no slapping of fingers, no mechanisms to prevent thumb-sucking or masturbation, no objects labeled in one way or another "mustn't touch." Toddlers even learned that a hot stove was "taboo" by being allowed to touch it rather than from restrictions enforced by parents.

Both from the literature and by observation and questioning in today's families, one learns that anal training was an unhurried and unharried process, neither associated with shame nor parental anger, and was, as a matter of fact, not taken very seriously by anyone concerned. A frequent Hopi statement in this connection is: "White people expect too much of small children." For bed wetting, punishment was employed only when a child continued the practice long after the normal period of establishing control. A mother's brother—or other member of the family—might hold the offender in the smoke of a juniper fire in front of his companions, even sometimes including others in the punishment. This was severe treatment but it was rarely utilized.

From a very early age the Hopi child had been given small bits of food which were chewed by various members of the family and placed in his mouth; in addition he had long been accustomed to sucking fruit, mutton and corn—anything that his elders ate which was small enough for him to handle. It would seem, therefore, that the transition from breast to solid food must surely have been a less critical period among the Hopi than among ourselves. As with anal training, there was normally no urgency in the process of weaning, which, by white standards, came at a late period in the infant's development. Moreover, even if there were another baby to nurse, it is of particular importance to note that the Hopi woman did not deprive a small child of breast *comfort* as abruptly and finally as do white mothers. The writer has sometimes observed a child who was not hungry quieted by the breast of a mother's mother or mother's sister who was unable to feed him thus, with no embarrassment on anyone's part

She has also seen other Hopi women "shame" a sleepy or perturbed child and say *kahopi* ("not Hopi," and therefore not good) when he tried to avail himself of this familiar panacea; but quite possibly this was done only because a *bahana* was present. In any case, from very early babyhood the breast was not the only source of food or oral satisfaction an infant knew, but was one of many sources, and it is very probable that it failed to acquire the overwhelming significance attributed to it in our society. In addition, a wide variety of caretakers relieved the weaning child of constant temptation in the form of an ever-present mother, so that it would seem logical to suppose that the child himself, as he grew older and was outside more of the time, developed new interests more or less naturally, as other foods made increasing appeal, and as other playthings and erotic satisfactions were substituted for the breast.

Of course the coming of other children frequently hurried this process and probably often caused some frustration, but by the time a mother was again pregnant the preceding child had long been familiar with many "mothers." In fact, the sooner a sibling appeared, the less severe would seem to have been the shock. It is perhaps easier to give up exclusive privileges one has had for only eighteen months than those which have been long reinforced. And since, as pointed out, the infant's dependence was not entirely upon his mother in any case, but included many relatives as well, his displacement at the time of the new child's arrival could not ordinarily have been so drastic as that of a comparable child in our culture. The conclusion is not to be drawn, however, that there was *no* sibling jealousy due to security displacement by other children. On the contrary, there is evidence of it in the personal history data gathered by the writer, and in published autobiographical materials, although in all cases it is strongly repressed and seems to be entirely subconscious. The implication, then, is not the elimination of a factor in the complex of Hopi psychology, but rather the modification of one which is being measured on the yardstick of white psychology.

Caution must be used particularly in applying the concept of a universal Oedipus complex to the Hopi. Here one must probe the affects inherent in relations within the maternal extended family and household and with paternal relatives as well. From the foregoing paragraphs, it could be predicted that both the concept of *extended* kinship and that of *sharing* one's loved ones would be *emotionally* very strong among adults. An individual would *feel*, as well as *say*, not only that he possessed many relatives but also that many others possessed them as well. There is much data which support such an assumption; whether in general conversation or formal interviewing, a Hopi must be continually interrupted and made to add the distinguishing "own" to kinship terms. Similarly, sharing relatives would seem to contribute to diffused affections and thus to a wider range of emotional security. The Hopi attitude in this regard has been strongly impressed upon those who have been formally adopted by various Hopi clans. In all else one's "relatives" are tolerant of *bahana* ignorance and bad manners, but the use of "my" instead of "our" before kinship terms brings a sharp reprimand to *bahana* and Hopi alike.

Many Hopi have asked me why such a distinction is so important, and in one instance, after many years of acquaintance with all three persons in the case, I discovered that the "mother" to whom an individual frequently referred was his mother's sister whose culinary habits he preferred, rather than his biological mother as I had supposed. Two out of twenty informants from whom dreams have been collected show symptoms of an "own" mother fixation, but even these have lived contentedly with other "mothers" for long periods and show much the same concern and affection for these as for their own. When asked why she added to her already heavy burdens by sewing for a "mother" who had several own daughters, my closest Hopi friend replied crossly: "Because she is *our* mother—a *bahana* wouldn't understand it." Another informant cried at my suggestion that she didn't "love" a certain "sister." The writer was not convinced that love between the two women existed; however, the strong affect of conditioned interdependence was illustrated well enough.

Indeed, it would seem that if an Oedipus complex developed in conservative, old-style Hopi families, it must of necessity have been built around a "composite mother" and a "composite father," for each child had several of both. Although most informants signified that they felt "closer" to their own parents than to the others, in only two cases known to the writer does this closeness tend to be approximately as exclusive as Western filial devotion. Of course the matrilineal organization of their society predisposed Hopi affection somewhat toward maternal relatives; but in actual habit, intense fondness for, and dependence upon, paternal relatives, ceremonial fathers, and particularly upon "doctor" fathers, to whom Hopi children were often "given" in illness, was by no means unusual.

In such a group, transference of father resentment to mother's brother would also seem to be an over-simplification of the actual problem. From birth, a child in this extended family normally slept in a room with many other individuals and close beside his mother, father, mother's mother, father, or siblings, as his fancy and their mutual convenience dictated. Grandparents, especially, liked their grandchildren to sleep with them. All of these individuals were a close part of the child's immediate environment and emotional cosmos; all indulged, cared for, and disciplined him as he became older and required it, although he perhaps had most to fear in this regard from his mother's brothers. During this early period his mother's brothers were frequently among the many sleepers in the room, but they were physically farthest removed from the female bodies from which the child drew his emotional and physical security. (At marriage, the mother's brothers went to live in other households, but still continued to regard their mother's or sister's homes as their own; thus they dropped in frequently to eat and to sleep and brought guests whenever they liked.) Sexual intercourse he witnessed, or, more exactly, was probably aware of if he happened to be awake, from the time he was dimly conscious of external stimuli. It was as natural and ordinary a part of these crowded sleeping quarters as the snoring or coming and going necessitated by other functions. Considering the soundness of Hopi sleep, it was quite possible for a child to be unaware of any of these things until he became sexually curious and purposely tried to stay awake.

In any case he was not pushed off to another bed, another room, or made to feel that he was being excluded from the security and easy companionship of this familiar room or these familiar bodies. On the contrary he occupied the *place of his choice* and his smallest whimper brought a quick response. As he grew older, if he imitated the sexual actions seen in the sleeping room or elsewhere, everybody laughed. He felt in their mirth the approval accorded to clowns; such laughter was not derision or reproval, and he knew it. From earliest babyhood, his genitals had been manipulated by adults and other children, not in deliberate masturbation, but playfully as a form of petting comparable to kissing children and ruffling their hair among ourselves. When he handled his own genitalia, no one fussed over it. By the time, then, that a Hopi child was aware that sexual intercourse had great value in the eyes of his elders, and also that members of his maternal family and clan were forbidden to him for this form of sexual indulgence, he was out of the house most of the time and had so many other objects for sexual experimentation and gratification that it is difficult to see how the few which were denied him could have assumed an immense importance. Moreover, the mother's brothers, whom neo-Freudians might expect to be substituted for the father in childhood resentment, were deprived of these same sexual objects more rigidly than was the child in question, who could at least sit on his mother's lap when in need of comfort, though he seldom did so after initiation.

There remain two factors which are significant in a discussion of childhood ambivalence toward adult males in any society, i.e., economic dependence and discipline. Among the Hopi every small child knew that his mother owned most of the fields from which his sustenance came, and that his father only tilled them for her. He was usually dependent upon her brothers only if his father died or deserted the family; and in either case the mother, more often than not, married again, and her new husband accepted the support of her children as a part of his obligations. The child's chief economic dependence upon his mother's brothers normally came at the time of his marriage, which probably seemed quite unimportant to him during the period in his life when the Oedipus complex was presumably being developed. During the same period, discipline was practically non-existent, or largely in the hands of his parents. There would seem to remain of enmity for mother's brothers, however, primarily that felt toward unaccustomed behavior restrictions, which increased rapidly through adolescence, but even this resentment must have been directed against several individuals. Not only were there often several mother's brothers who divided the responsibility somewhat, but in any case one of these was never the sole disciplinarian. All persons questioned on the subject agreed that ceremonial parents, adopted relatives, and paternal uncles, could and did scold them severely upon occasion; that *Soyoko* (a "giantess" wearing a hideous mask and women's clothes who was said to carry children away to her den and eat them) and her cohorts had been greatly feared during childhood, and that parents, including the father, were by no means helpless in administering physical punishment. Said one Hopi man, aged forty-two: "You have been reading books about the Hopi! I was 'licked' when I was a kid, by my father, mother, mother's and father's brothers, and even by an older sister sometimes. They licked us, but they didn't nag us like white parents do, and they didn't start as soon. They waited until we could understand what it was for."

It cannot be denied that affection for mother's brothers was often tempered by some animosity. Even so, the case for an Oedipus complex centering around the mother and father—or her brother—seems rather weak, whether considered in its narrow sense of sexual jealousy or in the wider one which includes rebellion against economic dependence and authority. And in any event, strong umbrage was as frequently directed towards the husbands of paternal aunts, who often carried their joking license to cruel limits and who were, from the child's infancy, jestingly named as his sexual rivals. Indeed since the paternal aunts did habitually engage in definite sex play with the boys, probably sometimes actually initiating them into sexual experience at an early age, as well as indulging them more and disciplining them less than anyone else in their world, a very real jealousy was probably felt for their husbands. This is particularly true in view of the castration threats, and actual simulation of castration, which often came from these husbands, who, as the boys grew older, quite possibly had reason to be jealous of them. When a boy married, his aunts' jealousy toward his wife was socially recognized and regulated by culturally patterned teasing and abuse. It has been admitted that "lovemaking with a *clan* aunt was the safest kind, because I could not marry her." We find, moreover, that "a boy is not supposed to tease *unmarried father's sisters*"—i.e., should not indulge in sexual banter or play with unmarried paternal aunts or cross-cousins, since he could not marry one if mutual teasing should result in the loss of self-control and pregnancy. The writer is unable to cite adequate evidence regarding boys who were initiated into heterosexual experience by paternal aunts or who later "abused" this joking privilege, but suspects that such evidence is so far withheld rather than non-existent. In any case, the verbal sexual incitation of a boy by the members of his adult world, using his paternal aunts as sexual objects, would unquestionably have directed his childish hopes and desires toward them and his jealousy toward their adult possessors—the father's sisters' husbands. And although he soon had a great variety of possible sexual objects, throughout life the mutual affinity between Hopi males and paternal aunts was very strong. Thus the subconscious affects here were quite probably *more nearly* equated with those of the original Oedipus concept than elsewhere in Hopi relations.

Somewhat of a parallel to the above constellation has been noted in the affectual attitudes of girls towards their mother's brother and his paternal aunts (her mother's father's sisters). The girl had less to fear than her brother from mother's brother, since her conduct was more directly regulated by her mothers. He might discipline her if she required it, was one of her chief advisors throughout his life, and in adult years— with her brothers—was a "strong arm" in marital and other crises. She might even joke mildly with him, though her joking was sexual in character only by analogy, in that she implied rivalry for his affection with his avowed sexual partners—his paternal aunts. It is quite possible that the parallel was stronger than it is here represented to be, but data assembled to date do not afford sufficient exemplification to reinforce it. In fact, it would seem that much more evidence will be required to establish reasonable proof that a definite Oedipus complex is present anywhere in Hopi relationships. Even if we assume that the Oedipus situation is universal, the Hopi seem to have resolved a great deal of its conflict by diffusing authority and transferring much of it outside the elementary family; and also by directing childhood sexual inclinations to "safe" objects outside the household—the father's sisters.

Examining now the general sexual development of Hopi children, we find a minimum of repression. The girls, of course, were more closely supervised than the boys, since their activities kept them nearer the house, especially if they were sibling nurses. But there is little doubt that they too managed to have ample freedom for sexual experimentation at an early age in spite of the "cultural ideal" to the contrary. In any event, none of these girls normally grew up in a state of fear or ignorance regarding sex. In all cases where friendly relations with the investigator have continued through several years and complete confidence has been established, women have admitted to both hetero- and homo-sexual play from early childhood through adolescence, and there seems no reason to suppose that boys were more backwards than girls. These informants insist, however, that homosexual behavior after marriage was "bad" and was never attempted; also that in no case did it ever exclude heterosexual indulgence or come to be preferred to it.

From the foregoing pages, there appears to have been little interference with most of the major security drives and primary needs of yesterday's Hopi children. What, then, can account for the fact that many adult Hopi exhibit signs of anxiety and are notoriously argumentative and hard to deal with? Ambivalence toward white rule is undoubtedly one factor which must be considered. The middle-aged Hopi of today, as children, bore the brunt of an inevitable acculturation impact. They were taught to give warning when a white person started up the mesa, heard endless arguments as to the best methods of placating these powerful intruders, and, if they belonged to "hostile" rather than "friendly" families, spent long hours in dark holes where their mothers hid them from policemen who came to take them to boarding schools. In the words of a Hopi:

> They have captured our war chiefs, imprisoned or enslaved our ablest men, lied to us without limit and like cruel giants, they have torn our children from their parents to make American citizens of them . . . but it was understood that we had to put up with them . . . until Hopi gods saw fit . . . to deliver us.

It would seem difficult to overestimate the effects of such impressions upon children. Normally in childhood—Hopi or white—parents and other persons of authority in the culture, are "gods" of a sort for a period at least. They are inevitably any child's conception of the "all-powerful," persons he looks up to, loves, or fears; but in any case he expects them to solve his problems. To find that these persons were themselves helpless in the face of a greater power must have shaken not only childish Hopi faith in their own parents and *racial group*, but their very foundations of security as well. And when they were finally caught by policemen, or sent by their parents to day or boarding school, their personal dignity was systematically offended: their hair was cut without permission, their clothes were often burned, and they were bathed by force. Worst of all, their hair was washed and their names were changed—items always included by the Hopi in *transition* rituals. In small Hopi eyes apparently nothing their parents had taught them or done for them was proper from the viewpoint of these arrogant beings. During their school years, they were bombarded by a confusion of ideas, manners, foods, trades, and "thou shalt nots," though they quickly found that the latter were very often translated by the whites in actual practice as "thou shalt not get caught." The more clever ones learned by the observation of *bahana* methods in dealing with each other that one could often coast on his reputation if he made a good first impression. Others reacted with what has been aptly characterized as "embarrassed stubbornness." It is small wonder, therefore, that a school sojourn did little to pave the way for adult adjustment. When the girls returned to their homes, they accepted for the most part their mother's standards rather than those advocated by white women, for the bond between Hopi women and their daughters was very strong. The failure to adopt white methods of child care and diet would seem to reflect the inability of white teachers to supplant the influence of the "educated" girls' mothers rather than a lack of love for children, as has sometimes been charged. The boys learned little in school which would prepare them for the hard life of a Hopi farmer and much which made them dissatisfied with it.

There remain for discussion two formative pressures which might be strong enough to account in part for the peculiarities in Hopi personality: fears of various kinds were systematically ground into childish minds from their first understanding of spoken words; and overt indulgence of aggression—which, in the form previously defined, is probably a universal tendency—was sharply and consistently restricted after the first few pampered years. It has been pointed out that anxiety and fear are both responses to real or fancied danger, the distinction being that anxiety is diffuse, uncertain, and helpless, while fear is positive and active in dealing with it. The physiological preparation of an organism for dealing with danger in response to fear and anxiety is too well known to require exposition here, but when the energy thus created is not expended in physical action, and particularly when this response is too frequent or too long sustained, the organism is in a more or less constant state of "alert." Such a condition could scarcely fail in toxic physical and psychic effects, and, whatever the reciprocity of these, it can hardly be denied than an "anxious" person is almost always a "touchy" one.

The inculcation of fear among the Hopi was a gradual but continuous process. First of all there were *Soyoko* and her ilk, who usually appeared once a year in connection with the *Powamu* ceremony in February, but might come more often if children misbehaved. Such fearful beings were a clever device for minimizing parent-child conflict but they brought nightmares throughout a lifetime, even though youngsters learned their identity at, or soon after, initiation. There was also the spider woman and *powakas;* there were owls, the dead, dread *Masau'u* (the God of Fire and Death), tales of fearful famine and droughts—a constant parade of horrors recited by the old and listened to with avid interest by the young. The Hopi use whatever cultural resources and personal stamina they possess in dealing with fear, and such resources naturally vary from Hopi to Hopi. As children, they became aware of various techniques for the displacement of anxiety, notably the "recitation of kinship terms to unite our hearts," dependence upon medicine men who could extract the poison arrows of *powakas*, and—after initiation—the faithful discharge of ritual obligations to gain favor with Gods and ancestors. Their conscious minds eventually learned that *Soyoko* did not eat children, but her damage had by that time been well done. Moreover, *Masau'u* (the dead) and *powakas* (famine and thirst) were not retired with *Soyoko;* rather the childish images of these fused with that of *Soyoko* to haunt Hopi dreams and the dark to the day of their death. As adults, these people dislike the dark and hate to sleep alone, while the frequent "bogey man" of their dreams is a vague shape which appears to symbolize from time to time all of these things.

Persons who have been long subjected to insecurity often fear to tempt fate by the acknowledgment of a temporary improvement in conditions, and undoubtedly the Hopi tendency to anticipate the worst is in part this not unusual reaction. Admittedly, these people had reason to be anxious over their food supply, for they had little control over it, and this basic fear is unquestionably linked with the others. For instance, a child who did not acquire the industrious and co-operative habits "recommended" by *Soyoko*, had small chance of survival as a Hopi farmer; furthermore, while the lot of invalids and aged among them was difficult, it would have been unbearable indeed had there been no afterworld from which these unfortunates might later return with blessings in the form of rain for the crops, or curses in withholding it.

The importance of aggression as a factor in personality development has been much stressed in recent years. Whatever the name or origin of this tendency, it cannot be denied that every normal individual strives to do something or other about it when his physical and emotional comfort is threatened. The energy thus created is normally directed against his total environment or some segment of it, from shortly after birth until he dies. Soon he ceases merely to react to comfort and discomfort, he anticipates them and devotes most of his energy for the remainder of his life to an effort to "seek pleasure and avoid pain." It will be seen, therefore, that aggression is not here interpreted as being always a consequence of frustration, but rather a response both to it and to the *anticipation* of it.

In considering aggression among the Hopi, we must remember that the *young* child was "lord of the manor" as far as his immediate surroundings were concerned. Numerous powerful and adept persons seem to have been put in his environment to administer to his wants; he had early found that aggressive behavior brought prompt and effective results. What little authority conflict existed was diffuse, and there was extremely little of it in early years. In his contacts with siblings and other boys and girls, a Hopi child took what he could get by any means at his disposal, as normal children probably do everywhere. But rather suddenly this pleasant state was interrupted. Society, represented by his parents, mother's brothers, and others, now began to mold this relatively unrestrained personality into its conceptually ideal pattern of smooth, selfless, co-operative effort which was necessitated by crowded villages, natural and human enemies, and an irregular food supply; the behavior which had been so pleasantly effective and easy during the first few years of his life was the antithesis of that which was now expected of him.

As he grew in understanding, the Kachinas slowly became real Gods, with power to punish as well as reward; no longer were they mere creatures of color dancing in and out of his consciousness while he lolled comfortably in a convenient lap. The first initiation usually came between the ages of seven and ten, and, for those who entered this next phase of their socialization by way of the Kachina society, it was a severe one. In any case, at initiation the child learned that the Kachinas were not *real gods* but merely representatives of them, and that they were an endless duty as well as a pleasure. The traumatic effect of this blow to a young Hopi's faith in his intimate world must be emphasized. All informants questioned by the writer have drawn the same picture of their reaction to initiation; their emphasis is rarely upon an anticipatory fear of it, nor upon the physical hardships endured during it. Rather they stress a previous struggle against disillusionment in which the hints—not very specific because of the severe penalty for betrayal—of earlier initiates were dismissed; and finally the intense disappointment in and resentment toward their elders which survived in consciousness for a long time.

There follows a direct quotation from one informant which differs from the others only in its better English phrasing:

> I cried and cried into my sheepskin that night, feeling I had been made a fool of. How could I ever watch the Kachinas dance again? I hated my parents and thought I could never believe the old folks again, wondering if Gods had ever danced for the Hopi as they said and if people really lived after death. I hated to see the other children fooled and felt mad when they said I was a big girl now and should act like one. But I was afraid to tell the others the truth, for they might whip me to death. I know now it was best and the only way to teach the children, but it took me a long time to know that. I hope my children won't feel like that.

It is true that childish conceptions of religion among ourselves undergo a radical change as we grow up, but rarely do these come about in one night. And for Hopi children there was the double burden of disenchantment and modified behavior, for while an altered concept of the Kachinas eventually became a vital part of their lives, excessive indulgence by their elders had disappeared never to return.

The mechanisms by which an individual learns the "best"—i.e., the least uncomfortable and most pleasurable—responses to his social or physical environment are many. Whether we call the entity which finally dictates the proper response a super-ego or conscience is here beside the point. The important thing is that by a system of rewards and punishments, imitation of approved cultural behavior, and the gradual repression of that which is not approved, "normal" children everywhere learn to make the socially accepted responses while others learn to cover the evidences of their failure to do so, or to take the consequences. But almost no one is ever completely reconciled to any of these three alternatives. The frustration thus produced is usually expended among ourselves by the socially accepted means of aggressive competition in schools, in work, and in games. In all of these, overt physical and verbal aggression is permitted within the none-too-stringent rules. But among the Hopi, competition is the worst of bad taste and physical aggression is rigorously suppressed. Outwardly, a Hopi learned to smile at his enemies, to use "sweet words with a low voice," to share his property, and to work selflessly with others for the good of the tribe. He learned, in short, that he must not engage in open conflict with other Hopi; but there remained another form of aggression open to him, which, judging by the low insanity incidence among these people, has some catharsis value: with a tongue as pointed as the poison arrow of a *powaka*, he carries on a constant guerilla warfare with his fellows.

In summary, this preliminary study of Hopi psychology suggests that the first five or six years of childhood among them may have a different significance than Freudian theory allows. If this should be verified with reference to their society and other societies, the relations of personality to culture will be even closer than we have generally assumed them to be. It seems evident that the Hopi were not subject to the same set of early pressures which encompass Western children, and even when the pressures were similar, they were much less intensely instigated. Hence it cannot be assumed that the stresses which in our society are said to play such an important role in personality formation are equally forceful among the Hopi. Of greater weight in Hopi personality development were the frustrating acculturation influences, the fears of various kinds which were ever present and inadequately sublimated, and the suppression of physical aggression. Probably the rather sudden shift from indulgence to control was also an important precipitating factor in personality formation.

17

Jules AND *Zunia Henry*

DOLL PLAY OF PILAGÁ INDIAN

CHILDREN

The following selection discusses hostility in still a third cultural context and extends the discussion to the relationship between hostility and sexuality. However, its primary importance for the purpose of this book is the contribution made to one of the most central questions that must be faced in a general theory of personality formation: Are certain psychological processes universal? Mrs. Eggan considered the Oedipus complex among the Hopi and also touched on the matter of rivalry among young brothers and sisters for the affection of the mother and other relatives. The Henrys' monograph is devoted to an examination of manifestations said by the psychoanalysts to be pan-human (sibling rivalry, penis envy on the part of girls, children's anxiety about the intercourse of their parents) among the Pilagá Indians of South America. Material on sibling rivalry in our own society is presented in the monograph by Dr. David Levy cited by the Henrys. Dr. Levy has also published data on sibling rivalry in a Guatemalan Indian community.[1]

Perhaps the best answer, in terms of evidence at present available, is that psychoanalysts are right in asserting the universality of certain

NOTE: Selections from this publication (*Research Monographs, No. 4, American Orthopsychiatric Association,* 1944, copyright, 1944, by the American Orthopsychiatric Association, Inc.) have been made with the kind permission of the authors and the American Orthopsychiatric Association. Most of the material reprinted here is from the last sections of the monograph, and a great deal of the Henrys' analysis of hostility patterns has not been included. Two brief descriptions of doll play have been appended as a sample of the material in the protocols.

[1] "Sibling-Rivalry Studies in Children of Primitive Groups," *American Journal of Orthopsychiatry,* Vol. 9 (1939), pp. 205–14.

emotional mechanisms of infancy and early childhood. Since all human children are born helpless and receive their early experience in a family made up of older persons of both sexes, it is plausible that there should be general themes that transcend culture and other effects of membership in a particular group. However, it also seems clear that the group-membership determinants bring about many variations on the basic themes. Among the Hopi, the boy's hostility is maximized against the husbands of his father's sisters, rather than against his father. Similarly, Malinowski claimed that among the Trobriand Islanders of the Southwest Pacific the form of the Oedipus complex was different—with the boy having incestuous wishes for his sister (not his mother) and hostility towards his mother's brother (not his father). The Henrys likewise document the reality of sibling rivalry among the Pilagá—but with some interesting and important differences in form. For example, the issue is not complicated by remorseful or defensive feelings, as in our society.

This selection has another significance for our interests. Limitations of space in this volume have prevented a section on methods for the study of personality in nature, society, and culture. In addition to observation, interviewing, the use of personal documents (e.g., the selection by Allport, Bruner, and Jandorf, p. 436), and the recording of case histories (e.g., the paper by Dollard, p. 532), these methods include tests, various sorts of experiments, drawings, play with dolls and plasticine. The monograph by the Henrys reports both free play with dolls and play in somewhat standardized situations with dolls similar to the "amputation dolls" used by Dr. Levy. Of all tests used thus far, the richest results have been obtained with the Thematic Apperception Test [2] and the Rorschach Test.

THE Pilagá Indians live near the Pilcomayo River in the Argentine Gran Chaco. Although they have a few small gardens, they depend for most of their food on wild fruits and fish. For about three months of the year, when these are very scarce, they suffer almost semi-starvation. The Pilagá live in small villages between which there are strong feelings of hostility and fear of sorcery. The population of our village was 127, of whom 37 were children under fifteen years of age. The eight houses of the village were arranged round a central plaza, which was used for drying fruit, as a playground for children, and for social gatherings. Almost as soon as one enters a Pilagá village one becomes aware of the children, for they are extremely violent and quarrelsome, and they are hungry for attention.

* * * *

Sibling rivalry is a characteristic feature of Pilagá familial relationships, and what might perhaps be called the "simple sibling rivalry situa-

[2] See William E. Henry, "The Thematic Apperception Technique in the Study of Culture-Personality Relations," *Genetic Psychology Monographs*, Vol. 35 (1947), pp. 3–135.

tion"—direct hostility toward the younger sibling because of displace‚ ment—emerges with great clarity in real life in the culture and in the ex‚ perimental play with dolls. In families in which there are several siblings, however, the conditions are complex, and where the Oedipus situation enters the entire conflict may be displaced in a different direction.

* * * *

In spite of complicating factors, sibling rivalry based on displacement may be detected in almost every case in which there is a younger sibling. The accompanying behavior patterns, however, are somewhat different from those in our culture. This is due in part to a difference in moral ideas and sentiments between the two cultures. Among the Pilagá there is no strong sense of guilt and no institutional support for guilt feelings. This does not mean that they do not experience guilt feelings, but rather that those feelings are different in some respects from what is experienced in our culture. An understanding of guilt in our culture is impossible without taking into consideration the feelings of remorse and loss of self-esteem which are so often part of it. Among the Pilagá, however, there is little of this. The daily experience of Pilagá children is not remorse or loss of self-esteem, but fear of retaliation. Self-punishment and self-accusation imply feelings of remorse, and they loom large in the Levy sibling rivalry experiments. *Self-punishment and self-accusation do not occur in any of the Pilagá material.*

* * * *

In studying the Pilagá child's behavior toward his peers, it is also important to consider the relevant love attitudes. Sibling love is not a trait for which a Pilagá child obtains the approval of his parents. The rewards for sibling love are *nil;* the punishment for sibling injury is mild and uncertain. Hence, it is not uncommon for a Pilagá child to detest his younger sibling, and he does not attempt to conceal it. The attitudes toward parents are also different among the Pilagá than among ourselves. The atmosphere of sanctity, impeccability, and compulsory love with which parents in our culture surround themselves is missing among the Pilagá. Pilagá parents are much more the equals of their children. They do not demand absolute love, honor, and respect as do parents in our culture, nor do they drill into children the "Oh, how much I have suffered for you" theme. If an angry Pilagá child hesitates to insult his erring parent, it is not for fear of disturbing some cloud-enveloped holiness, but rather to avoid a blow or a dousing with cold water. All of these contrasting attitudes should be borne in mind when considering the Pilagá material.

Field observation of sibling rivalry among the Pilagá has produced no evidence of self-accusation or self-defense. Inasmuch as those factors do not appear in the experimental material either, they may be said to be missing from the sibling rivalry syndrome among the Pilagá. Restitution, another factor in the guilt complex associated with sibling rivalry in our culture, does occur in theirs. This type of behavior has been described by

Levy [3] as "activities that restore or attempt to restore the attacked objects to their original state."

*   *   *   *

If all the children in a Pilagá family compete for the mother, the resulting hostility must become ever more intense. If C is the baby and B and A the older siblings, we have a situation something like this: B resents the intrusion of C, and C retaliates as best he can. But B, offended in his turn by the hostility of C, pays him back. Hence the hostility mounts. C, in his turn, resents B's competition for the mother too, and tries to keep him away from her. This gives rise to another cycle of hostility. Meanwhile the old strife between B and A, the oldest of the three, continues. A, however, also objects to the presence of C, and C to the presence of A. Thus C has to cope with A and B; A must struggle against B and C; and B must withstand A and C. Hence there is a constant, reversible, and clearly perceptible flow of blows from one sibling to another wherever strength permits. In all truth it may be said of a Pilagá baby that he learns to fight almost before he learns to walk.

*   *   *   *

Among the Pilagá, we do not find the condition, so common in our society, of rivalry of the younger sibling with the older *on the basis of parental preference for the older*. In our culture, the intelligence, beauty, or personality of the older sibling, or his resemblance to one of the parents may result in the concentration of attention on the older sibling to the exclusion of the younger. Such factors have little weight with the Pilagá. They do not make the fine distinctions we do, and, although they recognize resemblances to parents when the resemblances are called to their attention, these are of no importance to them. The whole weight of cultural interest is always thrown to the helpless infant. When it is no longer helpless, the Pilagá begin to lose interest in it.

In our culture "we see a diminishing percentage of overt rivalry frequencies by the next to the youngest child, with increased size of families." [4] Among the Pilagá, however, size of family has no effect on the development of overt rivalry, nor upon its intensity. In the six cases of intense hostility, the number of children in each family was as follows: three families with three children; two with two children, and one with four children. Since the average number of children per family in our village was three, the figures indicate that intensity of hostility is not determined by the number of children in the family. The two cases of least hostility occur in families of two or three children.

*   *   *   *

Why do Pilagá children inhibit their hostility at all; and what are the circumstances that assist this inhibition?

In another part of this monograph we pointed out that corporal sanc-

---

[3] David M. Levy, "Sibling Rivalry," *Research Monographs No. 2, American Orthopsychiatric Association*, 1937.

[4] Levy, *op. cit.*

tions and insults were the characteristic justice of the child world, where
the law of "an eye for an eye and a tooth for a tooth" prevails. It was
suggested, also, that it was the fear of retaliation rather than remorse
which motivated the attacks on the self. It is very likely, therefore, that
fear of retaliation is one of the factors that inhibits hostility and hence
keeps it well within the limits of the mild category with most of the
children.

Perhaps in no other aspect of child life is the fear of retaliation more
manifest than in the little girls' inter-village boxing bouts. In these bouts,
groups of embattled little girls fight each other with fists. Considering the
anger of the children, these bouts are remarkably mild. The girls face
each other in two long lines and pound each other on the arms. It is a
strikingly quiet give-and-take. But woe to the girl who strikes a blow that
is the least bit too heavy, or which falls on the body or face instead of
on the arms! She will be attacked ferociously by her opponent and
driven from the fight. It is no wonder that in these fights children are
rarely hurt. From time to time, one child or another goes too far and is
soundly whipped by her opponent; and there was the case of Chaupá
who attacked her opponent with four inch thorns and had to escape to
a relative's house to avoid a mob of infuriated little girls.

In addition to the fear of retaliation, which may be considered as a
direct influence on the display of hostility, we must also take account of
the sexual activity of the children as furnishing a diversion. Pilagá children
pass hours each day in violent sexual games of a rather simple character.
One of these is what we have called the genitalia-snatching game. In this
game the children go hurtling after one another trying to touch each
other's genital. Usually the boys chase the girls. Often they throw them
down, and try to touch the girl's genital, while she shrieks with excitement. Sometimes the fall hurts, and she cries. When the boys have caught
one girl, the other little girls dash up and beat off the boys. Then the boys
chase the other little girls, and the situation is repeated. The play may go
on for hours, and it is accompanied by the excited screams of the children.
This blending of violence and sexuality extends to many details of play.

*   *   *   *

Thus in everyday life *there is a constant veering between sexuality
and violence, and often the two are inextricably blended.* An examination
of the experimental material from this point of view seems pertinent.
Before going ahead, however, we should like to point out that it has always
seemed to us that for the children sexual intercourse itself has a certain
hostile component. This may readily be inferred from the foregoing data.
There are, however, many other indications of the way in which the
sexual act and hostility are blended in the minds of the children. A common threat is: "I'll have intercourse with you," and it passes from girl to
girl as well as from boy to boy and boy to girl. The boys speak of successful entry as: "I hit it" (as with a projectile), and failure to enter as "I
missed it." The little girls (and, indeed, the adults too) believe that men-

struation is caused by violent entrance of the vagina; and it can be seen that they scratch the little boys when they attempt intercourse. The great extent to which violence is associated with the sexual act leads to such angry remarks as that of the little foreign girl (i.e., from another village): "They wanted to mount us!"

In view of this situation, we have often been tempted to list every act representative of intercourse in the experiments as hostile. We have refrained, however, because we preferred to limit ourselves to acts obviously hostile in intent. Intercourse has been listed as a hostile act only where it is obviously so from the context, i.e., where it is accompanied by a derogatory remark or where the dolls are made to copulate very violently.

In our experiments there is a great deal of sexual as well as hostile activity. This is due to the fact that the experiment was set up not only in response to the needs of a specific problem, but also in response to the children's spontaneous demands on the material. In so far as experiments permitting spontaneous expression of drives are faithful mirrors of reality, we have much to gain by this procedure. Child life does not usually express itself in terms of one component only; besides, life is full of distractions and alternatives. One of these among the Pilagá is sexual activity. An analysis of a few experiments will illustrate the blending of sexuality and hostility.

Mátakana Trial III (see p. 251) puts the baby doll on top of the mother doll and says, "You have intercourse in the anus." By having the baby doll copulate in the anus with the mother doll, Mátakana gets the baby away from the breast; expresses contempt for her mother by referring to her vagina as an anus, and insults her baby brother by accusing him of anal intercourse. These are common forms of insult among young and old. A similar purpose is accomplished by placing the baby doll in a sexual position on the father doll, for it gets the baby away from the mother and scornfully implies homosexual activity.

The blending of hostility and sexuality is transparent in Yorodaikolík Trial IV. When Darotoyí puts the Tapáñi doll on the Yorodaikolík doll, Yorodaikolík promptly says: "She's deflowered" (i.e., her vagina has been attacked and injured). Later in the same trial, after hostility activity with the turtle, he puts the sister doll next to his own real penis and then puts the self and sister dolls in copulating position.

Yorodaikolík provides us with many interesting examples of the way in which overt hostility passes easily into overt sexuality and vice versa. Thus, in Trial III he is shaking his fist at the mother doll, but as soon as Simiti puts the Yorodaikolík doll between the legs of the mother doll he says: "Intercourse," and repeats the activity after her. In Trial VI he makes mother and baby dolls copulate; then he switches to murderous activity with a knife, but, as soon as it is taken away, he changes immediately back to sexual activity. In Trial XI, the falling of the sister doll on the self doll at once suggests her death and intercourse, and starts a sequence of frantic piling of the dolls one on top of the other. The inner significance of this piling is intercourse, and the piling terminates with the crowning of the heap with a scissors and the announcement that "it has teeth." In Trial XII (see p. 250), a long series of hostile acts, some of which have clear-cut sexual components, is followed by frank sexual activity and then a sudden switch to hostility again.

Maralú shows an almost constant alternation between transparently hostile acts and sexual ones. Thus, after putting the mother and father dolls in position of intercourse, he runs the toy tank over them. (We had explained to the children the function of a tank.) Then he amputates the father doll and, after replacing the parts, swings the doll in the way that is typical of the child custom of torturing the young of some animal —usually a bird—to death. Then, after improving the shape of the father doll's penis, Maralú again puts the mother and father dolls in position of intercourse.

Another interesting blending of hostility and sexuality is provided by Yatákana (Trial IV), where she pushes the father doll violently on the mother doll in the position of intercourse.

The experiment in which Nakínak, Kuwasiñítn, Wetél, and Yoro-daikolík take part is striking in its tremendous release of sexuality. The activity is about equally divided between hetero- and homosexual. Much of it is hostile in intent, that is, most of it consists of insulting references to sexual relations in Nakínak's family. This activity is interpreted as hostile also by Nakínak. Thus, when Wetél puts the brother doll on the mother doll, Nakínak takes it off scowling. Later, when Wetél puts the penis of the Nakínak doll into the anus of the brother doll, Nakínak pinches Yorodaikolík for saying: "They are having intercourse in the anus." Toward the end of the experiment, when Kuwasiñítn places the brother doll on top of the copulating mother and father dolls Nakínak shakes the box. Previously he had done the same thing himself.

These examples will suffice to indicate the general character of some of the psychic forces at work in the children's play. Hostile and sexual activity alternate with each other and are sometimes blended in the same act. In discussing inhibition of hostility in the Pilagá experiments, therefore, we must take into account not only the deference of Pilagá children to the dolls as our property, the sanctions against attacking relatives, and the absence of attempts to activate and encourage their attacks on the dolls; but also the fact that the children were free to divert their activity into other channels.

*　　*　　*　　*

Where children can see the adult sexual act, where sexual conversation and gestures are perfectly open, and child life is untrammeled by sexual taboos, it is reasonable to expect that the children should experiment with their sexual apparatus and attempt to imitate adult sexual behavior. This is the case among the Pilagá. Considering the extent of child knowledge about sex and the age at which this knowledge becomes articulate—three years—intercourse in a Pilagá household must not only be visible to the children, but carried on with little if any attempt to conceal the act from them. Absolutely no prohibition is placed on child sexual activity by the adults, so that the children are at liberty to do what they please. Under such circumstances, the only limits to the child's sexual activity are his physiological capacities and the tolerance of his companions.

It must be clear from the previous sections that sex is a strong and constant interest of Pilagá children. It is the purpose of this section to consider those aspects of sexual behavior which appear in the experiments and to relate them to day-by-day observation.

The commonest forms of sexual activity with the dolls—activity not directly related, however, to the standard play procedure of placing proper genitals on the dolls—in order of frequency are (1) placing the dolls in position of intercourse, (2) castration, and (3) change of sex. Other forms of activity are (4) piling, and (5) the making of penises that look like breasts. In one trial, Sorói repeatedly changed the form of the plastiline from penis to breast and back again. Throughout the play, verbalization often makes clear the intent of the activity listed as sexual.

The psychoanalyst would no doubt go much further than we have in analysis of this activity. Considering the newness of the material and our own lack of psychoanalytic preparation, we have thought it best to limit our discussion to the most obvious aspects of the material.

Intercourse position—The commonest form of sexual activity is placing the mother and father dolls in intercourse position. This was done by ten children. As being the most impressive act on the child's sexual horizon and the one most fraught with consequences and meaning for himself, it is to be expected that most of the children would perform at least this act even if they performed no other.

That the parents' sexual activity is disturbing to the child can be inferred from the following. In Darotoyí Trial I, Yorodaikolík says: "Simkoolí (Darotoyí's baby sister) is crying because her mother is having intercourse." Tanorow'í Trial I points to the self doll and says: "It's crying"; and in answer to the examiner's question, why, she says: "Because they (the parent dolls) are having intercourse." It will be observed that in this trial she stretches the arm of the father doll over the self while the father doll is lying in copulating position on the mother doll. Thus she signifies her desire not to be excluded during the sexual act. A similar and more intense expression of this attitude occurs in her second trial where, after maneuvering the mother and father dolls into a position which to her signifies intercourse, she puts the self doll in the arms of the father doll, the face of the self doll very close to the penis of the father doll. Thus, twice she placed the self doll in position of great intimacy with the father doll while she dramatized the sexual activity of her parents. The subsequent act, in which she places the self doll near the mother doll with the face of the self doll next to the anal region, seems to have been a move preparatory to further movements with the parent dolls.

Tanorowí's activity in these two instances demonstrates the desire to be included within the parents' sphere of interest at all times and the desire for sexual intimacy with the father. It is interesting that she does not put the self doll directly in position of intercourse with the father doll. The fact that the face of the self doll was placed close to the penis of the father doll may have been due to carelessness in arranging the dolls. Since we do not have in our notes that the move was deliberate, and since, being so rigid, the dolls are not always easy to manage, we cannot be sure that the position of the self doll in this instance really represents oral strivings.

Two other girls gave experimental evidence of desire for sexual relations with the father. In Trial II, Dañakána puts the baby doll on the father doll, close to the penis and says: "They're copulating." The examiner asks: "Is that its father?" and Dañakána answers, "Yes." The baby doll would, in this case, be Dañakána, because she is the youngest in her family. Besides, she occupies exactly the same position as Sorói, who had just been using the dolls, and who had been represented by the same doll. The case of her sister Simíti is clearer. Even though Simíti does not place the self doll in copulating position with the father doll, she spends practically the whole of her eighth trial manipulating and stroking the penis of the father doll, finally placing the father doll on the mother doll.

Considering the general sexual background and the way in which Pilagá men hold their children, it would be strange indeed if little girls did not view their fathers as sexual objects. The characteristic way in which a man holds his child is between his legs. Since often he wears only a breech clout, or sits with his skirt thrown back over his legs, exposing the breech-clout covered penis, it is natural that a little girl should develop an interest in her father's genital.

The desire of a little boy to have intercourse with his mother appears in Darotoyí II, and in Yorodaikolík V, VIII, and IX. Yorodaikolík was once observed to masturbate on the buttocks of his mother who was lying down. She did nothing to stop him. Darotoyí was never observed to do this. In the development of this kind of behavior in little boys, a number of factors must be taken into consideration. First is the pregnancy and lactation taboo on intercourse, and the long period of abstinence it imposes on the woman as well as on the man. During this period, two women were observed to use the feet of their infant sons to masturbate, i.e., they held the feet of the babies, eight and fifteen months respectively at the time of the observation, pressed against their genitals. Another important factor is that mothers manipulate the genitals of their infant sons. It is likely also that the straddled position of the child on its mother's hip in carrying would have some effect in focusing the child's sexual interests.

In the play of Yorodaikolík, one sometimes gets the impression that putting the self doll in copulating position with the mother doll really takes the place of putting the self doll to nurse. It is significant in this connection that Yorodaikolík never puts the self doll to nurse, although in other respects, as we have seen, he has given convincing evidence of his regression wishes. We have no evidence of regression from Darotoyí.

Yorodaikolík's sexual drives, however, are not limited to his mother. He also puts the self doll several times in position of intercourse with the sister doll. In everyday life, Yorodaikolík's sexual horizon is far from circumscribed. He utilizes a great deal of his environment for erotic purposes, masturbating not only against his sometimes compliant, sometimes rejecting sister, against Naichó, his housemate, against other little boys and girls in the village, but also against inanimate objects. He manipulates his penis constantly. A generalized erotic attitude toward the environment is also characteristic of the other little boys and girls.

In Trial III, Anetolí gives us another interesting insight into the child's attitude toward intercourse of the parents. After putting the mother and father dolls in position of intercourse, the father doll's penis falls off. She remarks: "It's good that way." Castration is a typical form of attack and may be conceived as serving three functions: the expression of general hostility, the expression of the wish to make sexual relations between the parents impossible, and the expression of penis envy. Sorói, in Trial IV, squashes the penis of the father doll and leaves it that way. Koitaháa, in Trial II, asks whether he may remove the penis of the father doll and, when told no, does it anyway. A few moments later in the same trial he says of the father doll: "I burn it." It will be remembered that Koitaháa has a three day old sibling.

Attacks on genitalia—Darotoyí, in a state of high exultation, attacks the genitals of his entire family. In Trial I, he attacks the genital of the baby doll and the penises of the father and brother dolls. He behaves similarly in Trial V, except that he attacks the genital of the older sister and the mother also. Yorodaikolík's remark in Trial IV that the sister doll is deflowered, and his attack on her vagina in Trial VI, are other aspects of the same drive toward attack on the genitals.

Attack on the genitals is the converse of *fear* of attack on the genitals. In everyday life attacks on the genitals—real, feinting, and verbal—are common enough. Below are a few examples.

1. Nagête, playing with Yorodaikolík, pushes him over and handles his penis, saying: "I'll cut it off."

2. Diwá'i screams and screams at Yorodaikolík: "Come here! Come here!" When he does go home, his mother, Nenarachí, and Piyärasáina scold him poisonously. Nenarachí says: "You will be killed." Piyärasáina says: "Your penis will be cut off."

3. Araná and Nagête put a dead nestling in Deniki's face. Deniki seems half frightened, half amused. He leans against Nagête's arm. Then she touches his penis and says: "It is bleeding." His penis, however, is not bleeding. Tapáñi says: "It's bleeding." Araná opens the bird's mouth and

puts Deniki's penis in it, saying: "It is devoured, it is devoured." Deniki starts to whimper, but does not touch the bird. Either Araná or Nagête at last takes the bird away. Deniki goes out.

4. Yorodaikolík is sitting near the fire. His mother says: "Your testicles." Yorodaikolík moves away. His mother picks up a firebrand and puts it between his legs, moving it towards his genitalia. He immediately bursts into tears and leans on his mother's breast. She strokes his head and he soon stops.

5. Nenarachí says to Simíti: "You'd better guard your vagina, because, if you don't, a man will come along, take you away, and rape you. Then you'll be a poor little thing when your vagina bleeds."

6. Sídingkí makes a jab at Simíti's vulva with a ten-inch hunting knife. She retreats. Sídingkí says to JH: "She's going to show you her genitals."

7. JH offers his watch to Dañakána so that she may listen to the tick. She withholds in embarrassment. JH withdraws the watch. Nagête, Nenarachí, and Piyärasáina insult her in a steady stream, until she weeps and goes home. The insult most often repeated is: "She is fed up with her big wormy vagina."

Change of sex—The next problem to consider is the change of sex of the dolls. The sex of the dolls was changed in the following trials: Kayolí V (mother), Dañakána II (father), Anetolí I and III (brothers), Naichó V (father?), Yorodaikolík XII (self and brother). In the cases of Kayolí and Dañakána, the activity following the change of genitals is too brief to permit an inference as to its possible meaning. In Anetolí's trials, the genitals of the dolls are not changed, but breasts are put on male dolls. This activity was carried on in perfect awareness of the sex of the dolls. It will be observed that in the first two trials Anetolí in collaboration with her playmates puts breasts on all the dolls. It is not until well on in the third trial that interest shifts momentarily away from breasts to genitals but ends up with breasts. Although it is true that girls of this age are extremely preoccupied with breasts and the possibility of their having them, it does not seem that this preoccupation is sufficient in itself to explain the persistent drive toward putting breasts on the male dolls. The authors have no specific explanation to offer for which they can produce supporting evidence from everyday life.

Yorodaikolík changes the penises of the self and baby dolls to vulvas in Trial XII. There is some indication that this change of sex is related to his fantasy life, for in another place he gave evidence of breast fantasies. In Tapáñi's Trial VI, when she puts large breasts on the self doll, he points to his own chest and says: "My breasts." Other little Pilagá boys have these fantasies too. Thus, while Darotoyí and Kapíetn were watching the little girls play a sexual game, they repeated over and over again to ZH: "My breasts hurt."

There is no direct evidence in the experiments that little girls have fantasies regarding their possession of the male organs. No little girl changed the vulva on the self doll to a penis. Kayolí is very interesting, not only because she was the only girl to change a female genital to a male one, but also because she is the one Pilagá girl we know who gave unmistakable and spontaneous evidence of desire to possess the male organ. Thus, one day, she improvised the following little song:

yóla	*laitá*
(my testicles)	(their odor)
ñikowí	*laitá*
(my penis)	(its odor)
yópi	*laitá*
(my vagina)	(its odor)
ñaté	*laitá*
(my buttocks)	(their odor)

Kayolí was also the inventor of the Matagói raping-game. In this game she impersonated Matagói, a maladjusted adolescent boy, who for some weeks was the hero of the children of our village because of his depredations on the mission store. But since he persisted in his robberies, and also quarreled with his family, the older Pilagá were frankly annoyed with him and called him *haláraik* (lunatic). Rape is a trait of *haláraiks*, and hence Kayolí called herself Matagói when she went on a "raping" rampage. Even though Matagói was never accused of rape, he was considered capable of it because he was *haláraik*. The following is an example of Kayolí's "raping behavior":

With her skirt parted in front and raised to expose her vulva, Kayolí chases the little girls around the village compound. When she catches one, she makes rhythmic movements of intercourse against her, lifting the girl's skirt and exposing her vulva, which the child then covers with her hands. JH goes into the house to write, and Kayolí comes in after him. Tapáñi starts to enter, and Kayolí makes a grab for her. Tapáñi first retreats but then comes in. Kayolí raises Tapáñi's skirt to the level of her vulva and makes rhythmic movements against Tapáñi's rump, as Tapáñi covers her vulva with her hands.

The following note is from an early stage in our field work, when the children were as yet unidentified. It is another example of a penis fantasy in a little girl. Unfortunately we do not know who it is.

A girl of seven is chasing two boys the same age. She is holding a piece of rubber tube which she calls a penis. When the ethnologist asks her whether it is her penis, she answers: "No." The following conversation takes place: Ethnologist: "Would you like to have one?" "No." E.: "Perhaps you have a vagina and a penis." "No."

Breast-like penises were made for male dolls by Wetél II, Sorói V, and Tanorowí II. Since Pilagá children are excellent modelers in clay, the breast-like appearance of the penis cannot be ascribed to lack of control of the material. It should also be noted that although Wetél put a breast-like genital on the father doll, he put a well-shaped penis on the self doll; and that after putting a breast-like penis on the brother doll, Soirói put a perfectly definite vulva on the self doll.

It is obvious that we are face to face with a definite problem, but one which, for reasons previously given, cannot be solved here. It is suggested that this behavior is functionally related to (1) phases of libidinal development described by Melanie Klein, (2) the desire to exclude the father from the circle of the mother's interest, and (3) hostile impulses centering around the removal of the dominating males. Soirói's repeated change of penis to breasts to penis in Trial VI is probably related to the first.

Spread of sexuality—Two children, Yorodaikolík and Mátakana, piled dolls on top of one another. In this activity we may see the kind of spread of a drive which Levy pointed out in another context. In Yorodaikolík's trials, the activity has a sexual meaning and may be related to the habit young Pilagá children have of piling one on top of the other in a frenzy of masturbation. In the case of Mátakana, it is not clear that the piling activity has a sexual meaning. In this type of behavior with the dolls, the child appears carried away by the power of the drive, so that he pushes on in the same way with all the objects at his disposal. Yorodaikolík even treats the scissors as something human.

Spread of the sexual drive can be seen not only in the play of Yorodaikolík and Mátakana, but also in that of Darotoyí. The spread, however, does not always consist in piling. It often extends laterally to other objects. Thus, for example, in the play of Yorodaikolík, he at times achieves a large number of lateral combinations. In Trial IX, the mother is placed face downward on all of the dolls at once, and Yorodaikolík says: "Look! Intercourse." Then he puts the self doll on the mother doll; then the baby doll on the sister doll and then on the mother doll, saying: "Has intercourse." He ends by placing the self doll on the sister doll and says: "Has intercourse." In XII, his placing of the sister doll on the mother doll is merely preliminary to a general piling whose significance is intercourse. A similar situation is presented in VI, where he places the baby doll under the mother doll, says "Intercourse," and at once places the self and sister dolls on the mother doll.

Lateral spread can be observed in Darotoyí II, where, after putting the self doll on the mother and then on the sister dolls, he puts the father doll on the brother doll and then on the sister doll. In Mátakana II, after putting the baby doll on the mother doll and saying: "The product of her vagina," she immediately begins to pile all the dolls except the father doll on top of the mother doll. The same move is repeated later in the same trial when, after putting the baby doll to nurse, she places the sister doll on the mother doll, and a few moments later she places the father doll on top of all. In Trial III we observe a lateral spread. After putting the mother doll on the father doll with the remark, "Has intercourse," she puts the baby doll on the mother with the derogatory remark, "You have intercourse in the anus," and transfers the doll to the father doll. After putting the mother doll to her mouth she again puts the baby doll on the father doll, penis to penis. Acts that are similar mechanically, however, are not always similar psychically. Thus, although in Trial II the moves with the baby doll have the meaning of getting the baby out of the way—sending it back where it came from—the movements with the other dolls cannot all be interpreted in that way because the self doll is included. On the other hand, in Trial III the intent seems the same in all cases, the contact of the penises of the baby and father dolls being related to the known homosexuality of little boys. In the play of Darotoyí and Yorodaikolík, however, the spreading has uniformly the significance of intercourse.

The phenomenon of spread of sexuality can be observed also in the experiment in which Nakínak, Kuwasiñítn, Wetél, and Yorodaikolík took part. Thus, Nakínak puts the brother doll on top of the copulating mother and father dolls, and Kuwasiñítn does the same thing immediately after him.

In the experiment last mentioned, there is so much homosexual activity that it seems worth while to include some more observations from every-day life of little boys.

Kapíetn pulls his penis into erection. Kuwasiñítn has been putting his fingers into Kapíetn's anus. Kuwasiñítn plays with Kapíetn's penis. Kapíetn goes away but keeps coming back to Kuwasiñítn. Kapíetn lies down on Kuwasiñítn's lap, face down, rubbing his penis on Kuwasiñítn's lap. Yorodaikolík runs over to him and, after pulling back his own fore-skin, thrusts his penis into Kapíetn's anus. Kapíetn goes off crying. Darotoyí is leaning on JH's lap, face down. Kuwasiñítn puts his finger into Darotoyí's anus, and he starts to cry. Kuwasiñítn goes away, and Darotoyí stops crying.

Darotoyí and Yorodaikolík are pressing their penises together. They are laughing and saying they are having intercourse. Meanwhile Kapíetn is pressing against Yorodaikolík's back.

Wetél pulls back Yorodaikolík's foreskin. Yorodaikolík appears to enjoy it.

CONCLUSIONS

The authors undertook their field research with the object of testing two propositions: (1) that, given similar intra-familial relationships in distinct cultures, the effects of these relationships on the children will be similar; (2) that, wherever sibling rivalry exists, the general symptom pattern will be approximately the same.

The study of Pilagá intra-familial relations shows that, under similar conditions of familial tension, children in distinct cultures will develop the same kind of symptom patterns.

The study of Pilagá children shows also that the patterns of behavior in sibling rivalry among them follow with little difference those found among children in our own society. The most important difference between the sibling rivalry patterns in our own society and those found among the Pilagá is that, among the Pilagá, remorse and self-punishment do not occur as consequences of hostility. Inasmuch, however, as remorse and self-punishment, while outstanding as general cultural sanctions in our own culture, do not occur as sanctions in Pilagá culture at all, it must be concluded that the difference in the sibling rivalry pattern between the two cultures is culturally determined. Remorse and self-punishment are not, therefore, fundamental to the sibling rivalry syndrome.

The problem naturally arises as to how far sibling rivalry is itself culturally determined. Since the loss, or fear of loss, of maternal affection is central to the problem, the question may very well be asked: Would sibling rivalry exist at all where not the mother but several aunts are the desirable objects? Or would it exist where the mother showered attentions equally on both children? Would sibling rivalry exist where the culture as a whole generated little anxiety? Finally, would sibling rivalry exist where membership in an important and wonderful secret society was absolutely contingent on the birth of a younger sibling, so that the presence of a new baby opened a new and wonderful life to the older sibling?

TWO SAMPLES FROM THE PROTOCOLS
OF THE EXPERIMENTS

The appendix of the Henrys' original monograph contains detailed recordings of the play behavior on which much of the foregoing analysis was based. A total of eighty-nine experiments, called "trials," were conducted with the children. Descriptions of two of these experiments have been selected to illustrate the play material.

Symbols. Arabic numerals in parentheses are used to indicate hostile acts. The hostile acts are numbered in succession as they occur in the trials. Italic numerals are used to indicate sexual acts.

YORODAIKOLÍK

Trial XII. 11–3–37. Points to sister doll and says "This one has no genital." Makes a very big one. Says: "Suña, Tapáñi says she has a big vulva" (1). He shows it to the examiner. Yorodaikolík remarks that self doll has no penis. Says: "Look, I know how to make a penis." Puts penis on self doll. Squashes it (2). Scratches it (3) and says: "The penis' opening." Says it looks like a vulva. Does the same to brother doll (4). Puts a new vulva on mother doll. Puts breasts on brother doll, and says that he's putting breasts on Deniki. Puts small blob of clay into vagina of sister doll, saying "The vagina's clitoris." Puts tiny breasts on sister doll and closes its eyes with clay (5). Says: "The covering of its eyes. *O way!* Blind!" He makes the abdomen, a clay blob on the abdomen. Says: "*O way!* Blind!" Asks for father doll. Picks up scissors. Lays them down. Picks up sister doll and says "*O way!*" Self doll and brother doll are side by side. Lays sister doll next to self doll, then brother doll, then mother doll. Pushes scissors closer to mother doll. Puts clay on head of sister doll and calls it "its hair." Puts extra nipple on mother doll's breast. Removes it and puts it on brother doll's head, calling it a hat. Handles scissors again. Asks for bubble pipe. Says he's thirsty. Drinks. Asks for hockey ball (a very heavy wooden ball about an inch and a half in diameter) and immediately puts it on mother doll's head (6). Picks up sister doll and puts hockey ball beside mother. Places sister doll beside mother doll. Places brother doll beside mother doll. Then places brother doll beside sister doll, self doll next to brother doll, then scissors. Treats the scissors like a doll. Pushes metal object into mother doll's foot (7). Picks up hockey ball. Yatákana looks in, and Yorodaikolík puts down hockey ball. He puts sister doll on genital of mother doll *1*, then across neck of mother doll. Puts brother doll on top of sister doll *2*, self on top of all *3*, and says "A heap." In answer to examiner's question, "Why is there a heap?" Yorodaikolík says "They are having intercourse." Yorodaikolík again puts sister doll across the breasts of the mother doll. Places self on sister doll *4*, and brother doll on top of all *5*. Changes brother doll to position on top of sister doll *6*. Then takes away self and brother dolls. Puts head of self doll into cup of water and says "I'm bathing." Puts legs of self doll into cup. Repeats. Puts self doll and brother doll (8) completely into cup. Takes out brother doll and says: "Denikí drowned (9). I, on the other hand, did not drown." Picks up scissors, looks angrily at Naichó at door, holds scissors up and puts them down again. Picks up hockey ball and rolls it on mother doll's face (10). Makes shoes for mother doll. Stands sister doll on head. Takes shoes off mother doll's feet and puts them on sister doll's head, saying: "The hat's brim." Puts sister doll aside on toy box within reach. Pulls up sleeves of self doll, drinks the water in which he had bathed the two dolls, and says: "I'll stop now."

* * * *

MÁTAKANA

Trial III. 10–14–37. After Yorodaikolík stops playing, Mátakana indicates that she wants to play by saying "I." She sits down and puts baby doll to nurse (mouth near breast of mother doll, body over genital of mother doll). She plays with joints of father doll. Arm comes off. Mátakana smiles and hands father doll to examiner. JH repairs it. Mátakana kisses sister doll on mouth. Puts mother doll on top of father doll *1* and says: "Has intercourse." Puts baby doll on top of mother doll *2* and says: "You have intercourse in the anus" (1) *3*. Says: "You have intercourse with milk" and puts baby doll on top of father doll's penis (2) *4*. Puts mouth of mother doll to her own mouth and says to mother doll, "Lie down, you poor little thing." Puts baby doll on top of father doll (3), baby doll's penis on penis of father doll *5*. Says: "At last her vagina is finished" (4), and puts baby doll next to mother doll. Leaves self doll next to baby doll and has sister doll next to self doll. Father doll is on other side of sister doll.

18

Allison Davis AND
Robert J. Havighurst

SOCIAL CLASS AND COLOR

DIFFERENCES IN CHILD-REARING

The selection by Davis and Havighurst is centered on the consequences of membership in special sub-groups within the larger society—class and caste. The importance of class differences, in spite of our traditional ideology, has been increasingly recognized in recent years. It has been said: "In the United States, the same pressures that are applied to make the child grow up are used to make him go up as an adult." The topic of class [1] has already been introduced by Ruesch. A caste is a class from which there is minimal escape through social mobility. Sometimes there are classes within castes. In this country, Negroes constitute the most obvious caste [2] group, though other racial or ethnic groups may have certain of the characteristics of castes.

We shall return to American culture and the class problem in the selections by Davis (p. 567) and Dollard (p. 532), and to other aspects

NOTE: Reprinted with abridgment from the *American Sociological Review*, Vol. 11 (1946), pp. 698–710, by kind permission of the authors and the American Sociological Society. (Copyright, 1946, by the American Sociological Society.)

[1] See also Bernice L. Neugarten, "Social Class and Friendship among School Children," *American Journal of Sociology*, Vol. 51 (1946), pp. 305–13.

[2] See also A. T. Childers, "Some Notes on Sex Mores among Negro Children," *American Journal of Orthopsychiatry*, Vol. 6 (1936), pp. 442–8.

*of Negro minority problems in those by Dai (p. 545) and Powdermaker
(p. 597).*

IN THIS report our purpose is to describe differences in the *cultural* training of children whose families are of different social and cultural status.

The nature of social stratification in the United States, and the cultural patterns of American social classes have been defined in extensive studies by Professor W. Lloyd Warner and his colleagues [3]. Furthermore, the powerful influence of this social-class system upon the American educational system, and upon what the child learns in school has been documented in a summary of recent studies [5].

To students of learning, and especially to those who wish to study the processes of socialization, a detailed understanding of American social-class cultures and motivational patterns is now a *sine qua non* of both research and therapy. For the social class of the child's family determines not only the neighborhood in which he lives and the play-groups he will have, but also the basic cultural acts and goals toward which he will be trained. The social-class system maintains cultural, economic, and social barriers which prevent intimate social intermixture between the slums, the Gold Coast, and the middle-class. We know that human beings can learn their culture only *from other human beings,* who already know and exhibit that culture. Therefore, by setting up barriers to social participation, the American social-class system actually prevents the vast majority of children of the working classes, or the slums, from learning any culture but that of their own groups. *Thus the pivotal meaning of social class* to students of human development is that it defines and systematizes different learning environments for children of different classes.

There are three major types of cultural systems in the United States. The first is (1) *the general American system of cultural behaviors;* these include some form of the American language, certain broad similarities, such as wearing "clothing," living in "houses," using machinery, etc.; the monogamous family; the prohibition of incest and murder; and certain democratic "ideals."

The other major cultural systems are (2) *the social-class cultures,* and (3) *the ethnic-group cultures.* So powerful are the social-class cultures that they tend to influence all the general *American cultural behaviors.* It is a fact that the specific form of the American language used, or of clothes, or of food, or of house, or even the social definition of a monogamous relationship, varies by social class.

It has been assumed generally that the basic areas of training young children were very similar in all social strata, including weaning, toilet-training, property-training, etc. This seems to be true. Our purpose, however, was to determine to what extent *the methods, the timing,* and *the pace* of this early training differed in the various social classes. We attempted to make the same comparison with regard to the training demands in middle childhood.

The primary questions which this research attacked were: (1) What are the training demands exerted upon the white and the Negro child in lower class and middle class in Chicago, and (2) What is the extent of the difference in the time of beginning, the length of, and the other conditions surrounding the training? The part of this study dealing with the differences between middle class and lower class white families in their child-rearing practices has been reported by Ericson [2].

I. PROCEDURE

The study consisted of holding guided interviews with mothers of young children, recording their responses on a schedule, and making a statistical analysis of the data from the schedules. All the mothers were residents of Chicago, and most of them lived on the South Side of Chicago.

There were fifty mothers in each of four groups, white middle class, white lower class, Negro middle class, and Negro lower class. Data on the ages of the mothers and fathers, the number of children, and the ages of their children are given in TABLE XXXIII:

TABLE XXXIII

Data on Families in the Study

	MIDDLE CLASS		LOWER CLASS	
	White	*Negro*	*White*	*Negro*
Number of Mothers	48	50	52	50
Median Age of Mother at Marriage	23	21	20	18
Median Age of Father at Marriage	25	25	25	21
Median Age of Mother at Interview	33	29	29	29
Median Age of Father at Interview	35	33	33	32
No. of Children at Time of Interview	107	109	167	184
No. of Families with 2 Children only	34	28	12	14
No. of Families with 3 or More Children	11	14	34	32
Median Age of Children at Interview	4	4	6	6

The interviewers were five women, who were trained specifically for the interviewing task in several sessions with the authors of this study. The interview was a lengthy one, lasting usually from two to three hours, and often taking place in two separate sessions.

The schedule for the interview was developed by one of the authors (A. D.) on the basis of previous work of this kind. It consisted of three main parts: The first and longest section dealt with the actual training of the child or children by the mother, and with individual differences in

TABLE XXXIV

Median Ages in Months for Various Aspects of Feeding and Toilet Training

	WHITE				NEGRO			
	MIDDLE		LOWER		MIDDLE		LOWER	
	N.	*Median*	*N.*	*Median*	*N.*	*Median*	*N.*	*Median*
Breast feeding finished	75	3.8	114	4.9	88	8.5	159	9.4
Bottle feeding finished	95	10.7	123	12.9	74	12.5	99	12.6
Sucking finished	99	10.5	147	12.8	104	12.0	177	12.2
Bowel training begun	99	7.5	158	10.2	105	5.5	172	8.5
Bowel training complete	91	18.4	152	18.8	95	13.4	160	18.6
Bladder training begun	93	11.2	156	12.2	102	9.2	124	11.1
Bladder training complete	81	24.6	139	24.0	76	18.0	143	19.0

personality among the children in the family. The second section dealt with the mother's expectations concerning the occupation, education, and responsibilities and privileges of her children, with the child's regimen (meals, sleep, recreation), relations with father, etc. The third section dealt with socio-economic data concerning the mother and father and their families.

The families were classified into middle and lower social classes by using data from the interview which have been found to be closely correlated with social class placement as defined and described by Warner and Lunt [4] and by Davis, Gardner, and Gardner [1].

The principal factors used in making the classification were occupation of parents and their siblings, education of parents, their siblings, and grandparents, property ownership, membership in churches and other associations, and section of the city. One of the authors (A. D.) discussed these data with the interviewer in each case, and made the classification. There was seldom any doubt as to the proper classification. For the Negro group, the criteria were parallel to those for the classification of the white families, but shifted systematically because of restrictions on opportunity for Negroes in American society. For example, where the occupation of

TABLE XXXV

Proportions of Children with Certain Kinds of Feeding and Toilet-Training Experience

| | WHITE | | | | NEGRO | | | |
| | MIDDLE | | LOWER | | MIDDLE | | LOWER | |
	No/Total	%	No/Total	%	No/Total	%	No/Total	%
Children breast fed only	5/106	5	28/163	17	32/107	30	80/179	45
Children bottle fed only	31/100	31	49/151	32	19/106	18	20/179	11
Children both breast and bottle fed	63/99	64	90/147	61	58/104	56	84/174	48
Children breast fed one month or more	76/106	72	118/163	72	90/107	83	164/179	92
Children breast fed longer than 3 months	34/106	32	66/163	41	63/107	59	145/179	81
Children sucking longer than 12 months	21/99	21	66/147	45	32/104	31	51/177	29
Children fed when they seemed hungry	3/106	3	53/153	35	6/108	6	87/175	50
Children having pacifiers	1/107	1	22/167	13	8/105	7	17/184	9
Children held for bottle or breast fed only	53/79	67	72/166	43	78/108	72	99/179	55
Children weaned sharply	20/101	20	23/154	15	7/105	7	39/182	21
Children who sucked thumb	54/105	51	30/166	18	50/104	48	54/183	30
Bowel training begun at 6 mo. or earlier	48/99	49	36/158	23	91/105	87	49/172	29
Bowel training complete at end of 12 mo.	25/91	28	31/152	21	46/95	49	37/160	23
Bladder training begun at 6 mo. or earlier	17/95	18	22/157	14	4/102	40	22/124	18
Bladder training complete at end of 18 mo.	26/81	32	67/139	48	51/76	67	73/143	51
No. of children who have masturbated	56/104	54	27/162	17	30/102	29	27/182	15

mail carrier would have suggested lower-class status for a white man, it suggested middle-class status for a Negro.

Since all the mothers were native-born, the factor of foreign parentage was ruled out of the study to a considerable extent. Nevertheless, this factor did appear in the white lower-class sample to a limited degree, due to the fact that about a third of this group were of fairly recent

TABLE XXXVI

Mothers' Reports on Strictness of Regime

| | WHITE | | NEGRO | |
| | Middle | Lower | Middle | Lower |
	%	%	%	%
Do Children Take Nap in Daytime?				
Yes	89	52	86	63
No	9	46	12	21
Sometimes	2	2	2	16
Total No. answering	46	50	50	48
Age at which Boys Go to Movie Alone				
5–7 years	17	35	35	70
Over 8	83	65	52	30
Doesn't approve of movies for child	0	0	13	0
Total No. answering	23	40	23	40
Age at which Girls Go to Movie Alone				
5–7	14	45	0	5
Over 8	86	55	72	95
Doesn't approve of movies for child	0	0	28	30
Total No. answering	28	42	32	37
Time Boys in House at Night				
5–6 o'clock	59	33	31	2
7–8 o'clock	32	33	63	57
9–10 o'clock	5	26	0	41
At will	4	8	0	0
Children not allowed on street	0	0	6	0
Total No. answering	22	39	32	44
Time Girls in House at Night				
5–6 o'clock	68	33	35	8
7–8 o'clock	29	39	56	63
9–10 o'clock	3	23	0	26
At will	0	5	0	0
Children not allowed on street	0	0	9	3
Total No. answering	31	39	34	38

foreign extraction, largely Italian, living in South Chicago. The remainder of the lower-class white group live in the Woodlawn and Hyde Park Areas, and are practically all of "old American" stock.

The middle-class sample is probably more representative of upper-middle than of lower-middle class people, with a high proportion of fathers in professions and managerial positions in business. The lower-class sample is definitely upper-lower rather than lower-lower class. The

TABLE XXXVII

Class Differences in Child Rearing

Feeding and Weaning

More lower-class children are breast-fed only.

More lower-class children breast-fed longer than 3 months (Negro only).

More lower-class children are fed at will.

Weaning takes place earlier (on the average) among middle-class children (white only).

More lower-class children suck longer than 12 months (white only).

More lower-class children have pacifiers (white only).

(c) More middle-class children are held for feeding.

(c) More lower-class are weaned sharply (Negro only).

Toilet Training

Bowel training is begun earlier (on the average) with middle-class children.

Bladder training is begun earlier (on the average) with middle-class children.

Bowel training is completed earlier by middle-class children (Negro only).

More middle-class parents begin bowel training at 6 months or earlier.

More middle-class parents begin bladder training at 6 months or earlier (Negro only).

More middle-class parents complete bowel training at 12 months or earlier (Negro only).

More middle-class parents complete bladder training at 18 months or earlier (Negro only).

(c) More lower-class parents complete bladder training at 18 months or earlier (white only).

Father-Child Relations

Middle-class fathers spend more time with children.

Middle-class fathers spend more time in educational activities with children (teaching, reading, and taking for walks).

Lower-class fathers discipline children more (Negro only).

Occupational Expectations for Children

Middle class expect higher occupational status for children.

Educational Expectation (Length of Education)

More middle-class children expected to go to college.

Age of Assuming Responsibility

Middle class expect child to help at home earlier.

Middle-class girls cross street earlier (whites only).

(c) Lower-class boys and girls cross street earlier (Negro only).

Middle-class boys and girls expected to go downtown alone earlier.

Middle-class girls expected to help with younger children earlier.

Middle-class girls expected to begin to cook earlier (white only).

Middle-class girls expected to begin to sew earlier (white only).

Middle-class girls expected to do dishes earlier (Negro only).

(c) Lower-class children expected to get job after school earlier.

(c) Lower-class children expected to quit school and go to work earlier.

Strictness of Regime

Middle-class children take naps in daytime more frequently.

Lower-class boys and girls allowed at movies alone earlier.

Middle-class boys and girls in house at night earlier.

TABLE XXXVIII

Color Differences in Child Rearing

Feeding and Weaning

More Negro children are breast-fed only.
More Negro children are breast-fed for three months or more.
More Negro children are fed at will (lower class only).
More Negro children have pacifiers (middle class only).
More white children are weaned sharply (middle class only).
Weaning takes place earlier (on the average) among white children (middle class only).
(c) More white children suck longer than 12 months (lower class only).

Toilet Training

Bowel training is begun earlier with Negro children.
Bladder training is begun earlier with Negro children.
Bowel training is completed earlier with Negro children (middle class only).
Bladder training is completed earlier with Negro children.
More Negro parents begin bowel training at 6 months or earlier (middle class only).
More Negro parents begin bladder training at 6 months or earlier (middle class only).
More Negro parents complete bowel training at 12 months or earlier (middle class only).
More Negro parents complete bladder training at 18 months or earlier (middle class only).

Father-Child Relations

White fathers spend more time with children (lower class only).
White fathers teach and play more with children (lower class only).
Negro fathers discipline children more (lower class only).

Educational Expectations (Length of Education)

More Negro children expected to go to college (lower class only).

Age of Assuming Responsibility

Negro boys and girls cross street earlier (lower class only).
(c) White girls cross street earlier (middle class only).
Negro boys go downtown alone earlier (lower class only).
Negro girls expected to dress selves earlier.
Negro girls expected to go to store earlier.
Negro girls expected to begin to cook earlier (lower class only).
(c) Negro children expected to quit school and go to work later.

Strictness of Regime

Negro boys allowed to go to movies alone earlier.
(c) White girls allowed to go to movies alone earlier.
White boys and girls in house at night earlier.

fathers were mainly steady, hard-working people at the semi-skilled and skilled levels.

The sample was not secured by a random procedure. Rather, it consisted mainly of people who had children in certain nursery schools, some private, and some war nurseries supported mainly by public funds. The South Chicago group consisted mainly of people who lived in the neighborhood in which one of the interviewers had grown up. The Woodlawn lower-class group was obtained by calling at random in certain areas where housing was obviously poor, and passing from one family to another with whom the person being interviewed was acquainted. Any systematic bias introduced by these procedures lay probably in the direction of getting a middle-class group which had been subjected to the kind of teaching about child-rearing which is prevalent among middle-class people who send their children to nursery schools.

2. RESULTS

There are a large number of reliable differences between classes and between colors. The accompanying tables report the differences between classes and colors on certain parts of the schedule. Those differences which are statistically reliable are summarized in TABLES XXXVII and XXXVIII.

Feeding, Weaning and Toilet Training—TABLE XXXIV shows the median ages for various aspects of feeding and toilet training, while TABLE XXXV shows the proportions of children in the various class and color groups who had certain kinds of feeding and toilet training experience.

Strictness of Regime—There were several questions in the interview which explored the degree of strictness of the regime set up by parents. These questions dealt with daytime naps, time children are due in the house at night, and age of going to the movies alone; all things in which the impulses of the child sometimes run counter to parental desires. TABLE XXXVI shows the results on these questions.

Age of Assuming Responsibility—A number of the questions dealt with the age at which certain things are expected, such as dressing oneself, helping with younger children, learning to cook and to sew; and with the age at which certain responsibilities would be permitted, such as crossing the street alone. In general these questions dealt with the theme of *assuming responsibility*. "At what age does your child assume responsibility for one thing or another?" was asked in a variety of forms. There is not space enough to report the results of this section in detail, but the reliable differences are summarized in TABLES XXXVII and XXXVIII. In general, it may be said that middle-class children are expected by their parents to assume responsibility earlier than lower-class children are expected to assume similar responsibilities by their parents.

Summary of Class Differences and Color Differences—TABLE XXXVII summarizes the class differences which are statistically reliable at the five per cent level, while TABLE XXXVIII summarizes the color differences which are statistically reliable at the same level. It will be seen that the differences tend to go together; a letter (c) indicates that the finding contradicts the general tendency of the results. For example, the general tendency is for lower-class children to be treated more permissively than middle-class children with respect to feeding and weaning. Contradictory to this tendency, however, more middle-class children are held for feeding.[3]

3. DISCUSSION

Personality Implications of Social Class Differences in Child-Rearing— Middle-class families are more rigorous than lower-class families in their training of children for feeding and cleanliness habits. They generally begin training earlier. Furthermore, middle-class families place more emphasis on the early assumption of responsibility for the self and on individual achievement. Finally, middle-class families are less permissive than lower-class families in their regimen. They require their children to take naps at a later age, to be in the house at night earlier, and, in general, permit less free play of the impulses of their children.

Generalizing from the evidence presented in the tables, we would say that middle-class children are subjected earlier and more consistently to the influences which make a child an orderly, conscientious, responsible, and tame person. In the course of this training, middle-class children probably suffer more frustration of their impulses.

In the light of these findings, the data with respect to thumb-sucking are interesting. Three times as many white middle-class children are reported to suck their thumbs as white lower-class children, and almost twice as many Negro middle-class children do likewise. Thumb-sucking is generally thought of as a response to frustration of the hunger drive, or of the drive to seek pleasure through sucking. Since middle-class children are fed less frequently and are weaned earlier, the higher incidence of thumb-sucking would be expected. The Negro middle-class children are treated much more permissively than the white middle-class children with respect to feeding and weaning, but much more rigorously with respect to toilet-training. Yet the proportion of Negro middle-class children reported as sucking their thumb is almost the same as the proportion of white middle-class children so reported. *Perhaps thumb-sucking is a response to frustration of any sort, rather than to frustration in the feeding area alone.*

The data with respect to masturbation are also of interest in this connection. Three times as many white middle-class as compared with lower-class children are reported as masturbating. Twice as many Negro middle-class children as compared with Negro lower-class children are reported as masturbating. The meaning of these findings is obscured by the possibility that some lower-class mothers may not have understood the ques-

---

[3] A table in the original article, which had to be omitted for reasons of space, compares the class and color differences which were statistically reliable and independent. On the basis of this the authors conclude: "There were more class differences than color differences, and the class differences occurred approximately equally often in Negro and white groups."

tion. Or it may be that some of them did not watch as carefully for masturbation as middle-class mothers do, or some lower-class mothers may have been more hesitant than middle-class mothers in admitting that their children followed this practice. Yet none of these explanations seems probable, and perhaps the data should be taken at their face value. Perhaps masturbation is much more common among middle-class infants than among lower-class infants. If this is true, it might be explained in terms of the hypothesis that masturbation is in part a palliative to frustration. Children who are frustrated more would masturbate more, according to the hypothesis.

It is a surprising fact that the middle-class mothers, in general, expected their children to assume responsibility earlier in the home, to help with the younger children, and to cook and sew at an earlier age. For it seems obvious that there is more actual need of the children's help in lower-class families, where the work of children to be cared for is greater and the mother has very little help with the housework. The explanation probably lies in a tendency on the part of middle-class people to train their children early for achievement and responsibility, while lower-class people train their children to take responsibility only after the child is old enough to make the effort of training pay substantial returns in the work the child will do. Middle-class parents can afford to use time to train children to dress themselves, help around the home, sew, cook, and so on, at such an early age that the children cannot repay this training effort by their actual performance, although they may repay it by adopting attitudes of self-achievement and responsibility.

In addition to training their children to take responsibility early and to adopt attitudes favorable to self-achievement, middle-class families attempt to curb those impulses of the child which would lead to poor health, waste of time, and bad moral habits, according to middle-class views. Therefore they require their children to take day-time naps longer, to come into the house at night earlier, and they do not permit their children to go alone to movies at an early age. Nevertheless, they encourage their children to be venturesome in the more "constructive" activities, from the middle-class point of view, of going downtown alone to museums, department stores, dancing lessons, and the like.

Whether the middle-class or the lower-class practices are preferable is, of course, largely a matter of private opinion. But it is significant that there is now a considerable body of scientific and lay judgment operating in the middle class to make child-rearing practices more permissive. It is contended that the orthodox middle-class practices make children too anxious, and frustrate them too much for the best mental health in later life. On the other hand, it may be contended that civilized life requires the individual to be tamed and to learn to take constructive control of his impulses. A certain degree of anxiety is valuable, in that it puts the individual on his toes to learn the lessons and meet the demands of modern society in order to win the very considerable rewards of modern civilized social life.

Our own view is that the better child-rearing practices can be drawn from both middle and lower-class life and made into a combination which is superior to both of the norms as they emerge in this study.

Personality Implications of Color Differences in Child-Rearing—The striking thing about this study is that Negro and white middle-class families are so much alike, and that white and Negro lower-class families are so much alike. The likenesses hold for such characteristics as number of children, ages of parents when married, as well as child-rearing practices and expectations of children.

There are, however, some very interesting color differences. The major color differences are found in the areas of feeding and cleanliness training.

Negroes are much more permissive than whites in the feeding and weaning of their children. The difference is greater in the middle class. Negro babies have a markedly different feeding and weaning experience from white babies.

The situation is reversed with respect to toilet-training. Here the Negro parents are much stricter than white parents, both in middle and lower-class circles. For example, 87 per cent of Negro middle-class mothers said they commence bowel training at 6 months or earlier, compared with 49 per cent of white middle-class mothers; and the comparable figures for bladder training are 40 and 18 per cent.

If feeding, weaning, and toilet-training have much influence on the personality, we should expect systematic differences between Negro and white people *of the same social class*, though it is not at all clear just what these differences should be, since one group is more rigorous in its training in one area while the other group is more rigorous in the other area.

There is another noticeable color difference. Negroes of both classes tend to give their girls an earlier training for responsibility in washing dishes, going to the store, and dressing themselves. This is probably traceable to the fact that Negroes of both classes have less outside help in the home than whites do, and consequently the help of the girls is more urgently needed. It is noticeable, also, that middle-class Negro girls are not allowed to play across the street or to go to the movies alone as early as white middle-class girls. This may be due to the fact that most middle-class Negroes are forced to live in much less desirable neighborhoods, from their point of view, than those in which middle-class whites live.

4. CONCLUSIONS

This study has given clear evidence of the following things:

1. There are significant differences in child-rearing practices between the middle and lower social classes in a large city. The same type of

differences exist between middle and lower-class Negroes as between middle and lower-class whites.

2. Middle-class parents are more rigorous than lower-class parents in their training of children for feeding and cleanliness habits. They also expect their children to take responsibility for themselves earlier than lower-class parents do. Middle-class parents place their children under a stricter regimen, with more frustration of their impulses, than do lower-class parents.

3. In addition to these social-class differences, there are some differences between Negroes and whites in their child-rearing practices. Negroes are more permissive than whites in the feeding and weaning of their children, but they are much more rigorous than whites in toilet-training.

4. Thus there are *cultural differences* in the personality formation of middle-class compared with lower-class people, *regardless of color*, due to their early training. And for the same reason there should be further but less marked cultural differences between Negroes and whites of the same social class.

5. In addition to the cultural differences between individuals due to early training experience, there are individual personality differences between children in the same family. These are probably due to physiological differences and to differences in emotional relationships with other members of the family.

REFERENCES

1. Davis, Allison, Gardner, Burleigh B., and Gardner, Mary R.: *Deep South* (Chicago, University of Chicago Press, 1941).
2. Ericson, Martha C.: "Child-Rearing and Social Status," *American Journal of Sociology*, Vol. 53 (1946), pp. 190–2.
3. *The Yankee City Series* (New Haven, Yale University Press). The first four volumes of this six-volume series have been published. See also reference 1, above.
4. Warner, W. Lloyd, and Lunt, Paul S.: *The Social Life of a Modern Community* (New Haven, Yale University Press, 1941).
5. Warner, W. Lloyd, Havighurst, Robert J., and Loeb, Martin B.: *Who Shall Be Educated?* (New York, Harper and Bros., 1944).

19

Benjamin D. Paul

SIBLING RIVALRY IN SAN PEDRO

In this analysis Dr. Paul describes the interrelationships between the kinds of strains imposed on personality in San Pedro, the ways in which these strains are expressed symbolically, and the means employed to cope with them. Culture, as a determinant of personality, is not merely a mould-ing agency, a die-stamp, or a cookie-cutter. Culture provides the arena in which personality is expressed, and in so doing provides symbols for its expression as well as techniques to meet the strains and conflicts which it creates.

PROJECTIVE tests have provided evidence for the existence of sibling rivalry among various nonliterate societies. By means of doll-play experiments, David Levy elicited hostile attitudes from Kekchi children of the Coban area in Guatemala. On the basis of this and other material, Levy asserts that the sibling rivalry situation "represents a universal experience when-ever a mother has more than one child in her own care [4]."

Jules and Zunia Henry likewise used doll-play to demonstrate the presence of sibling hostility in Pilagá children of the Argentine Gran Chaco. They found the pattern of sibling rivalry to be essentially the same as in our own society, the most important difference being that "remorse and self-punishment do not occur as consequences of hostility [1]," since

NOTE: Reprinted with slight abridgment of footnotes and the addition of one new paragraph from the *American Anthropologist*, Vol. 52 (1950), pp. 205–18, by permis-sion of the author and the American Anthropological Association.

these sanctions are not part of Pilagá culture. This finding demonstrates the importance of taking into account the specific cultural matrix which shapes and expresses basic psychological mechanisms. Relying on Thematic Apperception Test materials, William Henry writes of the Hopi: "Sibling jealousy seems particularly potent and would be expressed in indirect fashion, since there is some block, probably parental pressure, against the direct expression of sibling rivalry [2]."

Sometimes, however, the symbol system of a culture serves as the equivalent of a projective test, supplying evidence of sibling rivalry in belief and ritual. These symbolic condensations offer the investigator ready-made data comparable to projective test protocols. This seems to be the case in the agricultural village of San Pedro la Laguna, bordering Lake Atitlán in the mid-western highlands of Guatemala. Known as *Pedranos*, the inhabitants speak a variant (Zutuhil) of the Quiché language which belongs to the Mayan linguistic stock, practice Catholicism, and have a type of family and social organization based on seniority and male dominance. Marriage is within the community, most couples living with the parents of the boy until a child or two is born, at which time they set up separate housekeeping. Their culture is generally characteristic of the extensive highland Maya population, but since neighboring village communities often differ in important particulars [7] the following remarks should not be assumed to apply beyond San Pedro.

Pedranos believe that a child may destroy an infant sibling by "eating" its soul or spirit. To protect newborn babies from the appetite of such a child, parents occasionally resort to a well-patterned ritual which is supposed to "cure" the offender of his fratricidal craving. A ritual specialist, a midwife or shaman, grasps a chicken by the wings or legs, beating its body against the back of the culpable child until the fowl is dead. This is done privately and secretly and out of sight of the new infant which must not witness death. The sacrificial chicken is cooked in a savory broth and served exclusively to the older child who must finish the entire dish even if it takes three or four successive meals. While the chicken is being killed or when it is being eaten, the child is lectured on the meaning of the rite. The specialist or the mother warns: "Now that another little brother (or sister) is born you are not to eat him. This chicken will be your meal. Its meat is like the flesh of your little brother. You must take good care of him and never frighten him." If the child eats willingly it means that his predatory appetite is put at rest; reluctance would signify an unrepentant spirit.

This curing drama may be construed as a stylized statement of expected jealousy between siblings in San Pedro. A comparable type of interpretation may be found in Kluckhohn's analysis of Navaho witchcraft which associates cultural "projections" with underlying social strains; the tendency to attribute witchcraft to siblings reflects latent tensions arising from actual conditions of Navaho life [3]. As a system of belief, witchcraft likewise occurs in San Pedro but it is not a vehicle for expressing sibling rivalry, which finds symbolic outlet in other ways. The object of this paper is to place the San Pedro chicken-eating ritual in proper perspective by relating it to those social, situational, and cultural factors which lend it force and meaning. The problem may be phrased in the form of several questions. What is the context of socialization in which sibling attitudes are formed? What circumstances evoke performance of the sibling ritual? What cultural beliefs and dispositions underlie it?

SOCIALIZATION PATTERN

For convenience the last three children in a family may be regarded as occupying a series of statuses as follows: nursing baby, knee child, and yard child [5]. Pedranos treat nursing babies with warmth and affection. Infants are given the breast whenever they cry. If the mother is momentarily busy at the loom or at the grinding stone, an older sister or another woman of the household mollifies the infant by picking it up. Women who leave the house for long periods necessarily take along their nursing infants. The mother uses her shawl as a sling in which the infant is completely enveloped and held close to the breast. When time allows, mothers sit in the hammock rocking their babies in their arms. Fathers fondle their infants in the evening and laugh at their antics. Older sisters take good care of babies entrusted to their care and share their parents' pride over the accomplishments of the growing infant. Babies are given bits of solid food by the age of one, but they usually continue nursing until they are fifteen or twenty months old. Infants are weaned when the mother is pregnant with another child. Mother's milk is thought to be injurious to the nursing child when the woman is in the fourth or fifth month of pregnancy. To effect weaning, mothers often smear chicken feces or ground chili on their nipples. Babies of crawling age are carried about in a shawl rather than left for long periods on the floor where they may upset pots or get into the fire. First steps are applauded and encouraged, but children are not prodded into early walking. Little effort is made to teach them bowel or bladder control until they are old enough to walk and to understand instructions.

Though there is no institutionalized favoritism in San Pedro, the natives recognize the effect of differential infancy care on personality formation. The final child in a family is designated by a special native term which applies even when the individual has grown to adulthood. According to native accounts, terminal children are assertive, demanding, and quick to anger throughout their lives. Field observations support this view. A conspicuous difference in the experience of last-born children, and one recognized as influential by Pedranos, is the fact that they are allowed to nurse until the age of three or four years in the absence of succeeding siblings.

Final children excepted, transition from the status of nursing baby to that of knee child brings a reversal of treatment. The knee child is not neglected, but it can no longer be indulged in the degree to which it has grown accustomed. Endlessly busy preparing tortillas, carrying water from the lake, tending the breast baby, and performing other household routines, mothers have little time left for older children. The knee child reacts to the loss of the breast and curtailment of maternal attention with fits of petulance, temper tantrums, occasional trance-like withdrawals, and swift changes of mood. These demonstrations are half ignored and half humored by the mother, who is less tolerant of emotional outbursts on the part of the older yard child. San Pedro women explain that children become especially irritable and make more food demands during the several months before and after the birth of the next child.

The knee child is actively and effectively prevented from carrying out aggressive impulses against the new baby. This is accomplished by removal and direct punishment if necessary but more characteristically by deliberately fostering in the knee child a positive identification with the new sibling. When visitors come to see the newly born child, the mothers and elders of the household will announce in the presence of the knee child that the latter loves his little sibling and doesn't want anybody to take it away. The knee child acts out its aggression against the self (temper tantrums), against the older sibling (who does not retaliate on pain of punishment by the parents), or against the parent. The extract from field notes that follows indicates how the dispossessed child directs its anger at the mother. In this excerpt, Nicolasa is one of several women watching a religious procession; Petrona (knee child) is a three-year-old daughter and Bartolo a year-old infant.

> Petrona is crying loudly, her face buried in mother's lap. This interferes slightly with Bartolo who is nursing. Nicolasa is watching the procession and talking with the other women, but every once in a while shows a sign of affection for Bartolo. She seems to be paying absolutely no attention to Petrona, but she finally looks irritated, probably because of Petrona. I ask why she is crying. Nicolasa answers, "She hit Bartolo on the head." "Then why is she the one who is crying?" Nicolasa doesn't know why. After crying for about ten minutes Petrona stands up and starts slapping her mother with both hands in jerky and random fashion. Nicolasa wards off blows with one arm but doesn't do anything else. Finally as Petrona continues crying, Nicolasa orders her in an irritated voice, "Go, go home."

Another example of sibling behavior involves two sisters who were observed almost daily for nearly one year. Six-year-old Concepcion and Magdalena, age two and a half, were the only children of the family until a third girl was born. During the five months preceding the birth of this baby, Magdalena was observed to exhibit tantrums with increasing frequency. She was very "temperamental," one day boisterously joyous, the next day sullen and bursting suddenly into tears. This was attributed by women informants to the fact that her mother was pregnant, though children were supposedly ignorant of all matters pertaining to sex and reproduction. For the most part Magdalena's outbursts were ignored. She was not immediately placated nor was she strongly reproved. If she persisted long enough her mother would appease her with fruit or confections.

The yard child, Concepcion, was also disturbed by the impending birth but for a briefer period. Generally genial and even-tempered, Concepcion spent much of her time mothering Magdalena. She was generally undisturbed by Magdalena's occasional assaults upon her. Concepcion frequently asked the writer for fruit or sweets. These requests were always made in the name of Magdalena who did in fact receive most of the food granted. Concepcion's requests became increasingly insistent and annoying, reaching a peak during and shortly after the time her baby sister was born. Thereafter her attitude of aggressive demanding reverted to one of pleasant friendliness.

The evening her baby sister was born, Concepcion came or was sent into the writer's house to play. She was presumably unaware that her mother was about to deliver a baby. (Nor did we realize this at the time.) In the course of a long conversation she informed us with unusual frankness and feeling that she hated her mother and father. She disliked her grandparents less, she asserted, because they did not beat her. Next day, after the baby was born, Concepcion paid us another visit, this time to tell us that Mrs. Paul's house in the United States had burned to the ground and that her mother had perished in the flames. She repeated this story the following day, alleging that her own mother had heard the news from "a stranger." The conventional stork-story in San Pedro is that babies are purchased from foreigners. At the time, we were the only foreigners living in San Pedro. Apparently Concepcion's anxiety evoked by the childbirth was first channeled into bitter criticism of her parents and then displaced onto the foreigner held responsible for the intruding baby, not to mention the mother of the foreigner.

Verbal aggression, however, is exceptional on the part of the yard child, who learns obedience and industry by parental injunction and by whipping when necessary. Characteristically the yard child at the age of five years and upward runs errands and performs useful tasks for its parents, protects and instructs the knee child, and treats the nursing baby with the same devotion exhibited by the parents. The strength of the super-ego component at this stage is indicated by the response encountered when the observers introduced a doll into a children's play group. No parents were present. Assuming the doll to represent a nursing infant, children between the ages of five and adolescence uniformly displayed an admiring and solicitous attitude. The first reaction was to hold the doll tenderly and kiss it on the cheek. This gesture follows the conventional behavior of married women paying their first respects to a mother and her new baby. The second gesture on the part of yard children was to hand the doll to a younger sibling so that the latter might similarly kiss it.

To recapitulate the early socialization sequence in San Pedro, the nursing infant receives much care and affection and suffers little deprivation; actually dependent, it experiences so nearly complete gratification of its physiological demands as to leave it with an emotional impression of virtually unlimited mastery. The knee child, finding its imperious expectations thwarted by curtailment of attention, directs its anger not at the intruder but at the mother and the next older sibling. Although sometimes discouraged from aggressing too strongly against these objects, it is allowed to indulge in demonstrations of rage and is given substitute gratification in the form of food tidbits. The yard child finds security and peace of mind by inhibiting rather than deflecting its hostile impulses; under threat of physical punishment it renounces selfishness in favor of duty and compliance. Rather than resist authority, the yard child learns to identify with it. Under intense emotional strain, it resorts to aggressive imagery. If old enough to do so, the individual under stress may even resort to outright aggressive action.

Sibling competition does not end with early childhood, but in less direct form, competition for parental favor continues into adult life. From time to time one child will report the misdeeds of another to its parents. Parental favor is more than a matter of emotional gratification. The distribution of the family inheritance is a vital concern. While all children of both sexes are normally entitled to an equal share of agricultural land and other property, the possibility of favoritism always exists with its attendant hopes, fears, and efforts to curry favor. Not infrequently disputes over property break into the open after the death of the father, siblings bringing their arguments to the village court. The authority of the court resolves conflicts of interest between siblings, and between other contestants as well, but does not completely erase personal anxiety aroused by the overt manifestation of hostility. In some cases the intensity of feelings displayed during an argument in or out of the courtroom is followed either by an alcoholic spree or a dramatic fit of self-directed rage with a patterned set of symptoms which include suffocation and violent pains in the "heart." This type of demonstration appears to be an adult equivalent of the temper tantrum in the knee child.

Before turning to the circumstances under which the sibling ritual is performed the reader should be warned that preoccupation with the antisocial factors, while dictated by the nature of the problem under discussion, inevitably creates a distorted impression of San Pedro character structure which would emerge more favorably in a rounded treatment of the subject.

THE DEFINITION OF DANGER

Infant mortality is high in San Pedro as a result of dysentery, intestinal parasites, and other consequences of inadequate sanitation. Numerous children also die during epidemics of measles and chicken pox against which the Indian population has little immunity, but such deaths are usually attributed to natural causes or to supernatural punishment for general failure to heed the ways of the forefathers.

Conviction that a given child harbors a destructive spirit is founded on several types of evidence. A baby born with two hair whorls or with the cord wound round its neck is sometimes thought to have a baneful destiny. When it is two years older the curing rite may be performed to protect the lives of subsequent siblings. If the youngest of several children continues fretful and sickly for weeks or months after it is born, blame may center on the next older child; it is feared that he is slowly consuming the spirit of the vulnerable infant; or the ailing child may itself become suspect in due time and require ritual intervention to safeguard still younger babies. But the most typical situation leading to performance of the curing rite is one in which successive babies die while an older sibling remains alive. On the assumption that the living child has eaten its younger brothers or sisters and will do the same at the next opportunity, it is subjected to the magical cure a week after another baby is born, when mother and infant formally end their post-partum seclusion.

To reduce anxiety over the death of children, parents resort to various protective measures including the use of herbal charms to ward off "evil-eye," disparagement to avert the caprice of fate, burning candles before images of the saints and at hillside shrines, and similar acts. By itself the presence of death or disease need not direct suspicion against a child. Family misfortune is susceptible to a wide range of natural and supernatural explanations including fright, were-animals, domestic discord, and the malicious sentiments of envious neighbors. But the contrast between the favorable fate of one child and the unhappy fate of the next may easily arouse suspicion that one is surviving at the cost of the other. In any event the diagnosis is not made by the parents themselves. They bring their troubles to a shaman or the family midwife; both ritual specialists owe their calling to supernatural mandate. Like our own physicians they prescribe according to certain established doctrines, taking into account the peculiarities of the case, intimate knowledge of the client, and need for reassurance. If empirical measures prove ineffective they fall back on invocation, exorcism, and special ritual. Children may be credited with evil spirits when conditions fit the cultural "definitions of the situation" outlined above. In such event the shaman or midwife ascertains or assumes that the child had been born on an unlucky day, the native calendar being horoscopic. But any given situation permits a choice of interpretations depending on the intensity of concern, availability of alternate explanations, and the judgment of the consulting specialist.

Objectively considered, the individual faces unpredictable and uncontrollable hazards. In a cultural atmosphere which places a premium on raising a family, the death of children not only brings anguish to the parents but arouses intolerable feelings of inadequacy and self-blame. The anxiety of bereaved parents may be intensified by unconscious "guilt" arising from the ambivalence in parental attitudes toward children, who are a source of both hardship and gratification. To preserve the psychological integrity of the individual, San Pedro culture, like many others, supplies mechanisms of "prediction" and "control" which simultaneously lift the blame and restore normal motivation. Under acute stress people feel the need to "do something." Our interest, however, is not merely to show that the sibling ritual is one way of meeting this pressing need, but to analyze the factors which make this particular "solution" so acceptable to the average Pedrano.

To work properly, the procedural steps of the sibling ritual should conform to the accepted symbolic idiom of the culture. One rule is to keep the male principle and the female principle unmixed. If the subject of the ritual cure is a boy, the fowl should be a cockeral and the ritual specialist a shaman (who is always a male). In the case of a girl, a pullet is used and a midwife officiates. Symbolic differentiation of the sexes correlates with a clear distinction in social status between men and women in San Pedro. Another desideratum is that the chicken be black, a color connoting power, often evil. These are ideal rather than rigid requirements. Thus if a black chicken is not available another may be used, especially a red one (red is also a "strong" color).

SOME CULTURAL ASSUMPTIONS

More important than knowing the procedures is an appreciation of certain cultural concepts and implicit assumptions which impart conscious and unconscious meaning to the ritual. A crucial native conception is the duality of body and spirit. During consciousness the spirit is present, but it is not consciousness itself. It is temporarily absent during coma. At night it visits the places one dreams about. Fright induced by ghosts or were-animals results in sickness because the spirit is jolted out of the body. The spirit is the life principle, but it is conceived as a real entity rather than an abstraction. Like the wind, it is invisible but effective. It also has qualities. It can be strong or weak, benign or malignant; it is the equivalent of character. Since the spirit is at once real and incorporeal, the average Pedrano finds it plausible that a child is capable of killing its brother even in the absence of manifest aggressive behavior. The action occurs at the spirit level but the effect is as real as life and death.

The nearest corporeal counterpart of the spirit is the blood. Individuals are said to have "strong blood" if they are imperious or aggressive; but those who are submissive, hence possessed of "weak blood," are mildly disesteemed. The power of a person's blood varies with his physiological condition. Men who return from the field carrying burdens of corn or firewood on their backs are supposed to cool off before looking at their children. Exertion heats the blood and "hot blood" is injurious to young children. The danger is conveyed by "looking" rather than proximity. Pregnant women have hot blood and must likewise avert their glance in the presence of infants, or even in the presence of turkey chicks, which perish easily. This is the danger of the "evil-eye." The child that feeds on the soul of its younger sibling is also conceived as exerting the power of its strong blood, or to be strengthening its own blood by sapping the blood of its rival.

The meaning attached to the practice of killing the chicken by thrashing it against the offender must be inferred from native tenets of magic, discipline, and exorcism. It is fitting that the child's spiritual urge to kill its sibling should be vicariously fulfilled by having the child's body play a direct if passive part in destroying the life of the substituted chicken. The beating, moreover, magically links the child with the chicken. The human back has symbolic value with reference to ideas about the linked fates of family members, as indicated in idiomatic usage. Fathers are said to carry the "luck" of their daughters on their back, while the fate of sons rests on the back of their mother. If a family runs to boys it is said that the mother "carries her sons well." If only girls tend to be born or to survive, the father is credited with superior psychic strength. The beating is also a punitive measure; the child cries when the shaman or midwife strikes the fowl against its back. Physical punishment is a common corrective device in San Pedro, most generally incurred by children over 5 or 6 years old who disobey or disrespect parental authority. Such discipline, however, is not culturally defined as aggression.

But in addition to reprisal and correction, whipping in San Pedro has other implications. Before entrusting an infant to the hammock, the midwife routinely strikes the hammock, alternately warning and entreating its spiritual guardians to protect the child from harm and keep it from falling out of the hammock. This usage, which is paralleled in the ceremonial whipping of children, houses, and fruit trees on Holy Saturday, combines coercion with the exorcistic expulsion of evil. Use of the ritual chicken as a lash is thus rendered appropriate by the combination of magical, exorcistic, and expiatory connotations inhering in the whipping motif, in addition to the righteous means it represents of releasing parental anger.

The gesture of feeding the chicken to the child is both magical and propitiating. The magical aspect lies in the assumption that supernatural injury can be diverted from the intended victim by providing a substitute object. Another instance of this assumption is the belief that a person with "strong blood" can resist sorcery or supernatural punishment, which consequently falls upon his child or some other vulnerable kinsman. Propitiation consists not merely in giving food but in the generosity of setting before the child an entire chicken in contrast to the frugal serving he ordinarily receives. Thus the ritual that begins with punishment ends with appeasement. Force and entreaty are alternative techniques for influencing others in San Pedro. Unquestioning response to command, within the limits of established principles of authority, rests on the implicit threat of force. To win concessions from a superior, the individual is frequently reduced to the expedient of ingratiation. Supernatural powers are threatened and entreated in turn, as already indicated in the case of the guardians of the hammock. In a hierarchical setting in which parents dominate children and the fates dominate men, one can combat power by counter-force or by ingratiation. The chicken ritual tries both. The child is not a social superior but its voracious spirit is a kind of occult force.

We may now turn to the cultural motif which furnishes the very *raison d'être* of the sibling ceremony, that of symbolic cannibalism.

THE FEAR OF BEING EATEN

One of the most insistent symbolic themes in San Pedro culture is the fantasy of being devoured, which finds expression in a wide range of vivid imagery. It is scarcely true to claim that fanciful dangers interfere in any serious way with the daily round of San Pedro activities, which are pursued with energy and assurance, but it is hardly an exaggeration to assert that imagined dangers classically present themselves in cannibalistic shape.

The theme recurs in myth and legend. Long ago people lived on human flesh. For this they perished in the deluge which took place when the ocean rushed overland to have intercourse with Lake Atitlán, which is a female entity. But some people took refuge in safe places. These were changed into jaguars, coyotes, and other creatures. The black vulture was once an angel sent out to survey the toll of human life taken by the flood; assailed by irresistible odors, he fell to devouring the corpses. Thus present animals were once human. The sun disappeared one time when he was drawn into an argument with the moon, a deceitful woman. During the eclipse, a host of tigers, lions, and other fearful animals descended on the village to eat up the people, and this they will do again some day when the protective sun will disappear once more. These animals issue out of the volcano behind the village. Ordinarily they are chained down by their master, the lord of the volcano, who wields a writhing snake for a whip. All this has been seen by certain Pedranos who were lured into the volcano but allowed to leave when they resisted temptations. In times past, a monster came up from the depths of the lake. For his nightly fare he demanded the body of a child which the village authorities wrested from its parents, delivering it neatly combed and dressed to the monster at the edge of a local peninsula lest he carry out his threat to ascend into the settlement itself and eat up all the inhabitants.

To this day certain outlying points are said to be too dangerous to serve as way stations for overnight travelers; these are places or plantations reputedly run by "Moors" with a craving for human flesh. Negroes, who are observed on the coastal plane below San Pedro, are held to be fantastically powerful; in bygone times they were not above drinking the blood of ordinary people, though "this is now against the law." In the friendliest spirit and with no apparent fear, Pedranos would ask whether it were true that an Indian would jeopardize his life by visiting the United States; they had heard that we would eat him after roasting him to a turn. Women were inclined to give this fantasy an infanticidal twist, as illustrated in the following quotation from a middle aged mother·

They say that people in the United States don't die. When they get old they become young again. But I don't know whether this is true. They also say that when the Americans have 5 or 6 children they will eat one or two, because there are so many people. How terrible to eat people! Perhaps it isn't true. They say that they put the baby in an oven and toast it well and eat it.

Once the tables were turned on several women by improvising a "they say" story to the effect that Pedranos were in the habit of eating their children in the past. This they at first denied but in the same breath reversed this stand by adding: "They say that in the past people here didn't die because there was no sickness. They ate their babies because there were so many children."

Case material on insanity in San Pedro indicates that disturbed people may strike out against property and authority, or suffer hallucinatory aggressions, or attempt to drown themselves during periods of drunken depression. They apparently do not attempt to eat people. But this does not stop others from fancying that they do. Commenting on the case of a disturbed young Pedrano who was forcibly kept home to prevent him from fleeing or committing suicide, one adolescent female asserted that he wanted to eat people. At another point she said he wanted to drink their blood, adding that this craving was characteristic of insane people. A female informant with a more lurid imagination recounted that a woman who gave birth to twins some years ago suddenly went crazy; purportedly confusing them with chickens, she placed the babies in a vessel and killed them by scalding.

San Pedro culture provides a frame of reference which relates the regulation of hunger to the danger that a baby may become the victim of its sibling's appetite. According to local explanation the food cravings of a pregnant mother originate with the fetus, which will become unhappy and leave the womb prematurely if the cravings are not satisfied. The potential knee child, who is separated from the breast during the middle stages of pregnancy when the milk allegedly turns harmful, must likewise have its food whims satisfied; it too is "eating for the fetus." To forestall miscarriage, relatives and neighbors should appease the cravings of the expectant mother and of the demanding knee child. Little is said about the appetite of the expectant father, but it is sometimes said that the food he eats also aids the growth of the unborn infant.

Social relationships with the baby are thus established before it is born. The positive side of these relationships is regularly expressed in terms of eating behavior. The negative side is likewise expressed in eating terms, destructive eating replacing productive eating. The inversion on the part of parents is represented in fantasies that "other" parents do or did devour their own babies. On the part of the knee child, the inversion is represented by the belief that under certain circumstances it feeds upon its younger sibling. Blaming another child for the death of an infant conforms to a general Pedrano tendency to displace blame and aggression downward in the social scale. This tendency is congruent with a hierarchical authority system.

It is worth while to consider the bearing of cross cultural comparison upon our problem. The elements of procedure, belief, and assumption which enter into the ritual cure in San Pedro are not confined to that society alone. Taken singly or in partial combination, these cultural elements are common to other communities. Two comparative instances may be cited. The Mayan Indians of the Tzeltal-speaking village of Oxchuc in Chiapas believe that an individual can suffer from soul consumption and that whipping may facilitate recovery [9]. In that community all chiefs and elders are thought to receive the supernatural aid of an alter ego in the form of an animal (nagual) in effecting social conformity. To punish a social offender, the possessor merely allows his nagual to enter the victim's body and slowly eat its soul. Since the patient himself is at fault, he must confess his sins to recover; sometimes he has to be whipped to expiate his crime. In both Oxchuc and San Pedro one individual can kill another by eating his soul, but in the former case the aggressor acts in the public interest and the punishment is meted out to the victim rather than the aggressor, nor does there appear to be any sibling involvement.

In the Mayan town of Dzitas, Yucatan, a therapeutic ritual (kex) is believed to draw off the sickness of a patient. "The kex is the discharge of a promise, made when a patient is very sick, that cooked chicken will be offered. We did not find out to whom the offering is made. The context of the ritual suggests that the offering is made to the winds that caused the sickness. When the patient came—generally on a Friday—Aurelia [a female curer] prayed in Maya . . . holding the chicken over the patient's head and strangling it to death. For a male patient a pullet was used and for a female patient, a cockerel. . . . The chicken was immediately taken out in the bush and buried in its feathers. . . . When performed for a child . . . an egg was substituted for the hen [6]." Here again we encounter several familiar features but in contrast to San Pedro the symbolism of punishment and interpersonal aggression is absent.

Cultural elements such as the belief that human spirits can be eaten, and the practice of killing a chicken to draw off an affliction, may have spread from one group to another by contact and diffusion. But the distributions of the respective elements overlap rather than coincide. The culture of each community includes a somewhat different assortment of the common traits. These traits, moreover, are differently combined and construed to agree with other configurations or general orientations of the respective cultures. By the nature of our present interest we have analyzed certain aspects of San Pedro culture not in terms of their areal distribution but their meaning as parts of an integrated ritual pattern.

SUMMARY

To avert the death of a newborn infant the people of San Pedro occasionally kill a chicken by beating it against the back of an older child who is then served the cooked fowl with the warning that he desist from "eating" his little brother or sister. Taken at its face value, that is, at the level of manifest function, the ritual is an instrumental act—a means of influencing events—whose logic is part and parcel of the "primitive world view" which Guatemalan Indians combine with "civilized social relations," in Tax's words [8]. While it is a nonrational expedient from our point of view, it "makes sense" to the natives in terms of certain assumptions implicit in their culture. These assumptions have been stated above.

Analysis of the ritual on the level of latent function, however, indicates that it serves a number of purposes apart from its avowed function of forestalling misfortune. For one thing it is an institutionalized means of relieving the intolerable pressure to "do something" and of finding an acceptable scapegoat in time of emotional crisis.

Secondly, it gives symbolic expression to repressed antisocial sentiments which have their source in the social experience of the individual beginning in early childhood and continuing into adult life. Sibling antagonism is imputed in the form of an oral-aggressive projection, which is also indicative of parent-child tensions. These tensions are disallowed byproducts of a social system which emphasizes emotional constraint and respect for authority. An assumed magical connection between fetus, sibling, and parent provides a frame of reference for linking the fate of an infant with the appetite of its sibling. Belief that the death of a baby may be due to the appetite of another child is a specific instance of a more general disposition to entertain fantasies of being eaten.

Finally it should be understood that the sibling ritual not only *reflects* social strains. As part of the San Pedro culture pattern it also plays a role in *defining* the way in which sibling relations are perceived and experienced.

REFERENCES

1. Henry, Jules and Zunia: "Doll Play of Pilagá Indian Children," *Research Monographs*, No. 4 (New York, American Orthopsychiatric Association, 1944).
2. Henry, William E.: "The Thematic Apperception Technique in the Study of Culture-Personality Relations," *Genetic Psychology Monographs*, Vol. 35 (1937).
3. Kluckhohn, Clyde: "Navaho Witchcraft," *Papers of the Peabody Museum of American Archaeology and Ethnology*, Harvard University, Vol. 22 (1944), No. 3.
4. Levy, David M.: "Sibling Rivalry Studies in Children of Primitive Groups," *American Journal of Orthopsychiatry*, Vol. 9 (1939), pp. 205–14.
5. Mead, Margaret: "Age Patterning in Personality Development," *American Journal of Orthopsychiatry*, Vol. 17 (1947), pp. 231–40.

6. Redfield, Robert and Margaret P.: "Disease and Its Treatment in Dzitas, Yucatan," *Carnegie Institution of Washington*, Publication 523 (1940), pp. 49–81.
7. Tax, Sol: "The Municipios of the Midwestern Highlands of Guatemala," *American Anthropologist*, n.s. Vol. 39 (1937), pp. 423–47.
8. ——: "World View and Social Relations in Guatemala," *American Anthropologist*, n.s. Vol. 43 (1941), pp. 27–42.
9. Villa Rojas, Alfonso: "Kinship and Nagualism in a Tzeltal Community, Southeastern Mexico," *American Anthropologist*, n.s. Vol. 49 (1947), pp. 578–87.

CHAPTER

20

Dorothy Lee

ARE BASIC NEEDS ULTIMATE?

Imbedded in every culture is a system of values, standards which influence the kinds of choices men make when confronted with implicit or explicit alternatives. Dr. Lee illustrates the significance of values by showing that the concept of needs alone does not adequately explain the variation and variability in culture. Culture as a determinant of personality, then, must be seen as the context in which the developing personality takes on the values central to its organization. What the individual sees as the "right" and "proper" ways of behaving and believing are guided by what the culture defines as the "right" and "proper" things to do and to believe.

THE purpose of this paper is to urge a re-examination of the premise which so many of us implicitly hold that a culture is a group of patterned means for the satisfaction of a list of human needs. This is, of course, not a new issue either with psychologists or anthropologists. The concept of an inventory of basic needs rose to fill the vacuum created when the behaviorists banished the old list of instincts. Yet, in spite of dissatisfaction with this, many of us continue to think of cultural behavior in terms of some form of the stimulus-response principle. Anthropologists borrowed the principle from psychology, without first testing it against ethnographic material, so that often, when the psychologist uses anthropological material, he gets his own back again in a new form, and receives no new insights. There are two assumptions involved here: (1) the premise that action occurs in answer to a need or a lack; and (2) the premise that there

NOTE: Reprinted from the *Journal of Abnormal and Social Psychology*, Vol. 43 (1948), pp. 391–5, by permission of the author and the American Psychological Association. (Copyright 1948).

[335

is a list. In recent years, anthropologists, influenced by the new psychology, have often substituted *drives* or *impulses* or *adjustive responses* for the old term *needs*, but the concept of the list remains with us. We hold this side by side with the conflicting conception of culture as a totality, of personality as organismic, as well as with adherence to psychosomatic principles. We deplore the presentation of culture as a list of traits, yet we are ready to define culture as an answer to a list of needs.

This definition of culture has proved a strain on us. When we found that the original list of basic needs or drives was inadequate, we, like the psychologists, tried to solve the difficulty by adding on a list of social and psychic needs; and, from here on, I use the term *need* in a broad sense, to cover the stimulus-response phrasing of behavior. When the list proved faulty, all we had to do was to add to the list. We have now such needs as that for novelty, for escape from reality, for security, for emotional response. We have primary needs, or drives, and secondary needs, and we have secondary needs playing the role of primary needs. The endless process of adding and correcting is not an adequate improvement; neither does the occasional substitution of a "totality of needs" for a "list of needs" get at the root of the trouble. Where so much elaboration and revision is necessary, I suspect that the original unit itself must be at fault; we must have a radical change.

In applying the list of needs to different cultures, we found that modification was necessary. It was apparent that the need for food of a member of American society is far greater than that of the members of most other societies. Curiously enough, we also find that though a laborer on a New Guinea plantation needs a minimum diet of seven pounds of yams, plus a stated amount of meat, an Arapesh in his own hamlet, working in his fields, climbing up and down steep mountain sides, working hard at ceremonials, can live a good life and procreate healthy children on three pounds of yams a day, and almost no meat. Is further modification necessary here, or is there another factor at work? Faced with data of this sort, we have been tempted to apply a utilitarian calculus. We have said that when the Arapesh gardens inefficiently in company with his brother-in-law, and when he plants his fruit tree on someone else's distant land, he multiplies his exertions and minimizes his subsistence so as to achieve a maximum of social warmth. But is he really filling two distinct needs, slighting one at the expense of the other? When he takes his pig to another hamlet, and asks someone else's wife to feed and bring it up, what need exactly is he satisfying? And is this need greater than the general human need for food? And does its satisfaction supply a substitute for caloric intake? These questions are nonsense, but we do run into them if we carry the premise of a list of needs far enough.

The assumption of a list of needs was put under its greatest strain, I think, during the recent war. We had assumed that "the role of . . . needs in human behavior is that of first causes." Then how could we explain the behavior of certain small nations, who chose freely to lose necessary food, shelter, security, etc. rather than join the Axis? Why did whole nations court physical annihilation rather than subscribe to Axis doctrines? Why did fathers and husbands and daughters expose their beloved families to danger of torture or death by joining the underground? In this country, why did millions of people who had adequate food and shelter and "security" choose to jeopardize their lives? We can say, of course, that they were satisfying their need for emotional response, in this case the approval of others. One anthropologist did express it in this way. But why was it this particular course of action which was sure to bring them the approval of others? And how could these needs have been the cause of behavior whose goal was neither individual nor group survival? To my mind, this means that either needs are not the cause of all behavior, or that the list of needs provides an inadequate unit for assessing human behavior. I am not saying here that there are no needs; rather, that if there are needs, then they are derivative rather than basic. If, for example, physical survival was held as the ultimate goal in some society, it would probably be found to give rise to those needs which have been stated to be basic to human survival; but I know of no culture where human physical survival has been shown, rather than unquestioningly assumed by social scientists, to be the ultimate goal.

I believe that it is value, not a series of needs, which is at the basis of human behavior. The main difference between the two lies in the conception of the good which underlies them. The premise that man acts so as to satisfy needs presupposes a negative conception of the good as amelioration or the correction of an undesirable state. According to this view, man acts to relieve tension; good is the removal of evil and welfare the correction of ills; satisfaction is the meeting of a need; good functioning comes from adjustment, survival from adaptation; peace is the resolution of conflict; fear, of the supernatural or of adverse public opinion, is the incentive to good conduct; the happy individual is the well-adjusted individual. Perhaps this view of what constitutes the good is natural and applicable in a culture which also holds that man was born in sin, whether in Biblical or in psychoanalytic terms. But should we, who believe that other cultures should be assessed according to their own categories and premises, impose upon them our own unexamined conception of the good, and thus always see them as striving to remove or avoid ills? It seems to me that, when we do not take this negative view of the good for granted, other cultures often appear to be maintaining "justment" rather than striving to attain adjustment. For example, for the Hopi, the good is present and positive. An individual is born in hopiness, so to speak, and strives throughout life to maintain and enhance this hopiness. There is no external reward for being good, as this is taken for granted. It is evil which is external and intrusive, making a man kahopi, or unhopi.

In my opinion, the motivation underlying Hopi behavior is *value*. To the Hopi, there is value in acting as a Hopi within a Hopi situation; there is satisfaction in the situation itself, not in the solution of it, or the resolution of tension. I speak of value, but I am not prepared to define it; I shall try to indicate what I mean by presenting value situations. I am aware that when I introduce the subject of value here, I say nothing new to psychologists, who have been talking of value and defining value in recent years; however, I should like to add an anthropologist's contribution to their work. In addition to this, I want to point out that the notion of value is incompatible with that of a list of needs, or adjustive responses, or drives; so that, wherever it is held, the list must go.

Now, if we substitute the notion of value for that of needs, we are no longer troubled with the difficulty of trying to assess a totality in terms of an aggregate, since value is total and is to be found in a total situation. I do not know to what extent this statement is dogmatic, to what extent trite. When we listen to a symphony, we get satisfaction from a whole, not from eighteen thousand notes and a series of arrangements. I can give you an inventory of my daughter, her three teeth, her seventeen pounds, her inability to sit up yet, her mixed smell, her bald head; is it this which causes my behavior when I rush joyfully home to her as soon as I can leave my office? Again, we find that the Hopi like to eat corn; would we be justified in assuming that a Hopi would therefore find it good to work for wages so as to earn money to buy corn to satisfy his hunger? To the Hopi, corn is not nutrition; it is a totality, a way of life. Something of this sort is exemplified in the story which Talayesva tells of the Mexican trader who offered to sell salt to the Hopi group who were starting out on a highly ceremonial Salt Expedition. Within its context, this offer to relieve the group of the hardships and dangers of the religious journey sounds ridiculous. The Hopi were not just going to get salt to season their dishes. To them, the journey is part of the process of growing corn and of maintaining harmonious inter-relations with nature and what we call the divine. It is the Hopi Way, containing Hopi value. Yet even an ethnographer, dealing with Hopi culture in terms of basic needs, views the Salt Expedition as the trader did, and classifies it under *Secondary Economic Activities.*

So also with our earlier example of the Arapesh. Their eating is not a distinct act satisfying a single need. Food to the Arapesh is good; it incorporates intensive social intercourse; it is the medium of intimacy and identification with others, the symbol of human relations which to them are the primary good. It satisfies the total individual. When we analyze the mouthful of yams into so much nutrition plus so much social warmth, that is exactly what we are doing and no more; we do not find these distinctions or elements—we create them. What we find are aspects of a total situation without independent existence. Our impulse is to break up the situation because we are culturally trained to comprehend a totality only after we break it up into familiar phrasings. But in this way we miss the value inherent in it, since it disappears with analysis and cannot be re-created synthetically afterwards. Having created a series of elements, we then find no difficulty in motivating them according to a series of needs.

If needs are inborn and discrete, we should find them as such in the earliest situations of an individual's life. Yet take the Tikopia or the Kwoma infant, held and suckled without demand in the mother's encircling arms. He knows no food apart from society, has no need for emotional response since his society is emotionally continuous with himself; he certainly feels no need for security. He participates in a total situation. Even in our own culture, the rare happy child has no need for emotional response or approval or security or escape from reality or novelty. If we say that the reason that he has no need for these things is that he does have them already, we would be begging the question. I believe, rather, that these terms or notions are irrelevant when satisfaction is viewed in terms of positive present value, and value itself as inherent in a total situation.

On the other hand, it is possible to see needs as arising out of the basic value of a culture. In our own culture, the value of individualism is axiomatically assumed. How else would it be possible for us to pluck twenty infants, newly severed from complete unity with their mothers, out of all social and emotional context, and classify them as twenty atoms on the basis of a similarity of age? On this assumption of individualism, a mother has need for individual self-expression. She has to have time for and by herself; and since she values individualism, the mother in our culture usually does have this need for a private life. We also believe that a newborn infant must become individuated, must be taught physical and emotional self-dependence; we assume, in fact, that he has a separate identity which he must be helped to recognize. We believe that he has distinct rights, and sociologists urge us to reconcile the needs of the child to those of the adults in the family, on the assumption, of course, that needs and ends are individual, not social. Now, in maintaining our individual integrity and passing on our value of individualism to the infant, we create needs for food, for security, for emotional response, phrasing these as distinct and separate. We force the infant to go hungry, and we see suckling as merely a matter of nutrition, so that we can then feel free to substitute a bottle for the breast and a mechanical bottle-holder for the mother's arms; thus we ensure privacy for the mother and teach the child self-dependence. We create needs in the infant by withholding affection and then presenting it as a series of approvals for an inventory of achievements or attributes. On the assumption that there is no emotional continuum, we withdraw ourselves, thus forcing the child to strive for emotional response and security. And thus, through habituation and teaching, the mother reproduces in the child her own needs, in this case the need for privacy which inevitably brings with it related needs. Now the child grows up needing time to himself, a room of his own, freedom of choice, freedom to plan his own time, and his own life. He will brook no interference and no encroachment. He will spend his wealth installing private bathrooms in his house, buying a private car, a private yacht, private woods and a private beach, which he will then people with his privately chosen society. The need for privacy is an imperative one in our society, recognized by official bodies such as state welfare groups and the depart-ment of labor. And it is part of a system which stems from and expresses our basic value.

In other cultures, we find other systems maintaining other values. The Arapesh, with their value of social-ism, created a wide gap between own-ership and possession, which they could then bridge with a multitude of human relations. They plant their trees in some one else's hamlet, they rear pigs owned by someone else, they eat yams planted by someone else. The Ontong-Javanese, for whom also the good is social, value the sharing of the details of everyday living. They have created a system very confusing to an American student, whereby a man is a member of at least three ownership groups, determined along different principles, which are en-gaged co-operatively in productive activities; and of two large households, one determined along matrilineal lines, one along patrilineal lines. Thus, an Ontong-Javanese man spends part of the year with his wife's sisters and their families, sharing with them the intimate details of daily life, and the rest of the year on an outlying island, with his brothers and their families. The poor man is the man who has no share in an outlying island, who must eat and sleep only in a household composed of his immediate family and his mother's kin, when unmarried; and who must spend the whole year with his wife's kin, when married. He has the same amount and kind of food to eat as his wealthy neighbors, but not as many coconuts to give away; he has shelter as adequate as that of the wealthy, but not as much of the shared living which is the Ontong-Javanese good.

In speaking of these other cultures, I have not used the term *need*. I could have said, for example, that the Ontong-Javanese needs a large house, to include many maternally related families. But I think this would have been merely an exercise in analysis. On the other hand, when I spoke of our own culture, I was forced to do it in terms of needs, since I have been trained to categorize my own experience in these terms. But even here, these are not basic needs, but rather part of a system expressing our basic value; and were we able to break away from our substantival or formal basis of categorizing, I think we should find these to be aspects or stresses or functions, without independent existence. Culture is not, I think, "a response to the total needs of a society," but rather a system which stems from and expresses something had, the basic values of the society.

REFERENCES

In the body of the paper, I give no references, as the paper is intended to be expository, not polemic, and to deal with a general approach, not its specific expression. I have quoted, throughout, terms and phrases from several published items and have used material from several ethnographic works; I list all these below. I should add that what I say here has been said by G. W. Allport over a period of years, against a different frame of reference. My *value*, I believe, is the philosophic counterpart of his *functional autonomy* and his *intention*.

1. Beaglehole, E.: *Notes on Hopi Economic Life*, Yale Univ. Pub. in Anthrop., No. 15 (New Haven, Yale University Press, 1937).
2. Firth, R.: *We, the Tikopia* (New York: American Book Co., 1936).
3. Gillin, J.: "Cultural Adjustment," *American Anthropologist*, Vol. 46 (1944), pp. 429–47.
4. Hogbin, I.: *Law and Order in Polynesia* (New York, Harcourt, Brace, 1934).
5. Kardiner, A.: *The Individual and His Society* (New York, Columbia University Press, 1939).
6. Kluckhohn, C. and Kelly, W. H.: "The Concept of Culture," in Linton, R. (ed.): *The Science of Man in the World Crisis* (New York, Columbia University Press, 1945).
7. Lewin, K.: *A Dynamic Theory of Personality* (New York, McGraw-Hill, 1935).
8. Linton, R.: *The Study of Man* (New York, Appleton-Century, 1936).
9. Linton, R.: *The Cultural Basis of Personality* (New York, Appleton-Century, 1945).
10. Mead, M.: *Sex and Temperament in Three Primitive Societies* (New York, Morrow, 1935).
11. Mead, M.: "The Arapesh of New Guinea," in Mead, M. (ed.), *Cooperation and Competition among Primitive Peoples* (New York, McGraw-Hill, 1937).
12. Mead, M.: "The Mountain Arapesh: I. An Importing Culture," *Anthrop. Pap. Amer. Mus. Nat. Hist.*, Vol. 36 (1938), Pt. 3.
13. Murdock, G. P.: "The Science of Culture," *American Anthropologist*, Vol. 34 (1932), pp. 200–15.
14. Murray, H. A.: *Explorations in Personality* (New York, Oxford University Press, 1938).
15. Parsons, E. C., (ed.): *A Pueblo Indian Journal, 1920–1921, Amer. Anthrop. Assoc. Mem.* 32 (Menasha, Wis., 1925).
16. Simmons, L.: *Sun Chief* (New Haven, Yale University Press, 1942).
17. Watson, J. B.: *Psychological Care of Infant and Child* (New York, Norton, 1928).
18. Watson, J. B.: *Psychology from the Standpoint of a Behaviorist* (Philadelphia, Lippincott, 1919).
19. White, L.: "Energy and Evolution of Culture," *American Anthropologist*, Vol. 45 (1943), pp. 335–56.
20. Whiting, J.: *Becoming a Kwoma* (New Haven, Yale University Press, 1941).
21. Wissler, C.: *Introduction to Social Anthropology* (New York, Holt, 1929).

CHAPTER

21

Florence Rockwood Kluckhohn

DOMINANT AND VARIANT VALUE

ORIENTATIONS

A decade or two ago we might have argued that cultural variation was potentially boundless, and with it the variability of personality. That this is not so is cogently argued in this paper by Dr. Florence Kluckhohn. Certain problems are common to all mankind, and these problems demand solutions, although in the nature of the case it is clear that no final solution is possible. Some cultures have selected one solution, others have selected others. But there are boundries, and the types of solution to these problems are not limitless.

The nature of these problems might be called "philosophical," but abstract as the problems may be and abstract as the solutions may sound, they are of as vital concern in ordering the life of a community as they are of vital concern in the ordering of the life of an individual. The premises and assumptions which we make about ourselves, about our fellow men, indeed about the nature of man in general serve to guide our actions in dealing with ourselves, our fellow men, and men in general.

FOR all those concerned with the lives of individuals and the problems which arise in those lives there has long been a question—even an argu-

NOTE: Thanks are due the author for substantial reworking of two previously published papers especially for this volume. The present paper has been reworked from "Dominant and Variant Cultural Value Orientations," first published in the *Social Welfare Forum* (1951), pp. 97–113, and "Dominant and Substitute Profiles of Cultural Orientations," first published in *Social Forces*, Vol. 28 (1950), pp. 376–93. The reader is referred to these two papers for differing applications of the material here presented.

ment—as to how much the individual is a product of his biological heri-tage and how much the result of environmental forces. That either a strictly biological or an over-simply formulated environmental theory of human behavior is absurdly one-sided most of us have long ago accepted. Yet most of us also know that we are really only at the beginning of the time-consuming and arduous process from which we hope to derive an understanding of that infinitely complicated interplay of biological, psy-chological, social, and other factors which create the personality and character structure of individuals. We still display a tendency to concen-trate on certain factors and exclude others. Some are too much given to interpretations in strictly psychological terms, some too eager to use only the conceptual lenses provided by the sociologist, the anthropologist, or the economist, and others equally zealous with still other approaches.

Today the awareness is growing—rapidly growing—that anything like a full understanding of the concrete situations in which we see individuals requires a use of the explanatory concepts of more than one discipline and requires their use in other ways than an occasional "borrowing" to ac-count for the extraneous. The goal of an integration of our several ap-proaches to a study of human behavior and social situations is clearly be-fore us and recognized as a goal, yet no one can rightly claim that we are close to achieving it.

One specific example of this growing interest in the multiple rather than the unitary approach to individual and social problems is found in the current and frequent linkage of the terms "culture" and "personality." The practical reflection of this we find in the frequent queries of many in the fields of education, social work, colonial administration, and industry as to what a knowledge of cultural factors offers them for a better under-standing of the situations and the individuals with which they must deal. Even clinical psychologists and psychiatrists, whose theories have neces-sarily centered upon the psychological processes of the individual, are asking what the effect of varying cultural patterns upon the actions and motives of individuals may be.

Some real progress has been made, especially in the last decade or so, in bringing together psychological and socio-cultural theories. For those interested in a very brief sketch of the history of this development we would recommend Dr. Ralph Linton's introduction to Dr. Abram Kardi-ner's *The Psychological Frontiers of Society* [1].

Illustrative of the kind of conceptual integration which has to date been achieved, Linton points to the concept of "basic personality" which he himself and Dr. Kardiner developed in their collaborative work. Basic personality, he states, is a configuration involving several different ele-ments. It rests upon the following postulates:

1. That the individual's early experiences exert a lasting effect upon his personality, especially upon the development of his projective system.
2. That similar experiences will tend to produce similar personality configurations in the individuals who are subjected to them.

3. That the techniques which the members of any society employ in the care and rearing of children are culturally patterned and will tend to be similar, although never identical, for various families within the society.

4. That the culturally patterned techniques for the care and rearing of children differ from one society to another.

If these postulates are correct, and they seem to be supported by a wealth of evidence, it follows:

1. That the members of any given society will have many elements of early experience in common.

2. That as a result of this they will have many elements of personality in common.

3. That since the early experience of individuals differs from one society to another, the personality norms for various societies will also differ.

The *basic personality type* for any society is that personality configuration which is shared by the bulk of the society's members as a result of the early experiences which they have in common. It does not correspond to the total personality of the individual but rather to the projective systems or, in different phraseology, the value-attitude systems which are basic to the individual's personality configuration. Thus the same basic personality type may be reflected in many different forms of behavior and may enter into many different total personality configurations. [1, pp. vi–viii]

The studies of various other anthropologists and the collaborative work many of them have done with psychologists, psychiatrists, and sociologists have gone far in demonstrating many of the relationships between individual desires and group experiences—between culture and personality. Yet, for all of the valuable insights produced and the considerable progress thus far achieved, there have been some severe and—in the opinion of the writer—justified criticisms of many of the facile conclusions drawn by some anthropologists. Especially in some of the recent interpretations of so-called national character structure, one notes a repeated tendency to derive highly generalized and far sweeping conclusions from a few specific items of culture content. Sociologists and psychologists alike have cavilled at the apparent ignoring of interaction processes by some anthropologists and at the too deterministic effects often claimed for cultural factors.

Much of the difficulty in all attempts to use the cultural anthropologists' concepts and data arise from an absence of a systematic theory of cultural variation and from the tendency of most anthropologists to rely too much upon mere empirical generalizations. The most casual observer is aware that the customs of different societies vary. He knows, too, that the behavior patterns of individuals within a given society are often markedly different. Indeed, when dealing with variation at this level one cannot but be acutely conscious of the wide range of "individual differences." But it is not this plethora of specific content which is of the most critical importance if the aim is to understand better the relationship of cultural factors to either the structuring of social groups or the personalities of the individuals who comprise the social groups. It is rather the generalized meanings or values which should be the major, or at least the first, concern.

Specific patterns of behavior insofar as they are influenced by cultural factors (and few are not so influenced) are the concrete expressions reflecting generalized meanings or values. And to the extent that the individual personality is a product of training in a particular cultural tradition it is also at the generalized value level that one finds the most significant differences.

As Gregory Bateson has remarked: "The human individual is endlessly simplifying, organizing, and generalizing his own view of his own environment; he constantly imposes his own constructions and meanings; these constructions and meanings (are) characteristic of one culture as over against another [2]." Or, as Clyde Kluckhohn has stated: "There is a 'philosophy' behind the way of life of every individual and of every relatively homogeneous group at any given point in their histories [3]."

The writer agrees with these and many similar statements made by other anthropologists which emphasize the importance of "value orientations" in the lives of individuals and groups of individuals [4]. There is in many of them, however, too much stress—implied when not actually stated—upon the unitary character of value orientations. Variation for the same individual when he is playing different roles and variation between whole groups of persons within a single society are not adequately accounted for. More important still, the emphasis upon the uniqueness of the variable value systems of different societies ignores the fact of the universality of human problems and the correlate fact that human societies have found for some problems approximately the same answers. Yet certainly it is only within a frame of reference which deals with universals that variation can be understood. Without it, it is not possible to deal systematically with either the problem of similarity and difference as between the value systems of different societies or the question of variant values within societies.

Human behavior mirrors at all times an intricate blend of the universal and the variable. The universals and variations are of many kinds. All human beings have many and significant biological similarities as members of a particular species—*homo sapiens*—yet variability within the species is great. We frequently note both the similarities and differences which are psychological.

The problem of the dominant and variant in cultural patterning can, of course, be approached in different ways. Precisely which way will depend upon the specific type of investigation being made. The aim of the approach of this paper is a conceptual scheme which will permit a systematic ordering of cultural value orientations within the framework of common human—universal—problems. It is only when we have delineated the central types of value orientations and the ranges of possible variability in them that we can make systematic comparisons of either single orientations or the total *value orientation profiles* of whole societies or parts of societies.

The first fundamental assumption upon which the conceptual scheme is based is: *There is a limited number of basic human problems for which all peoples at all times and in all places must find some solution.* The five common human problems tentatively singled out as those of key importance can be stated in the form of questions:

(1) What are the *innate predispositions* of man? (Basic human nature)
(2) What is the relation of *man to nature?*
(3) What is the significant *time* dimension?
(4) What is the valued *personality type?*
(5) What is the dominant modality of the *relationship of man to other men?*

The problems as stated in these questions are regarded as constant; they arise inevitably out of the human situation. The solutions found for them are variable but not limitlessly so. It is the second major assumption of the conceptual scheme that *the variability in solutions is variability within a range of possible solutions.* The limits of variability suggested as a testable conceptualization are the three point ranges for each of the main orientations given in TABLE XXXIX. Nothing mystical is claimed for the number *three,* but as will be seen in the following explanations, such a breakdown seems, in almost all cases, to be both logically adequate and empirically sound.

TABLE XXXIX

Human Problems and Type Solutions

Innate Predispositions:	Evil (mutable or immutable)	Neither good nor bad (mutable or immutable)	Good (mutable or immutable)
Man's Relation to Nature:	Man subjugated to nature	Man in nature	Man over nature
Time Dimension:	Past	Present	Future
Valued Personality Type:	Being	Being-in-Becoming	Doing
Modality of Relationship:	Lineal	Collateral	Individualistic

To the question of what innate human nature is, there are the three logical divisions of evil, neither good nor evil (or mixed), and good. And such, in fact, seem to be the distinctions which have been made by societies. Variation within this range is, of course, possible. Human nature can be regarded as evil and unalterable or evil and perfectible. It can be good and unalterable or good and corruptible. It can be viewed as neither good nor evil and treated as invariant or as subject to influence. This kind of variability, however, falls within the basic threefold classification and is probably a result of the relationship of the human-nature orientation to other orientations.

Illustrations of these differences in the definition of innate predispositions are easily found. We have only to look about us to recognize that there is considerable variability in our present day American conception of human nature. The orientation we inherited from Puritan ancestors, and still strong in many of us, is that human nature is basically evil but perfectible. Constant control and discipline of the self are essential if any real goodness is to be achieved and maintained, and the danger of regression is always present. But some in our society today—perhaps a growing number—are inclined to the more tolerant view that human nature is a mixture of the good and the bad. These would say that control and effort are certainly needed, but lapses can be understood and need not always be severely condemned. Such a definition of basic human nature would appear to be a somewhat more common one among the peoples of the world—both literate and nonliterate—than the view we have held in our own historical past. Whether there are any total societies given to the definition of human nature as *immutably good* is to be doubted. The position is, however, a possible one and should be found ever present as an alternative definition within societies.

The three-point range of variation in the *man-nature* relationship—that of Man Subjugated to Nature, Man In Nature, and Man Over Nature—is too well known from the work of philosophers and culture historians to need a detailed explanation. Mere illustration will demonstrate the differences.

Spanish-American culture as I have known it in the American Southwest illustrates well the Man Subjugated to Nature position. To the typical Spanish-American sheep-raiser in that region there is little or nothing which can be done if a storm comes to damage his range lands or destroy his flocks. He simply accepts the inevitable as the inevitable. His attitude toward illness and death is the same fatalistic attitude. "If it is the Lord's will that I die I shall die" is the way he expresses it. Many a Spanish-American has been known to refuse the help of any doctor because of this attitude.

Another way of phrasing the *man-nature* relationship is to regard all natural forces and man himself as one harmonious whole. One is but an extension of the other, and both are needed to make the whole. Such was the attitude frequently found as the dominant one in China in the past centuries.

A third way of viewing this relationship is that of Man Against—or Over—Nature. According to this view, which is clearly the one characteristic of Americans, natural forces are something to be overcome and put to the use of human beings. We span our rivers with bridges, blast through our mountains to make tunnels, make lakes where none existed, and do a thousand and one other things to exploit nature and make it serve our human needs. In general this means that we have an orientation to life which is that of overcoming obstacles. And it is difficult for us to understand the kind of people who accept the obstacle and give in to it or even the people who stress the harmonious oneness of man and nature.

The possible cultural phrasings of the *man in time* problem breaks

easily into the three point range of Past, Present and Future. Far too little
attention has been given to this problem and its phrasings. Meaningful
cultural differences have been lost sight of in the too sweeping and too
generalized view that folk peoples have no time sense, and no need of one,
whereas urbanized and industrial peoples must have one. Whether days
are regarded as sunrise to sundown wholes or as split into hours and
minutes, and whether or not a clock is deemed a useful culture object are
not the critically important criteria for a consideration of the orientation
to time.

Spengler had quite another order of fact in mind than this when he
made, in his quite profound discussion of "time" in the *Decline of the
West*, this emphatic and categorical statement: "It is by the meaning that
it intuitively attaches to time that one culture is differentiated from another [5]." Time and Destiny were what were being related in Spengler's
conception. For the most part his concern was with the twofold division
of orientations into those which were the timeless a-historic present and
the ultra-historical projection into the future. Always on the plane of the
macroscopic and concerned with directionality as a cyclical unfolding, he
apparently did not feel a need to deal with the problem of the traditionalistic or past orientation which was so important a part of Max
Weber's treatment of moral authority. The threefold division proposed
for the cultural orientation schema has, therefore, its similarity to Spengler's conception in the distinction between a timeless, traditionless, future-
ignoring present and a realizable future. It differs in that it also differentiates from these an orientation which looks to the traditions of the past
either as something to be maintained or as something to be recaptured.
There is in the conception as here phrased an aspect of the orientation
which is relative to the standards and norms of authority which is not
explicitly included in Spengler's conception.

Obviously all societies at all times must deal with all the three time-
problems. All have some conception of the past, all have a present, and all
give some kind of attention to the future time-dimension. They differ,
however, in their emphasis on past, present, or future at a given period, and
a very great deal can be told about the particular society or part of a
society being studied, much about the direction of change within it can
be predicted, with a knowledge of where that emphasis is.

Illustrations of these different emphases are also easily found. Spanish-
Americans, whom we have described as having the attitude that man is a
victim of natural forces, are also a people who emphasize present time.
They pay little attention to what has happened in the past, and regard the
future as a vague and most unpredictable period. Planning for the future
or hoping that the future will be better than either present or past simply
is not their way of life. In dealing with Spanish-Americans one must always take into account the fact that they have quite a different time sense
from our own. Too often, we, who are in the habit of making definite

appointments for two or five o'clock and expect to keep and have them kept, are baffled by the Spanish-Americans to whom two or five o'clock means little or nothing. For an appointment made for two o'clock he may arrive at any time between one and five o'clock, or, most likely, he will not arrive at all. Something else which interests him more may well have turned up to absorb his attention.

China of past generations, and to some extent still, was a society which put its main emphasis upon past time. Ancestor worship and a strong family tradition were both expressions of this Past Time orientation. So also was the Chinese attitude that nothing new ever happened in the present or would happen in the future. It had all happened before in the far distant past. Thus it was that the proud American who thought he was showing some Chinese a steamboat for the first time was quickly put in his place by the remark: "Our ancestors had such a boat two thousand years ago." Many modern European countries also tended to stress the past. Even England—insofar as it has been dominated by an aristocracy and traditionalism—has voiced this emphasis. Indeed, one of the chief differences between ourselves and the English is to be found in our somewhat varying attitudes toward time. We have difficulty in understanding the respect the English have for tradition, and they do not appreciate our disregard for it.

Americans, more than most people of the world, place emphasis upon the future—a future which we anticipate to be "bigger and better." This does not mean we have no regard for the past or fail to give thought to the present. But it certainly is true that no current generation of Americans ever wants to be called "old-fashioned." We do not consider the ways of the past to be good just because they are past, and we are seldom content with the present. This makes of us a people who place a high value on change.

The fourth of the common human problems is called the *Valued Personality Type*. The range of variation in this case yields the Being, the Being-in-Becoming, and the Doing orientations. Since it is assumed that all the orientations are an aspect of the action and motivational systems of the individual personalities, *Valued Personality Type* is not the happiest of terms for designating this particular range of them. For the time being, however, we shall retain the term.

These orientations have been derived for the most part from the distinction long made by philosophers between Being and Becoming. Indeed, to a marked degree, the three-way distinction is in accord with the classification of personality components made by the philosopher Charles Morris—the Dionysian, the Apollonian, and the Promethean. The abstractly conceived component which he labels the *Dionysian*—the personality component type which releases and indulges existing desires—is somewhat what is meant by the Being orientation. His *Apollonian* component—the component type that is self-contained and controls itself through a meditation and detachment that bring understanding—is to some extent the Being-in-Becoming. His active, striving *Promethean* component is similar to the Doing orientation [6].

The accordance is, however, far from complete. As used in this schema the terms Being and Becoming, now made into the three-point range of Being, Being-in-Becoming, and Doing, are much more narrowly defined than has been the custom of philosophers. Furthermore, the view here is that these orientations vary independently relative to those which deal with the relation of man to nature, to time and innate predispositions. The tendency of the philosophers, writing with different aims, has been to treat these several types of orientations as relatively undifferentiated clusters.

The essence of the Being orientation is that it stresses the spontaneous expression of what is conceived to be "given" in the personality. The orientation is, as compared with the Being-in-Becoming or Doing, essentially *non*developmental. It might even be phrased as a spontaneous expression of impulses and desires; yet care must be taken not to make this interpretation a too literal one. In no society, as Clyde Kluckhohn has commented, does one ever find a one-to-one relationship between the desired and the desirable. The concrete behavior of individuals in complex situations and the moral codes governing that behavior usually reflect all the orientations simultaneously. A stress upon the "isness" of the personality and a spontaneous expression of that "isness" is not pure license as we can easily see if we turn our attention to a society or segments of a society in which the Being orientation is dominant. Mexican society, for example, is clearly one in which the Being orientation is dominant. Their wide-range patterning of *Fiesta* activities alone shows this. Yet never in the *Fiesta* or other patterns of spontaneity is there pure impulse gratification. The value demands of other of the orientations—the relational orientation, the conception of human nature as being good and evil and in need of control and others—all make for codes which restrain individuals in very definite ways.

The Being-in-Becoming orientation shares with the Being a great concern with what the human being is rather than what he can accomplish, but here the similarity ends. In the Being-in-Becoming orientation the idea of development so little stressed in the Being orientation is paramount.

Erich Fromm's conception of "the spontaneous activity of the total integrated personality" is close to the Being-in-Becoming type. "By activity," he states, "we do not mean 'doing something' but rather the quality of the creative activity which can operate in one's emotional, intellectual, and sensuous experiences and in one's will as well. One premise of this spontaneity is the acceptance of the total personality and the elimination of the split between reason and nature [7]." A less favorably prejudiced and, for our purposes, a more accurately limited statement would be: the Being-in-Becoming orientation emphasizes self-realization—self-development—of all aspects of the self as an integrated whole.

The Doing orientation is so characteristically the one dominantly stressed in American society that there is little need for an extensive definition. Its most distinguishing feature is its demand for action in the sense of accomplishment and in accord with standards which are conceived as being external to the acting individual. Self-judgment as well as the judgment of others is largely by means of measurable accomplishment through action. What does the individual do, what can he, or will he, accomplish are almost always primary questions in our scale of appraisal of persons. "Getting things done" and finding ways "to do something" about any and all situations are stock American phrases. Erich Fromm also recognizes this orientation as separable from that which he defines in his concept of spontaneity and which we have called the Being-in-Becoming, but he seems to view it as mainly compulsive. With this I cannot agree. Many persons in our society who follow patterns in accord with the Doing orientation are compulsive; many are not. Conformity, which is essential in all societies, whatever the arrangement of their orientations, should not be so much and so often confounded with compulsiveness.

The fifth and last of the common human problems treated in this conceptual scheme is the definition of man's relation to other men. This orientation, the *relational*, has three sub-divisions: the Lineal, the Collateral, and the Individualistic.

Sociologists have long used various types of dichotomies to differentiate homogeneous folk societies from the more complex urban societies. *Gemeinschaft-gesellschaft*, traditionalistic—rational-legal, mechanical-organic solidarity, or simply rural-urban—are the most familiar of the several paired terms. Anthropologists, who have for the most part studied *gemeinschaft* or folk peoples, have frequently in their analyses of kinship structure or social organization made much of the difference between lineage and a lateral extension of relationships.

The distinctions being made here obviously owe much to the concepts used in both these fields, but they are not identical with those of either field. The Lineal, Collateral, and Individualistic relational principles are analytical elements in total relational systems and are not to be confused with categories descriptive of concrete systems.

It is in the nature of the case that all societies—all groups—must give some attention to all three principles. Individual autonomy cannot be and is not ignored by the most extreme type of *gemeinschaft* society. Collaterality is found in all societies. The individual is not a human being outside a group and one kind of group emphasis is that put upon laterally extended relationships. These are the immediate relationships in time and place. All societies must also pay some attention to the fact that individuals are biologically and culturally related to each other through time. This is to say that there is always a Lineal principle in relationships which is derived from age and generational differences and cultural tradition. The fundamental question is always that of emphasis.

There will always be variability in the primacy and nature of goals according to which of the three principles is stressed. If the individualistic principle is dominant and the other two interpreted in terms of it—as is the case in the United States—individual goals will have primacy over the goals of either the Collateral or Lineal group. When the Collateral principle is dominant, the goals—or welfare—of the laterally extended group have primacy for all individuals. The group in this case is viewed as being moderately independent of other similar groups and the question of continuity through time is not critical. Where the Lineal principle is most heavily stressed it is again group goals which are of primary concern to individuals, but there is the additional factor that an important one of those goals is continuity through time. Both continuity and ordered positional succession are of great importance when Lineality dominates the *relational* system. Spanish-American society has been, until recently, one with a relatively strong Lineal stress, combined with a strong second order Collaterality.

How continuity and ordered positional succession are achieved in a Lineal *relational* system is separate from the principle as such. It does in fact seem to be the case that the most successful way of maintaining a stress on Lineality is through mechanisms which are either actual hereditary ones based upon biological relatedness or ones which are assimilated to a kinship system. The English, for example, maintained such an emphasis into the present time by consistently moving successful members of its more individualistic middle class into the established peerage system. Other societies have found other but similar mechanisms.

Thus far in the discussion of the major orientations the aim has been to show that different societies make different selections among possible solutions of common human problems. They raise to dominant position some one of the alternative principles. However, at no time has it been stated or implied that any society will or can ignore any of the dimensions. On the contrary, it is a fundamental proposition of this conceptual approach that all dimensions of all orientations *not only are but must be* present at all times in the pattern structure of every society.

However important it is to know what is dominant in a society at a given time, we shall not go far toward the understanding of the dynamics of that society without paying careful heed to the variant orientations. That there be individuals and whole groups of individuals who live in accordance with patterns which express variant rather than the dominantly stressed orientations is, it is maintained, essential to the maintenance of the society. *Variant values, are, therefore, not only permitted but actually required.* It has been the mistake of many in the social sciences, and of many in the field of practical affairs as well, to treat all behavior and certain aspects of motivation which do not accord with the dominant values as some kind of deviance. It is urged that we cease to confuse the deviant who by his behavior calls down the sanctions of his group with the variant who is accepted and frequently required as far as the total social system is concerned. This is especially true in a society such as ours, where beneath the surface of what has so often been called our compulsive conformity, there lies a wide range of variation. The dynamic interplay of the dominant and the variant is one of the outstanding features of American society, but as yet it has been little analyzed or understood.

Illustrations of variant value orientations, whether of individuals or whole groups, are numerous. Let us look at these three kinds: *ethnic difference* of which the United States has had, and still has, so much; *class difference; role difference* as it is seen in the role of the American woman.

The usual tendency of most observers has been either to view all ethnic groups as one undifferentiated whole or, with a concern for understanding better the problems of particular groups, to seek out the quite specific ways in which they differ. Attention is given to the kind of parental authority and to the attitude they have toward women, or perhaps we delve into the type of child-training patterns they follow. These specific patterns are, of course, important; but it is obvious that knowing them all is in most cases impossible. Furthermore, when there is no general framework within which to consider the specific differences noted, there develops so frequently a tendency to attribute too much to single items of cultural content.

We could know many such concrete patterns followed by the Spanish-Americans in the Southwest and still not know why after one hundred years within the borders of the United States their way of life has changed so little until very recently. We can know and have known many such patterns and yet are forced to admit that understanding between Anglo-Americans and Spanish-Americans is not, even now, very great.

When, however, we look to fundamental differences in value orientations we are led quickly to this proposition: The slow rate of assimilation of Spanish-Americans (and more recent Mexican immigrants), and the low level of understanding as well, are in large part attributable to a wide disparity in *all* the major orientations of Anglo-Americans and Spanish-Americans.

Illustrations of most of the Spanish-American orientations have already been given singly. Let us now take them as a whole system and again quickly compare them to the system of orientations dominantly stressed in American society. Where the Anglo-American stresses Individualism, the Spanish-American puts his primary emphasis upon a combination of the Lineal and the Collateral. The semi-feudal *patron-peon* system of Mexicans, both in the United States and in Mexico, has neither permitted nor required very much independent behavior of most people. Or, to phrase this another way, whereas the Anglo-American is quite systematically trained for independent behavior, the Spanish-American or Mexican is trained for dependence.

The American dominant *time* orientation has been noted to be Future, that of the Spanish-Americans, Present. We show a vague awareness of this difference when we so often refer to Mexicans in general as being a *mañana* people. Yet how very much bound by our own cultural values we are when we interpret *mañana* to mean that a Mexican will always put off until tomorrow what should be done today. Tomorrow in a highly specific sense is meaningless to the Spanish-American or Mexican. He lives in a timeless present, and as one Mexican scholar phrased it: "The Mexican never puts off until tomorrow what can be done *only* today."

Consider, too, the vast difference between the Spanish-American Being orientation and the American emphasis upon Doing or accomplishing. Doing things in the name of accomplishment is not usual Spanish-American behavior. That which "is" is in large part taken for granted and considered as something to be enjoyed rather than altered.

Our own and the Spanish-Americans' definition of the *man-nature* relationship are likewise poles apart. We set out to conquer, overcome, and exploit nature; they accept the environment with a philosophical calm bordering on the fatalistic. And seldom in Spanish-American culture does one find evidence of our own historical view of human nature as Evil but Perfectible. Their view would appear to be much more that of human nature as a Mixture of the Good and Bad.

With such differences as these it is not a cause for wonder that Anglo-American educators, social workers, politicians, and a host of others have been both deeply frustrated and quite unsuccessful in their many attempts either to alter rapidly the Spanish-American way of life or adjust Anglo-American values to it. Nor is it strange that the understanding between Anglo- and Spanish-Americans in an area (New Mexico and Arizona) where each group constitutes approximately a half of the total population has been inadequate throughout a whole century.

This highly dramatic and most extreme illustration of cultural variation within our own borders tells us a great deal if only we look at it in terms of major value orientations instead of always considering particular and specific bits of behavior. The specific to be very meaningful, we repeat, must be viewed as an item in the wider context of value orientations.

Generalizing this conception to all such groups as the Spanish-American we would first hypothesize that *the rate and degree of assimilation of any ethnic group into general dominant American culture will in large part depend upon the degree of goodness of fit of the group's own basic value orientations with those of dominant American culture.*

Class differences have greatly concerned American social scientists in the last two decades. The best known studies of the class structure of the United States are those of W. Lloyd Warner and his associates, but there are also many others with different points of view and in which conclusions very different from those of the Warner group are reached. From all the studies we have learned much that Americans have been unwilling to admit or discuss in past years. We know that there are great differences between the classes in attitudes toward education and politics, in association memberships, in family life, in occupational interests and opportunities, in reading habits, in recreational interests, and a host of other things.

Yet in spite of all the differences observed and recorded, there is a tendency in all the studies to assume that all the variation is variation on the same value theme—the so-called American Creed. What is remarked is that the behavior and attitudes of some classes are harmoniously in tune with the generalized creed, whereas those of other classes are off in pitch and limited in range. That the value themes themselves might be different is seldom suggested.

But according to the conceptual scheme of the dominant and variant in cultural orientations, it is assumed at the start that there is a dominant class—in the case of the United States, the middle class—in which adherence to dominant values is marked, but also that there are other classes which hold to variant values in much of what they do and believe. As I have suggested elsewhere, the observed behavior of an upper class in an old and declining community shows an adherence to lineality rather than individualism, to past time more than future, and to Being or Being-in-Becoming rather than to Doing personalities. Also in some parts of the lower class—a class so heterogeneous and diffuse that it should be "classes" and not just "class"—Present Time and Being orientations are often combined with either Individualism or Collaterality.

There has not been sufficient work done to date to state with certainty that this different approach will provide a more accurate knowledge and appraisal of class differences. One study now in progress—a study of the occupational aspirations of a large group of high school boys in a metropolitan community—shows some promise of demonstrating the existence of value orientation differences between classes and segments of classes [8]. In another study just completed by Dr. Charles McArthur it has been shown that, on the average, public and private school boys give predictably different responses to the pictures in the Murray and Morgan Thematic Apperception Test [9]. For example, on the assumption that in a public school boys are predominantly middle class and hence Future Time and Doing oriented, while private school boys—especially those of certain selected schools—are predominantly upper class with Past Time and either a Being or Being-in-Becoming orientation, these two predictions were made as to the variability of responses to the first picture of the test (a small boy is seated before a table on which there lies a violin; the subject is asked to invent a story about him):

1. That more public school boys (now students in a large university) will tell stories to the violin picture in which the parent demands work from the child.
2. That more private school boys will tell violin stories in which the music lesson is seen by the child as a way to create beauty and/or express or develop himself.

These predictions, and twelve others similar in type but different in content, were borne out by McArthur's data. And, as he himself said: "They constituted a neat demonstration that the attitudes of individuals, as measured by one of the psychologist's best projective tests of personality, can be predicted from a knowledge of the person's sub-cultural orientations profile."

The role of the woman in the United States is as little understood for its variant character as are some of the differences in social classes. Although it is frequently stated that the feminine role is poorly defined and full of contradictions, it has not been noted that the behavior expected of women in the wife-mother role is relative to value orientations which are markedly different from the dominant American values that are so well expressed in the man's occupational role.

In his occupational role the man—ideally at least—is expected to be autonomous and independent, whereas from the woman as a wife and mother we all expect a subordination of individualistic goals to those of the family as a group. Man's occupational role is also an action-oriented one which expresses well the Doing orientation, while in woman's role it is expected that much more attention will be given to all those things intellectual, aesthetic, and moral which busy men so often define as the nice but non-essential embroidery of American life. It is, in other words, more of the Being-in-Becoming orientation that women are supposed to adhere to. As for the *Time* orientation it is again man's role which most fully expresses American's dominant *future* emphasis. For the most part, daughters and wives are limited to a vicarious participation in the goals of future promise. The glory which comes from mounting success is largely a man's glory, the light from which is, for women, merely reflective.

Differences such as these are both considerable and troublesome in a society which theoretically stands for an equality of the sexes. When one studies carefully the history of the feminine role it becomes very clear that a central issue for many years has been woman's demand for the right to participate more fully in all those activities in which *dominant* American values are expressed.

In recent years the issue has become the more acute because of the kind of education the women of today receive. Girls are no longer trained in markedly different ways or for different things than boys are. Throughout childhood and youth the girl child goes to school with boys and is taught very much as they are taught. From babyhood on, she learns the ways of being independent and autonomous, and she is expected to know how to look after herself all through adolescence and beyond—forever if need be. The hope is expressed, of course, that she will not have to remain independent and will not, therefore, need to use much of what she has learned. Instead, and this is the truly great problem, she is expected, upon her marriage, or certainly after children are born, to give her attention to all those things which are defined as feminine and for which she has not been well trained at all. It is not to be wondered that the strains in the feminine role are numerous and make for serious personality difficulties in many women.

In other instances of variant roles the strains are fewer. The man who chooses to be a withdrawn scholar or an artist may often be made to feel that he is outside the main stream of life, but it is doubtful that he is subject to as many doubts about the value of what he does as are many women. In the main this is because he is creative, whereas the intellectual and aesthetic interests of women have been, to date, chiefly appreciative.

There is, of course, much more value variation within our own or any other social system than these few illustrations indicate. The varying roles any individual plays at different times and places may be, and often are, different in the value orientations they express. No individual any more than any whole society can live always in all situations in accord with patterns which allow for expression of only a single dimension of the orientations. One can also note many shifts of emphasis in the range of one orientation or another as between the different historical time periods of a social system. Indeed, whatever the type of problem or the situation one may choose for study, variations in value orientations are certain to be found. Thus we repeat: If a knowledge of major cultural value orientations is essential to the understanding of social situations and individual personalities (and I would say it certainly is), it is also necessary to know the variant value orientations and their relation to the dominant ones.

REFERENCES

1. Kardiner, Abram: *The Psychological Frontiers of Society* (New York, Columbia University Press, 1945).
2. Bateson, Gregory: "Cultural Determinants of Personality," in J. McV. Hunt (ed.): *Personality and the Behavior Disorders* (New York, Ronald Press, 1944), Vol. 2, p. 273.
3. Kluckhohn, Clyde: "Values and Value Orientations in the Theory of Action," in Talcott Parsons and Edward A. Shils (eds.): *Toward a General Theory of Action* (Cambridge, Harvard University Press, 1951), p. 409.
4. Kluckhohn, Ibid., p. 411, defines value orientation as follows: "a generalized and organized conception, influencing behavior, of nature, of man's place in it, of man's relation to man, and of the desirable and non-desirable as they may relate to man-environment and interhuman relations."
5. Spengler, Oswald: *The Decline of the West*, tr. Charles F. Atkinson (New York, Knopf, 1926–8).
6. Morris, Charles: *Paths of Life* (New York, Harpers, 1942), esp. chap. II.
7. Fromm, Erich: *Escape from Freedom* (New York, Farrar and Rhinehart, 1941).
8. This project is sponsored by the Harvard University Laboratory of Social Relations and is under the direction of Professors Samuel A. Stouffer and Talcott Parsons and the writer.
9. McArthur, Charles: Cultural Values as Determinants of Imaginal Productions, unpublished Ph.D. thesis, Harvard University, 1952.

Part Two

SECTION V

ROLE DETERMINANTS

INTRODUCTION

THE ROLE concept provides the principal theoretical point of articulation between analyses of the behavior of groups by anthropologists and sociologists and analyses of individual motivation by psychologists and psychiatrists.

Every culture provides not only a set of patterns that are applicable to all members of the society. It also formally recognizes age, sex, occupational, and other statuses and supplies blueprints for training individuals to play roles that correspond to these statuses. Each person's level of aspiration for success in playing his roles as culturally defined varies, but no personality escapes the molding influences of one set of roles as opposed to another. Even within the family at an early age,[1] the role of son as opposed to daughter makes a differentiating demand upon the developing personality. Once the child moves outside the family circle, membership in an age-sex category and later in an occupational group defines his relationship to the total society and delimits his cultural participation.[2]

The personality characteristics which the group's members regard as normal for the occupants of various status positions also play a part in personality formation through such activities as the mother's training of her children,[3] a chief's relation to his potential successor,[4] and the like.

The effects of occupational role playing have been impressionistically but suggestively discussed by Professor Waller in his chapter "What Teaching Does to the Teacher."[5] The relations between military status

---

[1] Frederick H. Allen, "Dynamics of Roles As Determined in the Structure of the Family," *American Journal of Orthopsychiatry*, Vol. 12 (1942), pp. 127–34.

[2] See Ralph Linton, "A Neglected Aspect of Social Structure," *American Journal of Sociology*, Vol. 45 (1940), pp. 870–87; and "Age and Sex Categories," *American Sociological Review*, Vol. 7 (1942), pp. 589–604.

[3] Margaret Mead, "On the Institutional Role of Women and Character Formation," *Zeitschrift für Sozialforschung*, Vol. 5 (1936), pp. 69–75.

[4] Clellan S. Ford, "Role of a Fijian Chief," *American Sociological Review*, Vol. 3 (1938), pp. 541–50.

[5] Willard Waller, *The Sociology of Teaching* (New York, V. Wiley and Sons, 1932), part 5.

and personality have been considered in a recent number of the American Journal of Sociology.[6]

The disrepancies between the expected personality manifestations in two or more different roles played by the same individual sometimes give rise to personal conflict. It is a frequent observation of ordinary experience that a Negro exhibits "one personality" in dealing with whites and quite another in dealing with his fellow Negroes. Some psychiatrists have speculated that neurosis often arises from the individual's demand for consistency in his behavior in all roles.

This Section includes only three selections, for, in spite of the obvious importance of role training and role playing for personality formation, this subject has barely been opened up at an empirical level.[7]

---

[6] Vol. 51, No. 5, March 1946. See especially the papers by Clark, Elkin, Hollingshead, and Stone.

[7] Two additional papers are now available in *Readings in Social Psychology* (T. M. Newcomb and E. L. Hartley, ed., 1947): Leonard S. Cottrell, "The Adjustment of the Individual to His Age and Sex Roles," pp. 370–3; and Theodore M. Newcomb, "Some Patterned Consequences of Membership in a College Community," pp. 345–57.

CHAPTER

22

Talcott Parsons

AGE AND SEX IN THE SOCIAL
STRUCTURE OF THE UNITED STATES

Professor Parson's paper deals primarily with age and sex roles in contemporary American society. He defines the structural situation that impinges on forming personalities and describes the role models towards which individuals consciously or unconsciously orient their strivings.

IN OUR society, age grading does not to any great extent, except for the educational system, involve formal age categorization, but is interwoven with other structural elements.[1] In relation to these, however, it constitutes an important connecting link and organizing point of reference in many respects. The most important of these for present purposes are kinship structure, formal education, occupation, and community participa-

---

NOTE: Reprinted from *American Sociological Review*, Vol. 7 (1942), pp. 604-16, with abridgment of footnotes, by permission of the author and the American Sociological Society. (Copyright, 1942, by the American Sociological Society.)

[1] The present paper will not embody the results of systematic research but constitutes rather a tentative statement of certain major aspects of the roles of age and sex in our society and of their bearing on a variety of problems. It will not attempt to treat adequately the important variations according to social class, rural-urban differences, and so on, but will concentrate particularly on the urban middle and upper middle classes.

[363

tion. In most cases the age lines are not rigidly specific, but approximate; this does not, however, necessarily lessen their structural significance.

In all societies, the initial status of every normal individual is that of child in a given kinship unit. In our society, however, this universal starting point is used in distinctive ways. Although in early childhood the sexes are not usually sharply differentiated, in many kinship systems a relatively sharp segregation of children begins very early. Our own society is conspicuous for the extent to which children of both sexes are in many fundamental respects treated alike. This is particularly true of both privileges and responsibilities. The primary distinctions within the group of dependent siblings are those of age. Birth order as such is notably neglected as a basis of discrimination; a child of eight and a child of five have essentially the privileges and responsibilities appropriate to their respective age levels without regard to what older, intermediate, or younger siblings there may be. The preferential treatment of an older child is not to any significant extent differentiated if and because he happens to be the first born.

There are, of course, important sex differences in dress and in approved play interest and the like, but if anything, it may be surmised that in the urban upper middle classes these are tending to diminish. Thus, for instance, play overalls are essentially similar for both sexes. What is perhaps the most important sex discrimination is more than anything else a reflection of the differentiation of adult sex roles. It seems to be a definite fact that girls are more apt to be relatively docile, to conform in general according to adult expectations, to be "good," whereas boys are more apt to be recalcitrant to discipline and defiant of adult authority and expectations. There is really no feminine equivalent of the expression "bad boy." It may be suggested that this is at least partially explained by the fact that it is possible from an early age to initiate girls directly into many important aspects of the adult feminine role. Their mothers are continually about the house, and the meaning of many of the things they are doing is relatively tangible and easily understandable to a child. It is also possible for the daughter to participate actively and usefully in many of these activities. Especially in the urban middle classes, however, the father does not work in the home and his son is not able to observe his work or to participate in it from an early age. Furthermore, many of the masculine functions are of a relatively abstract and intangible character, such that their meaning must remain almost wholly inaccessible to a child. This leaves the boy without a tangible meaningful model to emulate and without the possibility of a gradual initiation into the activities of the adult male role. An important verification of this analysis could be provided through the study in our own society of the rural situation. It is my impression that farm boys tend to be "good" in a sense in which that is not typical of their urban brothers.

The equality of privileges and responsibilities, graded only by age but not by birth order, is extended to a certain degree throughout the whole range of the life cycle. In full adult status, however, it is seriously modified by the asymmetrical relation of the sexes to the occupational structure. One of the most conspicuous expressions and symbols of the underlying equality, however, is the lack of sex differentiation in the process of formal education, so far, at least, as it is not explicitly vocational. Up through college, differentiation seems to be primarily a matter on the one hand of individual ability, on the other hand of class status, and only to a secondary degree of sex differentiation. One can certainly speak of a strongly established pattern that all children of the family have a "right" to a good education, rights which are graduated according to the class status of the family but also to individual ability. It is only in post-graduate professional education, with its direct connection with future occupational careers, that sex discrimination becomes conspicuous. It is particularly important that this equality of treatment exists in the sphere of liberal education, since throughout the social structure of our society there is a strong tendency to segregate the occupational sphere from one in which certain more generally human patterns and values are dominant, particularly in informal social life and the realm of what will here be called community participation.

Although this pattern of equality of treatment is present in certain fundamental respects at all age levels, at the transition from childhood to adolescence new features appear which disturb the symmetry of sex roles, while still a second set of factors appears with marriage and the acquisition of full adult status and responsibilities.

An indication of the change is the practice of chaperonage, through which girls are given a kind of protection and supervision by adults to which boys of the same age group are not subjected. Boys, that is, are chaperoned only in their relations with girls of their own class. This modification of equality of treatment has been extended to the control of the private lives of women students in boarding schools and colleges. Of undoubted significance is the fact that it has been rapidly declining not only in actual effectiveness but as an ideal pattern. Its prominence in our recent past, however, is an important manifestation of the importance of sex role differentiation. Important light might be thrown upon its functions by systematic comparison with the related phenomena in Latin countries, where this type of asymmetry has been far more sharply accentuated than in this country in the more modern period.

It is at the point of emergence into adolescence that there first begins to develop a set of patterns and behavior phenomena which involve a highly complex combination of age grading and sex role elements. These may be referred to together as the phenomena of the "youth culture." Certain of its elements are present in preadolescence and others in the adult culture. But the peculiar combination in connection with this particular age level is unique and highly distinctive for American society.

Perhaps the best single point of reference for characterizing the youth culture lies in its contrast with the dominant pattern of the adult male role. By contrast with the emphasis on responsibility in this role, the orientation of the youth culture is more or less specifically irresponsible. One of its dominant notes is "having a good time," in relation to which there is a particularly strong emphasis on social activities in company with the opposite sex. A second predominant characteristic on the male side lies in the prominence of athletics, which is an avenue of achievement and competition which stands in sharp contrast to the primary standards of adult achievement in professional and executive capacities. Negatively, there is a srong tendency to repudiate interest in adult things and to feel at least a certain recalcitrance to the pressure of adult expectations and discipline. In addition to, but including, athletic prowess, the typical pattern of the male youth culture seems to lay emphasis on the value of certain qualities of attractiveness, especially in relation to the opposite sex. It is very definitely a rounded humanistic pattern rather than one of competence in the performance of specified functions. Such stereotypes as the "swell guy" are significant of this. On the feminine side there is correspondingly a strong tendency to accentuate sexual attractiveness in terms of various versions of what may be called the "glamor girl" pattern. Although these patterns defining roles tend to polarize sexually —for instance, as between star athlete and socially popular girl—yet on a certain level they are complementary, both emphasizing certain features of a total personality in terms of the direct expression of certain values rather than of instrumental significance.

One further feature of this situation is the extent to which it is crystallized about the system of formal education. One might say that the principal centers of prestige dissemination are the colleges, but that many of the most distinctive phenomena are to be found in high schools throughout the country. It is, of course, of great importance that liberal education is not primarily a matter of vocational training in the United States. The individual status on the curricular side of formal education is, however, in fundamental ways linked up with adult expectations, and doing "good work" is one of the most important sources of parental approval. Because of secondary institutionalization, this approval is extended into various spheres distinctive of the youth culture. But it is notable that the youth culture has a strong tendency to develop in directions which are either on the borderline of parental approval or beyond the pale, in such matters as sex behavior, drinking, and various forms of frivolous and irresponsible behavior. The fact that adults have attitudes to these things which are often deeply ambivalent and that on such occasions as college reunions they may outdo the younger generation, as, for instance, in drinking, is of great significance, but probably structurally secondary to the youth-versus-adult differential aspect. Thus the youth culture is not only, as is true of the curricular aspect of formal education, a matter of age status as such, but also shows strong signs of being a product of tensions in the relationship of younger people and adults.

From the point of view of age grading, perhaps the most notable fact about this situation is the existence of definite pattern distinctions from the periods coming both before and after. At the line between childhood and adolescence "growing up" consists precisely in ability to participate in youth culture patterns, which are not for either sex, the same as the adult patterns practiced by the parental generation. In both sexes the transition to full adulthood means loss of a certain "glamorous" element. From being the athletic hero or the lion of college dances, the young man becomes a prosaic business executive or lawyer. The more successful adults participate in an important order of prestige symbols, but these are of a very different order from those of the youth culture. The contrast in the case of the feminine role is perhaps equally sharp, with at least a strong tendency to take on a "domestic" pattern with marriage and the arrival of young children.

The symmetry in this respect must, however, not be exaggerated. It is of fundamental significance to the sex role structure of the adult age levels that the normal man has a "job" which is fundamental to his social status in general. It is perhaps not too much to say that only in very exceptional cases can an adult man be genuinely self-respecting and enjoy a respected status in the eyes of others if he does not "earn a living" in an approved occupational role. Not only is this a matter of his own economic support but, generally speaking, his occupational status is the primary source of the income and class status of his wife and children.

In the case of the feminine role, the situation is radically different. The majority of married women, of course, are not employed, but even of those that are a very large proportion do not have jobs which are in basic competition for status with those of their husbands. The above statement, even more than most in the present paper, needs to be qualified in relation to the problem of class. It is above all to the upper middle class that it applies. Here probably the great majority of "working wives" are engaged in some form of secretarial work which would, on an independent basis, generally be classed as a lower middle class occupation. The situation at lower levels of the class structure is quite different, since the prestige of the jobs of husband and wife is then much more likely to be nearly equivalent. It is quite possible that this fact is closely related to the relative instability of marriage which Davis and Gardner (*Deep South*) find, at least for the community they studied, to be typical of lower class groups. The relation is one which deserves careful study. The majority of "career" women whose occupational status is comparable with that of men in their own class, at least in the upper middle and upper classes, are unmarried, and in the small proportion of cases where they are married the result is a profound alteration in family structure.

This pattern, which is central to the urban classes, should not be misunderstood. In rural society, for instance, the operation of the farm and the attendant status in the community may be said to be a matter of the joint status of both parties to a marriage. Whereas a farm is operated by a family, an urban job is held by an individual and does not involve other members of the family in a comparable sense. One convenient expression of the difference lies in the question of what would happen in case of death. In the case of a farm, it would at least be not at all unusual for the widow to continue operating the farm with the help of a son or even of hired men. In the urban situation, the widow would cease to have any connection with the organization which had employed her husband, and he would be replaced by another man without reference to family affiliations.

In this urban situation, the primary status-carrying role is in a sense that of housewife. The woman's fundamental status is that of her husband's wife, the mother of his children, and traditionally the person responsible for a complex of activities in connection with the management of the household, care of children, etc.

For the structuring of sex roles in the adult phase, the most fundamental considerations seem to be those involved in the interrelations of the occupational system and the conjugal family. In a certain sense the most fundamental basis of the family's status is the occupational status of the husband and father. As has been pointed out, this is a status occupied by an individual by virtue of his individual qualities and achievements. But both directly and indirectly, more than any other single factor, it determines the status of the family in the social structure, directly because of the symbolic significance of the office or occupation as a symbol of prestige, indirectly because as the principal source of family income it determines the standard of living of the family. From one point of view, the emergence of occupational status into this primary position can be regarded as the principal source of strain in the sex role structure of our society, since it deprives the wife of her role as a partner in a common enterprise. The common enterprise is reduced to the life of the family itself and to the informal social activities in which husband and wife participate together. This leaves the wife a set of utilitarian functions in the management of the household which may be considered a kind of "pseudo-" occupation. Since the present interest is primarily in the middle classes, the relatively unstable character of the role of housewife as the principal content of the feminine role is strongly illustrated by the tendency to employ domestic servants wherever financially possible. It is true that there is an American tendency to accept tasks of drudgery with relative willingness, but it is notable that in middle class families there tends to be a dissociation of the essential personality from the performance of these tasks. Thus, advertising continually appeals to such desires as to have hands which one could never tell had washed dishes or scrubbed floors. Organization about the function of housewife, however, with the addition of strong affectional devotion to husband and children, is the primary focus of one of the principal patterns governing the adult feminine role—what may be called the "domestic" pattern. It is, however, a conspicuous fact, that strict adherence to this pattern has become progressively less common and has a strong tendency to a residual status—

that is, to be followed most closely by those who are unsuccessful in competition for prestige in other directions.

It is, of course, possible for the adult woman to follow the masculine pattern and seek a career in fields of occupational achievement in direct competition with men of her own class. It is, however, notable that, in spite of the very great progress of the emancipation of women from the traditional domestic pattern, only a very small fraction have gone very far in this direction. It is also clear that its generalization would only be possible with profound alterations in the structure of the family.

Hence it seems that, concomitant with the alteration in the basic masculine role in the direction of occupation, there have appeared two important tendencies in the feminine role which are alternative to that of simple domesticity on the one hand, and to a full-fledged career on the other. In the older situation there tended to be a very rigid distinction between respectable married women and those who were "no better than they should be." The rigidity of this line has progressively broken down through the infiltration into the respectable sphere of elements of what may be called again the glamor pattern, with the emphasis on a specifically feminine form of attractiveness which on occasion involves directly sexual patterns of appeal. One important expression of this trend lies in the fact that many of the symbols of feminine attractiveness have been taken over directly from the practices of social types previously beyond the pale of respectable society. This would seem to be substantially true of the practice of women smoking and of at least the modern version of the use of cosmetics. The same would seem to be true of many of the modern versions of women's dress. "Emancipation" in this connection means primarily emancipation from traditional and conventional restrictions on the free expression of sexual attraction and impulses, but in a direction which tends to segregate the element of sexual interest and attraction from the total personality and in so doing tends to emphasize the segregation of sex roles. It is particularly notable that there has been no corresponding tendency to emphasize masculine attraction in terms of dress and other such aids. One might perhaps say that, in a situation which strongly inhibits competition between the sexes on the same plane, the feminine glamor pattern has appeared as an offset to masculine occupational status and to its attendant symbols of prestige. It is perhaps significant that there is a common stereotype of the association of physically beautiful, expensively and elaborately dressed women with physically unattractive but rich and powerful men.

The other principal direction of emancipation from domesticity seems to lie in emphasis on what has been called the common humanistic element. This takes a wide variety of forms. One of them lies in a relatively mature appreciation and systematic cultivation of cultural interests and educated tastes, extending all the way from the intellectual sphere to matters of art, music, and house furnishings. A second consists in cultivation of serious interests and humanitarian obligations in community welfare situations and the like. It is understandable that many of these orientations are most conspicuous in fields where, through some kind of tradition, there is an element of particular suitability for feminine participation. Thus, a woman who takes obligations to social welfare particularly seriously will find opportunities in various forms of activity which traditionally tie up with women's relation to children, to sickness, and so on. But this may be regarded as secondary to the underlying orientation which would seek an outlet in work useful to the community, following the most favorable opportunities which happen to be available.

This pattern, which with reference to the character of relationship to men may be called that of the "good companion," is distinguished from the others in that it lays far less stress on the exploitation of sex role as such and more on that which is essentially common to both sexes. There are reasons, however, why cultural interests, interest in social welfare, and community activities are particularly prominent in the activities of women in our urban communities. On the one side, the masculine occupational role tends to absorb a very large proportion of the man's time and energy and to leave him relatively little for other interests. Furthermore, unless his position is such as to make him particularly prominent, his primary orientation is to those elements of the social structure which divide the community into occupational groups rather than those which unite it in common interests and activities. The utilitarian aspect of the role of housewife, on the other hand, has declined in importance to the point where it scarcely approaches a fulltime occupation for a vigorous person. Hence the resort to other interests to fill up the gap. In addition, women, being more closely tied to the local residential community, are more apt to be involved in matters of common concern to the members of that community. This peculiar role of women becomes particularly conspicuous in middle age. The younger married woman is apt to be relatively highly absorbed in the care of young children. With their growing up, however, her absorption in the household is greatly lessened, often just at the time when the husband is approaching the apex of his career and is most heavily involved in its obligations. Since to a high degree this humanistic aspect of the feminine role is only partially institutionalized, it is not surprising that its patterns often bear the marks of strain and insecurity, as perhaps has been classically depicted by Helen Hokinson's cartoons of women's clubs.

The adult roles of both sexes involve important elements of strain which are involved in certain dynamic relationships, especially to the youth culture. In the case of the feminine role, marriage is the single event toward which a selective process, in which personal qualities and effort can play a decisive role, has pointed up. That determines a woman's fundamental status, and after that her role patterning is not so much status determining as a matter of living up to expectations and finding satisfying interests and activities. In a society where such strong emphasis is placed upon individual achievement, it is not surprising that there should be a certain romantic nostalgia for the time when the fundamental choices were still open. This element of strain is added to by the lack of clear-cut definition of the adult feminine role. Once the possibility of a career has been eliminated, there still tends to be a rather unstable oscillation between emphasis in the direction of domesticity or glamor or good companionship. According to situational pressures and individual character, the tendency will be to emphasize one or another of these more strongly. But it is a situation likely to produce a rather high level of insecurity. In this state the pattern of domesticity must be ranked lowest in terms of prestige but also, because of the strong emphasis in the community sentiment on the virtues of fidelity and devotion to husband and children, it offers perhaps the highest level of a certain kind of security. It is no wonder that such an important symbol as Whistler's mother concentrates primarily on this pattern.

The glamor pattern has certain obvious atttractions, since to the woman who is excluded from the struggle for power and prestige in the occupational sphere it is the most direct path to a sense of superiority and importance. It has, however, two obvious limitations. In the first place, many of its manifestations encounter the resistance of patterns of moral conduct and engender conflicts not only with community opinion but also with the individual's own moral standards. In the second place, it is a pattern the highest manifestations of which are inevitably associated with a rather early age level—in fact, overwhelmingly with the courtship period. Hence, if strongly entered upon, serious strains result from the problem of adaptation to increasing age.

The one pattern which would seem to offer the greatest possibilities for able, intelligent, and emotionally mature women is the third—the good companion pattern. This, however, suffers from a lack of fully institutionalized status and from the multiplicity of choices of channels of expression. It is only those with the strongest initiative and intelligence who achieve fully satisfactory adaptations in this direction. It is quite clear that in the adult feminine role there is quite sufficient strain and insecurity, so that wide-spread manifestations are to be expected in the form of neurotic behavior.

The masculine role at the same time is itself by no means devoid of corresponding elements of strain. It carries with it, to be sure, the primary prestige of achievement, responsibility, and authority. By comparison with the role of the youth culture, however, there are at least two important types of limitations. In the first place, the modern occupational system has led to increasing specialization of role. The job absorbs an extraordinarily large proportion of the individual's energy and emotional interests in a role the content of which is often relatively narrow. This in particular restricts the area within which he can share common interests and experiences with others not in the same occupational specialty. It is perhaps of considerable significance that so many of the highest prestige statuses of our society are of this specialized character. There is in the definition of roles little to bind the individual to others in his community on a comparable status level. By contrast with this situation, it is notable that in the youth culture common human elements are far more strongly emphasized. Leadership and eminence are more in the role of total individuals and less of competent specialists. This perhaps has something to do with the significant tendency in our society for all age levels to idealize youth and for the older age groups to attempt to imitate the patterns of youth behavior.

It is perhaps as one phase of this situation that the relation of the adult man to persons of the opposite sex should be treated. The effect of the specialization of occupational role is to narrow the range in which the sharing of common human interests can play a large part. In relation to his wife, the tendency of this narrowness would seem to be to encourage on her part either the domestic or the glamorous role, or community participation somewhat unrelated to the marriage relationship. This relationship between sex roles presumably introduces a certain amount of strain into the marriage relationship itself, since this is of such overwhelming importance to the family and hence to a woman's status and yet so relatively difficult to maintain on a level of human companionship. Outside the marriage relationship, however, there seems to be a notable inhibition against easy social intercourse, particularly in mixed company. The man's close personal intimacy with other women is checked by the danger of the situation being defined as one of rivalry with the wife, and easy friendship without sexual-emotional involvement seems to be inhibited by the specialization of interests in the occupational sphere. It is notable that brilliance of conversation of the "salon" type seems to be associated with aristocratic society and is not prominent in ours. In the informal social life of academic circles with which the writer is familiar, there seems to be a strong tendency in mixed gatherings—as after dinner—for the sexes to segregate. In such groups the men are apt to talk either shop subjects or politics, whereas the women are apt to talk about domestic affairs, schools, their children, etc., or personalities. It is perhaps on personalities that mixed conversation is apt to flow most freely.

Along with all this goes a certain tendency for middle-aged men, as symbolized by the "bald-headed row," to be interested in the physical aspect of sex—that is, in women precisely as dissociated from those personal considerations which are important to relationships of companionship or friendship, to say nothing of marriage. In so far as it does not take this physical form, however, there seems to be a strong tendency for middle-aged men to idealize youth patterns—that is, to think of the ideal inter-sex friendship as that of their pre-marital period.

Insofar as the idealization of the youth culture by adults is an expression of elements of strain and insecurity in the adult roles, it would be expected that the patterns thus idealized would contain an element of romantic unrealism. The patterns of youthful behavior thus idealized are not those of actual youth so much as those which older people wish their own youth might have been. This romantic element seems to coalesce with a similar element derived from certain strains in the situation of young people themselves.

The period of youth in our society is one of considerable strain and insecurity. Above all, it means turning one's back on the security both of status and of emotional attachment which is engaged in the family of orientation. It is structurally essential to transfer one's primary emotional attachment to a marriage partner who is entirely unrelated to the previous family situation. In a system of free marriage choice this applies to women as well as men. For the man there is in addition the necessity to face the hazards of occupational competition in the determination of a career. There is reason to believe that the youth culture has important positive functions in easing the transition from the security of childhood in the family of orientation to that of full adult in marriage and occupational status. But precisely because the transition is a period of strain, it is to be expected that it involves elements of unrealistic romanticism. Thus significant features in the status of youth patterns in our society would seem to derive from the coincidence of the emotional needs of adolescents with those derived from the strains of the situation of adults.

A tendency to the romantic idealization of youth patterns seems in different ways to be characteristic of modern Western society as a whole. It is not possible in the present context to enter into any extended comparative analysis, but it may be illuminating to call attention to a striking difference between the patterns associated with this phenomenon in Germany and in the United States. The German "youth movement," starting before the first World War, has occasioned a great deal of comment and has in various respects been treated as the most notable instance of the revolt of youth. It is generally believed that the youth movement has an important relation to the background of National Socialism, and this fact as much as any suggests the important difference. While in Germany, as everywhere, there has been a generalized revolt against convention and restrictions on individual freedom as embodied in the traditional adult culture, in Germany particular emphasis has appeared on the community of male youth. "Comradeship" in a sense which strongly suggests that of soldiers in the field has from the beginning been strongly emphasized as the ideal social relationship. By contrast with this, in the American youth culture and its adult romantization a much stronger emphasis has been placed on the cross-sex relationship. It would seem that this fact, with the structural factors which underlie it, have much to do with the failure of the youth culture to develop any considerable political significance in this country. Its predominant pattern has been that of the idealization of the isolated couple in romantic love. There have, to be sure, been certain tendencies among radical youth to a political orientation, but in this case there has been a notable absence of emphasis on the solidarity of the members of one sex. The tendency has been rather to ignore the relevance of sex difference in the interest of common ideals.

The importance of youth patterns in contemporary American culture throws into particularly strong relief the status in our social structure of the most advanced age groups. By comparison with other societies, the United States assumes an extreme position in the isolation of old age from participation in the most important social structures and interests. Structurally speaking, there seem to be two primary bases of this situation. In the first place, the most important single distinctive feature of our family structure is the isolation of the individual conjugal family. It is impossible to say that with us it is "natural" for any other group than husband and wife and their dependent children to maintain a common household. Hence, when the children of a couple have become independent through marriage and occupational status, the parental couple is left without attachment to any continuous kinship group. It is, of course, common for other relatives to share a household with the conjugal family, but this scarcely ever occurs without some important elements of strain. For independence is certainly the preferred pattern for an elderly couple, particularly from the point of view of the children.

The second basis of the situation lies in the occupational structure. In such fields as farming and the maintenance of small independent enterprises, there is frequently no such thing as abrupt "retirement," rather a gradual relinquishment of the main responsibilities and functions with advancing age. So far, however, as an individual's occupational status centers in a specific "job," he either holds the job or does not, and the tendency is to maintain the full level of functions up to a given point and then abruptly to retire. In view of the very great significance of occupational status and its psychological correlates, retirement leaves the older man in a peculiarly functionless situation, cut off from participation in the most important interests and activities of the society. There is a further important aspect of this situation. Not only status in the community but actual place of residence is to a very high degree a function of the specific job held. Retirement not only cuts the ties to the job itself but also greatly loosens those to the community of residence. Perhaps in no other society is there observable a phenomenon corresponding to the accumulation of retired elderly people in such areas as Florida and Southern California in the winter. It may be surmised that this structural isolation from kinship, occupational, and community ties is the fundamental basis of the recent political agitation for help to the old. It is suggested that it is far less the financial hardship of the position of elderly people than their social isolation which makes old age a "problem." That the financial difficulties of older people are in a very large proportion of cases real is not to be doubted. This, however, is at least to a very large extent a consequence rather than a determinant of the structural situation. Except where it is fully taken care of by pension schemes, the income of older people is apt to be seriously reduced; but, even more important, the younger conjugal family usually does not feel an obligation to contribute to the support of aged parents. Where as a matter of course both generations shared a common household, this problem did not exist. As in other connections, we are here very prone to rationalize generalized insecurity in financial and economic terms. The problem is obviously of particularly great significance in view of the changing age distribution of the population with the prospect of a far greater proportion in the older age groups than in previous generations. It may also be suggested that, through well-known psychosomatic mechanisms, the increased incidence of the disabilities of older people, such as heart disease, cancer, etc., may be at least in part attributed to this structural situation.

CHAPTER

23

Robert K. Merton

BUREAUCRATIC STRUCTURE AND

PERSONALITY

Certain roles (such as those of age and sex discussed by Professor Parsons) are ascribed on the basis of cultural patterns and biological facts. Other roles must be achieved by the individual's conscious choice and successful effort. In the latter case, the problem of the effect of role playing upon personality formation is complicated by the operation of selectivity. Do persons more or less voluntarily decide to become government officials because they sense that here their natural personality can most easily attain full expression? Probably. But Dr. Merton suggests that these tendencies that exist already at the time of the choice of a career are heightened by long experience of carrying out roles defined by cultural patterns and the informal rules that arise within interacting social groups. The personality manifestations that characterize bureaucrats, both on the job and away from it, result not merely from the individual characteristics of those who choose to become bureaucrats but are also the effects of the social groupings and of experience in standardized roles. Professor Hughes has also discussed the effects of occupational participation upon personality.[1]

NOTE: Reprinted, with slight abridgment, from *Social Forces*, Vol. 18 (1940), pp. 560–8, by permission of the author and the editors. (Copyright, 1940, by The Williams and Wilkins Company.)

[1] E. C. Hughes, "Personality Types and the Division of Labor," *American Journal of Sociology*, Vol. 33 (1928), pp. 754–68; and "Institutional Office and the Person," *American Journal of Sociology*, Vol. 43 (1937), pp. 404–14.

A FORMAL, rationally organized social structure involves clearly defined patterns of activity in which, ideally, every series of actions is functionally related to the purposes of the organization. In such an organization, there is integrated a series of offices, of hierarchized statuses, in which inhere a number of obligations and privileges closely defined by limited and specific rules. Each of these offices contains an area of imputed competence and responsibility. Authority, the power of control which derives from an acknowledged status, inheres in the office and not in the particular person who performs the official role. Official action ordinarily occurs within the framework of pre-existing rules of the organization. The system of prescribed relations between the various offices involves a considerable degree of formality and clearly defined social distance between the occupants of these positions. Formality is manifested by means of a more or less complicated social ritual which symbolizes and supports the "pecking order" of the various offices. Such formality, which is integrated with the distribution of authority within the system, serves to minimize friction by largely restricting (official) contact to modes which are previously defined by the rules of the organization. Ready calculability of others' behavior and a stable set of mutual expectations is thus built up. Moreover, formality facilitates the interaction of the occupants of offices despite their (possibly hostile) private attitudes toward one another. In this way, the subordinate is protected from the arbitrary action of his superior, since the actions of both are constrained by a mutually recognized set of rules. Specific procedural devices foster objectivity and restrain the "quick passage of impulse into action."

The ideal type of such formal organization is bureaucracy, and in many respects, the classical analysis of bureaucracy is that by Max Weber. As Weber indicates, bureaucracy involves a clear-cut division of integrated activities which are regarded as duties inherent in the office. A system of differentiated controls and sanctions are stated in the regulations. The assignment of roles occurs on the basis of technical qualifications which are ascertained through formalized, impersonal procedures (e.g., examinations). Within the structure of hierarchically arranged authority, the activities of "trained and salaried experts" are governed by general, abstract, clearly defined rules which preclude the necessity for the issuance of specific instructions for each specific case. The generality of the rules requires the constant use of *categorization*, whereby individual problems and cases are classified on the basis of designated criteria and are treated accordingly. The pure type of bureaucratic official is appointed, either by a superior or through the exercise of impersonal competition; he is not elected. A measure of flexibility in the bureaucracy is attained by electing higher functionaries who presumably express the will of the electorate (e.g., a body of citizens or a board of directors). The election of higher officials is designed to affect the purposes of the organization, but the technical procedures for attaining these ends are performed by a continuous bureaucratic personnel.

The bulk of bureaucratic offices involve the expectation of life-long tenure, in the absence of disturbing factors which may decrease the size of the organization. Bureaucracy maximizes vocational security. The function of security of tenure, pensions, incremental salaries, and regularized procedures for promotion is to ensure the devoted performance of official duties, without regard for extraneous pressures. The chief merit of bureaucracy is its technical efficiency, with a premium placed on precision, speed, expert control, continuity, discretion, and optimal returns on input. The structure is one which approaches the complete elimination of personalized relationships and of nonrational considerations (hostility, anxiety, affectual involvements, etc.).

Bureaucratization is accompanied by the centralization of means of production, as in modern capitalistic enterprise, or, as in the case of the post-feudal army, complete separation from the means of destruction. Even the bureaucratically organized scientific laboratory is characterized by the separation of the scientist from his technical equipment.

Bureaucracy is administration which almost completely avoids public discussion of its techniques, although there may occur public discussion of its policies. This "bureaucratic secrecy" is held to be necessary in order to keep valuable information from economic competitors or from foreign and potentially hostile political groups.

In these bold outlines, the positive attainments and functions of bureaucratic organization are emphasized, and the internal stresses and strains of such structures are almost wholly neglected. The community at large, however, evidently emphasizes the imperfections of bureaucracy, as is suggested by the fact that the "horrid hybrid," bureaucrat, has become a *Schimpfwort*. The transition to a study of the negative aspects of bureaucracy is afforded by the application of Veblen's concept of "trained incapacity," Dewey's notion of "occupational psychosis," or Warnotte's view of "professional deformation." Trained incapacity refers to that state of affairs in which one's abilities function as inadequacies or blind spots. Actions based upon training and skills which have been successfully applied in the past may result in inappropriate responses *under changed conditions*, as inadequate flexibility in the application of skills will, in a changing milieu, result in more or less serious maladjustments. Thus, to adopt a barnyard illustration used in this connection by Burke, chickens may be readily conditioned to interpret the sound of a bell as a signal for food. The same bell may now be used to summon the "trained chickens" to their doom as they are assembled to suffer decapitation. In general, one adopts measures in keeping with his past training and, under new conditions which are not recognized as *significantly* different, the very soundness of this training may lead to the adoption of the wrong procedures. Again, in Burke's almost echolalic phrase, "people may be unfitted by being fit in an unfit fitness"; their training may become an incapacity.

Dewey's concept of occupational psychosis rests upon much the same observations. As a result of their day to day routines, people develop special preferences, antipathies, discriminations, and emphases. (The term "psychosis" is used by Dewey to denote a "pronounced character of the mind.") These psychoses develop through demands put upon the individual by the particular organization of his occupational role.

The concepts of both Veblen and Dewey refer to a fundamental ambivalence. Any action can be considered in terms of what it attains or what it fails to attain. "A way of seeing is also a way of not seeing—a focus upon object A involves a neglect of object B." In his discussion, Weber is almost exclusively concerned with what the bureaucratic structure attains: precision, reliability, efficiency. This same structure may be examined from another perspective provided by the ambivalence. What are the limitations of the organization designed to attain these goals?

For reasons which we have already noted, the bureaucratic structure exerts a constant pressure upon the official to be "methodical, prudent, disciplined." If the bureaucracy is to operate successfully, it must attain a high degree of reliability of behavior, an unusual degree of conformity with prescribed patterns of action. Hence, the fundamental importance of discipline, which may be as highly developed in a religious or economic bureaucracy as in the army. Discipline can be effective only if the ideal patterns are buttressed by strong sentiments which entail devotion to one's duties, a keen sense of the limitation of one's authority and competence, and methodical performance of routine activities. The efficacy of social structure depends ultimately upon infusing group participants with appropriate attitudes and sentiments. As we shall see, there are definite arrangements in the bureaucracy for inculcating and reinforcing these sentiments.

At the moment, it suffices to observe that, in order to ensure discipline (the necessary reliability of response), these sentiments are often more intense than is technically necessary. There is a margin of safety, so to speak, in the pressure exerted by these sentiments upon the bureaucrat to conform to his patterned obligations, in much the same sense that added allowances (precautionary over-estimations) are made by the engineer in designing for a bridge. But this very emphasis leads to a transference of the sentiments from the *aims* of the organization onto the particular details of behavior required by the rules. Adherence to the rules, originally conceived as a means, becomes transformed into an end-in-itself; there occurs the familiar process of *displacement of goals* whereby "an instrumental value becomes a terminal value."

This process has often been observed in various connections. Wundt's *heterogony of ends* is a case in point; Max Weber's *Paradoxie der Folgen* is another. See also MacIver's observations on the transformation of civilization into culture, and Lasswell's remark that "the human animal distinguishes himself by his infinite capacity for making ends of his means." In terms of the psychological mechanisms involved, this process has been analyzed most fully by Gordon W. Allport, in his discussion of what he calls "the functional autonomy of motives." Allport amends the earlier formulations of Woodworth, Tolman, and William Stern, and arrives at a statement of the process from the standpoint of individual motivation. He does not consider those phases of the social structure which conduce toward the "transformation of motives." The formulation adopted in this paper is thus complementary to Allport's analysis; the one stressing the psychological mechanisms involved, the other considering the constraints of the social structure. The convergence of psychology and sociology toward this central concept suggests that it may well constitute one of the conceptual bridges between the two disciplines.[2]

Discipline, readily interpreted as conformance with regulations, whatever the situation, is seen not as a measure designed for specific purposes but becomes an immediate value in the life-organization of the bureaucrat. This emphasis, resulting from the displacement of the original goals, develops into rigidities and an inability to adjust readily. Formalism, even ritualism, ensues with an unchallenged insistence upon punctilious adherence to formalized procedures. This may be exaggerated to the point where primary concern with conformity to the rules interferes with the achievement of the purposes of the organization, in which case we have the familiar phenomenon of the technicism or red tape of the official. An extreme product of this process of displacement of goals is the bureaucratic virtuoso, who never forgets a single rule binding his action and hence is unable to assist many of his clients. A case in point, where strict recognition of the limits of authority and literal adherence to rules produced this result, is the pathetic plight of Bernt Balchen, Admiral Byrd's pilot in the flight over the South Pole.

> According to a ruling of the department of labor Bernt Balchen . . . cannot receive his citizenship papers. Balchen, a native of Norway, declared his intention in 1927. It is held that he has failed to meet the condition of five years' continuous residence in the United States. The Byrd antarctic voyage took him out of the country, although he was on a ship flying the American flag, was an invaluable member of an American expedition, and in a region to which there is an American claim because of the exploration and occupation of it by Americans, this region being Little America.
>
> The bureau of naturalization explains that it cannot proceed on the assumption that Little America is American soil. That would be *trespass on international questions* where it has no sanction. So far as the bureau is concerned, Balchen was out of the country, and *technically* has not complied with the law of naturalization.

Such inadequacies in orientation which involve trained incapacity clearly derive from structural sources. The process may be briefly recapitulated. (1) An effective bureaucracy demands reliability of response and strict devotion to regulations. (2) Such devotion to the rules leads to their transformation into absolutes; they are no longer conceived as relative to a given set of purposes. (3) This interferes with ready adaptation

---

[2] See Gordon W. Allport, *Personality* (New York, 1937), ch. 7.

under special conditions not clearly envisaged by those who drew up the general rules. (4) Thus, the very elements which conduce toward efficiency in general produce inefficiency in specific instances. Full realization of the inadequacy is seldom attained by members of the group who have not divorced themselves from the "meanings" which the rules have for them. These rules in time become symbolic in cast, rather than strictly utilitarian.

Thus far, we have treated the ingrained sentiments making for rigorous discipline simply as data, as given. However, definite features of the bureaucratic structure may be seen to conduce to these sentiments. The bureaucrat's official life is planned for him in terms of a graded career, through the organizational devices of promotion by seniority, pensions, incremental salaries, *etc.*, all of which are designed to provide incentives for disciplined action and conformity to the official regulations. The official is tacitly expected to and largely does adapt his thoughts, feelings, and actions to the prospect of this career. But *these very devices* which increase the probability of conformance also lead to an over-concern with strict adherence to regulations which induces timidity, conservatism, and technicism. Displacement of sentiments from goals onto means is fostered by the tremendous symbolic significance of the means (rules).

Another feature of the bureaucratic structure tends to produce much the same result. Functionaries have the sense of a common destiny for all those who work together. They share the same interests, especially since there is relatively little competition insofar as promotion is in terms of seniority. In-group aggression is thus minimized and this arrangement is therefore conceived to be positively functional for the bureaucracy. However, the *esprit de corps* and informal social organization which typically develops in such situations often leads the personnel to defend their entrenched interests rather than to assist their clientele and elected higher officials. As President Lowell reports, if the bureaucrats believe that their status is not adequately recognized by an incoming elected official, detailed information will be withheld from him, leading him to errors for which he is held responsible. Or, if he seeks to dominate fully, and thus violates the sentiment of self-integrity of the bureaucrats, he may have documents brought to him in such numbers that he cannot manage to sign them all, let alone read them. This illustrates the defensive informal organization which tends to arise whenever there is an apparent threat to the integrity of the group.

It would be much too facile and partly erroneous to attribute such resistance by bureaucrats simply to vested interests. Vested interests oppose any new order which either eliminates or at least makes uncertain their differential advantage deriving from the current arrangements. This is undoubtedly involved, in part, in bureaucratic resistance to change but another process is perhaps more significant. As we have seen, bureaucratic officials affectively identify themselves with their way of life. They have a pride of craft which leads them to resist change in established routines;

at least, those changes which are felt to be imposed by persons outside the inner circle of co-workers. This non-logical pride of craft is a familiar pattern found even, to judge from Sutherland's *Professional Thief*, among pickpockets who, despite the risk, delight in mastering the prestige-bearing feat of "beating a left breech" (picking the left front trousers pocket).

In a stimulating paper, Hughes has applied the concepts of "secular" and "sacred" to various types of division of labor; "the sacredness" of caste and *Stände* prerogatives contrasts sharply with the increasing secularism of occupational differentiation in our mobile society. However, as our discussion suggests, there may ensue, in particular vocations and in particular types of organization, the *process of sanctification* (viewed as the counterpart of the process of secularization). This is to say that through sentiment-formation, emotional dependence upon bureaucratic symbols and status, and affective involvement in spheres of competence and authority, there develop prerogatives involving attitudes of moral legitimacy which are established as values in their own right, and are no longer viewed as merely technical means for expediting administration. One may note a tendency for certain bureaucratic norms, originally introduced for technical reasons, to become rigidified and sacred, although, as Durkheim would say, they are *laïque en apparence*. Durkheim has touched on this general process in his description of the attitudes and values which persist in the organic solidarity of a highly differentiated society.

Another feature of the bureaucratic structure, the stress on depersonalization of relationships, also plays its part in the bureaucrat's trained incapacity. The personality pattern of the bureaucrat is nucleated about this norm of impersonality. Both this and the categorizing tendency, which develops from the dominant role of general, abstract rules, tend to produce conflict in the bureaucrat's contacts with the public or clientele. Since functionaries minimize personal relations and resort to categorization, the peculiarities of individual cases are often ignored. But the client who, quite understandably, is convinced of the "special features" of *his* own problem often objects to such categorical treatment. Stereotyped behavior is not adapted to the exigencies of individual problems. The impersonal treatment of affairs which are at times of great personal significance to the client gives rise to the charge of "arrogance" and "haughtiness" of the bureaucrat. Thus, at the Greenwich Employment Exchange, the unemployed worker who is securing his insurance payment resents what he deems to be "the impersonality and, at times, the apparent abruptness and even harshness of his treatment by the clerks. . . . Some men complain of the superior attitude which the clerks have."

Still another source of conflict with the public derives from the bureaucratic structure. The bureaucrat, in part irrespective of his position with*in* the hierarchy, acts as a representative of the power and prestige of the entire structure. In his official role, he is vested with definite authority. This often leads to an actual or apparent domineering attitude, which may only be exaggerated by a discrepancy between his position within the hierarchy and his position with reference to the public.[3] Protest and recourse to other officials on the part of the client are often ineffective or largely precluded by the previously mentioned *esprit de corps* which joins the officials into a more or less solidary in-group. This source of conflict *may* be minimized in private enterprise since the client can register an effective protest by transferring his trade to another organization within the competitive system. But with the monopolistic nature of the public organization, no such alternative is possible. Moreover, in this case, tension is increased because of a discrepancy between ideology and fact: the governmental personnel are held to be "servants of the people," but in fact they are usually superordinate, and release of tension can seldom be afforded by turning to other agencies for the necessary service. (At this point the political machine often becomes functionally significant. As Steffens and others have shown, highly personalized relations and the abrogation of formal rules [red tape] by the machine often satisfy the needs of individual "clients" more fully than the formalized mechanism of governmental bureaucracy.) This tension is in part attributable to the confusion of status of bureaucrat and client; the client may consider himself socially superior to the official who is at the moment dominant.

Thus, with respect to the relations between officials and clientele, one structural source of conflict is the pressure for formal and impersonal treatment when individual, personalized consideration is desired by the client. The conflict may be viewed, then, as deriving from the introduction of inappropriate attitudes and relationships. Conflict with*in* the bureaucratic structure arises from the converse situation, namely, when personalized relationships are substituted for the structurally required impersonal relationships. This type of conflict may be characterized as follows:

The bureaucracy, as we have seen, is organized as a secondary, formal group. The normal responses involved in this organized network of social expectations are supported by affective attitudes of members of the group. Since the group is oriented toward secondary norms of impersonality, any failure to conform to these norms will arouse antagonism from those who have identified themselves with the legitimacy of these rules. Hence, the substitution of personal for impersonal treatment within the structure is met with widespread disapproval and is characterized by such epithets as

---

[3] In this connection, note the relevance of Koffka's comments on certain features of the pecking-order of birds. "If one compares the behavior of the bird at the top of the pecking list, the despot, with that of one very far down, the second or third from the last, then one finds the latter much more cruel to the few others over whom he lords it than the former in his treatment of all members. As soon as one removes from the group all members above the penultimate, his behavior becomes milder and may even become very friendly. . . . It is not difficult to find analogies to this in human societies, and therefore one side of such behavior must be primarily the effects of the social groupings, and not of individual characteristics." K. Koffka, *Principles of Gestalt Psychology* (New York, 1935), pp. 668–9.

"graft," "favoritism," "nepotism," "apple-polishing," etc. These epithets are clearly manifestations of injured sentiments. The function of such "automatic resentment" can be clearly seen in terms of the requirements of bureaucratic structure.

Bureaucracy is a secondary group mechanism designed to carry on certain activities which cannot be satisfactorily performed on the basis of primary group criteria. Hence behavior which runs counter to these formalized norms becomes the object of emotionalized disapproval. This constitutes a functionally significant defence set up against tendencies which jeopardize the performance of socially necessary activities. To be sure, these reactions are not rationally determined practices explicitly designed for the fulfilment of this function. Rather, viewed in terms of the individual's interpretation of the situation, such resentment is simply an immediate response opposing the "dishonesty" of those who violate the rules of the game. However, this subjective frame of reference notwithstanding, these reactions serve the function of maintaining the essential structural elements of bureaucracy by reaffirming the necessity for formalized, secondary relations and by helping to prevent the disintegration of the bureaucratic structure which would occur should these be supplanted by personalized relations. This type of conflict may be generically described as the intrusion of primary group attitudes when secondary group attitudes are institutionally demanded, just as the bureaucrat-client conflict often derives from interaction on impersonal terms when personal treatment is individually demanded. (Community disapproval of many forms of behavior may be analyzed in terms of one or the other of these patterns of substitution of culturally inappropriate types of relationship. Thus, prostitution constitutes a type-case where coitus, a form of intimacy which is institutionally defined as symbolic of the most "sacred" primary group relationship, is placed within a contractual context, symbolized by the exchange of that most impersonal of all symbols, money.)

The trend toward increasing bureaucratization in Western society, which Weber had long since foreseen, is not the sole reason for sociologists to turn their attention to this field. Empirical studies of the interaction of bureaucracy and personality should especially increase our understanding of social structure. A large number of specific questions invite our attention. To what extent are particular personality types selected and modified by the various bureaucracies (private enterprise, public service, the quasi-legal political machine, religious orders)? Inasmuch as ascendancy and submission are held to be traits of personality, despite their variability in different stimulus-situations, do bureaucracies select personalities of particularly submissive or ascendant tendencies? And since various studies have shown that these traits can be modified, does participation in bureaucratic office tend to increase ascendant tendencies? Do various systems of recruitment (e.g., patronage, open competition involving specialized knowledge or "general mental capacity," practical experience) select different personality types? Does promotion through seniority lessen competitive anxieties and enhance administrative efficiency? A detailed examination of mechanisms for imbuing the bureaucratic codes with affect would be instructive both sociologically and psychologically. Does the general anonymity of civil service decisions tend to restrict the area of prestige-symbols to a narrowly defined inner circle? Is there a tendency for differential association to be especially marked among bureaucrats?

The range of theoretically significant and practically important questions would seem to be limited only by the accessibility of the concrete data. Studies of religious, educational, military, economic, and political bureaucracies, dealing with the interdependence of social organization and personality formation, should constitute an avenue for fruitful research. On that avenue, the functional analysis of concrete structures may yet build a Solomon's House for sociologists.

24

David M. Schneider

THE SOCIAL DYNAMICS OF PHYSICAL
DISABILITY IN ARMY BASIC TRAINING

*In this paper Dr. Schneider shows how a host of different motivations
may bring a person to play a particular role, and how, despite the variety
of motivational forces involved, the very participation in the role engenders
and enforces certain uniformities in the personalities of the players.*

I

IN THIS paper I present a brief description and analysis of the role of the
sick in a particular milieu, that is, the role of those soldiers in army basic
training who are not hospitalized for physical disabilities and who remain
in their training group.

In basic training a soldier with a physical disability can be hospitalized,
treated on an out-patient basis, or ignored by the medical officer.

Where the disability requires extended hospitalization, the soldier's
relationship with the group is terminated and, as far as we are concerned,
needs no further comment. But where the medical officer allows the soldier to remain a part of the group, then it is possible for him to develop
a special relationship to it.

The kinds of disabilities which permit the soldier to remain a part of

NOTE: Thanks are due the author for revising and rewriting portions of this paper
especially for this volume. It first appeared in *Psychiatry*, Vol. X (1947), pp. 323–33.
Thanks are due the William Alanson White Psychiatric Foundation for permission
to reprint.

the group are, most frequently, rheumatic, asthenic, gastro-intestinal complaints, and occasionally cardiovascular complaints and "chronic headaches." As the group sees them, they are pains in the leg or arm or back, "trick knee," chronic "dislocations," sore joints, stiffness of joints, weakness of muscles or limbs, cramps, nausea, stomach pains, frequency of bowel movement, dizziness, palpitation, "murmur," shortness of breath, rapid pulse rate, "pounding heart," "sinus trouble," "migraine," and "awful headache." Tooth extractions, athlete's foot, mild nearsightedness or farsightedness, minor abrasions, and so forth are never observed to function socially as is described below.

In certain cases these physical incapacities are symptomatic of the failure of the civilian to adjust to his new role as a soldier. This adjustment depends on the ways in which the new recruit meets three major problems: (1) the disposition of the affectional relationships and needs which characterized his civilian life; (2) the method of meeting the heightened demands on his masculinity; and (3) the establishment or re-establishment of brother-brother-like and father-son-like relationships. When there is a failure in adjustment, that failure may be referred to the kind of response which the soldier makes to one or all of these problems.

In certain other cases, however, the physical incapacity may exist prior to the adjustment failure and may account for it by precluding adjustive responses, particularly to the second problem listed above, and permitting only kinds of responses likely to prove maladjustive.

In brief, any particular case of physical incapacity seen in the early stages of basic training may be a developed symptom of maladjustment or it may be, psychologically speaking, a relatively uncomplicated organic malfunction which will necessarily develop psychic correlates as the training period proceeds. This point will be more fully explored in section II below.

The development of the relationship between the physically incapacitated and the group takes place as follows:

Physical disability is defined as something which is external to the person, over which he can ordinarily be expected to have no control, something which happens to him through no fault of his own, like the "hand of fate." The idea of incapacity which comes from outside the self correlates closely with the idea that help from the outside is necessary in dealing with it. This idea is supported, of course, by a specially trained group of technicians toward whom one can turn for help. Physical disability is a weakness; that is acknowledged. But it is a kind of weakness that does not usually arouse blame or condemnation. It is not the same kind of weakness as chronic alcoholism or masturbation or cowardice. One can "see" no reason for alcoholism, but one can "see" and "feel" pain and discomfort. As a weakness, externally caused physical incapacity is neither condemned nor punished in army culture. Rather, it arouses sympathy and beneficent feelings ideally, and strong negative sanction is directed toward any action which aggravates or exploits such disability.

Initially the physically incapacitated, whom I shall call "the sick," define their own position and role in the group. They point out their particular complex of symptoms when an occasion warrants and make clear exactly why they should not do many of the things which others do. The initial response to them is sympathetic, and support is given to them in their efforts to avoid the kinds of activity which they feel are injurious to their health. They are urged not to do anything that will hurt them. It is remarked and repeated that they have no business in the army; the army has no business trying to force the impossible out of them, that they can count on "the rest of us" to support them in a show-down should any officer or noncommissioned officer demand their full participation.

At this time, the beginning of the training period, the non-sick members extending sympathy are having their own troubles accustoming themselves to the hard work, the hours, the food, the discipline. The army seems to them harsh and brutal. Their extensions of sympathy to the sick are a protest of their own and at the same time define a wished-for role of exemption. The protest amounts to a convenient channel for directing group-sanctioned hostility toward the agent of their own distress, and the role of exemption shines brightly as an escape from the very things which are forcing adjustment from them. To the sick this sympathy represents a perfect kind of social sanction. Not only is sickness culturally sanctioned as an excuse for avoiding undue physical effort, but the whole-hearted group expression and reiteration of the sanction leaves them doubly secure in their role; the sick know that they cannot be expected to do things, and when the group agrees with them and tells them so they are sure of their role.

But as time goes on, fewer and fewer expressions of sympathy are forthcoming, and the sick participate less and less in the normal everyday group activities. The gap between the sick and the group gradually widens. The group has begun to make its own adjustments, and the social distance between the sick and the group becomes the initial ground for future friction.

The final relationship is reached when the group makes its stable adjustment, and the sick find themselves cut off from the group. Group life continues to encompass more and more new experiences in which the sick are not included. By this time the group members have come to accept the masculinity competition and have learned to handle their recently truncated affectional needs and responses by suppression, so that toughness standards are functioning, emotional expressions are suppressed, and the constant need to secure one's masculine self-conception is operating at full force. By this time, the sick are strangers. Clashes occur in any of the following situations: when the whine of the sick demands sympathy, pity, emotional identification; or when the picture that the sick present arouses anxiety on the part of the masculine self-conception; or when the simple "unable" existence of the sick impedes the tough role of the others. The sick man is then actively despised and aggressed against; he is brushed from the field. He retires, securing himself with the previously accepted sanction that the sick cannot be expected to do things, and he does not do them. He continues to attempt to define the situation as he had at the outset and only gradually abandons his attempts to force his definition of the situation onto the others. The rest of the group rolls onward past this casualty, without troubling itself over him except for the brief, isolated encounters in which their conflicting views clash. Illustrating this final adjustment is the following short extract from my field notes:

T. sweeping down the middle of the floor. K. sitting on his foot locker. T. asked K. if he'd swept under his bed yet; K. said No, he felt sick. (T. had either to sweep under K's bed for him or wait until K. swept it himself so that all the dirt could go down the aisle at once.) T. shouted at the top of his lungs, "You bastard! You're always sick—now get the hell on the ball, goddam it, and sweep under your bed!" K. silently rose and slowly swept under his bed.

Although a large majority of those who define their role as "sick" play it as described above, deviations from it are occasionally observed. Some few become even keener participants in the masculine competition, forcing themselves and exploiting the common fund of experience to gain full participation. A few find difficulty in defining their role, first excluding themselves on the basis of their ailments, then ignoring and denying these ailments to force ahead, and eventually adjusting either as full-blown sick or as non-sick members of the group. And some maintain their sick role in action but attempt to deny the fact that they are omitted from the group by forcing vocal and verbal interaction: by boasting, loud talk, and emphasis on other areas such as gambling or drinking. These people realize that they are losing status, that being omitted from the competition also means that they cannot win, and for them the ego hurt involved demands compensation.

II

The existence of the sick role and the patterned ways of responding to it; the course of the relationship and its final equilibrium; the sanctions which support it, including the specially tempered kinds of punishments and deprivations; the institutional provision of doctors, dispensaries, and avenues for establishing relationship; and the existence in the army of heightened demands on masculinity and physical activity—all these are functionally interrelated.

Physical disability, as we have seen, disqualifies the suffering soldier from situations which are physically competitive. The competition is between "brothers," and participation in the competition aims at securing a satisfactory masculine self-conception. It is competition to "keep up with the boys," to be tough and strong. The old army cliché, "make a man of you," is perhaps the most accurate way of saying this. The army tries to "make a man of you," but if one cannot, or does not want to, be a "man," one does not have to be. One calls on some external, hand-of-fate kind of disqualification. For those persons who cannot operate within the preferred cultural channels, the existence of this institutionalized arrangement both *allows* disqualification and *patterns* its operation.

The physical disabilities of those who play the sick role may have any one of a wide variety of origins. But no matter what the origin, the soldier with a physical disability is faced with the psychological problem of how to treat it, how to behave, what role to play. Some attempt to compensate for their disabilities by intense conformative behavior, pushing themselves to the limit to gain full participation. Others try simply to ignore the disability, saying nothing of it and acting as if it did not exist; others play the sick role. The determinants of this choice lie in the psychic make-up of the individual and in the available, permitted alternatives of the culture. But when the sick role is played it does affect the disqualification of the individual from the masculinity demands of army life.

This disqualification may initially be either the direct response to a specific motivation to evade masculinity demands or an indirect consequence of behavior based on other motivations. It is clear, however, that no matter what motivations are originally involved, the same psychological consequences prevail for all who play that role. Initially the soldier is supported in his sick role. Later he finds himself omitted from the group, thereby losing those satisfactions which membership brings. He loses a common ground of experience upon which feelings of belonging are based. He loses the sense of achievement which the others have, and he feels the loss. He is himself responsible for his omission from the group, for he has taken the initiative in avoiding the bulk of the activities, largely physical, in which the group participates. Isolated from the group and unable to participate in its physical activities because he has defined himself as unable, he then experiences a sense of inadequacy, of failure, of inability, of being left-out. This feeling of inadequacy, supported by the various motivations which, initially, made the sick role psychically necessary, provides additional motivation specifically for the evasion of masculinity demands.

Hence, whether the need to evade masculinity demands is antecedent to the playing of the sick role or derived from the playing of the sick role after it has been embarked upon for other reasons, both kinds of people have this need once the sick role is embarked upon.

This is an account of the primary function of this pattern of behavior in terms of the motivations of people. But such regularization of behavior patterns never arises and endures simply to fulfill one person's need. It seems evident, then, that the number of men who must evade the masculinity demands in this basic training situation is great enough so that some provision for their problem has to be made—provision for how they are to act and how they are to be treated. One of the provisions found in American army basic training culture is this sick role.

It should be explicitly noted that physical disqualification is not the only provision in the army for persons who need to evade masculinity demands. It is simply the particular one which I am discussing.

The existence of this kind of cultural provision for the evasion of masculinity demands (disqualification without consciously directed punishment) means that the society treats this syndrome as a legitimate, though not specially honored, one. In other societies, or even other times within the history of this society, other provisions exist, that is, severe punishment, public shaming, and so on. Why this kind of cultural provision should exist rather than some other is a question which cannot be dealt with here.

III

The readily visible and vocal disabilities are physical rather than intellectual, asthenic rather than affectional. Hardly anyone goes to any army dispensary to complain that he is not thinking clearly or efficiently, or that he does not enjoy sexual relations as much as he once did. Instead he complains that he has sacrioliac strain or stomach trouble or chronic lumbar pain or a "trick knee." This does not mean that these are the only disabilities which occur. Rather, it reflects the comparatively high social valuation and emphasis which is placed on the masculine physical capacities. It is hardly necessary to point out that this is quite consistent with the whole orientation of army basic training as training for combat.

Consistent with this too is the content of the sick role, which emphasizes physical disability as the grounds for disqualification, and it is in the light of these observations that the *selective* and *suggestive* functions of the sick role become apparent.

The sick role, as a pattern of behavior in operation, is selective insofar as it emphasizes and makes socially and psychically effective for the soldier those of his disabilities which are appropriate to the army situation —the physically disabling rather than the emotionally or intellectually incapacitating. It is suggestive in that by its form and existence, the sick role provides a model for unconscious adoption.

From the point of view of the general question of symptom formation, these data suggest that in those cases where the psychic aim of a disability is to disqualify, to evade, to remove the subject from a psychologically intolerable situation, then the kind of disability he develops must be patterned from the social situation if it is to be effective. The cultural definition of effective is not specific, but may provide a range of alternatives. When an individual needs a disqualifying somatic condition in the army, for instance, this condition does not have to be specifically lumbar sprain or chronic sinusitis, but it cannot be hypertension or hay fever unless it is severe. Where a physical complaint has symbolized psychic conflict and has also functioned as a disqualifier in civilian life, the change to the army situation, because of the change in social valuation, may have necessitated a shift to a somatically synonymous symbol bearing the new socially dis-qualifying evaluation. Or where such a need has given rise to a range of hypochondriacal physical complaints, then the effect will be a selection and emphasis of some of them and a negation of the importance of others; this selection and emphasis is reflected in the soldier's behavior, for he comes to feel more discomfort from the socially relevant aches and pains and less discomfort from the others.

It is perhaps in this area, the area of masculinity demands, that the social importance to psychic security of the kind of symptom can be seen most dramatically. Where induction into the army is interpreted as an im-mersion into a traumatic, threatening situation; where the social demands on the person are neither intellectual nor affectional but explicitly physi-cal; where there are channels of escape, avenues of relationship leading to these channels, doctors and hospitals and certificate-of-disability discharges or at least limited service, and where others are seen or heard about who demonstrate the pattern which effectively removes them; where initial self-disqualification is greeted with such strong group support; then the development of some kind of physical disability is almost inevitable. And if these soldiers develop some ineffective symptom first, perhaps because it is more precisely expressive of their conflict, then its very social ineffec-tiveness, which also means psychic ineffectiveness, soon necessitates further efforts at tension reduction. This effort results in the genesis of an effective symptom, modeled from the situation which is so readily apparent. The development of a socially irrelevant symptom simply cannot relieve the anxiety sufficiently. It can reduce it, it can be of some comfort, but the bulk of the anxiety will still be there, still demanding relief.

IV

The initial support of the sick in their role has the psychological effect of reinforcing their disqualification by rewarding it. This reinforcement pro-vides, along with the person's internal tensions and motivations, one of the bases for the compulsiveness of the behavior which follows—the sick role.

The socially isolating effect of the sick person's role arouses strong feelings of isolation. He is cut off from the group and, feeling cut off, the depth and satisfaction of his interpersonal relationships suffer. More and more of his satisfactions have to come from within himself and less and less can he depend upon others for his psychic wants. This makes his dis-ability more important than ever for him and so contributes to the sec-ondary gain.

The neurotic compulsion is further reinforced through the social isola-tion which has a fixing effect, once the soldier has embarked upon a career of disqualification. Although there is no institutional provision making the process irreversible, all factors militate against turning back. The longer he is out of the group, the harder it is to get back in. The greater the initial satisfaction from the social sanction supporting his disqualification, the less likely is it that it will be relinquished later. The feelings of isolation—consequent on divorce from the group and barren and punishing relationships—are consolidated but minimized by the secondary gain. The personality which adjusted to its civilian masculinity anxieties by avoiding competition and accepting a passive role, has its whole adjustment shattered by a traumatic inundation of a violently threatening situation and new fears are added to old ones, consolidating the fears and the imperative need for avoidance. Since he is cut off from the group and hurt by it, he may well develop feelings of persecution which consolidate his isolation.

It should be noted that the group makes no conscious effort to punish the sick except where the situational convergence brings the group and the sick into direct conflict. Yet the net effect is a punishing one for the sick, as a result of the periodic clashes and the fact that they are cut off from the group, with a dearth of rewarding human relationships. But this is more a mechanical consequence than a premeditated effort on the part of the group. The overt pattern is one of mutual avoidance and withdrawal; the sick avoid participation and the well find themselves isolated from the sick and keep it that way.

The fixing effect plus the nonpunitive nature of the role are two factors which tend to support and perpetuate both the patterned behavior of the sick and the masculinity-evading motivation with which it is associated. If one were to pose the question, Why does the sick role as a patterned behavior persist?, the answer could be made, at least in part, in the following terms: first, the need to evade masculinity demands in a sufficiently large portion of the population requires some kind of provision; second, for reasons which I have not explored, this role is met by a nonpunitive, noneliminative kind of behavior by the rest of the group; third, one of the consequences of this role is its fixing effect upon the person. These last two factors mean that the anxiety which has been aroused by masculinity demands in those who play the sick role is reduced to a tolerable degree but never eliminated, cured, or punished out of existence at any point in its development. And since the anxiety is reduced but neither altered nor eliminated in the soldier, he perpetuates both the anxiety and the adaptation to it.

It seems clear, then, that for any given person the existence and perpetuation of this kind of neurotic response—the sick role—cannot be ascribed exclusively to the peculiar or accidental psychic history of that person, but is rather a function of the interaction of the particular psychic situation with the social situation in which the neurotic response takes place. Other cultures, or different social responses to the same psychic needs, may have different psychic consequences for the person and he in turn may respond differently. Not the least important of the conditions accounting for the particular neurotic response is the existence of permissive cultural definitions—definitions which function so as to allow and perhaps even facilitate the perpetuation of the response. It is to these functionally permissive conditions of genesis that one must look for a part of the answer to the question of why different cultures exhibit different kinds of neuroses.

V

If the perspective is shifted from the individual in the society to the society alone, it can be seen that the following situation exists: a minority of the group deviates from the cultural role defined for it. The deviation is neither random nor erratic, but regular both in its occurrence and in its pattern. The role of the group toward the deviant develops into a regular and consistent one, and it too cannot be described as erratic or random. The conditions for the transmission and perpetuation of the deviant role exist as part of the definition of both the sick role and the role of the group. This meets Linton's definition of cultural behavior [2].

The role of the deviant negates the major orientation or goal of the culture. Army basic training is centered on the efficient execution of the masculine role of toughness, ruggedness, ability to "take it" (stamina), which has the aim of successful combat; the sick evade this specific role by acting in precisely the contrary fashion; they are neither tough nor rugged but weak and unable to "take it."

As Benedict [1] and Mead [3] have pointed out, deviance of this sort is *abnormal* only in the cultural context which disallows and disvalues it, rather than institutionalizing and sanctioning it. The disvalued definition of the sick role in army basic training culture, however, is not the arbitrary judgment of a malevolent society; rather, the disvaluation of the sick role follows logically from the functional requisites of a culture oriented primarily toward one major goal.

Where a culture is largely oriented toward one major interest or goal, any role in that culture which is defined in antithesis to it is potentially disruptive or threatening. If the major goal is to be maintained, then the threat to it must be neutralized. The neutralization of the threat does not mean, however, that the whole role must be eliminated. In the case of the sick, the threatening behavior becomes *isolated* from the group and the group *insulated* from the threat by the minimal and brief contact with it. This isolation is enforced by defining the deviant role as disvalued, by punishing the actor through the kind of ostracism previously described, and by the punishing effect of the brief encounters.

But these devices which effectively isolate the deviants are first made possible by the structure of the culture. The basic training situation is one in which the whole group participates in the same kind of activity at the same time. Since it is the evasion of these activities that is the distinguishing feature of the sick role, it follows that by evading that participation which is shared, the sick are separated from the group. It is this which makes isolation possible. Although the separation of the sick from the group is made possible by their separate spheres of experience, it would still be feasible, within this same structure, to integrate the two. Instead, because the sick role constitutes a threat against the major goal of the culture, this initial separation is capitalized upon, and the devaluation of the sick role, the punishment of the sick through withholding sympathy, and the aggression of the brief encounters effectively enforce the isolation of the sick role.

Although, as pointed out above, the definition of the sick role allows for its perpetuation, its existence also has positive social functions. Since the personnel of this society, the army, are drawn from a different society with a different culture, army basic training society has the problem of selecting only those personnel who meet its specifications. Insofar as its selective techniques do not prevent the introduction of potential deviants and insofar as deviance can be learned after induction, the sick role permits the group to become homogeneous by draining off deviants and isolating them, thus strengthening the group. By the same token, any given person who was maladjusted has the possibility of escape from what may be for him an intolerable situation.

VI

One of the more important conclusions of this analysis is that a necessary functional interdependence exists between both social and psychological functions within any system of interacting people. It seems quite obvious that if the assumption of the sick role were not to entail a psychologically punishing price for the soldier, the social organization of army basic training—as an integrated unit training for combat—could not be maintained; on the other hand, if the social organization of army basic training were oriented by, let us say, principles of "fun-for-all," as in a boys' summer camp, a psychological consequence of the sick role could not have been severe punishment. Similarly, the *psychological* isolation of the individual is functionally coordinate with his *social* isolation, which is necessary in maintaining a homogeneous group and weeding out the incompetents; psychological isolation has the social effect of creating people who, from the point of view of the group, "don't really belong" and so can be easily weeded out. Again, the channeling of symptom formation into physical and not intellectual disabilities is functionally congruent with the necessity, defined by the social organization, for selecting physically fit fighters. Psychological needs which are in implicit conflict with the social functions of training for combat might not be detected at that time. But when these needs become expressed in physical terms, they do come into conflict with the social value and thus make possible the selection of psychologically as well as physically fit combat men.[1] So, too, the fact that the sick are not actively and aggressively sought out and attacked

---

[1] That this is not a perfectly efficient system, measured in terms of actual combat in the future, is clear from the appearance of combat neuroses later. But *within* the basic training situation it is effective and it is integrated.

by the group is functionally congruent with the psychological orientation of aggression in army basic training. The social value is on aggression against the enemy, not amongst the group. If the group members attack each other indiscriminately, this may have the effect of heightening the general level of aggression, but it will also militate against co-operative team work. The failure of the group to attack the sick supports the soldier's displacement of his aggression onto the enemy, and permits team activity by minimizing intragroup aggression.

VII

To summarize: one finds in army basic training a patterned sick role for those who have physical disabilities but who must remain with the group. The sick role develops first as a group-sanctioned behavior, focusing the resentment of the group against the army, providing a *cause célèbre* for aggression against the army. But the group sanction lasts only until the group has begun to make its adjustment to aborted affectional needs, heightened masculinity demands, and brother-like relationships. When the group begins to make its adjustment, its need for rationalization of its aggression against the army diminishes, so that it no longer supports the sick in their role. At this time the divorce between the sick and the group takes place, and the sick find themselves isolated from the group. The final adjustment is reached with the stable isolation of the sick from the group, with occasional, isolated clashes when the sick impede the others. The group, however, usually makes no special, conscious effort to punish the sick, but only maintains the social distance between them. Whether or not the original motivation for any given soldier is the specific evasion of the heightened masculinity demands which army life imposes, this becomes the effect once the sick role is played. It is apparent from the existence of this patterned sick role that a sufficiently large portion of the population either comes into basic training with this motivation, or acquires it shortly thereafter, and that this motivation has such social relevance as to warrant the regularization of the sick role. Hence the sick role has as one of its cultural functions provision for the social problem of those who, for one reason or another, are forced to evade masculinity demands. From the point of view of the question of symptom formation, the social valuation of the particular symptom stands as one of the most important criteria determining the choice and manifestation of a symptom. For if the symptom is to be effective in relieving this anxiety, it must function socially so as to permit disqualification in meeting masculinity demands. For the sick soldier, the adoption of the sick role reinforces narcissistic trends and fixes almost irreversibly his neurotic role. The nonpunitive, noneliminative kind of role which the sick role represents, plus the fixing effect on the person, are two among many factors which tend to reduce the anxiety to a tolerable degree but at the same time perpetuate it *in toto* rather than eliminate it at its source—in the family relationships of early life or later, as the case may be—or at any point in its development.

The existence and perpetuation of this neurotic response—the sick role —cannot be ascribed only to the peculiar or accidental psychic history of the person but must be considered also as a function of the cultural situation in which that psychic history takes place. One important condition which must be taken into account is the existence in the cultural structure of permissive rather than eliminative role definitions.

From the cultural perspective alone, the sick role is the direct negation of the major goal of army basic training culture. As a potential threat to the major goal, the disvaluation of that role is a functional necessity and not an arbitrary assignment of cultural value. This threat is neutralized by isolating it and insulating the group from it. This isolation of the sick is possible only because the cultural structure allows for its requisite conditions, namely, the separate areas of common experience for the group on the one hand and the sick on the other. The separation is enforced by defining the deviant role as disvalued, and by punishing the sick by withholding sympathy, and in brief hostile encounters.

Major cultural functions of the sick role are, first, the support of the major goal of the culture by draining off deviants and allowing the group to become more homogeneous and, second, providing the individual soldier with an avenue for escape from, for him, a psychologically intolerable situation.

The necessary integration of the social and the psychological functions of the sick role is understood in terms of the following: first, the punishing price paid by the soldier directly supports the organization of the society around the primary goal of training for combat, and conversely, such an organization, if it is to survive, must exact just such a price from any soldier who deviates from its major goal. Second, the psychological isolation of the soldier gives rise to his social isolation, thus permitting the deviant to be weeded out of the group and the group to maintain its homogeneity, and conversely, the social isolation of the soldier consolidates his psychological isolation. Third, the channeling of psychological symptoms into physical manifestations permits their ready recognition and elimination, thus supporting the homogeneity of the group, while conversely the major emphasis on physical fitness as the socially defined value enables the individual to express his psychic problems in those socially recognized terms.

REFERENCES

1. Benedict, Ruth: *Patterns of Culture* (Boston, Houghton Mifflin, 1934), pp. 237–57.
2. Linton, Ralph: *The Cultural Background of Personality* (New York, D. Appleton-Century, 1945), pp. 32–8.
3. Mead, Margaret: *Sex and Temperament* (New York: William Morrow, 1935), pp. 290–322.

Part Two

SECTION VI

SITUATIONAL DETERMINANTS

INTRODUCTION

WE NOW come to those determinants of personality that result from "accidents" of the individual's life history or from relatively temporary situations in which a number of individuals are placed by external events. The child may grow up under family circumstances of economic security[1] or in a situation where frustrations for all members of the family are for some years maximized by war.[2] He may "happen" to be an only child[3] or to have to adjust to sudden blindness[4] or to obesity[5] (which may originally have been caused by a constitutional factor or by previous situational factors or both).

Western science (at least during the past two centuries) has abhorred "chance." All events would be seen as determined by mass regularities. Hence the popularity of such slogans as "Culture (or "the times") makes the man"; "Geniuses are the products of their age." There is undoubtedly a measure of truth in such formulations, but this element of correctness has been hypertrophied and given a one-sided over-estimation in the eagerness to derive generalizations of the beautiful neatness and simplicity which brought the classical mechanics such fame. But the new physics fails to support such expectations. In physical events involving the intervention of a small number of quanta the principle of indeterminacy reigns. A leading physicist has recently suggested that human acts are likely to belong to the class of divergent rather than convergent phenomena. In other words, students of personality formation must study particular individuals and all the unpredictable events of their life experience as well as general "problems."

[1] H. A. Meltzer, "Economic Security and Children's Attitudes to Parents," *American Journal of Orthopsychiatry*, Vol. 6 (1936), pp. 590–608.

[2] M. E. Wright, "The Influence of Frustration Upon the Social Relations of Young Children," *Character and Personality*, Vol. 12 (1943), pp. 111–22.

[3] Louis Taylor, "The Social Adjustment of The Only Child," *American Journal of Sociology*, Vol. 51 (1945), pp. 227–32.

[4] Dorothy Burlingham, "Psychic Problems of the Blind," *American Imago*, Vol. 2 (1941), pp. 43–85.

[5] George H. Reeve, "Psychological Factors in Obesity," *American Journal of Orthopsychiatry*, Vol. 12 (1942), pp. 675–8.

In no other way than this, unless we resort to mystical reliance upon biological determinism, can we hope to explain observed facts in culture change. Creative individuals develop to make new social inventions. General biology and each culture place important limitations upon the actions of people, but occasional persons are able to organize and direct social trends in ways which are by no means wholly determined by preexistent culture patterns.

Without invoking any exaggerated "great man theory of culture," it may be submitted that the course of the history of various societies was unmistakably different because at a crucial period there "just happened" to be a leader who was an innovator or who showed unusual capacity or foresight. Various American Indian tribes were exposed to pressures of about equal intensity. Some disintegrated almost completely; others didn't. In some cases the strategic factor may have been some difference in the internal organization of the aboriginal culture or some slight but crucial variation in environmental position. However, it seems most unlikely that such factors would cover all the cases. Indeed in the case of the Yaqui village of Pascua [6] we have the documentation of a remarkable individual, Juan Pistola, who appears to have made all the difference. His existence and his qualities could not have been predicted by any amount of knowledge of the culture, of the history of the society, or of the environmental press on the whole group. His personality was created by some special combination of constitutional and situational determinants.

[6] Edward Spicer, *Pascua, a Yaqui Village in Arizona* (Chicago, University of Chicago Press, 1940).

CHAPTER

25

Ira S. Wile
AND Rose Davis

* THE RELATION OF BIRTH
TO BEHAVIOR

*All men are born, but they are born in different ways. At the mo-
ment preceding the beginning of the birth processes, each fetus has its
own constitutional predispositions dependent upon genetic structure and
the particular intra-uterine environment in which it has developed. But
birth is an event and a somewhat violent event. One group of psychoanalysts (the Rankians) have built a whole psychological system around
the conception of the "birth trauma." Wile and Davis agree with other
empirical investigators that the speculations of Rank and his followers
have little foundation. However, if the birth process in general cannot
be considered of such crucial importance for personality formation,
there are some indications that the mode of birth—whether speedy or
slow, spontaneous or by instruments—may be related to such personality characteristics as hyperactivity and aggressiveness. It is obvious
that "accidents" in delivery that result in actual malfunction are momentous for personality. Some investigators have also claimed that, even
when there is no organic injury at birth, the psychological effects may
be related to the later development of the disease. It is said, for example,
that asthmatic cases have often experienced birth strangulation.*

*Wile and Davis examine also the influence of birth order upon per-
sonality. They consider, however, only the situational (i.e., noncul-*

NOTE: Reprinted from the *American Journal of Orthopsychiatry*, Vol. 11 (1941), pp.
320–34, by permission of the American Orthopsychiatric Association. (Copyright,
1941, by the American Orthopsychiatric Association.)

turally patterned) aspects. Birth order is also a role determinant. Especially in cultures emphasizing the first-born or last-born position, the personality is affected by playing the role of eldest or youngest child or, more specifically, of oldest or youngest son or daughter.

IN A PREVIOUS investigation, one of us [15] presented as a conclusion: "Birth order does not determine behavior or personality characteristics, but the family constellation does provide a different environment for each individual. In specific cases, this special environment may affect individual behavior disadvantageously, but birth order lacks the effect of a determinism."

The act of birth itself is an environmental factor, the effect of which theoretically should vary in terms of relative spontaneity. We have, therefore, sought differences in behavior that might be proven the result of instrumental or operative delivery. Behavior variations are frequently attributed to birth traumata of a physical or psychological nature, without adequate evidence that the behavior is the direct consequence of the birth.

While the behaviorist might assert the absence of any factual hindrance of capacity, talent, temperament, mental constitution and characteristics at birth, he would admit that structural defects can limit function. It has also been urged that there is no type of characteristic in which individuals may differ that has not been dependent upon the genes. These views stress the effects of environment and heredity, regardless of birth pattern. A birth injury, however slight, shifts the corporeal environment which enters into the personality. The existent genetic forces may thus be limited in structure and function so as to modify physical, intellectual, and emotional phases of behavior.

Undeniably pre-natal elements are affected by natal experience, and natal factors may affect post-natal activity and reactivity of a personal and social type. The life of a child is conditioned by the factor of birth. The innate individual equipment, including all adaptive mechanisms, is subjected to different pressures under conditions of spontaneous birth or instrumental, including operative, delivery.

To study the problems directly, we have selected 500 children consecutively admitted to the Children's Health Class at Mount Sinai Hospital. For purposes of checking, we are comparing them with 500 children observed in private practice, with the records of 10,000 children delivered in a Bronx Hospital, with the general figures for New York City in 1933, and for those of Philadelphia during 1931–33. We have classified the children generally into those with spontaneous and instrumental birth.

Of the 500 Mount Sinai Hospital children, 380 (76 per cent) were born spontaneously, 120 (24 per cent) were assisted into the world by instrumental and other operative methods. The spontaneous group will be termed Group A, and those born instrumentally, Group B. The Caesarian section group is also segregated, although included in Group B.

The following table indicates that our data are in harmony with general experience.

TABLE XL

Birth Methods in Sample Groups

	MT. SINAI HEALTH CLASS	PRIVATE PRACTICE	BRONX HOSPITAL	NEW YORK 1933	PHILA. 1931– 33
	500	*500*	*10,000*	*21,095*	*68,733*
A. Spontaneous	380(76%)	401(80%)	7,650(76.5%)	15,896(75.4%)	72%
B. Instrumental	115(23%)	91(18%)	2,272(22.7%)	4,612(21.9%)	25.3%
Caesarian	5(1%)	8(1.6%)	279(2.79%)	575(2.7%)	2.5%

The clinic group parallels the figures for Bronx Hospital and New York City in the percentage of spontaneous births, but contains a smaller proportion of children born by Caesarian section, the reason for which will shortly appear. It is also evident that the number of Caesarian children is lower than the percentage seen in private practice.

Of the 120 instrumental cases, 5 were Caesarian, 8 were breech, and 5 were prematurely delivered during the eighth month or earlier. While the Caesarian group formed only 1 per cent of the total clinic cases, it constituted 4.2 per cent of those instrumentally delivered. In Group A, breech cases were 1 per cent, and in Group B, 5.6 per cent; premature births numbered 1.6 per cent of Group A and 5 per cent of Group B.

There is marked difference between the percentage of Caesarian-born treated in personal and clinic practice as compared with the number born in the city and in Bronx Hospital. This is due to the fact that guidance clinics see only survivors while hospitals report all. The fact that a higher percentage of Casearians is noted in private than in clinic practice arises from the greater frequency of this operation among private patients than among ward patients. This is exemplified by Kushner's figures from Bronx Hospital [7] where the number of children born by Caesarian sections to ward mothers was 1.6 per cent, whereas among private patients it was 3.3 per cent.

There is a difference between birth rates and natal survival rates. Thus the birth mortality of ward infants born operatively was 8 per cent, while for similarly born private patients it was 5.5 per cent. This is significant, since ward patients numbered only 3,166, while private patients were

6,834 in number. The effect of birth upon patients is also indicated by the fact that while infant mortality in Bronx Hospital for all births was 1.38 per cent, mortality for those born through operative procedure was 6.1 per cent, and mortality for children born by Caesarian section was 10 per cent.

Further explanation for difference between clinic or private practice and the general city data concerning births is found in the neo-natal mortality during the first month which, in 1933, was 29.9 in New York City [9]. The main etiological factors were birth injuries, asphyxiation, prematurity, and congenital malformation. Clearly the rates of spontaneous and instrumental births must vary definitely from the data for children presenting themselves at a clinic because of varying figures for survival of birth, the first month, first year, etc.

Significantly, the more severe the operative interference, the higher the infant mortality rate. Thus in Bronx Hospital the infant mortality of those delivered by low forceps was 1.6 per cent, mid-forceps 4 per cent, high forceps 20 per cent. Our clinic cases represent the effects of selective survival which thus minimized the presence of many possible sequelae from instrumental delivery.

Children born by Caesarian section have a high mortality rate at birth due to causes ante-natal rather than natal, and survivors may later show hemorrhages or other factors incidental to conditions antecedent to birth.

The mere fact of spontaneous birth does not definitely indicate that the child may not have been subject to various injuries incidental to protracted compression, to forcible contractions under the action of pituitrin despite a structural inadequacy of the mother, or even to precipitate birth.

Spontaneous birth affords no insight concerning dysgenic ante-natal factors. Adequate data are not available to indicate the extent to which labor was protracted, or its degree of difficulty, even though resulting in unaided expulsion of the infant. The fact remains, however, that lack of birth palsies, plegias, and spastic paralyses indicates that our spontaneous group have had some measure of physical protection, although there is evidence that the intelligent use of low forceps may be highly protective against even such disabilities. The effects of spontaneous birth cannot be wholly separated from the effects of precipitate delivery, prematurity, or the pressures incidental to birth positions other than that of the vertex.

Our instrumental group does not reveal an increased number of mental defectives attributable to birth injuries, but then one finds high intelligence quotients among children with spastic paralysis, showing thus diverse rather than characteristic effects of birth injury. The fact remains that every non-spontaneous delivery involves some trauma and compression regardless of ordinal position. Among the spontaneously born, the first born is particularly subject to undue pressure and compression and even to some definite asphyxia. But one finds little specifically characteristic of birth trauma, as suggested by Schroeder [12]. Reporting upon clinical material, he attributed the behavior problems of children with a history of difficult labor mainly to their mental retardation. Our experience does not confirm this observation. With him, however, we found hyperactivity and distractibility an apparent characteristic following birth trauma, with no other specific behavior syndromes save those due to structural impairment.

The term "spontaneous birth" is an indefinite one and is modified in meaning and effect because analgesia or anaesthesia are so commonly employed, and their effects upon the child are practically unknown save in extreme instances. These effects, however, are far more prolonged and more profound when instrumental procedures are adopted. Apnea from any cause may be regarded as a hazard to the brain, and asphyxia at birth may cause degenerative changes of the brain cells, especially in the higher centers; but this is more likely to be associated with actual birth injuries. Trauma and asphyxia are two major causes of cerebral injury, but the reasons for them are varied for the spontaneous as well as for the instrumental cases. It is important to remember that intracranial injuries have been frequently reported among children delivered by Caesarian section, although, according to Brander [3], they are more likely to occur when the infant is premature.

The selective mortality rate is very important because it eliminates many children who otherwise would constitute serious problems in behavior. The neo-natal mortality rate at one month was 29.4, while at one year it was only 46.5 over a period of years [1]. This neo-natal mortality included prematurity, congenital debility, birth injuries, and congenital malformations.

Order of birth and birth factors have some relationship. According to Goldstein [5] there is a tendency for the later born to have shorter duration of labor, a greater incidence of ante-partum complications, but a lesser incidence of labor complications, a higher percentage of spontaneous deliveries, and a greater percentage of normal births. The extent to which our figures present a larger number of first born indicates some effects of birth, but it reflects many other conditions well known in connection with children brought to guidance clinics. As Rosenow [11] has noted, in considering the records in Cleveland and Philadelphia: "We may conclude definitely that first born children from small families present problems to child guidance clinics more frequently than do other children from such families." Our tabulations do not permit very many definite conclusions concerning ordinal position. Whether age as a selective factor accounts completely for this preponderance of first born remains an open question. We concur in Rosenow's statement: "But nothing whatever can be concluded on the basis of primogeniture alone." Rosenow's studies make no reference to specific problems or to the type of birth.

Wetterdal [14] indicates the varying effects of forceps deliveries which we have tabulated to show natal mortality and general condition at age 12.

TABLE XLI

Effects of Forceps Deliveries

	SPON-TANEOUS	LOW FORCEPS	MIDDLE AND HIGH FORCEPS
Died in Hospital	2.55	3.6	6.2
Mental Defective at Age 12	6.6	6.7	9.9
Physical Defective at Age 12	5.3	4.5	6.2

Of 2,000 children delivered with instruments, 536 had asphyxiation, of whom 16.8 per cent died in hospital, and 5.6 per cent of survivors were mentally defective and 4 per cent were physically defective at age 12. The devastating effect of middle and high forceps delivery is quite obvious, but it is evident that delivery with low forceps is not productive of an increased amount of mental or physical deficiency and may indeed even be a protective factor against them. As Wetterdal remarks: "In the cases of low forceps operation, excluding those for asphyxia in the fetus, the primary mortality exceeds very slightly (about 1 per cent) that associated with spontaneous deliveries." Further, he states: "When compared with those spontaneously delivered, no inferiority can be detected in those delivered by forceps as far as the mental and physical conditions at age 12 are concerned." These findings are exceedingly significant considering the increased tendency to employ forceps in obstetrics.

The clinic children in both groups ranged between 13 and 15 years of age, as indicated by the following table. This does not permit any comparison of our data with those of Wetterdal, because he began his studies from the birth data, while such were secondary to our study. The chronological age alone is not wholly indicative of the problems, if birth

TABLE XLII

C. A. Range 13–15 Years

C. A.	A	B
3– 6 years	30 (7.9%)	11 (9.1%)
6–13 "	328 (86.4%)	100 (83.3%)
13–15 "	22 (6%)	9 (7.5%)

order is to be regarded as a definite factor. Birth order in our groups was as follows:

TABLE XLIII

Birth Orders

	A	B
First Born	88 (23%)	47 (39.2%)
Second Born	114 (30%)	31 (25.8%)
Only Children (first born)	95 (25%)	33 (27.5%)
Only Children (not first born)	26 (7%)	0
Youngest in Family	38 (10%)	9 (7.5%)
Other Positions [4, 5, 7]	19 (5%)	0

It is noteworthy that the operative or instrumental cases (B) contain a much larger percentage of first born children, which only suggests a greater hazard to the surviving first born. It may also suggest that the problems presented by the first born, instrumentally delivered, give rise to more parental concern. There is comparatively little difference between the percentages of only children (first born) and the youngest in family in the two groups. The only children (not first born) in Group A and the youngest in family in Groups B and A showed decidedly low percentages, thus giving support to the findings of Goldstein, even though our data may be too limited to constitute a specifically valid sample.

Our experiences with birth order are usual, but we wish to call attention to the fact that our data represent children coming from incomplete families, and therefore there is probably overweighting of children in the earlier ordinal categories. It is also true that infant mortality increases with the order of birth, but our groups are too small to segregate respective age groups for statistical purposes. Admitting the higher neo-natal mortality of first born children, some of the later problems of those surviving, spontaneously delivered, may be due to the fact that the duration of labor is 50 per cent higher for primipara, and, therefore, they were subject to a higher degree of cerebral compression. This factor undoubtedly enters into the neo-natal death rate, and it may have possible results upon the psychology of the survivors, especially those subjected to severe compression, as during instrumental delivery.

Thurstone and Jenkins [13] reporting on clinic cases 17–21 years of age, found 38.6 per cent first born, 23.8 per cent second born, and 12.9 per cent third born. Their figures are not differentiated, but they are not dissimilar to our figures for first and second born in our Group B. The disparity probably arises from a difference in the selection of the age groups.

This excess of the first born in guidance clinics has been generally reported from Philadelphia [11] and Minneapolis [6]. In our group there are certain outstanding figures concerning first and second born children, first children including the only child as first born. The A group of 380 contained 183 (48 per cent) first born children, including only children; the B group of 120 children included 80 (66 per cent) first born and only children. The A group contained 114 (30 per cent) second born and the B group, 31 (25 per cent) second born children. These figures are more striking if one considers the general family structure in the communities from which our patients came. According to the United States Bureau of Census, 32.2 per cent of Manhattan and Bronx families were one-child families and 28.8 per cent had two children. It is now evident that our groups had a tremendous excess of first born children, which was particularly noticeable in the B group, instrumentally delivered. It must be borne in mind that the census calculation merely indicated the number of children in the family at time of census and gave no data concerning ordinal position of the one or two children enumerated. The fact remains that the birth order is not indicated in the census figures, but the families represented at the clinic showed fewer completed families than those reported in the census.

Despite these data, we agree with Goodenough and Leahy that: "There is probably no position in the family circle which does not involve as a consequence of its own peculiar nature certain problems of adjustment." Their material showed that the greatest number of problems were presented by the oldest child. The mere fact of position in the family is not in itself the determiner of the particular form of behavior to any great extent, as will be illustrated when we present the problems of the children in various ordinal positions. That permanent effects attributable to ordinal position factors are uncommon has been shown by Malzberg [8]. After studying manic-depressive and schizophrenic patients, he reports: "I conclude that there is no undue representation of the earlier born in our samples. The early-born are undoubtedly at a disadvantage with respect to many physical and mental characters most of which appear at birth." Mental diseases practically do not exist among the very young, and mortality through the years would alter all relationships between mental disease and form or order of birth.

Without reference to selective survival values, our patients demonstrate that despite possibly severe trauma, for some spontaneously delivered as well as for those instrumentally born, there is little evidence of great lowering of mental level in the two groups. The number of intelligence inferiors for both groups is many times more numerous than expected on the basis of their presence in the general population, which is understandable because the children were referred to the clinic because of behavior difficulty. This fact constitutes a further selective element to be recognized in considering the data specifically. Our groups were constituted as follows:

TABLE XLIV

Stanford-Binet IQ Range 50–149

		A	B
Inferior	IQ 24– 74	15%	15.83%
Dull and Average	IQ 75– 89	20–	12.5–
		60	42.5
	IQ 90–109	40–	50.0–
Superior	IQ 110–149	25	41.67

It is noteworthy that there is practically no difference between the inferior A and B groups of low IQ's. The sum of the dull and the average is 60 per cent for the A group and 42.5 per cent for the B group, which emphasizes the fact that in the B group superiors are represented to the extent of 41.67 per cent, as compared with only 25 per cent for the A group. This shows a very definite selective factor in the survival rate of those subjected to instrumental delivery. It may also suggest that other elements in the personality conduce to more disturbing behavior in those instrumentally born than in those spontaneously born.

The median IQ for groups A and B lay between 90 and 110, thus proving again the reasonable normality of the selection. The data for the prematures and those born by Caesarian section are too few for any judgments, but it is worth noting that in the premature, Group A, IQ's varied from 69 to 121, and in the B group, from 74 to 149, while among the Caesarian group there was one IQ of 65, and four above 110.

Plass and Jeans [10] reported non-instrumental birth associated with intracranial damage, especially when delivery was rapid, as after the administration of pituitary extract. More important, however, are their comments upon the late consequences of intracranial hemorrhage as evidenced by epilepsy, mental deficiency, behavior difficulties, athetosis, and spastic paralysis. They found mental deficiency occurred in two-thirds of their cases and considered it responsible for the behavior difficulties. They emphasize the presence of athetoid movements as a result of injuries to the basal ganglia. Our groups did not show such high percentages of mental deficiency, and we are of the opinion that the mental deficiency is not always the element responsible for the deviating and undesired behavior.

Bostroem [2] refers to cases which he believes attributable to early organic brain injury, pre-natal or occurring through accident at birth. He speaks of social maladjustment incidental to attitudes of lack of reserve, accentuated by vanity and search for self-satisfaction. He observed physical and neurological symptoms that suggested athetosis and signs of infantile motor activity. The important fact to us is that a psychic condition, not truly athetotic but representative of great irritability, characterized many of our children. This may be incidental to some basic brain injury particularly under conditions of instrumental birth or severe and essentially traumatic protracted labors, or those made precipitate by special medication. This will be further evident in discussion of the special problems presented.

The relation of the problems to birth order involves the relation of intelligence status to ordinal position.

This tabulation of intelligence quotients and ordinal position indicates definitely that our group manifests a number of selective elements: survival, parental interest, and undesirability of behavior leading to clinical attendance. We have already called attention to the fact that the superior group constituted a larger percentage (41.67 per cent) in Group B than in Group A (25 per cent). While only 25 per cent of the A group were in the superior range, it is significant that 34 per cent of the first born and 34 per cent of the youngest were in this category and were, therefore, in outstandingly high proportions, but they were by no means as relatively frequent as in the B group, where 64 per cent of the first born and 48 per cent of the only-child group were in the superior IQ range.

The instrumental group shows definitely smaller proportions (42.5 per cent) in the dull and average group as compared with Group A (60 per cent). Noteworthy, however, is the similarity of the percentages for the inferior Group A (15 per cent) and B group (15.8 per cent). This suggests that instrumental delivery has not a devastating effect upon the mentality of children who survive. In the A group we note the high percentage (17 per cent) of inferior mentality in the first-born only child. It is, of course, possible that the family is incomplete or that the mental inadequacy of the child has led to contraception. This suggestion becomes less cogent because in the instrumental group, children from families of more than three children, but 12 per cent of the only children and 11 per cent of the youngest children, were amongst those in the lowest range of IQ. The idea finds some support, as 47 per cent of the children in Group A from families with more than three children were in the inferior group. Instrumental birth appears to be a definite factor in causing behavior related to inferior mental status but no more frequently than does spontaneous birth. Survival selectivity operates, as is implied in problems presented to the clinic, when inferior mentality plays a part in disturbing behavior, but less so than when high intelligence is equally significant, because, obviously, 25 per cent of A and 41.67 per cent of B are far out of harmony with any expectancy of superior intelligence.

The dull and average groups are far below the theoretical expectancy on the basis of common experience in the population, because the group between IQ 75 and 109 should approximate 75 per cent of the children. The superior group above 110 should approximate 20 per cent. Groups A and B vary from the expected norms of intelligence quotients. In Group B the differential between the first and second born is well illustrated by the fact that 74 per cent of the second born, as opposed to only 17 per cent of the first born and 40 per cent of the only children, almost reached the expected 75 per cent in the dull and average group. In general, the instrumental group shows larger numbers of superiors as compared with the spontaneous birth group.

We recognize that IQ's are not true guides of behavior. We concede that in Group B our numbers in each one of the ordinal positions are distinctly limited. Our data merely represent a special group of children brought to the clinic for personality and behavior disorders. The figures

TABLE XLV

Intelligence Quotients and Ordinal Position
Spontaneous Delivery (A) 380 Cases

	TOTAL	SUPERIOR GROUP		DULL AND AVERAGE GROUP		INFERIOR GROUP	
		IQ 110–140		*IQ 75–109*		*IQ 24–75*	
		95 cases— 25%		*228 cases— 60%*		*57 cases— 15%*	
		Per Cent	Number	Per Cent	Number	Per Cent	Number
First Born	88	34	30	52	46	14	12
Second Born	114	17	20	74	84	9	10
Only Child (first born)	95	24	23	59	56	17	16
Only Child (not first born)	26	27	7	58	15	15	4
Youngest	38	34	13	50	19	16	6
Other Positions	19	11	2	42	8	47	9

are highly selective because they cannot be compared directly with data based upon the large numbers of instrumentally born children who so behave as not to require specific attention. This criticism applies likewise for Group A, as there are large numbers of children spontaneously born who are lost in the general population because of adequate adaptation. High intelligence, low intelligence, or, indeed, average intelligence, in the presence of restlessness, irritability, and annoying activity, would invite inquiry as to its relation to the disturbing behavior. The types of behavior reported for the two groups possibly indicate a reason for the overweighting of our clinic with children in the superior and inferior groups.

In grouping the problems presented by the 500 children, we omit specific categories in which the occurrences were very few or where percentage differences were wholly negligible. Behavior is grouped to show definite patterns of reaction. In the following tabulation, the aggressive phases of personality are definitely lower in the B than in the A group, despite a markedly increased general motor activity and restlessness in the B group which otherwise evidences a negative trend in personality energy.

Group A shows an increased proportion, almost to the extent of 100 per cent or more for every problem except general hyperactivity in which the B group has the 100 per cent superiority in frequency. General hyperactivity appears to be almost the characteristic problem offered by those in Group B and so dominates all other types of behavior, including the aggressive, and the tic, nail biting, food fad variety of behaviors as to

TABLE XLVI

Intelligence Quotients and Ordinal Position
Instrumental Delivery (B) 120 Cases

	TOTAL	SUPERIOR GROUP		DULL AND AVERAGE GROUP		INFERIOR GROUP	
		IQ 110–150		IQ 75–109		IQ 50–74	
		50 cases— 41.67%		51 cases— 42.5%		19 cases— 15.8%	
		Per Cent	Number	Per Cent	Number	Per Cent	Number
First Born	47	64	30	17	8	19	9
Second Born	31	10	3	74	23	16	5
Only Child	33	48	16	40	13	12	4
Youngest	9	11	1	78	7	11	1

suggest it as the most significant single problem manifested in the survivors of instrumental birth. Strangely, the percentage of school difficulties reported by Group B was only half of those offered by children in Group A. This may be explained by the high percentage of children with IQ above 110 in Group B. Another explanation lies in the fact that the instrumental group is more selective because of its higher natal and neo-natal mortality, while the children spontaneously delivered include many minor grades of mental impairment, represented in the much larger dull and average group.

There is an additional implication in the greater probability of survival of children with superior endowment. This is in harmony with the admittedly higher mortality rates of children with lower mentality. It is none the less significant that while children with low IQ's in the B group are relatively only slightly more numerous than in Group A, they apparently report relatively half as many maladjustments at the school level, even though the median C. A.'s of the two groups are practically identical.

It is possible, of course, that schools create fewer difficulties for children who manifest definitely little aggressive behavior, even though they manifest restlessness. The survival factor is particularly evident in the marked differentiation between the state of physical health, which apparently is much better for those in Group B than in Group A.

The smaller proportion of intersibling conflicts in Group B has numerous reasons but has definite relation to the fact that among the B group there were no families of more than three siblings, and 48 per cent were only children, while more than 11 per cent of Group A were from families with more than three children. As 51 per cent of the children in Group A were also in the only-child position, the intersibling conflicts may be more related to the very aggressive behavior. This is evident in comparing Group A with Group B.

TABLE XLVII

Problems Presented—Total: 380(A), 120 (B)

	A	B
Aggressive Types of Behavior (rages, tantrums, pugnacity)	65%	33.3%
General Hyperactivity (restlessness, irritability, distractibility)	25	50
Submissive Types of Behavior (fears, unhappiness, fantasy life, no friends)	40	25
Tics, Nail Biting, Food Fads	70	33.3
Peculation	12	6.7
Infantile Home Relationships	55	20
School Difficulties	45	22.3
Intersibling Conflicts	30	15
Physical Ills	10	3.3

Ordinarily, school success or difficulty is not regarded as a matter of ordinal position. To indicate all of the generic effects of type of birth and ordinal position we rechart these problems in terms of ordinal position.

It is noticeable that in Group B aggressive behavior is most marked (43 per cent) in the first born, with the second frequency amongst only child; whereas among Group A, 96 per cent of only (but not first) children, and 68 per cent of only children, were brought for aggressive behavior, the first born having the lowest frequency rate.

Under general hyperactivity in Group A one notes the highest frequency among only child (not first born), followed by youngest child; whereas among the B group the percentage of first, second, and only born was approximately the same for this problem, with 44 per cent for the youngest child. While there is a suggested relation to ordinal position among those spontaneously born, it is lacking for the instrumentally delivered, thus leading to the inference that this behavior is related to the manner of birth rather than to ordinal position, further substantiated by the data for submissive behavior. In Group A, the greatest frequency was for only child (not first born), followed by youngest and children in other positions, whereas among children of Group B the maximum was among only child followed by the first born and the youngest, but the total variation in percentages for the B group shows little effect of ordinal position.

The various habits are entirely out of harmony, and it is informative that in Group A, 95 per cent of the youngest children offered these problems with the lowest frequency among the only child group, whereas in Group B, the highest proportion was manifest in the first born followed by the only child.

Infantile home relationships in Group A showed the highest percentages in only child (96 per cent and 95 per cent), followed by youngest child (92 per cent). There is some harmony of data in Group B, where 45 per cent of only child offered this difficulty and there were not more than 10 or 11 per cent for first born, second born, or youngest. This suggests that infantile home relationships probably have a greater dependence upon only-child status than on ordinal position or method of birth, unless it can be urged that the traumata even of normal delivery for the only child was a factor affecting adjustment within the home.

Strikingly, in both groups the youngest child showed the highest proportion of peculation, and the only child the lowest percentage. The proportion among the instrumental (44 per cent) is almost twice that found among the spontaneously born (24 per cent) which suggests that there may be some birth factor operative in reducing inhibitions as well as a possible factor due to ordinal position.

Both Groups A and B indicate that the only child and second born have the smallest proportion of school difficulty, while the youngest child apparently is more likely to be brought to the clinic because of school maladjustment. The high proportion of youngest children in the B group (33 per cent) cannot be explained by low or high IQ. That the maladjustment is not wholly due to IQ is evidenced by the fact that only 16 per cent of the youngest children in the A group were in the inferior IQ range, while 34 per cent had superior IQ, but 79 per cent of the youngest children presented school difficulties.

While method of birth may play some part in school maladjustment, it is clear that ordinal position has no uniform effect, particularly as the only child is subjected to the same forces as a first born child, and there is a marked difference between the percentage of school difficulties presented by the only child and the first born child. A comparison of only child, and only child (not first born), reveals a percentage difference of only 3 per cent. Further, the difference in relative frequency of school difficulties among first born and second born amounts to 19 per cent in Group A and 16 per cent in Group B. Among the B group, school difficulties were slightly more among the youngest children than among the first born; the percentage of inferior mentality was only moderately higher among the first born, and the superior mentality was exceedingly low among the youngest, and high among the first born. These facts suggest that possibly low forceps or careful instrumentation or operation are

TABLE XLVIII

Problems Presented (in Percentages) in Terms of Ordinal Position for Group A (380) and Group B (120)

	FIRST BORN	SECOND BORN	ONLY CHILD	ONLY CHILD (NOT FIRST BORN)	YOUNG-EST CHILD	OTHER POSI-TIONS
Aggressive Types of Be-havior						
A, 65%	50%	60%	68%	96%	79%	79%
B, 33.3%	43	20	36	0	22	0
General Hyperactivity						
A, 25%	28	9	16	77	47	37
B, 50%	51	50	50	0	44	0
Submissive Behavior						
A, 40%	56	26	14	88	63	63
B, 25%	26	20	30	0	22	0
Habits, Tics, etc.						
A, 70%	74	76	44	77	95	84
B, 33.3%	43	26	30	0	22	0
Infantile Home Relation-ships						
A, 55%	19	22	95	96	92	89
B, 20%	11	10	45	0	11	0
Peculation						
A, 12%	20	9	5	7	24	11
B, 6.7%	6	3	0	0	44	0
School Difficulty						
A, 45%	63	44	16	19	79	84
B, 22.3%	32	16	12	0	33	0
Intersibling Conflicts						
A, 30%	28	31	0	0	100	84
B, 15%	21	20	0	0	22	0
Physical Ills						
A, 10%	8	2	10	35	18	16
B, 3.3%	2	0	3	0	22	0

actually protective of cerebral structure. This is somewhat supported by such facts as: 34 per cent of first born in Group A had IQ above 110 and 64 per cent of first born in Group B were in this category; 24 per cent of the only child (first born) in Group A as compared with 48 per cent of the only child category in Group B had IQ above 110. This is more striking as 34 per cent of the youngest in Group A and only 11 per cent of the youngest child in Group B had IQ in the superior IQ range.

Intersibling conflicts must in part reflect ordinal position, as obviously they cannot occur in only child families. But it is interesting that 100 per cent of the youngest child in Group A offered this problem, while the minimal presentation was by the first born child. In Group B there was practically identical frequency for first born, second born, and youngest child. Inasmuch as Group A showed comparatively little difference between first born and second born children, this behavior can scarcely be related to ordinal position in families of three or fewer children. It is possible, of course, that the generally reduced aggressiveness in Group B may partially account for the comparative infrequence of sibling conflict in Group B as compared with Group A.

Physical ills were most marked in the only child (not first born), 35 per cent in Group A, and in the youngest 22 per cent in Group B. The variations in figures for the various other ordinal positions is not pronounced. Incidentally, the only child in Group A offered this problem with the mild frequency of 10 per cent, as compared with 35 per cent of only child (not first born). In Group B, the youngest showed the most physical ills (22 per cent) but in Group A, the distinction between the youngest (18 per cent) and those in other positions in the family was not sufficient to indicate to what extent ordinal position itself was a factor. The much lower percentages in Group B are probably functionally related to the survival value of the physically superior child exposed to instrumental birth and the survival of many who, despite spontaneous birth, had been exposed to marked dysgenic factors.

Apparently the data from instrumental deliveries suggest that personality is somewhat contracted in the presence of a general hyperactivity. This leads us to believe: (1) that the hyperactivity and restlessness are probably related to the greater pressures involved in protracted and especially instrumental delivery; (2) that there are changes due to asphyxia, since the brunt of the cerebral disturbance arises on the basis of a compression from above with the maximum pressure falling upon the basal ganglia; (3) that residual elements of varying degree cause inadequate co-ordination, expressed as restlessness, irritability, and lowered inhibition for muscular activity. Extreme effects of pressure would be expressed as paralyses or spastic plegias.

Hyperactivity is not to be regarded in the same category as the birth trauma from a psychological standpoint. Our experiences lead us to agree with Campbell [4]: "As to claims about the imprint left on the personality by the devastating experience of birth, these constructions are hypothetical, of dubious value and subject to no control." The trauma during birth varies for spontaneous and instrumental deliveries, and it is difficult to evaluate the differentials save in terms of survivals, and the known effects of pressure which, unfortunately, are lacking.

It is practically impossible to generalize concerning any group of activity trends, particularly in relation to specific conditioned behavior patterns. Obviously the total personality is affected by any changes in structure and function, the physiological and psychological, and these are related to the complete organization of the personality.

It is still impossible to analyze the habit systems of the child at birth, nor can one determine exactly what happens to them or what interferes with basic patterns of conditioning. Neither can we state with exactness that children are only relatively bright or dull wholly as a result of the mode of birth. We cannot be certain whether infantile needs are altered or whether modes of response only are changed. Certainly our instrumental group shows alterations in personality as compared with the children of spontaneous birth, but with the exception of the general hyperactivity we are aware only of a qualitative reduction in meeting conflicts with the environment. Behavior reveals no particular foci of personality disorder. We cannot be certain about the effect of birth compressions upon infantilism, the resultant effects upon feelings or the fundamental needs of the personality. Admittedly the process of birth offers environmental factors whose impact upon personality may modify structure and possibly may shift inherent trends and even alter some of the genetic bases of accomplishment by limiting their growth.

Certainly the birth trauma, in the Rankian sense, has comparatively little meaning in the light of the behavior exhibited by our two groups.

There can be little doubt that the immaturity of the psyche and the central nervous system at the time of birth enables the personality to escape a larger measure of injury than would be anticipated in the light of the severity of the process of birth that required 24 per cent of operative interference. There are no adequate terms or statements to indicate the degree to which psychological factors in personality are definitely modified by the mode of birth, except as they are exhibited in general behavior reactions. Again mental reservations must be made because of the marked differences in survival values as instruments may destroy many of the unfit who are often saved in spontaneous birth.

It is interesting, however, that the rage and tantrums type of aggressiveness, as well as the fear and fantasy submissive groups, were respectively 65 per cent and 40 per cent in Group A, as compared with 33.3 per cent and 25 per cent in Group B, while general hyperactivity was only 25 per cent in Group A, as opposed to 50 per cent in Group B. The children with instrumental birth appeared to show a general reduction of personality energy rather than a mere increase in sensitivity and irritability, although the physical element was generally noted as a restless, distractible, irritable hyperactivity. We may conclude, therefore, that behavior reactions are not to be interpreted as shocks of birth, with a persistence of fear and pain and an anxiety state that later might become a source of neurotic behavior. Sensitivity does not appear to be especially related to the form of birth, nor can strict evaluations be made because the differential infant mortality rate, natal and neo-natal, precludes the tabulation of the psychological effects of precipitate or protracted birth, spontaneous or instrumental delivery. A physical trauma factor is evident in the heightened restlessness, poorly controlled hyperactivity, and the contraction of personality forces. The instrumental factor in birth deserves greater consideration from the standpoint of prophylaxis as well as diagnosis.

REFERENCES

1. *Births, Stillbirths, Maternal Mortality, and Infant Mortality in New York State, 1934–1938* (Bull. N. Y. Dept. of Health).
2. Bostroem, A.: "Special Form of Inhibition of Psychic Development Related to Athetoid or to Infantile Motility," *Arch. f. Psychiat. u. Nvkrht.*, Vol. 75, No. 1 (1925), p. 1.
3. Brander, T.: "Cerebral Defect in Children Born by Caesarian Section," *Acta Paediat.*, Vol. 23 (1938), pp. 145–164.
4. Campbell, C. MacFie: *Human Personality and Environment* (New York, Macmillan Co., 1934), p. 92.
5. Goldstein, H.: "The Relation of Order to Birth Factors," *Child Devel. Abst.*, Vol. 9 (1938), p. 127.
6. Goodenough, F. L., and Leahy, A. B.: "The Effects of Certain Family Relationships upon the Development of Personality," *J. of Genet. Psychol.*, Vol. 34 (1927), pp. 45–71.
7. Kushner, J. Irving: "Obstetrics in a General Hospital," *N. Y. State J. of Med.*, Vol. 40 (1940), p. 195.
8. Malzberg, Benjamin: "Is Birth Order Related to the Incidence of Mental Disease?" *Am. J. Physical Anthropology*, Vol. 24 (1938), pp. 91–104.
9. "Maternal Mortality in Philadelphia 1931–33," *Quart. Bull. N. Y. Dept. of Health*, Vol. 2, No. 4 (1934), p. 91.
10. Plass, Edward, and Jeans, P. C.: "Intracranial Hemorrhage in the New Born: Early and Late Manifestations," *J. Iowa State Med. Soc.*, Sept., 1932.
11. Rosenow, Kurt: "The Incidence of First Born among Problem Children," *Pedag. Sem. and J. of Genet. Psychol.*, Vol. 37 (1930), pp. 145–51.
12. Schroeder, Paul L.: "Behavior Difficulties in Children Associated with the Results of Birth Trauma," *J. A. M. A.*, Vol. 92 (1929), pp. 100–4.
13. Thurstone, L. L., and Jenkins, R. L.: *Order of Birth, Parentage, and Intelligence* (Chicago, University of Chicago Press, 1931).
14. Wetterdal: "Diagnosis for Children Delivered by Forceps: A Re-Investigation of 2,000 Children Delivered by Forceps and 2,000 Spontaneously Delivered," *Acta Obst. et Gynec. Scandinav.*, Vol. 6 (1927), p. 349. (*Child Devel. Abst. and Biblio.*, Vol. 3 [1929], p. 149.)
15. Wile, Ira S., and Jones, A. B.: "Ordinal Position in the Behavior Disorders of Young Children," *J. of Genet. Psychol.*, Vol. 57 (1937), pp. 61–95.

CHAPTER

26

Franz Alexander

EDUCATIVE INFLUENCE OF
PERSONALITY FACTORS IN
THE ENVIRONMENT

Although the reaction of human beings to a more or less private situation will occur in terms of individual constitution and the individual's previous experience (including cultural training and role playing), such situations—whatever the responses made by the individual—can have momentous effects upon personality formation. Of these, none are more clear-cut and certain than those that derive from participation in particular family units.

The composition of families varies within every society and within every class and caste sub-group of each society.[1] The size of the family may range from three to twelve or more, or to still larger numbers in extended families where there is little distinction between members of the immediate biological family and the mother's or father's sisters or brothers and their children. In one family, the father is considerably older than the mother; in another, both parents are unusually old or unusually young at the time of the birth of the first child. In some cultures, the immediate family may consist either of a single husband and wife, or of a wife and two or more husbands. In the psychological sense, even

NOTE: Thanks are due to the University of Chicago Press and to the author for permission to reprint, with slight abridgment, from "Environment and Education," University of Chicago, *Supplementary Educational Monographs, No. 54* (1942), pp. 29–47. (Copyright, 1942, by the University of Chicago Press.)
[1] See Dorothy M. Spencer, "The Composition of the Family as a Factor in the Behavior of Children in a Fijian Society," *Sociometry*, Vol. 2 (1939), pp. 47–55.

[421

brothers and sisters are hardly members of the same family. For the eldest child spent the first three years of his life without the situation being complicated by the presence of other children, whereas the youngest child had to react not only to his parents but also to ten brothers and sisters; the eldest child had a young mother, while the youngest had an experienced mother of more than forty years of age.

No family lives up perfectly to the code of behavior which the culture prescribes for child training and family life in general. The domestic environment is molded not merely by group standards, not merely by the roles played by family members, not merely by the numerical composition of the family at any given time point; it is also influenced by the constitution and the total personality of each individual in the family and by the "accidents" that have happened to each family member and to the family as a unit. The effects of the family upon the formation of each child's personality cannot be predicted from knowledge of the group to which the family belongs, nor the roles, nor the constitutions of family members. The combination of these influences is in some sense unique for each family—and indeed for the same family at different time points. The family situation is therefore a situational determinant par excellence.

Psychoanalysis has demonstrated the crucial contribution to personality formation of the child's early emotional relationships within his own particular family. The basic emotional patterns are universal, but the variations on these basic themes are manifold. As Dr. Alexander has written elsewhere: "For the finer individual differences between people, constitution and the early and highly specific family influences together must be held responsible."

THE development of scientific thought follows the dialectic principle. The complexity of natural phenomena permits the observer to select and emphasize one set of factors, to neglect others. No field offers greater opportunity for one-sided analysis than personality research, particularly the genetic problem of personality development. What we term "personality" is the total expression of the integrated activity of a complex biological system which is subject to the laws of heredity and at the same time is molded by post-natal experiences. The deep-going influences exerted on personality formation by the emotional attitudes of the parents, such as ambition invested into the child, maternal rejection, and overprotection, have been well demonstrated by psychiatric and particularly psychoanalytic experience. Partially these influences are highly specific—different in each case, dependent on the particular features of the family in which the child spends its first years, as well as on accidental occurrences. Partially, however, they are uniform—typical for the society of which the family is a part.

Any of these factors may legitimately become the focus of interest:

the biological influences, particularly the role of heredity; the highly specific features of the immediate environment; or, finally, those more general influences which are characteristic of a particular culture. Yet one-sided emphasis on any of these factors necessarily leads to false conclusions.

I. CRITIQUE OF THE HEREDITARY POINT OF VIEW

In the second half of the past century, the hereditary factor alone was considered. Typical of this view in anthropology was the Lombroso school; in philosophy, Nietzsche, Chamberlain, and Gobineau; and in history, Carlyle. While Lombroso considered the criminal a victim of heredity, Nietzsche saw the future of humanity in the breeding of a super-race of supermen, who by reason of their hereditary qualities would be destined to rule over the inferior slave races. Carlyle believed that history is shaped by superior individuals, who are not so much the products as the creators of the age in which they live.

Psychiatry, too, was completely dominated by the hereditary point of view. Mental diseases, whenever they could not be retraced to infectious origin, such as the postsyphilitic conditions, were explained on a hereditary basis. Genetic etiological studies were limited either to histopathological search after morphological tissue changes or to statistical studies of hereditary traits. Brain research and statistics ruled over psychiatry with an amazing neglect of the personality of the patients and of the environmental influences.

In the course of the last forty years, the interest in environmental factors, particularly under the influence of Freud's principles, has gradually permeated psychiatry, although hereditary considerations still prevail. One of the most common diagnostic labels is still that of the "constitutional psychopath," comprising all those cases in which a behavior disturbance appears early in childhood and, stubbornly resisting all therapeutic efforts, persists throughout life. It is much easier for the physician to blame the great idol of inalterable heredity than to admit the ineffectiveness of his own therapeutic measures.

However, at the turn of the century, the pendulum in scientific thought began definitely to swing to the opposite direction. The most favorable soil for this reversal was the scientific scene in the United States, particularly the field of the social sciences.

In Europe the hereditary point of view was determined, if not solely, yet to a high degree, by cultural conditions. The aristocratic structure of society, in spite of the French Revolution and the Industrial Revolution which gradually invaded the whole continent, survived to some degree even in economic life and much more powerfully in traditional ideology. The Industrial Revolution did not transform the class system into a society of great class mobility. The individual's fate remained largely determined by birth. Heredity as a social factor loomed large, and accordingly it also retained its distinguished position in scientific thought. The division of the comparatively small and overpopulated continent into small national and linguistic groups had a further powerful influence on theoretical thought. These racial and linguistic differences remained of primary importance in the shaping of Europe's destiny at a time in which changing technological conditions required the creation of greater political and economic units. This ideology—that there is an unbridgeable gap between racial groups—remained untouched even in the face of the fact that actually the personalities of the Western and even the Central Europeans became more and more similar.

It is most remarkable that another, even more antiquated expression of hereditary ideology, the institution of hereditary monarchy, was dethroned only after the first World War, after it had ruled over Europe for centuries. The political philosophies of hereditary monarchy and of national states are merely different gradations of the same all-pervading hereditary ideology. Its scientific expression is the hereditary explanation of personality and mental disease—a theory which is obviously not the cause but the result of the political conditions in Europe.

All these historical side lights may appear to be a detour in the approach to the topic of my presentation. I feel, however, that without a full appreciation of these cultural determinants of scientific thought, the shifting fashions in the field of personality research remain unintelligible. We render lip service today to the cultural determinants of scientific ideas, but, when it comes to applying this principle to our own field, we prefer to remain blind and consider our own point of view as the absolute and the only possible truth.

2. CRITIQUE OF THE ENVIRONMENTAL POINT OF VIEW

To understand the extremes to which certain environmental views of personality development go, we must consider their cultural sources with the same unbiased scrutiny which has made us recognize how deeply rooted the hereditary concept was in the traditions of European civilization. Just as the European cultural and political traditions favored the stress on the hereditary factor, so was the American scene ripe for the recognition of the environmental factors. American science readily assimilated the environmental concepts of Freud because it was already oriented in the same direction in the views represented by Sumner and particularly by Boas and his followers. Here the emphasis was on the determining influence exerted on personality by cultural factors, with a consequent underevaluation of heredity.

The most radical blow to heredity is contained in the tenet of the Declaration of Independence "that all men are created equal." It is obvious that this view represented a revolt against the class system of the mother-country. Its scientific expression is the belief that man's fate is determined after his birth and not before. How could any other theory survive in a culture in which the supreme ideal is the man who achieves everything by his own efforts and who owes everything to himself? The ideology of free opportunity given to men, all born equal, does not "jibe" with too much consideration for heredity. The most extreme scientific expression of this view is the concept that man at birth is a *tabula rasa* and that all his later personality traits are molded by external circumstances. These circumstances may be more or less favorable, but at least the unjustness of inequality is not so great as if heredity were determining everything. Life-conditions, unlike heredity, are changeable; and, even though society does not give everyone exactly the same chances, one can improve conditions and can strive to approach an ideal system. Indeed, the American culture, by its political democracy and particularly by its extension of education, opened up opportunities to everyone more equally than had ever been done before.

The environmental theory found an even stronger ally than political theory which, after all, always contains many wishful elements. This ally was sound, well-founded, everyday observation. Here was a civilization in which people belonging to all the different races of Europe were, in a brief period, transferred into a new and different type of personality. The second-generation immigrant, under favorable conditions, lost all distinguishable racial or national characteristics and assumed features common to the inhabitants of this country. As so often happens, scientific thought here only lagged behind everyday observation, which in this case indeed could not easily be disregarded. No wonder that heredity became dethroned as the all-important factor and gave way to cultural environment. Thus the pendulum of thought development reached the other extreme.

The study of the influence of the intimate personal environment, as distinguished from the broader cultural environment, is likely to restore the equilibrium between these two extreme swings in opposite directions and to lead to a more seasoned evaluation of the determinants of personality formation.

3. THE IMPORTANCE OF EARLY ENVIRONMENTAL INFLUENCES

Early environmental influences, both those which are uniformly characteristic for the culture in which the child grows up and also those specific influences which vary from family to family according to the personalities of the parents, act through the family milieu. Following the development of a child as it actually takes place, step by step, we find that the decisive factors are the emotional relationships between the child and other members of the family. In later years, the child is exposed to the cultural milieu outside his family. There is little doubt that these later influences, so far as the basic personality structure is concerned, are less deep-going and their effects more reversible than are those earlier influences which the child receives in contact with his parents and siblings. We know that these more superficial character traits are changeable because we can readily observe how adolescents and even some mature persons adjust themselves, often remarkably well, to a new cultural milieu and assume many traits characteristic of their new habitat. Just as people may learn a new language at a relatively advanced age, they can also learn new customs and even a new outlook toward goals and ideals. What remains more unchangeable is the groundwork of the personality, something to which Hippocrates referred in his theory of temperaments. Whether or not a person will become timid or out-going, cautious or enterprising, self-confident and optimistic or self-critical and pessimistic, aggressive or submissive, dependent or independent, generous or withholding, orderly or careless—all these personality features which make a person a well-defined individual, different from others—depends on the influence of the intimate personal environment on a hereditary substratum.

All these different personality types, although possibly in different statistical distributions, we find in all cultural environments. There is good evidence that they are formed in the early years of growth, when the main problem confronting the child is not yet to adjust to a cultural milieu but primarily to adjust to the rapidly changing phases of his biological growth, which, in contrast to the external, we may call by the expression of Claude Bernard, the "inner environment." Later we shall return to this fundamental aspect of personality development. Before we proceed, however, it will be necessary to clarify certain current dogmatic and one-sided exaggerations concerning cultural influences—exaggerations which come from a lack of precise distinction between specific influences of the human environment, as represented by the human material which served to support these extreme views: I mean the study of the static primitive civilizations. For the sake of brevity I shall refer to this error as "the ethnological bias."

The most recent emphasis on the cultural factors in personality development in general, and in the causation of psychoneurosis in particular, comes from a group of critics of Freud, to whom occasional reference is made as "neo-Freudians," represented by authors such as Horney, Kardiner, Fromm, and others. That sometimes I also seem to be included in this group comes from the fact that I, too, recognize the need for re-evaluation of cultural factors in personality development and share the views of this group concerning certain gaps in traditional psychoanalytic formulations.

The essence of the argument runs about as follows: Freud, although the greatest psychologist of all times, under the influence of the nineteenth century's scientific tradition, was too biologically oriented. He postulated a too elaborate, biologically predetermined instinctual structure which, in its main features, unfolds in a more or less autochthonous manner, like a flower. He recognized, possibly even overemphasized, the importance of those early experiences which arise in family life, but he overlooked the fact that the parental attitudes themselves are strictly determined by cultural factors. This neglect is the basis of a significant error. Those psychological factors which he discovered as all-important, such as the Oedipus complex, the castration complex, the masculinity trend of women, the repression of sexual and aggressive trends, are not necessarily universal but are of prime importance only in our present culture and may even be lacking in other cultures. Freud, not noticing the cultural determination of family attitudes, assumed that all these psychological factors are biologically determined; he overlooked their local nature. In this respect Freud's attitude was somewhat like the geocentric theory in cosmology, which maintains that the earth is the center of the universe. Also, Freud declared the personality structure of the European and American of the nineteenth century to be the universal human nature, not noticing that what he dealt with was nothing but a specific edition of human nature—if there is such a thing—an alternative among unlimited possibilities.

This is not the place to evaluate in detail the merits of this criticism, but there can be no doubt that much of this contention is valid. However, as so often is the case, the critic, in his dialectic fervor runs into errors which are precisely the opposite of those which he attacks. Most of the basic psychodynamic constellations, such as the Oedipus complex and the castration fear, have a more fundamental foundation than these critics of Freud want to admit; they have sources which are based on certain biological factors overreaching specific cultural variables. The emotions involved in the Oedipus constellation are the most direct expression of the biological helplessness of the human infant. It is the expression of the possessiveness which the little child feels toward the main source of his security and pleasure, the mother. In this respect, it is a parallel manifestation of sibling rivalry. The significance of certain premature, genitally tinged, sexual interest of the little son in his mother is, according to my experience, overrated. Among the sources of this interest, it is difficult to distinguish between infantile forms of pleasure sensations involved in all phases of the nursing procedure and some early appearance of rudimentary genital desires. The jealousy aspect of the Oedipus complex is certainly universal and is based on the prolonged dependence of the human infant on its mother's care. Even among animals, similar phenomena can be readily observed; even the puppy, attached to a single person from whom it receives food and care, will show hostility and aggressiveness against any supposed competitor. It is, then, of secondary importance and is dependent on cultural conditions whether hostility is directed against the father or against some other member of the family— the maternal uncle, for instance, who in certain cultures happens to be the head of the family and, as such, the main disciplinarian and obstacle to the child's sole possession of the mother. It is, of course, true that, according to accepted standards of education, this attitude may be strengthened or weakened in different cultures. Where the child is encouraged toward independence, the attachment to the mother may be of shorter duration; where the mother as a rule has a rejecting attitude toward the child, as in the Marquesan culture, the hostility and fear of the mother may overshadow the attachment, which the child may soon transfer to less prohibitive members of the family. In all these cases, the essence of the Oedipus complex remains the same, namely, the possessive attachment to a person upon whom the child depends for his gratifications and security, with jealousy and hostility against competitors. The distribution of attachment and hostility among the different members of the family may vary according to traditional customs and parental attitudes. Critics of Freud correctly question the universality of the specific form of this emotional constellation, which Freud called the Oedipus complex. They correctly point out its cultural variations, but they err when they do not recognize its biological origin in the prolonged post-natal dependence of the human infant upon parental care, not necessarily maternal exclusively.

The same argument applies to another fundamental psychological factor known as the "castration complex." There is no question that parental attitudes toward the early genital activities of the child vary according to cultural conditions. Children who exercise early sexual freedom probably have less anxiety in connection with their genital pleasure sensations than do children of our Western civilization. On the other hand, Stärcke and I [2] succeeded in demonstrating the early biological precursors of the castration fear. We showed that it is based on the repeated experience of the emotional sequence: pleasure followed by pain and frustration. The castration threat, customary in the sexually repressed Victorian era, only reinforces the universal emotional attitude: to expect pain and evil after pleasure and gratification. This is expressed in so universal a human attitude as the expectation of misfortune at the peak of good luck. "Knocking on wood" and the fear of envious gods when fortune is too generous are symbolic of the deeply rooted experience of the continuous sequence of pain and pleasure during the biological process of life.

We come to the conclusion that cultural constellations can reinforce and bring into the foreground certain emotional mechanisms but cannot introduce any fundamental dynamic principles into human nature.

The new and deserved emphasis on the cultural factors of personality development should not make us overlook the fact that the first and fundamental task which the child's ego must accomplish is the continuous adjustment to those biological changes which follow one another in rapid succession during the process of maturation in the first years of life. Though these changes may be influenced by external conditions, in their

---

[2] August Stärcke, "The Castration Complex," *International Journal of Psychoanalysis*, Vol. 2 (June 1921), pp. 179–201; Franz Alexander, "Concerning the Genesis of the Castration Complex," *Psychoanalytic Review*, Vol. 22 (1935), pp. 49–52.

main features they are universal for the human species. From an endo-parasite and after birth an ectoparasite, dependent for its basic biological needs on the mother-organism, the child develops into a biologically in-dependent being. This change involves the mastery of diversified func-tions of body control and of intelligence, and there is a marked emotional resistance in the child against this process of maturation. Psychiatric ob-servations offer the most convincing evidence for the strong resistance which the ego puts up against accepting the gradually increasing inde-pendence which biological maturation brings. Psychopathological phe-nomena reveal a regressive urge to return to earlier, more dependent phases of life. This regressive urge is mobilized whenever external condi-tions become difficult beyond the organism's capacity to cope with them. Cultural factors as they appear in the methods of child-rearing and in parental attitudes undoubtedly have a great influence on the child's readi-ness to accept the process of maturation. But also the individual attitudes of the parents have an equally powerful influence. There are overpro-tective and rejecting mothers in every culture, although either type may be the rule in one and the exception in another civilization. In our own culture there is a tremendous variability of parental attitudes, as I shall try to demonstrate in the following paragraphs.

4. THE IMPORTANCE OF SPECIFIC CHARACTER TRAITS OF PARENTS

This leads us back to the error into which the so-called "neo-Freudians" fall when they underestimate the specific influence of the parental per-sonalities and overstress the cultural factor—an error to which I have re-ferred as the ethnological bias. We saw that the denial of the biological foundation of the fundamental dynamics of personality was the outcome of the dialectic overemphasis on the environmental factor. The lack of emphasis that the so-called "neo-Freudians" place on the specific per-sonal influences results from the particular nature of the ethnological source material which is used to demonstrate the cultural determination of personality. The primitive civilizations which serve as material are, in comparison with our Western civilization, characterized by their static nature and their highly rigid organization. These societies undoubtedly offer most valuable material for psycho-sociological studies. However, they are, in comparison with our world, well-nigh petrified, unchanged for considerable periods, and consequently are not suitable for the study of social dynamics as these manifest themselves in historical development. Their history, as compared with that of the Western world, is relatively uneventful, does not contain deep-going changes of the social structure, and, above all, is almost completely unknown. The static nature of these cultures manifests itself in a stable structure which precisely defines the individual's social place, functions, and attitudes with a rigidity un-known in the dynamic societies of the Greek-Roman type or the European and American civilizations. The role of tradition in these static societies is incomparably greater than in our culture.

The dynamic nature of our culture manifests itself primarily in the fact of social change. In the nineteenth century, this change assumed almost astronomic proportions. This change best reflects itself in the literature of the period, one of the most favored topics of which was the clash of ideological attitudes between two succeeding generations. The ideological clash between fathers and sons is something which cannot be overlooked, and it does not allow explanation of personality attitudes by tradition alone. What was right and valid a generation ago is outmoded today. In our age, customary attitudes rapidly lose their dynamic power and are replaced with new trends in a fast adaptation to an ever changing world.

Another factor contributing to the dynamic nature of our society consists in its extremely complex structure, based on a far-reaching division of labor, that is to say, on the differentiation of social functions. This differentiation leads to a diversified stratification of the population; it is futile to speak of any homogeneous attitudes. Almost every generalization is likely to be fallacious. Even such basic attitudes as those toward property, discipline, and sexuality vary tremendously according to social stratification. One often speaks of the repressive attitude toward sexuality characteristic of the Victorian era. Even this attitude, however, was found mainly among the bourgeois and was absent to a high degree among peasants or in the proletariat. With some neglect of individual differences, one might justifiably speak of the personality structure of a Zuñi Indian. It is more difficult to speak of the typical character of the Swiss, although these mountaineers represent possibly the most uniform nation of the Western world. It is even more difficult to speak of a typical Italian cultural attitude without first defining whether one means the attitude of the Italian peasant, laborer, manufacturer, or artisan.

The heterogeneity of ideological trends in present-day America is its most outstanding feature. Certain traditional attitudes of the frontier, of course, still prevail, such as the ideal of the self-made man—self-reliant, enterprising, with a sense of freedom in economic activities and his supreme ideal that of success. But even these trends are mingled with new ones, with a certain longing for order and security, with a new evaluation of purely scientific and artistic accomplishment in contrast to merely technical and practical achievements. And all the old and new attitudes are blended in varying proportions in different groups—different in the East, in the Middle West, and in the South; different among farmers and industrial wage-earners; different among intellectuals, "white-collar" workers, and executives. From psychoanalytic case histories in our culture we learn that every attempt simply to deduce attitudes or a basic personality structure from social determinants is futile. Even in a single group, individual differences are more impressive than conformities, the exception more frequent than the rule. In every other family we see a dissenter, a peculiar type. The cross-currents of cultural fertilization offer a kaleidoscopic picture. In a solid middle western family of farmers, through three generations there appears a freethinker and radical who becomes a school teacher; the son of an immigrant labor leader becomes a staunch Republican; the son of an orthodox rabbi, a frivolous columnist of a New York periodical. Orientation toward goals in life, sex, property, and children appear in a chaotic confusion and defy every attempt at generalization. Under the microscope of psychoanalytic case studies the reason becomes evident: it is the highly individual determination of personality development by the reaction of the child's inherited constitution to the immense range of individualities as represented by the different personalities of the parents in their relationships to each other and to their children. All this, of course, does not mean that the cultural structure has no influence on personality formation. It means that the cultural structure in our dynamic civilization is complex and changing and accordingly offers an infinite selection to individuals of how they will be influenced by the culture. From the great variety of cultural trends in a changing civilization in which the past, the present, and the future are woven into an intricate pattern, the child's ego selects what it needs. This selection is determined by those emotional needs which develop in early family life under the influence of the particular personalities of the parents. A few brief examples may illustrate this thesis.

A number of my patients have belonged to a similar, rather well-defined cultural background. The patients were second-generation Americans, members of immigrant families, and belonged to a racial minority group. The father's role in the family was of determining significance. His success in adapting himself to his new environment had a great influence on the son's development. In most of these cases the father became a more or less depreciated figure. He struggled for his existence, could not speak good English, was a mediocre provider, and sometimes even the target of ridicule of the son's contemporaries. The mother's attitude was likewise typical. She set all her hopes in the son, particularly in her oldest. The husband could no longer fulfil her ambitions. His position in the family was definitely on the decline. The mother's attitude was to sacrifice everything for the son's sake. The son would make up for everything; he would become a real American in a free country in which all possibilities were open for him. Thus the father became dethroned at home, and the whole life of the family revolved around the eldest son. From the meager earnings of the father a disproportionate amount was spent for the education of the son. The father's own little hobbies and customs, brought over from the old country, his weekly card game or some other inexpensive entertainments, were looked on jealously as unnecessary expenditures. His slow ways, particularly if he had some intellectual interests, were considered laziness. The family's future was bound up in the son. Everything must be sacrificed for his sake. This description, with minor variations, shows a typical culturally determined family situation, recurring in thousands and thousands of instances. In the few cases I have had the opportunity to study, I have observed a variety of outcomes, dependent on the specific human characters involved.

One common outcome is that the son, usurping father's place in mother's affections as well as in many material respects, develops tremendous ambition. He wants to justify all the hopes and sacrifices of the mother and thus appease his guilty conscience toward the father. There is only one way to accomplish this end. He must become successful, whatever the cost. In the hierarchy of values, success becomes supreme, overshadowing everything else, and failure becomes equivalent to sin (the greatest sin consists in a senseless and unjustified human sacrifice: the father). Consequently all other vices, such as insincerity in human relationships, unfairness in competition, disloyalty, disregard for everyone else, appear comparatively as nothing; and there emerges the formidable phenomenon of the ruthless careerist, obsessed by the single idea of self-promotion, a caricature of the self-made man, a threat to Western civilization, the principles of which he reduces to an absurdity.

Much depends, however, on the father's character and attitude. If he accepts the situation graciously and shares the mother's ambitious attitude toward the son, the guilt reaction is likely to become stronger than in cases in which the father responds with aggression or a spiteful neglect of his family. In the latter case the son may feel justified in his attitude. It is more difficult to accept sacrifice from a magnanimous, self-offering father than from an aggressive, rejecting parent. I have seen both constellations. In one case the father was not very effective, but he was an amiable man who fully accepted the mother's attitude toward the son. The outcome for the son, who became socially successful, was a neurosis which manifested itself in severe depressions and anxiety states mixed with a driving ambition. He tried always to discredit the father as much as possible. This deprecatory attitude was then transferred to all authority; a suspicious personality developed, inclined to find fault with everyone, unable to acknowledge real values or generous trends in anyone else. In the other case the father withdrew and neglected his family, obviously resenting the mother's exclusive concern with her son. This son developed a morbid self-confidence, formidable ambition, but comparatively little guilt. The only thing which caused him real concern was his unpopularity, which was the natural reaction of the environment to his personality. Those who did not dislike him feared him because of his uncanny driving power. In this case all the aggressions remained directed toward the environment, whereas in the former case they were turned back against the self. The good father thus proved to be a greater load than was the neglectful parent. The first patient, of course, became a more socialized individual, since he was unable to free himself from the influence of his conscience.

In another case the guilt reaction was so severe that its alleviation required paranoid projections. This patient, in order to justify his ruthlessness, had to see in everyone an aggressor. This view made him feel that his war against the world was waged in self-defense.

In yet another case of similar background, the mother's attitude was not only ambition for the son but also overprotection. Here the father was a better provider and gave no cause for depreciation. The son developed a tremendous lack of confidence, nurtured by his mother's overprotectiveness and by the feeling that he would never be able to accomplish what was expected of him. The father, a self-made man, appeared to him like a giant. He tried to drive himself forward and developed the compulsive urge to face dangerous situations. Gradually, however, all this effort became so much of a strain for him that he withdrew toward a more contemplative, artistic outlook, with a tendency to accept a dependent but secure position. On the whole, the outcome in this case was a more continental type of personality which, in another environment, would have received more appreciation.

In one case there was developed, as a reaction to the same family situation, a real productive urge, the wish to create something of lasting value. In this family, the father was himself a productive type who did not lose his social status and did not give up his creative ambitions after his immigration. In this case also the mother's exaggerated ambition was transferred to the boy and created a neurotic conflict between a thirst for external success and real productive tendencies.

These few sketchy examples may illustrate the overwhelming significance in personality development of the specific character traits of the parents and their relations to each other and to their children. Cultural environment accounts only for certain similarities in persons of the same cultural group. For the tremendous individual differences among human beings living in the same group, constitution and the specific human influences are responsible.

5. THE TASK OF THE PSYCHIATRIST AND THE EDUCATOR

The primary concern of the psychiatrist, as well as of the educator, is the human being as an individual, with all his peculiarities and specific makeup. Both deal with personalities and must understand their development in the most specific terms and not in generalities. Psychiatrists and educators cannot be satisfied with recognizing in their patients or pupils the exponents of cultural configurations; they must understand each on his own merits in terms of his own highly individual life-history.

The admonition of the so-called "neo-Freudians" that psychiatrists are not yet sufficiently aware of the cultural determinants in personality development is valid in a sense, but the implication that this cultural point of view will increase their therapeutic acumen is unfounded. For the social planner or reformer, this cultural point of view is of primary importance.

The therapist will have to continue to deal with his patients, and the educator with his pupils, as individual cases whose personalities must be understood as the result of specific factors, different in each case. The significance of the cultural point of view lies in another field, which undoubtedly is more significant than individual therapy: in the preventive or mental-hygiene aspects of psychiatry and education. Recognizing the influence of cultural factors to which everyone is equally exposed may contribute to the prevention of certain personality disturbances by social reform.

On the other hand, the introduction of the study of case histories is the most important methodological contribution of psychiatry to the social sciences. Social life is lived by individuals. In each person the prevailing social trends necessarily reflect themselves. The recognition of certain common features in individual case histories is thus one valuable method of establishing the ideological trends in a given society. This method will also contribute to the understanding of the dynamics of social change, which is foreboded in the emotional tensions of individuals—emotional tensions which are the result of the discrepancies between traditional attitudes and a changing social structure. Traditional attitudes represent adjustment to earlier, abandoned phases of historical development. It is my conviction that rapid social change is, as a rule, accompanied by a spread of neurosis which results from these discrepancies. In such historical periods, neurosis becomes the rule and the normal individual the exception.

In our dynamic civilization, however, recognition of the specific influences of the personal environment, as distinct from the more general cultural factors, will remain an indispensable point of view. The most characteristic feature of our cultural life lies in its great emphasis on the individual. Its complex structure, its highly differentiated distribution of social functions, would be impossible without variegated human material. Through the psychological influence of individualities upon one another, new differences are created, and thus is provided that highly diversified human material which is the secret of our rapidly changing world.

From a distant point of view, the present cultural period, beginning with the Renaissance, might appear in the history of the human race as somewhat similar to a period of mutation in biological evolution. The breeding of highly differentiated human material is the most characteristic manifestation of this period of social change, in which the traditional influences give place to rapid and enforced adjustments. The accent is on the individual. This period, thought of as beginning with the Renaissance, produced the most individualistic type of art and literature, as represented by the great reproducers of personality: the Italian painters of the fifteenth and sixteenth centuries; Rembrandt and Shakespeare in the seventeenth century; and the great novelists of the past century, such as Stendhal, Balzac, and Dostoevski. In the past hundred years, this period of rapid transformation has assumed the almost explosive qualities of biological mutation. The latest fruit of this development is the scientific mastery of the human personality, the psychological study of the human being, not only as the exponent of a biological specimen or of a social group, but as an individual person, molded by the specific personality influences of his early emotional environment.

CHAPTER

27

G. W. Allport, J. S. Bruner, AND E. M. Jandorf

PERSONALITY UNDER SOCIAL

CATASTROPHE

NINETY LIFE-HISTORIES OF THE NAZI REVOLUTION

There are times in world history when personal tragedy is endemic. The physical suffering and uprooting endured by many under the Nazi regime gave Allport, Bruner, and Jandorf an opportunity to study the effect of catastrophe as a situational determinant. Their subjects were all adults, and their investigation confirms the position of the psychoanalysts that the basic pattern of personality tends to become quite set in early life. Adjustments are made to a new situation—but ordinarily not so rapidly as reality demands—and in the course of adjustment there may be some variation in the manifestations of personality.[1]

Individuals of any group in any environment meet with frustration of some of their impulses. When there is frustration, the impulse may be modified or directed toward other objects or turned inward. Where there is relatively little frustration, the impulse may carry through to a modification of the situation external to the individual. The impulse may be manifest as a somatic process of observable behavior or as a subjective process reported verbally to an observer. Any physical modification of the environment will be accompanied by a subjective pattern,

NOTE: Reprinted with slight abridgment from *Character and Personality*, Vol. 10 (1941), pp. 1–22, by permission of the authors and Duke University Press. (Copyright, 1941, by Duke University Press.)

[1] For another illuminating analysis of the effects of political events upon personality, see A. M. Meerloo, "Psychopathic Reactions in Liberated Countries," *The Lancet*, April 7, 1945.

and sometimes these patterns crystallize in symbolic cultural or sub-cultural patterns. Sometimes the environment cannot be modified physically, only subjectively. When these subjective modifications are expressed, they may be termed "pseudo-objectifications." Aggression is only one of many possible alternative modifications of an impulse under the press of frustration. The authors of this paper rightly reject the over-simple formulation that sees frustration inevitably followed by aggression.

We must likewise beware of the tacit assumption that frustration occurs only in terms of deprivation imposed by persons. Children, for example, are also prevented from goal responses by illness, inability to walk, inaccessibility of desired objects, lack of the skills requisite to manipulate gadgets. While children sometimes blame other persons for these deprivations, this surely is not always the cause. The impersonal environment also frustrates and has implications for personal development, even though it must be freely granted that those aspects of the environment that are mediated by persons have more vivid emotional and hence structural consequences.

The following paper illustrates another method used by students of personality—that of the personal document.[2]

IN RESPONSE to an announcement of a prize competition, over two hundred life-histories on the subject "My Life in Germany Before and After January 30, 1933" were received. They averaged one hundred pages in length. All were read and analyzed by a corps of psychologists and sociologists working in close collaboration. Of these documents, ninety were submitted to particularly detailed psychological analysis. As a means of guiding and standardizing this analysis, a schedule comprising the following spheres of information was employed:

> Identifying data
> Development of attitudes toward National Socialism
> Frustrations in various spheres of activity
> Sources of frustration
> Reactions to suffering and cruelty
> Changes in mental activity with respect to
>   fantasy, planning, and valuation
> Conformity behavior
> Feelings of insecurity and their alleviation
> Fluctuations in the level of aspiration
> Identifications with groups

---

[2] For a survey and critical discussion of the use of personal documents from non-literate societies, see C. Kluckhohn, "The Personal Document in Anthropological Science," in L. Gottschalk, C. Kluckhohn, and R. Angell, *The Use of Personal Documents in History, Anthropology, and Sociology* (Social Science Research Council, Bull. 53, 1945), pp. 79–173.

Aggressive behavior
Regressive reactions
Defeat and despair responses
Suicide attempts and fantasies
Expression of wit and humor
Plans and achievement of emigration
Post-emigration adjustment
Thumbnail summary of case

The schedule called for detailed information on each of these topics and wherever possible made use of rating scales and other quantitative devices. The nature of the categories of analysis employed will become clearer as our results are presented.

The ninety cases are classified in TABLE XLIX in respect to sex, age, religious affiliation, socio-economic status, place of residence, and occupation in Germany. By the nature of circumstances, they were all individuals living outside of Germany at the time of writing, and all were strongly anti-Nazi in their political persuasion. The manuscripts include documents received from the United States, China, Palestine, Switzerland, Colombia, France, and England.

In spite of the extensiveness of the material and the complexities of analysis, the principal findings of psychological interest can be presented in the form of answers to five questions:

1. How does the persecuted adult defend himself psychologically against catastrophe?
2. How far does catastrophic social disorganization disrupt basic personality integration?
3. How do political attitudes change under the impact of catastrophe?
4. What responses appear under conditions of extreme frustration?
5. What is the psychological plight of the persecutor?

I. PSYCHOLOGICAL PROTECTIONS AGAINST CATASTROPHE

Several lines of evidence force us to the conclusion that our subjects actively resisted recognition of the seriousness of the situation or, in cases where the seriousness was realized, failed at first to make a realistic adjustment to it. The evidence is of six types.

1. *Date of recognizing the threat of National Socialism*—In the course of analysis, each investigator made a record of the date at which writers first showed clear awareness of the menace that National Socialism might contain for their own personal lives. Of the 48 German cases for which this information was recorded, 29 per cent showed their first recognition of the threat in 1933, when the revolution was already upon them. An additional 15 per cent of the cases did not recognize the threat until 1935 or after. Of the 12 Austrian cases reporting, 7 did not recognize the threat until 1938.

TABLE XLIX

Classification of the Ninety Writers

CATEGORY	PERCENTAGE OF CASES
Age	
20–30	10.1
30–40	30.0
40–50	32.2
50–60	15.5
60 and over	12.2
Religious affiliation	
Jewish	67.7
Protestant	26.6
Catholic	4.6
Undetermined	1.1
Class membership	
Upper middle	26.5
Middle	36.6
Lower middle	14.5
Intelligentsia	8.9
Uncertain	13.5
Occupation	
Lawyer	18.8
Businessman	13.3
Physician	10.0
Writer	10.0
Housewife	10.0
Teacher	7.7
Student	4.0
Uncertain and other	26.2
Place of residence	
Berlin	38.9
Vienna	18.9
Munich	7.7
Hamburg	5.5
Frankfurt	2.1
Dresden	3.4
Other	23.5
Sex	
Male	76.0
Female	24.0

A German visiting Vienna in 1937 records his impressions as follows: "People were living there in oblivion and quite unconcerned. They ridiculed anyone who even suggested similarities between the German and Austrian political developments." The situation was the same in Germany in 1933 and after. One subject writes: "This sort of attitude found its extreme expression in one of my relatives who reacted to all new decrees, irrespective of their content, stoically with the same sentence, which was designed to calm him and others: 'They'll make exceptions, yes, they'll make exceptions.' " Again, a Jewish woman quotes her husband's reaction to her fears for their safety: " 'You're a child,' said he, as he made himself comfortable in his luxurious bed. 'You mustn't take everything so seriously. Hitler used the Jews very skillfully as propaganda to gain power. Now that his goal is achieved you'll hear nothing more about the Jews.' "

These instances may seem to imply nothing more than an inability on the part of the individuals concerned to evaluate properly and appraise correctly the current political scene—a display of an all too common lack of political foresight. In the context of the Germany of 1929–33 or the Austria of 1933–38, however, such inability must have been sustained not merely by ignorance but by dynamic psychological mechanisms of a protective nature. For in Germany these were years of severe social unrest marked by rising Nazi vote, riots, strikes, and disorderly political events. The situation was even more disturbed in Austria, with assassinations and political chaos the order of the day. To account for the persistent indifference and apathy of our subjects in the face of such violent disorder, one may reasonably infer the presence of active resistance on the part of these writers, an unwillingness to perceive the anxiety-ridden reality of the situation. Additional data support this hypothesis.

2. *Interval elapsing between recognition of danger and first thought of emigration*—Of the subjects who provided the relevant information on this point, we find that 63 per cent waited for two or more years after Hitler's successful conquest before considering emigration as an adjustment to the situation. The mean interval is three years, with 39 per cent of the subjects reporting intervals exceeding the mean, ranging from four to more than six years. What happened during the period of time between recognition of the threat and first thoughts of emigration will be seen in the discussion of responses to frustration. The point here is that, before most of our writers could break their psychological ties by such a step, they had to undergo the trials of extreme suffering and deprivation.

3. *Intensity of shock required to bring realization of catastrophe*— One Jewish bookseller persisted in her attempt to maintain her business against all manner of indignities until 1938, when a new release of anti-Jewish legislation made it totally impossible for her to continue. She writes: "It became horribly clear to me that everything was over now, that this was only the beginning of a long chain. I had made my decision, I had to give up. To fight was senseless, the opposition was too overwhelming." In this case we see clearly the prolonged delay in recognizing the simple fact of defeat. Of the 67 German subjects for whom reports are available, 65 per cent of the cases did not admit defeat until 1934 or after, and 39 per cent did not give up psychologically until 1938 or 1939.

The fact that often as many as five years of intense suffering have to be experienced before an individual gives up his struggle for the fulfilment of long-established needs is first and foremost a tribute to the persistence of goal-directed behavior. One is led to the conclusion that reinforcement of response is not the whole story in accounting for maintenance of goal-striving. Certainly no incentives for such persistent activity were present in the social situation. Indeed, the opposite was the case.

Probably one of the most important reasons for remaining in Germany was the unstructured nature of the emigration situation. While consummation of basic desires was becoming increasingly less likely there, nevertheless emigration—so uncertain did it appear to our subjects—contained no improvement of prospect. The conflict was one between the positive valence exerted by a familiar, though frustrating, environment and the disagreeable uncertainties awaiting them in an unknown land.

4. *The lure of the familiar*—How strong was the pull of the familiar and how effective its influence in preventing realization of, and response to, the catastrophic social change, is revealed in the two following excerpts. The author of the first quotation is a writer who, unable to sell his manuscripts in Germany after 1933, went abroad to find a publisher in England, France, or Austria.

> Finally I returned to Germany and decided, after much wavering, to try it once more. Today I know that this was the greatest mistake of my life; but at that time there were enough weighty arguments for it. Abroad I had spoken to many emigrants who were completely ruined and who were leading a desperate life. I had my mother, whom I had to care for; I had all my relatives, all my friends—many non-Jews among them—who were faithful to me; I had my comfortable home. And I was given the opportunity to enter a commercial enterprise where I could earn a living for myself and my family.

A woman who, though a refugee herself, was in charge of a committee to aid refugees in Switzerland, reports the following:

> The young actor Meier rushed into the office, one morning, excitedly waving a letter that he held in his hand: "Friends, a very fine position was offered to me; it is as good as in my hands," he said enthusiastically. "Where, how, when?" exclaimed those who happened to be in the office at the time. It was a rare occurrence indeed for a refugee to find a position and a good one at that. "With the Jewish Kultur Bund. I must leave immediately because I was requested to sing before them prior to making definite arrangements. But that is just a matter of form."
>
> It was quite unusual for anyone who escaped Germany to return there, but it seemed that all the horror, all the suffering and humiliation that caused him to leave his country were forgotten and forgiven as soon as the opportunity of making a livelihood presented itself. In vain did they remind him of his own tale of woe when he arrived in Switzerland; in vain did they point out to him the dangers of a return visit to Germany—the young actor seemed like one possessed. A few hours later he boarded the train to Germany.

5. *Active avoidance behavior and suppression*—Evidence of suppression is abundant in our data.

We tried to escape into another sphere by reading novels, studying foreign languages, doing different more or less irrelevant things in order to save our energies. I remember the bitter contrast such activity once showed. One day the mobs were howling at the near-by synagogue. Hysteric yells shrieked, the sky was radiant with the flames of the burning temple. The smell of the burned wood and smoke penetrated the windows. We were at the piano and played a Mozart concerto. Often our eyes went to the window, but we did not stop. We played Mozart, we made music, we did not want to admit disturbing reality. We wanted to spare our nerves.

6. *Physical escape*—Physical escapes from the field also occurred. Some subjects traveled to secluded country places where Nazi influence had not yet made itself felt. One subject, whenever the going got too rough, betook himself to the cottage of a friend in the Black Forest, where he managed to lose himself in the "grandeur of nature." This device proved adequate until the pogrom of 1938 made it all too clear that escape was not to be had in such a manner. Temporary holidays from the catastrophe solved no problems. Eventually all of our writers were forced by the logic of events, and in many cases by extreme physical suffering, to escape permanently from the field.

Summary—Many psychological mechanisms seemed to play a part in this intricate story of protection against catastrophe. To summarize them briefly: (1) *Persistent goal striving* is the indispensable postulate. Families to defend, children to educate, businesses to foster, friends to help—in short, the conservation of personality structure and all the major values of life call for tenacity. (2) Such differentiated goal-striving demands the *retention of a structured field*. Migration into a new and strange life-space removes the "behavior supports" essential to the pursuit of long-established goals. In adulthood, our data show, such a catastrophic change in frames of reference, values, and supporting habits meets with active resistance. (3) One of the factors in this resistance is the *pull of the familiar*. The dynamic character of the familiar—as yet, an inadequately developed chapter in psychology—will probably be found to involve at least two interrelated mechanisms. The first, cross-conditioning, accounts for the incidental reinforcement of behavior through collateral stimulation by familiar objects and events. The second, less understood, mechanism implies that familiarity *qua* familiarity acquires positive valence. (4) Resistance is aided further by the operation of *unconscious mechanisms of defense* frequently discussed in the psychoanalytic literature. Among these our evidence points to the presence of suppression, rationalization, fantasy, and isolation.

2. ENDURING CONSISTENCY OF PERSONALITY

Very rarely does catastrophic social change produce catastrophic alterations in personality. Neither our cases nor such statistics as are available reflect any such number of regressions, hysterias, or other traumatic neuroses as the gravity of the social crisis might lead one to expect. On the contrary, perhaps the most vivid impression gained by our analysts from this case-history material is of the extraordinary continuity and sameness in the individual personality. A light-hearted, optimistic, expansive advertising man, affable and adventurous, encounters the revolution and is ruined. His attempts at readjustment for five years within Germany reflect his expansive, sociable, extroverted nature. Finally, without discouragement and in a wholly adventurous spirit, he emigrates to take up the same type of life in a new country, continuing his cheerful, unworried, and bouncy career. In contrast, an individual of negligible social appeal, after a long and colorless career as gymnasium teacher, encounters unemployment and persecution. But his very negligibility saves him from severe treatment. His emigration is not planned; it takes place in a curiously accidental way. In the New World he continues his colorless existence, no more and no less submerged in manner and outlook, persisting somewhat vaguely in his efforts to be a routine teacher as he was in his homeland. Or, a highly competent Jewish woman, of means and position, with much sparkle of manner, leaves Austria well supplied with parting gifts from her non-Jewish friends and admirers, and soon assumes the same social position in her new home, writing a distinguished and characteristic volume of memoirs.

To those who hold naïvely to the view that "personality is the subjective side of culture," this conclusion may be startling. The magnitude of the cultural disruption should, according to this theory, dismember the personalities of its victims almost beyond recognition. Yet it is, on the contrary, *resistance* to social catastrophe that is the outstanding characteristic of our cases.

The conclusions of the analysts of our life-histories and the testimony of those who have known refugees before and after their immigration are not the only evidence available. An experiment employing ten of our cases confirms the point. Interviews with these ten cases, conducted by an investigator who had no knowledge of their life-histories, were focused primarily on the present adjustment of personality. When thumbnail sketches based on these interviews were matched by twenty judges with thumbnail sketches written by the analysts of the same persons based on their life-histories, the result was a figure of 95 per cent agreement. The conclusion is that personalities studied by independent investigators, working a full year apart in a period of serious disruption, are seen by a new set of judges to be recognizably the same.

Where change does take place, it seems invariably to accentuate trends clearly present in the pre-emigration personality. Radical transformations do not occur, selective reinforcement and partial inhibition accounting for what change there is. In no case does the alteration correspond to the complete upset in the total life-space.

3. THE EFFECT OF THE REVOLUTION UPON POLITICAL ATTITUDES

Whereas temperament, basic traits, and forms of expressive behavior showed resistance to catastrophic change, there were inevitably alterations in attitude and behavior. Through these, the subjects were enabled to make partial adjustment to the catastrophic social change going on about them.

Prominent, for example, are shifts in attitudes toward the National Socialist movement. An analysis of attitudes as reported in the life-histories is found in TABLE L. In this table, Period I covers the time

TABLE L

Attitudes Toward National Socialism

ATTITUDE	PERCENTAGE PER PERIOD*			
	I	*II*	*III*	*IV*
	$N = 39$	$N = 59$	$N = 81$	$N = 51$
Tolerant	5	2	0	0
Neutral	36	15	5	4
Antagonistic	56	61	78	80
Fearful	21	47	70	37

* Percentages in TABLE L and all succeeding tables are based on the number of cases supplying sufficient data for a judgment about the phenomenon in question rather than on the total population.

from the subject's first knowledge of National Socialism to the year 1929; Period II from 1929 to January 30, 1933, the date on which the Nazis seized power in Germany; III from 1933 to the time of emigration; IV from emigration to the date of writing, 1939 or 1940. For convenience, the attitudes are grouped under four principal divisions: tolerant (from strong to slight favor); neutral (indecision and indifference); antagonistic (disgust, hatred, anger, revolt); and fearful (anxiety, fear, despair).

Two things stand out especially in TABLE L: the inevitable desertion of neutral attitudes as the social crisis verged into personal catastrophe; and the concomitant adoption of fearful and antagonistic attitudes. Among our subjects only three showed a somewhat favorable attitude toward National Socialism in its early stages; none at all after 1933. A large number, on the other hand, were at first indifferent or neutral regarding the movement. With very few exceptions, all these cases of original indifference shifted to a negative position by 1933. We may assume that, had we a representative sample of the German population, a proportionally large group would have been found which shifted from indifference to positive affiliation.

The social forces in Germany which seemed to push people off the fence into the ranks of the pro-Nazis or anti-Nazis are intimated by one of our subjects: "Everything was appraised according to pro- or anti-Nazi; there was simply no possibility of being objective. You had to belong to one group. We individualists instinctively felt the loneliness of our position outside any group." In Germany, as in Austria, when the battle lines were drawn, political opinions departed from a normal distribution wherein the Great Undecided sits comfortably on the fence, to a state of bimodality in which one had either to be for or against "The Revolution." Here we have evidence of a general law of public opinion: in times of grave crisis, the central tendency in attitude-distribution disappears; its place is taken, as opinion is forced to the extremes, by a U-shaped curve.

A striking difference in the persistence of anger and fear after emigration is also evident in TABLE L. Even after emigration, our subjects were as angry and disgusted with the Nazis as they had been while living beneath the heel of their persecutors. Of their anxiety and fear over the Nazi menace, on the other hand, half our subjects were freed soon after they reached a new haven. Anger and hatred survive two thousand miles of open water; to a much less degree so do fear and anxiety.

In the analysis schedule, analysts were asked to record the range of objects over which the subject's attitude toward National Socialism extended. Data on attitudinal range, though not amenable to strict quantitative treatment, seem to indicate a growth in generality as time went on. Thus, if in Period I Hitler and his circle were dismissed as a crack-brained group worthy only of contempt, in Period III not only this cluster of individuals but the entire population of followers, all governmental bureaus, all laws, and even, in some cases, the entire German people became objects of disgust or rejection. As the attitude grew more intense, the range of equivalent stimuli arousing it increased.

But even though the degree of generalization increases with intensification, there is evidence that with time discrimination likewise occurs. Although subjects professed to a hatred of all Nazis, many of them would make exceptions of particular Nazis. When, as frequently happened, they found surprising instances of kindness and compassion within the Nazi ranks (see below), over half the subjects accepted the favors gratefully, willingly making exceptions in their attitude of condemnation for Nazism and all things pertaining thereto. The general principle involved here seems to be that while intensity leads to a widening of the range of equivalence, experience and discrimination lead to specialization in the attitude, just as irradiation in the Pavlovian experiments is succeeded by progressive differentiation. There were, of course, die-hards who to the end looked upon all good deeds of individual Nazis with suspicion, cynicism, and contempt; it was thus they kept intact their generalized hatred.

4. FRUSTRATION AND ITS CONSEQUENCES

That extreme frustration in the lives of our subjects was the consequence of the advent of Hitler is obvious. Important here are the specific kinds of frustration experienced and the specific types of response elicited. TABLE LI contains a listing of the spheres of activity in which our subjects suffered severe restriction, together with a report of those cases in which some expansion of freedom was noted by the analysts.

TABLE LI

Percentage of Subjects Reporting Restriction and Expansion of Freedom in Various Spheres of Activity
(Percentage of Cases Reporting on Personal Freedom = 92)

SPHERE	CONSTRICTION OF FREEDOM	EXPANSION OF FREEDOM
Family	42%	4%
Friends	55	4
Occupation	84	2
Recreation	44	1
Religion	22	6
Politics	24	3

Instances of constriction and frustration were almost universal in the life-histories. The following example of constriction in friendship-relations appears in the manuscript of a Viennese *Privatdozent*:

> With two Jewish friends I had planned a trip to Zell am See. At the railroad station, I met quite unexpectedly two of my colleagues from the university office. They were both gentiles. . . . I distinctly remember the disagreeable feeling each of us five had in the compartment. The two Jews sat there—polite but reserved. The two gentile officials felt not less uncomfortable. We all tried to overcome the situation. . . . But when I stepped out to have a cigarette, one of the gentiles said to me: "I hope you did not mind our joining your company." I had to answer: "Oh, not at all, it was only by chance that I met the two." This sounds ridiculous today; but it was very serious in 1937. I had to avoid any suspicion that I was enjoying Jewish company. We all felt relieved when the two gentiles changed in Salzburg. One of my friends remarked, "I hope we did not cause you any trouble," and I felt pretty bad for some time.

On the subjective side, the life-histories report many experiences of suffering. These are summarized in TABLE LII.

TABLE LII

Percentage of Subjects Reporting Suffering
(Percentage of Cases Reporting = 87)

CAUSE OF SUFFERING MENTIONED	PERCENTAGE OF CASES
Physical	73
Social	69
Financial	64
Threats	63
Insults	56

Great anxiety and feelings of insecurity were practically universal. The sources from which they derive are summarized in TABLE LIII.

TABLE LIII

Percentage of Cases Experiencing Insecurity for Various Reasons
(Percentage of Cases Reporting = 92)

BASIS OF INSECURITY	PERCENTAGE OF CASES
Uncertainty of future status	76
Collapse of previously reliable habits and expectations	62
Fear of physical harm	64
Financial uncertainty	59
Secret police activities	57

Of physical torture our manuscripts tell many a grim and almost incredible tale. Readers of a single autobiography, such as Valtin's *Out of the Night*, sometimes incline to accuse the writer of exaggeration and romancing. The two hundred life-histories submitted to our competition unexpectedly provided a test of reliability for many statements of Nazi barbarity. In several cases it so happened that more than one of our contributors had been prisoners in the same concentration camp (photostats of releases from the camp sometimes being included with the life-histories). Although entirely unacquainted with each other, the separate writers many times reported events, sometimes of a routine order and sometimes bizarre, which agreed fully with each other. The following excerpts from three separate ex-prisoners at one concentration camp illustrate the point.

The hands are bound together on your back with a chain which is then fastened to a hook in a tree, so that the tips of your toes don't touch the ground.

. . . His hands were bound together on his back, and he was hung from the tree by a chain. . . .

I was bound to a tree and had to hang there for two hours. . . . The arms are bound together on the back by the wrists with a chain. Then one is hung from a hook so that, at the most, you touch the ground with the tips of your toes. The whole body weight hangs on the drawn-back arms. The pain is excruciating.

Before listing its psychological consequences as revealed in these life-histories, it is well to examine the nature of the frustrating situation. In many discussions of the psychology of frustration, the stimulus situation presents nothing more than a barrier to some goal. The normal response to such frustration is alleged to be aggression directed against the barrier. If punishment is threatened for aggressive attacks upon the barrier, it is held further, the subject will turn either to substitute activity or to displacement of his aggression. In the case of our subjects, the barrier and the threat of punishment both inhered in the same agents (representatives of the Nazi regime). For this reason frustration was extreme, and any aggression subject to severe and instant punishment.

Applied to our material, the frustration-aggression hypothesis, even when modified by the subsidiary concepts of displacement and substitution, falls far short of accounting for the complexity of responses brought on by extreme frustration. So diverse were these responses that they fail to fit the three categories provided by the Yale hypothesis: aggression, displaced aggression, substitute activity. The responses we find demand rather a ninefold classification.

1. *Resignation and other defeat reactions*—One of the possible responses to frustration not mentioned by the Yale authors is resignation. Not infrequently our writers reported giving up in their struggles to resist Nazi intrusions into their lives. With a minimum of action and no aggression whatsoever, they seemed to surrender to the irresistible tide. In many cases, to be sure, this surrender was accompanied by feelings of depression and a bitter consciousness of defeat. In all these cases, however, we have evidence of a yielding to circumstance that may be considered an instance of adaptation through completely submissive behavior. In very few instances were these submissive adjustments the only or the permanent form of response to frustration; but that they were present in the majority of our sample is seen in TABLE LIV, which presents the various spheres of activity in respect to which defeat reactions occurred.

TABLE LIV

Percentage of Cases Reporting Defeat Reactions
in Different Fields of Activity
(Percentage of Cases Reporting Defeat = 61)

FIELD OF ACTIVITY	PERCENTAGE OF CASES
Occupation	61
Family	52
Community	29
Financial	19
Political	49
Church	18

2. *Adoption of temporary frames of security*—Another common response to the insecurity engendered by frustration is the adoption of temporary islands of security. TABLE LV lists some of their forms.

TABLE LV

Percentage of Cases Adopting Various Temporary
Frames of Security
(Percentage of Cases Reporting Temporary
Frames of Security = 74)

FRAME OF SECURITY	PERCENTAGE OF CASES
Faith in underground movement	23
Hope of emigration	75
Official conformity	26
Occupational continuity	47
Continuation of routine activities	32

Some women found a sense of safety in the routine of housework, some men in continuing day by day in occupational activities; many in thinking of the day when they might escape. Others show in their accounts how they shopped around for islands of security, sampling in turn political Zionism, rumors of underground movements, religious activities —anything, in short, that would provide distraction and hope. In the case of one family worrying about the perils of transportation to a concentration camp (many prisoners were shot for the slightest offense), the writer records with grim humor the following episode: "My mother, too, had heard about these things, and now the reader will no longer be astonished that my mother cried happily when, after three weeks, we received a letter from my father: 'Otto, thank God, father is in Dachau.'" To this agonized wife even a concentration camp became a frame of security.

3. *Heightened in-group feelings*—Another response to frustration, possibly an instance of substitute satisfaction for thwarted desires, was the strengthening of ties within already established groups. Most dramatic are the many instances of return to the healing intimacy of the family after bitter experiences of persecution on the street, in the office, or in prison. Others turned with new zest to their remaining friends as a means of alleviating frustration, still others to clubs or religious groups (see TABLE LVI). In some of the secondary in-groups (occupation, neighborhood school, and club), to be sure, a decreased cohesion was inevitable because of the enforced dichotomies (Aryan–non-Aryan, Party–non-Party).

4. *Shifts in the level of aspiration*—Still another consequence of frustration was a marked lowering of ambitions. For the 44 subjects on whom relevant data are available, appreciable diminishing of the level of aspiration is found in respect to occupational status (29 cases), community status (13 cases).

TABLE LVI

Percentage of Cases Reporting Change in In-Group Feeling
(Percentage of Cases Reporting on In-Group Feeling = 66)

GROUP	PERCENTAGE OF CASES		
	Decrease	*Same*	*Increase*
Family	16	26	56
Ethnic group	1	18	54
Friends	18	16	48
Church	1	19	34
Occupation	38	18	19
Club	19	13	9
Neighborhood	44	16	9
School	16	9	6

5. *Regression and fantasy*—The two following excerpts reflect a type of emotional exhaustion that led in various ways to primitive forms of psychological response.

> Brief statements in my notebook also show these reactions to the general atmosphere; they tell of nervous irritation, restriction, nervous crisis, and a somatic longing for sleep. . . . A friend told me, "But now I enjoy sleeping. A deep night's sleep is the only relief one can have in these days."

> This certain state of mind, feeling secure only alone at home, possibly covered up in a dark room, was like a psychotic disease. I met hundreds of people showing these symptoms. Fright and terrified nervousness were the first reactions of indifferent or anti-Nazi circles.

Some of the cases of heightened in-group feeling within the family might safely be classified likewise as instances of regression, possibly too some of the cases of return to religion.

Twenty-five per cent of the cases on whom there were data respecting change in higher mental processes reported an increase in the amount of daydreaming, and none a decrease. Twelve per cent reported experiencing suicide fantasies. In only two cases had there been a mental breakdown. As compared with later evidence on the increasingly realistic character of thought and planning under conditions of frustration, these figures respecting fantasy life are notably low.

6. *Conformity to the regime*—That a severe frustrating situation will not gain the voluntary assent of the sufferer is obvious; and yet a kind of segmentalized conformity to the demands of the situation serves as a means of avoiding punishment and further frustration. Evidence of such partial conformity is found frequently in our manuscripts. The writers' stories seem an endless procession of petty conformities to the harrowing demands of the Nazi persecutors. As for conformity-behavior, virtually

all the cases for whom we have data justified themselves in terms of the avoidance of punishment. A few gave as their reasons the safeguarding of emigration plans or of schemes to bore from within.

7. *Changes in philosophy of life*—As the months of persecution went on, 42 per cent of the cases for whom data are available betrayed an increased fatalism in outlook. This finding is related, no doubt, to the occurrence of defeat reactions and resignation previously recorded. Others made attempts to come to terms with the situation through alterations in their scale of values. Thus, 33 per cent of the cases reporting became more religious, and an equal number increased in their concern for the welfare and safety of others. In none of these instances was the change revolutionary; rather, it took the form of an accentuation of the dominant pre-existing philosophical tendency. Such alterations as occurred are seen to best advantage in a qualitative analysis of the documents. In some cases for example, moral standards were disturbed.

> Do we know what the next day will bring? A girl whom I begged to give herself to me, who had consistently said no, now said: "Perhaps soon." Esther, whom I was still in love with, had left without my being able to see her. Would I ever see her again? What of principles! Principles? What for.

In other cases, hedonic scales of valuation were markedly shifted. "Since we have expected the worst and most dreadful at all times, things struck us Jews as pleasant which among people of culture are taken for granted and do not command particular attention." On the whole, however, our evidence does not uncover drastic alterations in the philosophy of life, but rather a modulation here and an intensification there.

8. *Planning and direct action*—Intelligent planning and action to overcome obstacles are also stimulated by frustration. In Section 5, above, we noted that of the cases reporting on "changes in higher mental processes," 25 per cent revealed an increase in daydreaming. For this same group of cases, the percentage increase in "planning" activity is 77. These figures confirmed the strong subjective impression of the analysts that, as the extent of the catastrophe became realized, our subjects developed an astonishing amount of ingenious, adaptive, and realistic planning. One author tells how infinitely inventive he became in getting rid of dangerous books banned by the Nazis. Others tell similar tales.

> There was only one general topic we all thought of: how to outlast the almost superhuman strain of the situation; and how to rid oneself of these surroundings by getting into another country, and how to find a new place in a new world which at least would not bar one's right to live.

> The mind was constantly on the alert to find out "what to do." One thought and thought, called friends, discussed secretly, listened to rumors. . . .

Problems of adaptation arising from severe personal frustration beset our subjects, calling for realistic ingenuity. That ingenious solutions occurred, culminating in emigration, is clear proof of one primary psychological consequence of frustration: non-aggressive planning.

9. *Aggression and displaced aggression*—With the threat of punishment immediate and severe, there was, naturally, little evidence of direct aggression against Nazi persecutors. Only three physical attacks against Nazis are reported. Antagonism toward the regime, to be sure, was expressed by a majority of the cases among friends in secret. In 45 per cent of the cases for whom there are adequate data on fantasy-life, aggression against the regime played a prominent part in the content of dream and fantasy. Occasional clear instances of displaced aggression were recorded. "Hatred rose within me. Hatred for this symbol, swastika: hatred for this flag and everything connected with it. I could not look at it without becoming furious. I became incensed and would take it out even upon my poor mother in rudeness."

Mentions of desire for revenge against Hitler occurred in 14 per cent of the total sample, against the Nazis in general in 22 per cent. Two or three of the authors report at the time of writing that revenge is now their sole motive in living.

Although aggression and displaced aggression in various forms are frequently encountered in our cases, they must be considered merely one mode of response to frustration, and by no means the most frequent nor most prominent. Nor is it possible to subsume under the concept of "substitute activity" such responses as *resignation, defeat, conformity, direct action,* and *planning.* It is questionable, moreover, whether appreciable psychological insight is to be gained by classifying as "substitute activity" such responses as *regression, adoption of temporary frames of security,* or *lowering of the level of aspiration.*

Having encountered these difficulties with the frustration-aggression hypothesis as it is stated by the Yale group, the writers suggest three problems upon which further research on the matter might profitably be undertaken: (1) *Response as a function of the intensity of frustration.* Our data revealed, for example, that excruciating frustration—physical torture, even—is most likely to elicit either crafty planning for escape or resignation, rather than aggression. (2) *The relation between ego-involvement and response to frustration.* It is possible that deep involvement of the ego in the circumstances surrounding frustration can inhibit or abrogate aggressive impulses. (3) *The effect of threat of punishment upon aggressive responses.* The threat of punishment may have a different effect upon the response to frustration if it is inherent in the frustrating situation (as in Germany) or if it is administered by an outside agent (as in the laboratory situation). It may be that the former is the more effective in inhibiting aggression.

Even if improvements and extensions are made in the frustration-aggression hypothesis, it seems unlikely that its basic conception will ever be adequate to cover all the consequences of frustration in human behavior.

5. THE PSYCHOLOGICAL PLIGHT OF THE PERSECUTOR

An incidental, though very important, finding concerns the mental conflicts of the persecutors. Of considerable significance for the psychology of the Nazi, this finding came to light as a by-product of analysis of our ninety cases. It was noted that some 53 per cent reported sporadic kindness from Nazis. Of the 61 Jewish cases, 66 per cent were treated kindly by the Nazis at some time.

One of the subjects related the following incident:

One day several SA men and policemen came to the house to undertake a search. After turning the entire house topsy-turvy they helped themselves to valuables and left with their pockets stuffed to capacity. Late that night the bell rang again, and one of the policemen who had participated in the "search" entered. He put a package on the table and explained that he had stolen the property in the morning in order that he might save something for them. The package represented his share of the loot.

Another case, often duplicated, shows how compassion manifests itself in the very process of torture:

I had been standing erect, with my face to the wall, for several hours, waiting to be called and questioned by the Gestapo. I dared not move. The young man who had been standing next to me had been manhandled badly because he had moved ever so slightly. But I was growing weak. A policeman sitting in the room must have noticed this. When one of the Gestapo men came in, he said to him that I was in the way where I was standing, took me out of the room and gave me a glass of water to drink.

That some Nazis were consistently cruel and sadistic there seems to be no doubt. Others, as in the instances cited above, seemed to be capable of participating in persecution and yet at the same time of alleviating it at certain times when their socialized impulses had the upper hand in their natures. Countless Germans, of course, were only surface Nazis, and underneath worked consistently to redress the wrongs the party members were inflicting. The following case is typical:

We made the amazing discovery that we had many more friends than we had ever been aware of. A high official, for example, with whom my husband had no relations other than business ones, and with whom he communicated in no other way than over the telephone or in writing, suddenly revealed himself as a friend who dared warn us that we were in danger of immediate arrest. One can hardly estimate the risk that this man took in this unselfish act which surprised us greatly.

One of our contributors, a typical burgher who had at first ardently favored National Socialism because it aspired to a restoration of German honor, put the matter this way in his life-history: "But this revolution is being fought out in the dark of night by each German in the privacy of his own room." For many a German, as in the case of this burgher, pride in certain Nazi accomplishments was turned to bitter shame by the excesses, broken promises, and indecencies. It is questionable whether the morale of a country can be high whose citizens, whether Nazis or not, betray so much and so agonizing a conflict within themselves.

6. SUMMARY

Personal documents reporting the experiences of ninety individuals during the Nazi Revolution have provided a means whereby certain prevailing theories of personality can be tested and amplified. These documents have likewise yielded insights that no other method of investigation could have yielded. In this article we report five especially significant findings:

1. Several lines of evidence demonstrate the inability of the individual to realize the imminence of catastrophic change; adjustment to such change is not made with appropriate swiftness. Rather, mechanisms of self-protection against full realization of danger are most conspicuous. Outstanding are the victims' adherence to familiar scenes and activities and their persistence toward established goals, even though the familiar has become fraught with danger and the attainment of established goals is no longer possible. The significance of such a finding for Americans in their present national emergency is apparent. Constant efforts to awaken realization of imminent danger are demanded.

2. Even catastrophic social change does not succeed in effecting radical transformations in personality. Before and after disaster, individuals are to a large extent the same. In spite of intensifications of political attitudes (toward extreme opposition), and in spite of growing awareness and criticism of standards of judgment and evaluation, the basic structure of personality persists—and, with it, established goal striving, fundamental philosophy of life, skills, and expressive behavior. When there was change in our subjects, it did not seem to violate the basic integrations of the personality, but rather to select and reinforce traits already present.

3. In times of deepening crisis, the middle of the road is forsaken. Attitudes fall into an asymmetrical bimodal distribution. The majority swings to the party in power, while the minority, which composes our sample, is forced to the position of extreme opposition. In this position it is persecuted, frustrated, and of necessity must conform a hundred times a day to one type or another of painful coercion. In the process, it indulges in a certain amount of unrealistic escape-activity, but at the same time sharpens its planning powers and inventive efforts. If not wholly destroyed or whipped into an apathetic submission, members of the minority group may ultimately break out of the field entirely—as did our subjects—and escape to a new homeland.

In the course of their development, attitudes toward National Socialism showed a marked tendency to broaden as well as to become extreme. Not only did our subjects hate Hitler and his immediate circle more intensely, but they grew to despise and reject everything connected with the movement, in extreme cases even the German *Volk* itself. But although they came to abhor the movement and its adherents, many of our subjects were still able to make exceptions in the case of individual Nazis who showed compassion and kindness to them in their personal conduct.

4. That aggression is neither the most conspicuous nor the most natural consequence of frustration is abundantly demonstrated in our data. Although there is some evidence of aggression, displaced aggression, and of substitute activity, these responses fail to cover the rich variety of behavior stemming from the acute frustration caused by the Nazis. Besides aggressive responses, direct or displaced, we find defeat and resignation, regression, conformity, adoption of temporary frames of security, changes in standards of evaluation, lowering of levels of aspiration, heightened in-group feeling, increased fantasy and insulation, and, above all, increased planning and problem-solving. It is obvious that not all these forms of response can be listed under "substitute activity" as the frustration-aggression hypothesis would be forced to maintain.

5. An important by-product of our analysis is insight into the psychological plight of the persecutor. Frequent were instances of kindness shown our subjects by members of the Nazi party in the course of persecution. The inevitable conclusion must be that these men suffered acutely for their activities with pangs of conscience and conflict. Whether a regime demanding from its adherents so schizoid an adjustment can survive is the fateful question before the world today.

28

J. McV. Hunt

AN INSTANCE OF THE SOCIAL ORIGIN OF CONFLICT RESULTING IN PSYCHOSES

As Dr. Hallowell's paper dealt with situational determinants arising from conflict between two cultures, Dr. Hunt's deals with the effects of a situation wherein the behavioral practices of a small informal social group are in conflict with certain cultural patterns to which some of the group came to subscribe. The pressure of situation on each different individual in the clique is also brought out by presentation of the factors that prevented some members from exposure to one or both of the conflicting influences.

At the end of his paper, Dr. Hunt intimates that constitutional factors may also have been operative. This provides a nice transition to papers of the next section illustrating the interrelationship of the determinants.

AN INTIMATE relation between social conditions and the organization of the motives and attitudes of the individual (personality) is widely recognized. Socialization is fairly accepted to be learning, broadly speaking, where those members of society who come in contact with the developing individual are the teachers. Neuroses, moreover, are rapidly coming to

NOTE: Reprinted from the *American Journal of Orthopsychiatry*, Vol. 8 (1938), pp. 158–64, by permission of the author and the American Orthopsychiatric Association. (Copyright, 1938, by the American Orthopsychiatric Association, Inc.)

be regarded as the issue of unfortunate attitudes or tendencies and conflicting tendencies inculcated by society, if not precisely those attitudes and tendencies themselves. The case material presented in this paper illustrates concretely this relationship between social conditions and the personal disorganization of individuals. Furthermore, this material appears to indicate that socially induced conflicts may have considerable etiological importance for subsequently appearing psychoses, and suggests certain additions to the current theory just mentioned.

The individuals concerned here were taught sexual perversions in one social setting and given a set of values incompatible with these practices through the medium of religious conversions in another setting. It was during an experiment at St. Elizabeth's Hospital that one of the writer's subjects (*Ww*) remarked that the boys of his neighborhood had lacked a fair chance in the world. He pointed out that five of his contemporary and intimate neighborhood group of fifteen boys had been committed to St. Elizabeth's Hospital. When hospital records regarding the other four patients appeared to agree generally with his reports, further consideration of this patient's story appeared warranted.

The informant—*Ww* was a manic-depressive patient, born in Maryland in 1898. He had only a common school education, but fairly wide reading and almost continuous study of the Bible had made of him something of a "philosopher." Although his Biblical interpretations were made in terms of his own personal needs and problems, his insight and understanding were rather remarkable. His history contained a long story of chronic alcoholism, and the alcoholic indulgences, according to his own report, were attempts to escape from his feelings of guilt. He had suffered guilt continuously since early adolescence over his overt homosexual practices. At the time this material was reported, *Ww* had recently recovered from several months of depression—his form of mental disorder. He was attempting to integrate himself by writing short stories, and some of these were creditable.

The method—The description of the community and the information concerning the individuals involved were obtained by having *Ww* write his own personal history and one for each of the boys in his neighborhood group. These histories were then amplified by questioning to obtain information on as many aspects as possible for each boy. Care was taken, of course, to shield the purpose of the queries. *Ww* could recall the relative economic status of each family and the pertinent social situations in which each of the group had taken part. Thus were obtained reports of: (1) which boys had entered into the neighborhood athletic games (usually baseball), (2) which had entered into the perversions, (3) which had attended revivals to any extent, (4) which had taken part in the alcoholic indulgences, and (5) which had been committed to the hospital (checked with hospital records).

The community—The neighborhood in question was in a suburb of northeast Washington, D. C. Family providers, except for a doctor, were laborers. Economic status varied from low average to outright poverty. In several of the families Ww knew of constant friction. Alcoholism in one or both parents was common. Many of the children grew up with little or no parental guidance.

Near this neighborhood were a slaughterhouse and barns in which race horses were wintered. When they were about ten years of age many of the boys were accustomed to play about these places. Some of the group were initiated into bestiality by the laborers around the slaughterhouse. Some were also seduced into homosexual practices by men about the barns. Before Ww was thirteen a ring of the boys from this neighborhood had formed in which these perversions were continually indulged. The homosexual practices were continued longer, and the ring was not dissolved until about the end of the war, when most of the group were nearing the age of twenty.

Near this same community was a Pentacostal church where emotional revivals were frequently conducted. Ww and certain others attended regularly, entering violently into the experience of conversion. Ww reported that on one evening he and Ho "went forward" four times, crying each time, until they were finally informed that once each evening was sufficient. After such experiences, the boys present promised each other to give up their perversions. Only a little later, however, as their sexual appetites mounted and following the merciless taunts of Wl and sometimes of Wa, who were the neighborhood bullies, the offspring of anti-religious parents and thus outside the influence of the revivals, these resolves were shattered. Every broken resolve resulted in feelings of guilt which increased with time so that in later adolescence the boys were almost continuously miserable. Most of this ring lost interest in school. They quit, did odd jobs, or "mooned about" in their misery.

All this is seen through the eyes of Ww. The subjective reactions of the other individuals are not available. Ww reported that in himself there developed a tendency to oscillate between two attitudes, one in which he attempted "to pull himself up and make good," and another in which he "didn't give a damn." Temptations soon altered the former, and guilt feelings changed the latter. For several years he had a girl, but he could never allow or, perhaps, bring himself to marry her. He found no way to escape from the conflict between his homosexual drive and his social and religious values.

THE SIGNIFICANT DATA

1. *Apparent importance of the two social situations*—So far we have merely a rehearsal of an unfortunate community situation. Significance appears only when the relationship between the social situations in which the various individuals were associated and the incidence of later mental disorders is noted (see TABLE LVII). All those, save two, whom Ww remembered as having entered into both the pervisions and the revivals were

later committed to St. Elizabeth's Hospital. One of these two (*Ho*) died on the streets of Washington in the winter of 1919, in what appears from *Ww*'s description to have been either a depression complicated by alcohol or a catatonic stupor. Although he was *Ww*'s best friend, he described him as having become "queer" and given to going off by himself on "moons." For several weeks he was severely "depressed," but he had appeared to have "come out of it" for a few days before his death. *St*, the other exception, took part only in the practices of bestiality and in those only during the earlier years. He went to church only "socially," so attended very few of the revivals. Thus these two individuals may be fairly considered no exceptions in view of these circumstances, and we might say that all those who entered vitally into both of these conflicting social settings succumbed to a subsequent psychosis.

Ha also might be thrown out of the group as an exception at first sight, for one would ordinarily expect no relationship between social factors and a commitment years later with a diagnosis of general paresis. It is interesting, therefore, to note that *Ww* reported him sexually unattractive to other members of the "gang." This was supposed to have caused him some embarrassment. He was the first to go to the "red-light districts," and he made himself objectionable to these old associates by bragging of his purchased heterosexual exploits. Viewed from one angle, and an important one, *Ha*'s associations in this neighborhood group could be considered a determinant of his contracting syphilis which subsequently became neurosyphilis as a result of other factors.

One should note that five boys of the group entered into the sexual perversions without suffering subsequent commitment to a mental hospital. Two of these (*Wa* and *Wl*), in fact, were ring leaders. On the other hand, one boy (*Gr*) attended the revivals frequently, but, it must be added, at his mother's behest, without any dire results. It appears, then, that the conflict induced into these persons by these two antithetical influences may be the key to their subsequent disorders.

2. *Age of onset and type of psychosis*—The age at which a psychosis appeared and the form of the psychosis varied considerably for those who were committed to hospitals. The significance of *Ho*'s death on the streets of Washington is dubious, although some disorder appears to have been involved. *Ww* was the first member of the group actually committed to St. Elizabeth's Hospital in 1920, with a diagnosis of manic depressive psychosis, depressed type. He recovered and joined the Navy, only to be dismissed during his second year for overt homosexual practices. His dismissal caused him considerable shame. For a number of years he bummed about the country, working as a waiter in restaurants. During these years he was committed several times to hospitals, once in Utah, and at least once more in Baltimore, commitments which do not appear in TABLE LVII.

The younger *Ri* (1) was committed second, and also at the age of twenty, with a diagnosis of paranoid state. He had, however, deteriorated

TABLE LVII

Participation of Individuals of Neighborhood in the Various Social Situations

INDI-VIDUALS	SOCIAL SITUATIONS IN WHICH ASSOCIATED				HOSPITAL RECORDS (ST. ELIZABETH'S)
	Neighbor-hood Athletics	*Perversions*	*Revivals*	*Alcoholic Indulgences*	
En	Occ.				
Gr			Freq.		
Ha	Occ.	Reg.	Occ.	Occ.	Adm. 4/4/33 General paresis
Ho		Reg.	Freq.	Freq.	Died on streets, winter, 1919
Jo	Freq.	Reg.	Freq.	Occ.	Adm. 1/18/32 Schizophrenia
Le	Occ.	Occ.*			
Ol	Occ.	Occ.*			
Ri-1	Freq.	Reg.	Freq.	Occ.	Adm. 11/18/21 Paranoid state
Ri-2	Occ.	Reg.	Occ.	Freq.	Adm. 6/20/34 Schizophrenia (catatonic excitement)
Sa-1	Freq.				
Sa-2	Freq.				
St	Occ.	Occ.*	Occ.		
Wa	Occ.	Freq.**		Occ.	
Wl	Freq.	Reg.			
Ww	Occ.	Reg.	Freq.	Freq.	Adm. 9/9/20, 11/25/32, 9/25/33, 9/30/35 Manic depressive psychosis (depressed type)

Occ.—Occasionally
Freq.—Frequently
Reg.—Regularly

* in bestiality only
** in bestiality primarily

so much that he was inaccessible for questioning at the time these reports were obtained. One would conclude that his psychosis had taken the classical course of dementia praecox.

Jo survived until 1929 when he was sent to Walter Reed for mental observation. His marriage had been unsuccessful, and his history refers to a marked change in his personality following marriage. While he confessed to homosexual practices, he reported a compulsion to look at the genital region of every man on the street. And, following his discharge from Walter Reed, he is reported to have attempted a compensatory flight into excessive heterosexual activity. Threats against a girl with whom he was living resulted in his commitment to St. Elizabeth's Hospital in 1932, diagnosis schizophrenia.

The older *Ri* (2) managed reasonably well until the depression. He had been married 7 years and fathered several children. Before he was committed, it is interesting to note, he returned to the Pentacostal church, ceased sexual relations with his wife, and became maniacally religious. At the hospital, his diagnosis was schizophrenia, catatonic excitement. These facts tend to support the hypothesis that the influences during adolescence inculcated his fundamental conflict, but that other factors have determined the age of onset and the form of his psychosis.

3. *Factors determining the escape of the other boys*—One wonders what kept the other boys of the neighborhood away from these antithetical influences. *Ww*'s reports contain some of the answers. *En* came from a family relatively better than most of the others. Being somewhat older, and under more parental supervision, occasional participation in athletic games was his only contact with the group. *Gr* was the "goody-goody" boy of the neighborhood and seldom associated with the group at all. Although the economic status of *Gr*'s family was marginal, his mother watched him closely, taking him regularly to the revivals where he had almost his only common influence with other boys of the neighborhood. *Le* and *Ol* both participated in the bestiality, but never became part of the homosexual ring. Their families moved away at approximately the outbreak of the war so that they were removed from the unfortunate scene relatively early. The *Sa* brothers were the doctor's sons. They lived on the edge of the neighborhood. While they were leaders in athletic games, parental supervision kept them from the play about the barns when younger, and they never knew of the perversions, so far as *Ww* could remember. These brothers also went to high school and on to college, so that they, too, were separated from neighborhood associations relatively early. *Wa*, according to *Ww*, was of "tougher stuff" than most, and he matured very early. He married at 16, and left the neighborhood. As *Wa*'s father was a practising atheist, and *Wa* aped his father in this respect, he never came under the religious influence. He took part in the practices of bestiality at the slaughterhouse, but seldom in the homosexual perversions, and in those only during the earliest years. Social reasons for his failure to enter into homosexuality are not evident. *Wl* was the bully

of the group. He was a leader in both the athletic games and in the perversions. *Ww* considered him "hard inside." He "didn't get worried and depressed, just took everything easy." *Wl's* parents were also antireligious, so that he also took no part in the revivals and usually teased and taunted those who did. As he also went to high school, he, too, left the neighborhood gang relatively early. He is now a policeman, and apparently successfully married. Thus fairly specific social factors appear to have prevented these other members of this neighborhood group from entering into the situations which appear to have been crucial for those who became disordered.

One would like further verification of *Ww's* description of this community. The writer found one individual who remembered the existence of the slaughterhouse, the horse barns, and of the Pentacostal church, but he recalled very little of the boys in the community. The whereabouts of the doctor is not known. Hospital histories were of little use, for the individuals committed had moved to other parts of the city before they were committed. It was impossible to check even the story of the homosexual perversions, for only *Ww* had frankly confessed them. There were, however, indications of homosexual conflict already noted in the histories of two of the other patients. Nevertheless, the data reported carry considerable weight. *Ww* had no reason to invent his descriptions. Except for realizing that his community had been unfortunate, he had noted none of the relationships pointed out here until they were pointed out to him after the material had been obtained.

4. *Theoretical considerations*—Such case evidence as this is not to be considered proof of any hypothesis. It does, however, since this community situation is certainly not unique in its implications, lend further weight to the conception of neuroses and even functional psychoses as issues of conflicting attitudes and tendencies learned from and reinforced by members of an individual's social milieu. Differential social conditions responsible for the two forms of psychosis (manic-depressive psychosis and schizophrenia) are not evident. Some factor inherent in the constitution of the individuals in question, or at least socio-psychological influences occurring earlier than these described, may determine the form of disorder.

The fact of onset of psychosis at varying ages in the individuals concerned is suggestive of the operation of still another variable. One may assume from *Ww's* description of his own internal reactions to these adolescent influences (alternating between two impulses) that those individuals who later became psychotic were more or less continually "in conflict." There is also suggestive evidence for this in the psychiatric histories of the other three patients. (In view of the fact that *Ha's* disorder resulted from syphilitic infection, he must be excluded from this part of the discussion.) It follows, then, that those who became psychotic at approximately twenty years of age (*Ri*-1 and *Ww*) were less resistive to conflict than those in whom the psychosis did not appear until after the age of thirty (*Ri-2* and *Jo*). Taken in this light, a constitutional factor appears in the form of *frustration* or *conflict tolerance*. The effects of the conflict may be assumed to accrue until the tolerance of the individual is passed. This tolerance may not be entirely constitutional, in the biological sense, for it is likely that childhood influences may affect it considerably. The fact that psychological disorders are not unique in man allows such a hypothesis as this to open a new experimental approach to the disorders with animal subjects where conflict can be experimentally induced and controlled. It should be possible to measure this tolerance and to determine the effects of various types of training, of diets, etc., upon it.

5. *Summary*—A neighborhood group of boys who were taught sexual perversions and given a set of values incompatible with these practices through religious conversions are described. Those members and only those members of the neighborhood group who experienced both these antithetical influences were later committed as psychotic. Social factors are described which prevented the other members from participating in one or both of these influences. The facts lend additional weight to the hypothesis that socially induced conflicts are of etiological importance in mental disorders. The fact that the individuals concerned succumbed to psychoses at varying ages suggests that a constitutional factor in the form of frustration tolerance may be operative, an hypothesis which affords another experimental approach to the disorders.

29

Saul Rosenzweig

THE GHOST OF HENRY JAMES:
A STUDY IN THEMATIC APPERCEPTION

This paper might equally well be considered as a case study or as a study in method. On the one hand it traces what it meant to be born Henry James, son of the famous Henry James, Sr. and brother of the equally famous William James, situational determinants of a very particular kind. On the other hand the personality of the subject is explored through his creative productions, his novels and stories and autobiographical writings. The procedure is similar to that employed by interpreters of projective protocols. An example of a similar device will be found in the paper by Dr. Leites in Part III of this volume.

IN THEIR original paper on the Thematic Apperception Test Morgan and Murray [17, p. 289], after pointing out the intrinsic psychological rationale of their procedure, add: "Another fact which was relied on in devising the present method is that a great deal of written fiction is the conscious or unconscious expression of the author's experiences or fantasies." It thus becomes natural to seek occasional validation of the technique by applying it to apt examples from belles-lettres. One such opportunity is afforded by an author of both American and English renown. Henry James is the

NOTE: Reprinted by permission of the author and the editor of the journal *Character and Personality*, Vol. XII (1943), pp. 79–100, with slight revision and incorporation of footnotes into the body of the text. (Copyright 1943.)

writer here under discussion. While his longer novels continue to be a notorious stumbling-block for the average reader, his short stories are by no means difficult. It is these latter tales that play the larger part in the present attempt to interweave the author's life experience with certain of his productions in the manner of a thematic apperception analysis. The result should, if successful, contribute both to the literary criticism of James's work and to the psychology of his biography.

Among the tales of Henry James is a supernatural series composed during the final third of his life and peopled by ghosts of a character utterly Jamesian. It is the peculiarity of these wraiths which merits special attention at this time; for, unlike the ordinary creatures of their kind, they fail to represent the remnants of once-lived lives but point instead to the irrepressible unlived life. To consider these ghosts in their significance for him is a fitting expression of interest in James's immortality.

Weirdly enough, these apparitions lead back to their point of origin in James's first published tale, "The Story of a Year," which appeared in the pages of the *Atlantic Monthly* for March, 1865. Here at the very outset was announced that "death" which spoke more elusively in his earlier writings and more explicitly provided toward the end the basis for his literary specters. It was, in short, the story of his own life—written prophetically and published at the early age of twenty-two, complemented in too perfect a fashion for other interpretation by the tales of his later years, and clarified autobiographically in the last full book he lived to complete. Singularly this tale has never been reprinted, though, as every reader of James is aware, nearly all of his other short stories have appeared in collected form once at least, and many of them more than once.

The story may be more significantly reviewed after some facts of James's early life have been recalled. The second child of Henry James, the theological and semiphilosophical writer (William James, the famous psychologist, having been by but a little over a year the first), Henry James, the novelist, spent his earliest days in a household richly gifted with intellectual fare and gracious cheer. The father [cf. 18] was a strongly individual student of cosmic problems which for a period brought him into close association with the transcendentalist group of Concord and Boston. Emerson was a close friend, as were also many other literary and scholarly figures of the time. His books dealt with religious questions, such as the nature of evil, and with social problems, like those of marriage and divorce, in which the relation of the individual to society occupied a central place. His views were distinctly unconventional. Though he was at various times an enthusiastic student of Fourier and of Swedenborg, he was never a mere disciple—the individualistic stamp was too strong on all that he thought and wrote. Indeed, this markedly idiosyncratic bias made his books, despite their vivid language and command of style, accessible to a very limited audience. The majority tended to be of a mind with the reviewer who said of "The Secret of Swedenborg" that the elder James had not only written about the secret of Swedenborg but that he had kept it. One is inevitably reminded of the similar quips with which the works of his son and namesake were later received: for example, the comment in *Life* expressing the hope that Henry James would sharpen his point of view and then stick himself with it, and Mark Twain's avowal that he would rather be damned to John Bunyan's heaven than have to read *The Bostonians*.

The early life of the elder James is not without interest in the present context, especially as concerns an accident which befell him at the age of thirteen and which left its mark upon him for the rest of his life. While a schoolboy at the Albany Academy, he formed one of a group who used to meet in a near-by park for experiments in balloon flying. The motive power for the balloon was furnished by a ball of tow soaked in turpentine. The ball would drop when the balloon caught fire, and the boys would then kick the ball around for their amusement. During one of these experiments, when Henry's pantaloons had by chance got sprinkled with turpentine, one of the balls came flying through the open window of a stable. The boy, in an attempt to put out the fire, which would otherwise have consumed the building, rushed to the hayloft and stamped out the flame. In doing so he burnt his leg severely and had to remain in bed for the next two years. A double amputation above the knee proved necessary. He had a wooden leg in later years and was prevented by his infirmity from leading a very active life. Fortunately he had inherited sufficient money from his father—an influential and wealthy citizen of Albany—to obviate any routine means of earning a livelihood. Accordingly, the children of Henry James were much more closely companioned by him than would otherwise have been possible, and it is thus easier to understand that the strength of character which he had should have left so strong an impression upon their young personalities. In the *Notes of a Son and Brother* [7, p. 192], the son Henry refers to his father's handicap and couples the latter's acceptance of it with a further resignation to the lack of worldly recognition the message of his books received. The similarity to the son's own fate is again noteworthy, not merely for their both having been neglected by the general public—a circumstance already mentioned—but for their common lot of infirmity.

The particular infirmity of the son must at the very outset be recognized as having established itself upon fertile soil. Henry was apparently always unsure of himself. As a boy his incapacity for athletics and for schoolwork equally stood out in his impressions although he occupied the place of favorite in his mother's affections. He was especially aware of a certain inferiority to his older and more energetic brother William—he has said as much—and William has in counterpart written in one of his letters [cf. 16, Vol. I, p. 288] about "innocent and at bottom very powerless-feeling Harry." But the accident which offered this general orientation a specific date and place for its disclosure is still inescapably important.

No better description of it could possibly be given than that which the victim has himself provided. He is speaking in *Notes of a Son and Brother* of his year at the Harvard Law School, and of the inception of his literary career. He continues [7, pp. 296 ff.]:

Two things and more had come up—the biggest of which, and very wonderous as bearing on any circumstance of mine, as having a grain of weight to spare for it, was the breaking out of the [Civil] War. The other, the infinitely small affair in comparison, was a passage of personal history the most entirely personal, but between which, as a private catastrophe or difficulty, bristling with embarrassments, and the great public convulsion that announced itself in bigger terms each day, I felt from the very first an association of the closest, yet withal, I fear, almost of the least clearly expressible. Scarce at all to be stated, to begin with, the queer fusion or confusion established in my consciousness during the soft spring of '61 by the firing on Fort Sumter, Mr. Lincoln's instant first call for volunteers and a physical mishap, already referred to as having overtaken me at the same dark hour, and the effects of which were to draw themselves out incalculably and intolerably. Beyond all present notation the interlaced, undivided way in which what happened to me, by a turn of fortune's hand, in twenty odious minutes, kept company of the most unnatural—I can call it nothing less—with my view of what was happening, with the question of what might still happen, to everyone about me, to the country at large: it so made of these marked disparities a single vast visitation. One had the sense, I mean, of a huge comprehensive ache, and there were hours at which one could scarce have told whether it came from one's own poor organism, still so young and so meant for better things, but which had suffered particular wrong, or from the enclosing social body, a body rent with a thousand wounds and that thus treated one to the honour of a sort of tragic fellowship. The twenty minutes had sufficed, at all events, to establish a relation—a relation to everything occurring round me not only for the next four years but for long afterward—that was at once extraordinarily intimate and quite awkwardly irrelevant. I must have felt in some befooled way in presence of a crisis—the smoke of Charleston Bay still so acrid in the air—at which the likely young should be up and doing or, as familiarly put, lend a hand much wanted; the willing youths, all round, were mostly starting to their feet, and to have trumped up a lameness at such a juncture could be made to pass in no light for graceful. Jammed into the acute angle between two high fences, where the rhythmic play of my arms, in tune with that of several other pairs, but at a dire disadvantage of position, induced a rural, a rusty, a quasi-extemporised old engine to work and a saving stream to flow, I had done myself, in face of a shabby conflagration, a horrid even if an obscure hurt; and what was interesting from the first was my not doubting in the least its duration—though what seemed equally clear was that I needn't as a matter of course adopt and appropriate it, so to speak, or place it for increase of interest on exhibition. The interest of it, I very presently knew, would certainly be of the greatest, would even in conditions kept as simple as I might make them become little less than absorbing. The shortest account of what was to follow for a long time after is therefore to plead that the interest never did fail. It was naturally what is called a painful one, but it consistently declined, as

an influence at play, to drop for a single instant. Circumstances, by a wonderful chance, overwhelmingly favoured it—*as* an interest, an inexhaustible,
I mean; since I also felt in the whole enveloping tonic atmosphere a force
promoting its growth. Interest, the interest of life and of death, of our national existence, of the fate of those, the vastly numerous, whom it closely
concerned, the interest of the extending War, in fine, the hurrying troops,
the transfigured scene, formed a cover for every sort of intensity, made
tension itself in fact contagious—so that almost any tension would do,
would serve for one's share.

Two points stand out in these stirring words: first, James did not
doubt from the beginning that his hurt would involve much and last long;
and, second, he could not view it as a merely personal experience but
found it indissolubly united with the war which was at that moment en-
gulfing the entire nation. The former consideration shows clearly that
somewhere in his personality the seed had been sown for what had now
transpired, despite the appearance of mere accident, just as in his later
tale, "The Beast in the Jungle," the hero knew without any statable basis
that he would one day suffer some extremity of disaster from which his
life would acquire its significance. It may be conjectured from what has
already been said regarding the accidental crippling of the father at the
age of thirteen that the dire experience of the son at eighteen was in some
sense a repetition—that by one of those devious paths of identification
which creates strange needs in sensitive personalities, Henry James, the
son, while likewise engaged in extinguishing a fire may, if only for a moment, have suffered a lapse of attention or alertness, due possibly to some
glimmering association about his father's accident on a so similar occasion; and that thus favored, the accident took effect. It seems not unlikely
that but for this momentary inco-ordination, the injury—described somewhere as a sprain—would not have been sustained. Such a psychological
moment can surely not be underestimated by any sympathetic reader
of James since he himself made of just such minutiae the essence of his
art. How much the proximity of the rhythmically moving men may
have contributed to the mental association with the father and the
"lapse" must remain like the lapse itself a matter of conjecture. But the
presumed relationship to the father's accident seems to explain the
son's avowed receptivity for the event and his certainty as to its consequences.

The coincidence between the accidents of Henry James, Sr., and his
son Henry is amazingly paralleled by a similar duplication of experience
between the father and the son William. In this latter case a psychological
catastrophe rather than a physical injury is involved, but the powerful relationship between father and son is again inescapable. (For a description
of the experiences, compare William James, *The Varieties of Religious
Experience*, New York, 1902, pp. 160–161, with the footnote reference to
the work of Henry James, Sr., *Society: The Redeemed Form of Man*,
Boston, 1879, pp. 43 ff., where the father's case of equally sudden terror is
recounted.)

These considerations also shed some light upon the nature of the injury to Henry James, Jr., especially in its psychological significance. James himself describes it as "the most entirely personal" and as "a horrid even if an obscure hurt." It is known also that it in some way affected his back. But the physical aspect which has on occasion been stressed is of purely secondary importance. Paramount is the subjective depth of the injury as James experienced it. Occurring at the very outbreak of the war, the event may well have caused him to suspect himself as an unconscious malingerer. A complex of guilt could thus have remained. Coming as it did at a time when *men* were needed by the country and were, like his own brothers Wilky and Robertson, answering the call, the injury even more surely constituted a proof of his powerlessness and crystallized a sense of impotence from which he never fully recovered. The avoidance of passion and the overqualification in his later writings are largely traceable to such an implicit attitude of combined guilt and inferiority, as are also some of his subsequent actions including as will be pointed out presently, his participation in World War I.

In such a context is understandable also his consternation after having revealed his problem at finding it "treated but to comparative pooh-pooh —an impression I long looked back to as a sharp parting of the ways, with an adoption of the wrong one distinctly determined [7, p. 330]." The great surgeon to whom his father conducted him for consultation in Boston might at least have offered some warning of what was in store. Obviously, the surgeon, with a not uncommon lack of interest in psychological implications, did not even begin to fathom the depths to which the experience of his patient reached, and the patient, feeling much from these depths, was the more appalled at the medical advice he received. Corresponding to this negative aspect is a positive one—the orientation which James tells of adopting toward his injury in trying to come to terms with it. In the late summer of 1861 he visited a camp of invalid and convalescent troops in Rhode Island. He had here his "first and all but sole vision of the American soldier in his multitude, and above all—for that was markedly the colour of the whole thing—in his depression, his wasted melancholy almost; an effect that somehow corresponds for memory, I bethink myself, with the tender elegiac tone in which Walt Whitman was later on so admirably to commemorate him [7, pp. 310 f.]." James tells of talking with the soldiers and comforting them as he could, not only with words but by "such pecuniary solace as I might at brief notice draw on my poor pocket for. Yet again, as I indulge this memory, do I feel that I might if pushed a little rejoice in having to such an extent coincided with, not to say perhaps positively anticipated, dear old Walt—even if I hadn't come armed like him with oranges and peppermints. I ministered much more summarily, though possibly in proportion to the time and thanks to my better luck more pecuniarily; but I like to treat myself to making out that I can scarce have brought to the occasion (in proportion to the time again and to other elements of the case) less of the consecrating sentiment than he [7, pp. 314 f.]." As he sailed back to Newport that night feeling considerably the worse for his exertion in his "impaired state," there established itself in his mind, "measuring wounds against wounds," a correspondence between himself and the soldiers "less exaltedly than wastefully engaged in the common fact of endurance. [7, p. 318]."

Another heartening aspect presented itself at the Harvard Law School, which he at this time attended for some months, where the "bristling horde of . . . comrades fairly produced the illusion of a mustered army. The Cambridge campus was tented field enough for a conscript starting so compromised; and I can scarce say moreover how easily it let me down that when it came to the point one had still fine fierce young men, in great numbers, for company, there being at the worst so many such who hadn't flown to arms [7, pp. 301–02]."

His new orientation entailed a constructive step forward. In the months which followed, James turned to the art of fiction. His first published tale was, as has already been mentioned, "The Story of a Year." Needless to say, the story should be read in its original form to be fully appreciated. Unfortunately it is not easy of access since it was never reprinted, but a synopsis, however lacking artistically, may convey certain essentials of the plot which are needed here.

John Ford has a second lieutenancy in the Northern Army and is about to leave for the war. On a long walk just before his departure he proposes marriage to the ward of his widowed mother. The girl is named Elizabeth, or Lizzie, Crowe. She is a simple, pretty creature who is overjoyed at the prospect of marriage, but he exacts from her the promise that if anything should happen to him in the war, she will forget him and accept the love of another. He also cautions that, to avoid gossip, it may be better to keep the engagement a secret, but he does not bind her on this point. On getting home the girl goes to her room, while he tells his mother of the engagement. Mrs. Ford is definitely against the match because she thinks Elizabeth shallow and not good enough for him. (The author suggests that, having been a good mother, Mrs. Ford would have liked her son to choose a woman on her own model.) He refuses to accept his mother's judgment about the girl, but tries to avoid contention on this last night of his stay at home. He asks his mother not to discuss the matter with Elizabeth.

After he is gone the two women say nothing about the engagement to each other at any time, but the mother has her secret plans. When Elizabeth's first blush of excitement is over, she is sent by Mrs. Ford on a visit to a friend in another city and there, decked out in finery of her guardian's making, she soon wins another suitor—Bruce. When she leaves for home, he comes to the train to see her off and accidentally shows her the newspaper which contains the announcement of Ford's having been severely wounded. Elizabeth is in great conflict and now avoids Bruce, who would accompany her to the next station. When Elizabeth reaches home, Mrs. Ford states her intention of going to nurse her son—the very thing the girl had planned to do herself; but Lizzie is strangely relieved by this shift of responsibility. She stays at home, while Mrs. Ford goes off.

Elizabeth now dreams one night that she is walking with a tall dark man who calls her wife. In the shadow of a tree they find an unburied male corpse covered with wounds. Elizabeth proposes that a grave be dug, but as they lift the corpse it suddenly opens its eyes and says "Amen." She and her companion place it in the grave and stamp the earth down with their feet.

Various changes occur—Ford gets better, gets worse, etc., and at one point when it seems he is dying, Elizabeth accepts Bruce, who is visiting in the town at the time. But Ford unexpectedly has a turn for the better and is brought home to be nursed. His mother manages to keep Elizabeth away from him for some time. When first rejected at his door, Lizzie wraps a blanket around herself and goes out on the steps. Bruce comes by, but she will not talk with him and leaves him standing there stupified. The next day she manages to get into Jack's room, and he appears to recognize her. When Mrs. Ford learns of this visit, she is very angry, but Jack asks for Elizabeth to come again. This time he explains to her that he knows he is going to die. He is, however, glad that she has found someone else and blesses them both. He asks Elizabeth to be kind to his mother. He dies. The next day Elizabeth encounters Bruce, but she is willing only to say farewell. She says she must do justice to her old love. She forbids Bruce to follow her. "But for all that he went in [12, p. 281]."

The story has today a timely interest of a general sort since it embodies a type of problem confronting many young men and women in the confused contemporary world. But it obviously goes deeper by bringing home the manner in which events on a national, or even international, scale may have a peculiarly personal significance for the individual which is timeless in character. Thus James says [12, pp. 263 f.]: "I have no intention of following Lieutenant Ford to the seat of war. The exploits of his campaign are recorded in the public journals of the day, where the curious may still peruse them. My own taste has always been for unwritten history, and my present business is with the reverse of the picture."

As a first step in interpretation must be noted the facts that the hero foresees his own death, is wounded, and dies. In foreseeing his death, he makes his sweetheart promise that she will forget him and choose another if need be. The mother opposes the match, and the girl in the case is represented as abandoning the hero well before his own physical wounds have doomed him. It is hardly possible to escape the conclusion that those wounds were not meant to be fatal without the contribution of the unhappy love motif. The wounds of the hero are, in other words, not those of a patriot who dies of what befalls him in military combat. They are those of a lover forsaken by his psychological fate. The personal significance of the war as opposed to its national or external one is emphasized. The war serves merely as a screen upon which the deeply private problem can be projected.

At this point one comes readily to see that this tale of the Civil War and the author's description of his civilian injury at the time of its outbreak are closely related. The correspondence which had established itself in his mind between the wounded soldiers in Rhode Island and his own impaired state is expressed imaginatively in the first story. But the author's view takes precedence over the soldier's even in the fiction, since it is the implication of the wounds for love rather than for war that is stressed. The death of the hero in "The Story of a Year" is thus a representation of James's own passional death as implied in the *Notes of a Son and Brother*.

The dream of Elizabeth is one of the high lights of the tale and clearly illustrates James's early mastery of certain psychological processes which have been more formally described by professional psychologists only recently. Not only is the dream prophetic—that would be banal—but it portrays in clear images the conflict in the dreamer's mind and the inevitable solution she will adopt in keeping with her deepest wishes. Such a reading of the dream indicates unmistakably that the hero's fate was sealed not by the wounds he sustained in battle, but by the psychological forces in the situation. Among such forces were not only the faithlessness of the girl but also the opposition of the mother and the hero's own self-doubt. The banter with which his early conversation with the girl is embellished—the references to the possibility of the hero's looking like a woman instead of a man after he returns with his wounds—is of considerable interest as indicating the presence of certain feminine elements in his personality to which his self-doubt and the anticipated injury may bear some relationship.

Certain details of the story might with further knowledge of James's own life lend themselves to a fuller interpretation. Thus, for instance, the possessive character of the mother in relation to the son; the heroine's being a ward of the mother—a "cousin" of the hero; and the personality of the successful rival Bruce—all raise interesting problems regarding possible intimates in James's environment. Perhaps even more significant is the absence of a father—the widowed state of Mrs. Ford. If Henry James's own father is here in question, the filial relationship may have been sensed as "too sacred" for exposure. The depth of identification between father and son in terms of their common infirmity, already discussed, agrees with such a view. This construction is, moreover, borne out by the fiction itself if the father's death existing as a given fact when the story opens is taken as corresponding to the death of the son at its close. The identity of their fates may be regarded as symbolizing their psychological identification. Paramount, however, is the other equivalence of Henry James's blight and John Ford's death. For from this "death" came the ghost which was to appear again and again in the later tales.

Before turning to the subject of this specter, attention must be paid to some intermediate stages of development. As if to materialize the "death," James actually left America to take up residence in England in 1875. The fantasy had for a time been adequate as a form of adjustment, but in the end it yielded as a forecast to the actual physical withdrawal. For this often discussed self-exile seems to have represented an escape from a world disagreeable before and now no longer tolerable. Most of James's tales and novels were written while he was living abroad, and a great number of them, from *A Passionate Pilgrim* (1871) to *The Ambassadors* (1903), present the problems of the expatriate and the allied contrast between Old and New Worlds. He returned twice to his native land in the early eighties. His mother died during the first visit, and his father's sudden and final illness brought him back almost immediately. He then remained away again for over twenty years. After a decade, however,—in the early nineties—he began writing a series of supernatural tales to which allusion has already been made. "Sir Edmund Orme" [11], which was copyrighted in 1891 and appears to have been the first, concerns the fate of a lover who as a prerequisite to his marriage must rid himself of a ghost that represents an early jilted suitor of his prospective mother-in-law. The uncanny relationship between the older woman and young man, with the apparition of an unloved youth as intermediary, unmistakably revives the situation in "The Story of a Year." There, it will be recalled, the subordination of John Ford to his mother's judgment eventually coincided with his own presentiment of death and made together for what could on the surface well be taken for a jilting by his sweetheart. The supernatural tale, however, records the triumph of the hero over the ghost and thus sounds the keynote of James's new orientation. The same restorative tendency is even more obviously at work in "Owen Wingrave" [9], which appeared in 1893. Owen has been preparing for a military career—the traditional profession of his family—when at the eleventh hour he decides to brave every misunderstanding, even that of cowardice, and keep faith with his deepest convictions by giving up his plans. In the stormy days that follow he accepts the challenge of the girl who, somewhat like Elizabeth Crowe in the case of John Ford, had been his childhood playmate and a dependent of his family. To prove his courage, he allows her to lock him for the night in a haunted chamber where his great-great-grandfather had mysteriously died after having accidentally caused the death of his own young son by an angry blow. Like his ancestor, Owen *wins* his *grave* in that room. "He looked like a young soldier on a battle field [9, p. 220]." In this instance the relationship of the hero to the paternal figure, rather than the maternal one—as in "Sir Edmund Orme"—is portrayed and, similarly, the emphasis is laid upon aggression (or war) rather than upon love. The other aspect of John Ford's problem seems thus to be bared—the role of the father figure as an inhibitor of aggression. The integrity of the hero is in the end established even if, like his sire before him, he has to yield his life to the ghost of an accidental violence.

As a presentation of James's personal problem at the outbreak of the Civil War, including even the relation of his injury to that of his father, this tale is once more clearly autobiographical. As in "Sir Edmund Orme," the vindicating theme is again dominant. For the ghosts which haunted their author from the undying past (as a return of the repressed) could only be exorcised by the achievement of some solution.

After a series of such tales had for over ten years proclaimed his deep preoccupation with the past, James began to plan eagerly for an American visit of six or eight months. His supernatural fantasies had foretold this revisit even as "The Story of a Year" had previously forecast the departure for Europe. The counterpart to the defensive escape was to be a compulsive return. The need he felt was strong. His brother William tried to dissuade him, in order doubtless to spare him the pain which the exposure would inflict upon his sensitive nature. But Henry insisted that he actually needed "shocks." How he experienced these in 1904–05 is vividly recorded in *The American Scene* [2], which he wrote on his return to England.

The itinerary of his American trip as reflected in the chapters of this book is in itself instructive. The "repatriated absentee" or "restless analyst," as he variously styles himself, went first to New England. He then saw New York, Newport, Boston, Concord, and Salem; Philadelphia, Baltimore, Washington, and the South—Richmond, Charleston, and Florida. He traveled also to the Far West, but his book concludes with Florida. The sequel he planned was never written, and it strikes one that, with the South accounted for, the rest was to him merely appendix.

At any rate, his visits to Richmond and Charleston—where he had never been before—stand out as especially significant. He says, in speaking of his "going South," that it somehow corresponded now to what in ancient days the yearning for Europe seemed romantically to promise. Early in the chapter on Richmond he alludes to the outbreak of the Civil War and describes the city almost purely in terms of its having been the Confederate capital. He characterizes it [2, p. 358, *passim*] as "the haunted scene" and "the tragic ghost-haunted city," full of an "adorable weakness" that evokes a certain "tenderness" in the visitor. The gist of his impression he gives in an image: "I can doubtless not sufficiently tell why, but there was something in my whole sense of the South that projected at moments a vivid and painful image—that of a figure somehow blighted or stricken, discomfortably, impossibly seated in an invalid-chair, and yet fixing one with strange eyes that were half a defiance and half a deprecation of one's noticing, and much more of one's referring to, an abnormal sign [2, p. 362]." A strong suspicion arises that the image here projected is that of James himself in 1861. For confirmation one need only recall his own description of the manner in which his youthful injury had united itself indissolubly in his mind with the Civil War. As his own inner turmoil had corresponded then to the internal conflict of the country, so now his highly sympathetic and tender response to the vanquished faction seems builded on an understanding of his quite similar fate.

This view is borne out by his impressions of Charleston. Here, again, the war of North and South dominates his field of vision, but, unlike the Northern friend who accompanied him, he finds himself concentrating on the "bled" condition and his heart fails to harden even against the treachery at Fort Sumter. Once more a synoptic image emerges, this time one of feminization: "The feminization is there just to promote for us some eloquent antithesis; just to make us say that whereas the ancient order was masculine, fierce and moustachioed, the present is at the most a sort of sick lioness who has so visibly parted with her teeth and claws that we may patronizingly walk all round her. . . . This image really gives us the best word for the general effect of Charleston . . . [2, pp. 401 f.]." One recalls almost with a start the bantering conversation in "The Story of a Year" of forty years earlier between the hero and the heroine as to the possibility of his looking like a "lady" when he returns from the war with his wounds. John Ford carries on the figure by saying that even if he grows a moustache, as he intends to do, he will be altering his face as women do a misfitting garment—taking in on one side and letting out on the other—insofar as he crops his head and cultivates his chin.

In general, then, the impression seems sustained that Henry James's visit to America in 1904–05, after twenty years of absence, was largely actuated by an impulse to repair, if possible, the injury and to complete the unfinished experience of his youth. He was, as it were, haunted by the ghost of his own past and of this he wished to disabuse his mind before actual death overtook him. Since the Civil War had played so vital a part in his early blight, he now visited the South for the first time and received there those impressions which bear so strong a mark of personal projection.

The plausibility of this reconstruction and of the preceding interpretation of "The Story of a Year" is strengthened by a psychological reading of the later supernatural tales, especially "The Jolly Corner" [5]. This short story was first published in the *English Review* for December, 1908, shortly after his visit to the United States. It is the story of Spencer Brydon who as a man of fifty-six returns to America after many years of residence in Europe. He has come to look at his property—the house on the Jolly Corner—where he was born and grew up. Before long he becomes absorbed in the old house to the point of visiting it nightly in the strange hope of encountering there his own alter ego—the ghost of his former self. When he finally does succeed and is confronted by the specter he has been seeking, he notes among other things that two fingers on its right hand are missing. He cannot endure to face the image before him— he refuses to recognize himself there—and overwhelmed by the extremity of his emotion, he falls unconscious. When he revives, Alice Staverton, whom he had known in his early days before taking up residence abroad and whom he has been seeing since his return, is standing over him. She, too, has seen the ghost—in a dream—and thus knew that Spencer had made the encounter. He protests to her that the shape he has met was not himself till she simply declares, "Isn't the whole point that you'd have been different? [5, p. 483.]" It is clear from the context that the heroine could have been in love with the rejected personality (the ghost) since she understood it. She is, however, equally ready to accept Brydon as he is today and reconcile him, if may be, to those unacknowledged aspects of himself which have kept him from her all these years—which have driven him abroad to escape himself.

The specter in this tale is typical of Henry James. Unlike the ghosts of other writers, the creatures of James's imagination represent not the shadows of lives once lived, but the immortal impulses of the unlived life. In the present story the ghost of Spencer Brydon is obviously his rejected self. Moreover, an injury—the two lost fingers—here stands in some relation to the fact that the life was not lived or that, in other words, a kind of psychological death had occurred. Finally, the injury and the related incompletion have entailed an unfulfilled love. The hero has fled the heroine because he could not face himself.

At this point one is obviously but a step from "The Story of a Year," written forty years earlier than "The Jolly Corner." To repeat what has more than once been implied: with the death of John Ford the ghost of Spencer Brydon came into existence. The story of the latter is a complement to that of the former. As Henry James—or Ford—left America to reside abroad, Brydon returns to confront his former self. The identity of the characters is established by the injuries each suffered—James's "obscure hurt," Ford's wounds, and Brydon's missing fingers. But like James during his visit in 1904–05, Brydon is obviously attempting to rectify the past—to face it again and test the answer previously given. There is thus represented here not merely a harking back with vain regrets but an obvious effort to overcome old barriers and pass beyond them. It is in this spirit that the woman in the case, Alice Staverton, now likewise appears as a complement to Elizabeth Crowe. Whereas Elizabeth had been faithless, Alice is ever faithful and still ready to accept her lover both as he was and as he is. Even the device of the dream recurs—the dream of Elizabeth having presaged her abandonment of Ford, while that of Alice brings her through her empathy to the scene of Brydon's overwhelming encounter with his ghost.

The complementary relationship of these two tales, standing at the very beginning and at all but the end of James's creative work, is so striking that one is impelled to believe that the second was intentionally written as a counterpart to the first. This conjecture is supported by chronological considerations. When James toured America in 1904–05, memories of the Civil War were vividly revived for him, as has already been mentioned. "The Story of a Year" must surely have been recalled at that time in sharp relief. But one of the more practical reasons for the journey was to arrange for the publication of the definitive New York Edition [8] of his collected fiction. After completing *The American Scene* on his return to England, he spent the next two years in rereading, selecting and meticulously revising his novels and tales. Critics have assailed the rigorous censorship to which the earlier writings were subjected in this process, but James's action is understandable if one compares the revision and the revisit as attempts equally to reclaim the past and reshape it while there was yet time. In the careful review of all his past work which the preparation of the collected edition entailed, James must again have come upon "The Story of a Year"—this time paginally. But he did not include this tale. What one does find there—psychologically instead—is a new story, "The Jolly Corner," which was first published in 1908 and was probably written during the arduous process of the collective revision. This tale was plainly based on the American visit, yet it no doubt also represented a retelling of the omitted "The Story of the Year"—the most radical revision of them all. For in "The Jolly Corner" one finds a coalescence of revisit and revision which satisfactorily explains the complementary relationship of this story to the first ever written. Through marking the persistence of the trend one comes to see that, despite the wishful rewording, "The Story of a Year" was nevertheless the story of a life.

Towards the end of 1909 and for nearly a year thereafter, James suffered from a severe nervous depression which completely incapacitated him for work. This illness must in the foregoing context be taken as a reaction to the failure of his restitutive efforts. Neither the supernatural tales nor the American return nor the definitive revision of his works had achieved the solution he desperately sought, and despair overtook him. His brother William's death toward the end of 1910 removed a mainstay of his life and deepened his misery. Further illness in 1912 made the end seem tragically near.

But through everything he held on, actuated still by the same forward impetus that had unfailingly declared itself before. He was unwittingly preparing for the final and highest adventure of his life. For with the outbreak of World War I in 1914, this reticent man of seventy-one, until now without any obvious interest in political affairs, of a sudden identified himself with *social action*. He recognized the cost that might be involved when he compared himself to the quiet dweller in a tenement upon whom the question of "structural improvements" is thrust and he feared for his "house of the spirit" where everything had become for better or worse adjusted to his familiar habits and use. But this "vulgar apprehension" could not deter him; and, as he says, "I found myself before long building on additions and upper storys, throwing out extensions and protrusions, indulging even, all recklessly, in gables and pinnacles and battlements—things that had presently transformed the unpretending place into I scarce know what to call it, a fortress of the faith, a palace of the soul, an extravagant, bristling, flag-flying structure which had quite as much to do with the air as with the earth [13, pp. 19 f.]."

His efforts for the Allied cause knew no bounds. He visited army hospitals and refugee encampments (as he had on a certain earlier occasion visited a military camp of invalids in Rhode Island); made pecuniary contributions and wrote articles for war charities; supported movements like the American Volunteer Ambulance Corps; and performed a host of lesser tasks as a daily routine from the beginning of the war until his death. His friends were amazed—even as they were inspired—by the fervor of this notoriously passionless writer. As Percy Lubbock, the editor of James's *Letters*, well says: "To all who listened to him in those days it must have seemed that he gave us what we lacked—a voice; there was a trumpet note in it that was heard nowhere else and that alone rose to the height of the truth. For a while it was as though the burden of age had slipped from him; he lived in the lives of all who were acting and suffering—especially of the young, who acted and suffered most. His spiritual vigour bore a strain that was the greater by the whole weight of his towering imagination; but the time came at last when his bodily endurance failed. He died resolutely confident of the victory that was still so far off [6, Vol. II, p. 379]." Edmund Gosse, among others, expressed the opinion [14] that James's death early in 1916 was definitely hastened by his profligate expenditure of energy in war service.

The significance of his death during World War I well lends itself to further examination. Without detracting in the least from the positive significance of the contribution, one may still trace the line of its descent from the earlier record already revealed. Is it too much to suggest that the unparalleled fervor of his actions is to some extent explained by a belated compensation for his failure at the time of the Civil War? In favor of such a view is the fact that the last book he lived to complete—*Notes of a Son and Brother*—and the one in which he recounted the memories of his youth, including his injury, was published in 1914. His early experiences were thus unusually fresh in his mind at the outbreak of the war. But to this inference may be added his own testimony as found in the opening sentences of the little volume, *Within the Rim* [13], in which are collected his wartime essays:

> The first sense of it all to me after the first shock and horror was that of a sudden leap back into life of the violence with which the American Civil War broke upon us, at the North, fifty-four years ago, when I had a consciousness of youth which perhaps equalled in vivacity my present consciousness of age. . . . The analogy quickened and deepened with every elapsing hour; the drop of the balance under the invasion of Belgium reproduced with intensity the agitation of the New England air by Mr. Lincoln's call to arms, and I went about for a short space as with the queer secret locked in my breast of at least already knowing how such occasions helped and what a big war was going to mean. [13, pp. 11 f.]

The analogy of the wars in his own consciousness thus attested, it is not difficult to believe that a common motivational tie was at least implicitly at work. He might have been found wanting in 1861, but he would not be found so on this second and doubtless final occasion. At that earlier time he had adjusted to his personal wounds by withdrawal and by such constructive acts as the art of fiction permitted. But now a positive participation in real social action would provide the solution for the problem which had haunted him through life. Instead of hanging his head as a war disability, he would stand forth as a war hero; England, which had been for him a refuge of escape, would become a citadel of his true assertion; and America, which had exhibited him as weak, would now be exhibited by him as weak.

In this setting becomes intelligible the mooted question of James's assumption of British citizenship a few months before his death. He had, on the one hand, been adding to his numerous activities in the Allied interest repeated statements of his consternation that America did not enter the war at once. On the other hand, his fervent identification with the English cause increased daily. Thus in July, 1915, he at last became a naturalized British subject. By this stroke he changed for himself the orientation of a lifetime. His haven of refuge was transformed into the many-flagged and turreted embattlements of which he well might write with a surge of liberated passion. From these heights he could in the end look down upon America hanging back in the distance. His own words—in a letter to his nephew—again at this point offer direct confirmation:

> I have testified to my long attachment here in the only way I could—
> though I certainly shouldn't have done it, under the inspiration of our
> Cause, if the U. S. A. had done it a little more *for* me. Then I should have
> thrown myself back on that and been content with it; but as this, at the
> end of a year, hasn't taken place, I have had to act for myself, and I go
> so far as quite to think, I hope not fatuously, that I shall have set an ex-
> ample and shown a little something of the way. [6, Vol. II, p. 491.]

It seems not improbable that this excessive expenditure of energy in a man over seventy brought on a death premature by some months or even years. But he must surely have felt that the reward had been worth the cost. And regarding these final events in the terms not of what they may have been surmounting in his past, but, as from the vantage point of the present, they appear progressively to mean, one can respond in full accord: since James by the active assertion of that period re-established vital contact with contemporary social realities.

So at last the pattern of the genius which was Henry James emerges. Suffering from childhood with a keen sense of inadequacy, he experienced in his eighteenth year an injury that sharply crystallized this attitude into a passional death. The ghost which as an apotheosis of his unlived life appears repeatedly in his later tales was liberated from this "death." Many aspects of his experience and work up to the very time of his actual death were oriented as movements back to and forward from this nucleus.

The broader application of the inherent pattern is familiar to readers of Edmund Wilson's recent volume, *The Wound and the Bow* [19]. This title paraphrases the *Philoctetes* of Sophocles in which the hero's rare skill with the bow is portrayed as having a mysterious, if not supernatural, relationship to his stubbornly persistent wound—a snake bite to the foot. Abandoned in his illness for years on the island of Lemnos, Philoctetes is finally conducted to Troy, where he fights and kills Paris in single combat, thus becoming one of the great heroes of the Trojan War. Reviewing the experience and works of several well-known literary masters, Wilson discloses the sacrificial roots of their power on the model of the Greek legend. In the case of Henry James the present account not only provides a similar insight into the unhappy sources of his genius but reveals the aptness of the Philoctetes pattern even to the point where the bow of the wounded and exiled archer is at the last enlisted literally in a crucial military cause.

PSYCHOANALYTIC EPICRISIS

In the jargon of psychoanalysis the story just sketched could be retold as follows. The Oedipus situation of Henry James included a highly individualistic father—a cripple—and a gifted sibling rival (William) who together dwarfed the boy in his own eyes beyond the hope of ever attaining their stature. A severe inferiority complex resulted. The problematic relationship to father and brother was solved submissively by a profound repression of aggressiveness.

At the age of eighteen, in the earliest days of the Civil War, Henry sustained a persistent physical injury. A keen sense of created impotence, combined with a possible suspicion of unconscious malingering, now crystallized his early sense of inferiority into "castration anxiety." The "obscure but intimate hurt" was experienced as involving not only the manliness of war, then socially so moot, but also the virility of love, which was focal in the adolescent stage of his individual development. Identification with the crippled ("castrated") but powerful father could have figured in the trauma both through the son's remarkably similar accident and in their common incapacitation. At the same time, the injury, interpreted more deeply, may have been unconsciously embraced as a token of filial submission: the acknowledged weakness was at once peculiarly appropriated as "an inexhaustible interest." Introversion in which both aggression and sexuality were repressed was now established as a *modus vivendi*.

The possible role of constitutional bisexuality should be noted in passing, even if only speculatively. Injuries like the one experienced by James may be conceived to subdue the more active and masculine components of personality and accentuate as a counterpoise the more passive feminine ones. The creative drive of genius seems often to be enhanced even as its capacity is paradoxically also limited by such a destiny.

It was at any rate after his injury that James turned to the art of fiction. His writing served him both as an escape from frustration by way of fantasy and as a partial means of solving his problems through sublimation. But the fantasied escape proved insufficient, and he therefore soon abandoned the American scene that had become to him intolerable. During most of his life he lived in England. His various novels and tales written both before and after the departure from America acquired their notorious peculiarities—precious overqualification of style and restraint of sexual passion—from the repressed pattern of his life. The acute psychological insights in which his work abounds sprang in part, however, from the introspective vigilance allied with these defects.

As James began to enter the final third of his life, a resurgence of his buried drives occurred. The supernatural stories which began to come from his pen during this period testify to this "return of the repressed." His ghosts consistently represent an apotheosis of the unlived life. This fictional attempt to face again the early unsolved problems was followed compulsively by an actual revisit to America. As the criminal returns to the scene of his crime, James now went back to the haunts of his catastrophe. But the neurotic repressions failed to yield, and a severe nervous depression that expressed his sense of defeat ensued.

With the outbreak of World War I soon following, when he was already over seventy, came a final effort at solution—now not by sublimation in fiction, by escape or return, but in relationships to the real social world. It is not surprising that a note of overcompensation was present in these war activities, especially in the assumption of British citizenship, and that his end was probably hastened by his profligate expenditure of energy. But in large measure he re-established contact with the realities of his environment by these acts and in the same degree he thus succeeded in laying the ghost of his unlived past before death overtook him.

Three wars are thus spanned by the ghost of Henry James: the Civil War, which evoked it mortally in his youth; World War I, which permitted it to be laid before his death; and World War II, which, occurring during the centenary of his birth, recalled it anew in the immortal sense.

REFERENCES

Works by Henry James

1. *The Ambassadors* (New York, Harpers, 1903).
2. *The American Scene* (New York, Harpers, 1907).
3. "The Beast in the Jungle," in *The Better Sort* (New York, Scribners, 1903), pp. 189–244.
4. *The Bostonians* (New York, Macmillan, 1886).
5. "The Jolly Corner," in *Novels and Tales*, New York Edition (New York, Scribners, 1909), Vol. XVII, pp. 433–85).
6. *Letters*, Edited by Percy Lubbock (New York, Scribners, 1920), 2 vols.
7. *Notes of a Son and Brother* (New York, Scribners, 1914).
8. *Novels and Tales*, New York Edition (New York, Scribners, 1907–17), 26 vols.

9. "Owen Wingrave," in *The Wheel of Time* (New York, Harpers, 1893), pp. 147–220.

10. *A Passionate Pilgrim and Other Tales* (Boston, James R. Osgood, 1875).

11. "Sir Edmund Orme," in *The Lesson of the Master and Other Tales* (New York, Macmillan, 1892), pp. 266–302.

12. "The Story of a Year," *Atlantic Monthly*, Vol. 15 (1865), pp. 257–81.

13. *Within the Rim* (London, Collins, 1918).

Works by Others

14. Gosse, Edmund: *Aspects and Impressions* (New York, Scribners, 1922), pp. 17–53.

15. James, Henry, Sr.: *The Secret of Swedenborg* (Boston, Houghton Mifflin, 1869).

16. James, William: *Letters*, Edited by his son Henry James (Boston, Atlantic Monthly Press, 1920), 2 vols.

17. Morgan, C. D. and Murray, H. A.: "A Method for Investigating Fantasies: the Thematic Apperception Test," *Arch. Neurol. and Psychiat.*, Vol. 34 (1935), pp. 289–306.

18. Warren, Austin: *The Elder Henry James* (New York, Macmillan, 1934).

19. Wilson, Edmund: *The Wound and the Bow* (Boston, Houghton Mifflin, 1941).

Part Two

SECTION VII

INTERRELATIONS BETWEEN THE DETERMINANTS

INTRODUCTION

WHILE FEW of the preceding selections have failed to take account, explicitly or implicitly, of more than one class of determinants, each has been primarily concerned with the influence of a single class. The conceptual framework of the papers in the final Section of Part Two gives more prominent recognition to the many-sidedness of the problem of personality formation, though one set of factors may be emphasized. Each contribution deals explicitly with the interrelationship between at least three classes of determinants.

Only by considering the interrelationship of all the determinants can we account for that total configuration which distinguishes each individual from all others. Personality formation occurs through time as the individual organism (with biologically given capacities and limitations) organizes his experience to meet the demands of the immediate environment as structured by group membership, role, and situational factors. No two persons (save one-egg twins) have identical hereditary equipment, and, in detail, the sequence of life experiences of one person is not paralleled precisely by those of another person. The essential uniqueness of each personality is thus understandable.

The uniqueness of any person may be recognized on the basis of a combination of physical characters. Individuality is also manifested by the system of needs or goals which mark one organism's strivings as distinct from the strivings of another. Likewise characteristic are the set of means that an individual chooses for the attainment of his goals. But no list of traits, whether physical or psychological, will serve satisfactorily to describe a personality. For personalities are distinguished by organization, integration, and structure as much or more than by isolate characteristics. If we are told only what notes are played in a measure of music, we can't go very far toward predicting the effect this music will have upon others. We need to know the rhythm, order, duration, and intensity of the several notes. Personality, like music, has a pattern. More exactly, a personality—like the more complex musical forms—is composed of many major and minor patterns more or less perfectly integrated in terms of a single master pattern.

The form of a symphony, however, is a fact. The notes are all there before the student, and qualified analysts will reach the same results from their dissection. Personality, on the other hand, is an abstraction. It is, as was pointed out in Part One, a hypothetical formulation, not directly verifiable. The correctness of the formulation can be tested only to the extent to which it makes sense of the life history of the subject and, more importantly, makes possible correct prediction of his future acts. Prediction is, to be sure, dependent upon the observer's realization of the situation as well as upon knowledge of the subject's motivations and habit systems. The observer must also have information about the subject's knowledge or ignorance of the factual side of the situation in which he finds himself—his cognitive orientation to the field.

Personality manifests itself in a continuous series of activities or proceedings. These proceedings are ephemeral. There may be consistent repetitions in form or in content, but each proceeding is a separate event. Regularities are obtained by abstraction of the common elements. Each proceeding, however, has some element of novelty, small though it may be. Even the tiniest variations in experience may modify the personality in ways that finally become perceptible. The study of personality necessarily, therefore, involves not only the interrelationship of the determinants but that interrelationship in a time dimension. Some proceedings are sufficiently dramatic or traumatic to effect an immediate and lasting change in the total personality. In other cases the perduring aspects of the constitutional determinants gradually obliterate the effects of cultural or situational determinants that have brought about a temporary change in personality traits or organization.

A personality is an organization of 1) the actions and manners of behavior and perception characteristic of a person's proceedings under certain circumstances or in general and 2) the motivations underlying such proceedings. The former is a generalization on the basis of observation of his proceedings through time. The latter is inferred through one or another method of psychodynamic analysis. To study personality we must, on the one hand, observe the individual in action and get his account in conscious and unconscious terms of the intention of these activities; on the other hand, we must observe others' reactions to him and get their account of those reactions. Personality is thus a two-sided concept, including both the individual's reactions and his stimulus value for others. But when we speak of a person's social effectiveness the idea of conation is immediately introduced. The individual is trying to accomplish something; therefore, a person's social stimulus value is not the same as the stimulus value of a natural object. Hence for an understanding of social stimulus value we must study two orders of data: the individual's actions with his account of them and the reactions of others.

Personality is the unique organization of relatively persistent action-systems which attempt to satisfy the biologically and socially determined needs of the individual. The action-systems may operate in relation to the physical world, the socio-cultural world, the self, or any combination of these. The needs may be conscious or unconscious. The recognition of unconscious needs assumes the necessity to explore unconscious aspects through such media as projective tests, dream analysis, and the like.

The foregoing review of some basic theoretical considerations that have been treated elsewhere in this volume may assist the reader in orienting himself to the highly complex interrelationships considered in the papers that follow.

CHAPTER

30

Lester W. Sontag

SOME PSYCHOSOMATIC ASPECTS

OF CHILDHOOD

Dr. Sontag's article deals with the complicated interconnections be-
tween constitutional factors, group-membership determinants (the bio-
logical environment of the group and "accepted social custom"), and sit-
uational determinants. The main stress is upon constitution and upon
situational events (illness, deformity, an "abnormal" parent) in the lives of
particular children.[1] However, the place of socio-cultural factors in the
environment to which a given constitutional type must adjust is clearly
recognized. The development of psychosomatic manifestations[2] of per-
sonality is seen as understandable only in this multi-dimensional matrix.

"The specific autonomic symptoms vary from individual to individual. For
example, failure of mastery in similar situations may be accompanied in
one individual by diarrhea, in another by salivation, in another by con-
junctival congestion, etc. These differences may be related to constitutional factors or physiological states prevailing at the moment which not
only play a part in determining whether mastery shall fail in a given instance but also in determining the nature of the autonomic response."
—Albert Kuntz, *The Autonomic Nervous System*

NOTE: Reprinted from *The Nervous Child*, Vol. 5 (1946), pp. 296–304, by permission
of the author and the editor. (Copyright, 1946, by Dr. Ernest Harms, editor.)
  [1] See also Margaret E. Fries and Beatrice Lewi, "Interrelated Factors in Development: A Study of Pregnancy, Labor, Delivery, Lying-In Period and Childhood,"
American Journal of Orthopsychiatry, Vol. 8 (1938), pp. 726–52; and Margaret E.
Fries, "Psychosomatic Relationships between Mother and Infant," *Psychosomatic
Medicine*, Vol. 6 (1944), pp. 159–62.
  [2] Cf. Margaret Mead, "The Concept of Culture and the Psychosomatic Approach,"
Psychiatry, Vol. 10 (1947), pp. 57–76.

A DISCUSSION of the psychosomatic and somatopsychic aspects of child-hood really should include every aspect of the behavior, health, growth, nutrition, and care of children, for certainly the child's body, his emo-tions, and his environment are even less divisible than are those of an adult. There is almost no aspect of any of them which may be considered separately from and without regard to the others. A paper on the psy-chosomatics of childhood might, therefore, assume the proportions of a text integrating the various aspects of the child and his world into a significant whole. Infancy and childhood are the periods of rapid change —change in body size and maturity of structure; expansion in experience, in breadth of social contact and adaptation, in the acquiring of new knowledge and concepts; interpreting and understanding of the world, and adoption by the child of attitudes toward it. The child's physiological functions are developing a more adequate homeostasis, but they are, nevertheless, in a relatively immature state. No one of these changes is a unit in itself. Each is a part of a total developmental pattern and does not vary independently. Impingement upon the organism from any point distorts the shape of the whole, just as surely as will pressure on one point of a water-filled rubber balloon. Perhaps the same statement might be made about the adult as a mind-body-environment unit. Yet, the processes of growth, social adaptation, and many other features of child-hood make its psychosomatic aspects different from those of later life. Therefore, a paper devoted to the psychosomatics of childhood and infancy can perhaps most profitably become a discussion of some of the more important points of environmental pressure and vulnerability of this whole organism, the child, and which are, to a greater or lesser de-gree, peculiar to that period.

Since the birth of the child or the beginning of his post-natal life follows immediately the termination of his intra-uterine life, the young infant may be subject to a unique source of psychosomatic disturbance, one resulting from a disturbed fetal environment. A modification of a new born infant's overt behavior pattern, together with a disturbance of his various body functions, may arise from disturbances of his mother's emotional or metabolic processes. There are, of course, no communicating fibres between the mother's and the fetus' nervous systems; therefore the fetus cannot experience an emotional episode of the mother in the usual sense. The fetus is, nevertheless, an intimate part of a total psychosomatic organism. To the degree that the mother's somatized anxieties or fears modify the function of her endocrine organs, cell metabolism, etc., and therefore change the composition of the blood momentarily or over longer periods, the fetus is able to experience certain aspects of this emo-tional state. Such "blood borne" anxieties or stimulations may, and in certain instances do, prove irritating to the fetus, as evidenced by his immediate increase in bodily activity. A fetus subject to repeated stimuli of this sort may maintain throughout his later intra-uterine life a state of irritability and hyperactivity. As a newborn infant, his muscular activity level is high, as is the level of certain other of his physiological functions. He is the infant who is prone to have an exaggerated bowel activity and a higher fluctuation of heart rate. Such disturbances of somatic function may include cardiospasm. Infants who do not tolerate their feedings, regurgitating them or passing them as undigested curds, often have a history of such disturbing prenatal environment.

There are many other aspects of ante-natal environment which may or may not be of importance in a similar way. Excessive fatigue of the mother, exposure to loud and prolonged noise and vibration, the possibility of the development of fetal allergies are all factors of problematical importance. For example, one woman recently complained of her fetus' violent reaction to the vibration of the washing machine motor which was transmitted directly from the apparatus to her abdomen on those days when she was working in the laundry.

This whole problem of fetal environment offers, then, the possibility of lessening the adaptibility of the infant to his new environment at birth, rendering him less able to utilize food successfully. In addition, his behavior may be further modified by causing him to be an irritable, nervous, and crying infant, a less desirable child in the eyes of the new mother.

The physiological immaturity of infancy and early childhood is in itself a factor somewhat unique to the psychosomatics of that period. A newborn infant is limited in his ability to maintain in a fairly stable state his various physiological processes. He will have a normal body temperature so long as he is in a room the temperature of which is not much above or below that of his own, but the degree of differential necessary to destroy this stability of temperature control is much less during infancy than during later life. Disturbances of and motor control of the bowels and of the peristalsis of the stomach are much more readily accomplished during infancy than later, when maturational processes have established more resistance or stability of the gastro-intestinal tract. This same immaturity of homeostasis is a part of the picture of every physiological function of infancy, almost any of which may be more readily disturbed during childhood than in later life. There is a greater somatic component of anxiety during infancy when even mild environmental stress situations will produce significant degrees of somatic dysfunction.

J. B., an eighteen-months-old child, developed diarrhea which lasted with varying severity for six weeks. No fever was present nor any laboratory nor physical findings which would explain the condition on an infectious basis. The child, son of a soldier, was born two months after his father sailed for England, only ten months after his parents' marriage. His mother, an extremely insecure woman, had always expressed her anxiety over her husband's absence and safety quite overtly. During the child's eighteen months of life, he had had every opportunity to feel very strongly his mother's insecurity. The period of his diarrhea coincided with anxiety on his mother's part resulting from her conviction that her husband was a member of the invasion troops then landing on the Normandy beach head. Premonitions that her husband would never return, plus the conviction that he was participating in a terribly hazardous operation, raised her expression of anxiety to a high level. Her child, too young to understand the dangers, and without any direct emotional attachment to the father, expressed this reflected anxiety somatically through hyperactivity of his colon.

The process of acquiring and utilizing food necessary for growth and energy has psychosomatic implications peculiar to the period of early childhood. Nutrition during infancy and childhood is, of course, not merely a matter of an individual's selecting for himself what he needs or wants to satisfy his appetite or his concepts of what his body needs in terms of vitamins, proteins, etc. It involves, rather, a combination of the appetites and desires for food and emotional needs of the infant plus the anxieties and concern of his parents for his proper nutrition and welfare. Such anxieties are to a degree an expression of the parents' desire for the child to grow into a strong and beautiful creature, reflecting the virtues and beauties of the parents themselves. The child's failure to cooperate with his parents in this process of gratifying the parents' ego is, of course, a parental frustration and a source of tension to the infant. The somatic effect of this psychosomatic situation may be important both in terms of the child's optimum processes of growth and in the maintenance of a maximum level of energy with which to meet competitive situations and to adjust to the various environmental impacts on his expanding social world. Here again, the particular and peculiar circular aspects of certain of the processes of the child's reaction to his environment are important. Limitations on growth and energy level as a result of poor nutrition increase the mother's anxiety which in turn further disturbs the child and makes him less satisfactory to his parents.

R. H. is a slender, stringy, underweight girl, the daughter of a prominent manufacturer and his former secretary, whom he had married when she was forty. The mother's doubt as to her ability to have a child at her somewhat advanced child-bearing age, plus her determination that the child by its perfection should justify her own elevation from the status of secretary to that of wife of her wealthy employer, were important factors in her handling of the child. Her concern for the child's nutrition was marked, and R. H. appeared to learn rapidly that refusal of food was a potent tool and bargaining weapon for privilege. Poor eating habits and food refusal were followed by underweight, which in turn increased the mother's anxiety and made her most solicitous of her child. As a result of this poor nutrition, R. H.'s energy and strength were probably less adequate and her adjustment to competitive situations less aggressive.

One of the aspects of the psychosomatics of infancy and childhood which deserves mention is the limited social environment of the child. The adult is the product of an adjustment process in which the social group in which he moves has been enlarged from one which at birth may have had only two important components, father and mother. to one which includes innumerable people in vocational, social, and family groups. The intensity of this interpersonal relationship of the infant and immaturity of social adaptation is in itself a hazard. The growth of an anxiety or resentment on the part of either parent for the child leaves little opportunity for compensation or amelioration through identification with others. A situation in which the mother rejects the child leaves him without resource for substituting other relationships for this maternal affection which he so badly needs. Furthermore, the intensity of his relationship to parents is such as to provide opportunity for anxieties and resentments from types and degrees of emotional exchange which might be unimportant at a later age. The child tends to respond to every fluctuation in this intensive emotional interplay between him and his parents in terms of changes in overt behavior and, organically, as changes in the motor and secretory functions of the gastro-intestinal tract, vasomotor control, and other body functions. It is an example of a somatopsychic situation in which a child's physical peculiarities or lack of conformity to certain physical standards modify his environment, which in turn modifies his social-emotional adjustment and is a potent factor in the final delineation of his personality.

The use of attention-getting devices by infants and children has psychosomatic aspects peculiar to their early age level. As mentioned above, the high degree of dependence of an infant upon one individual as a source of his emotional gratification and need leads to an intensity of relationship between mother and child, or child and other members of the immediate household. Therefore, if a child's emotional needs are not being adequately or properly met by this relationship, he is left to adopt measures or find devices of his own with which to secure the attention and concern which he desires. In addition to the fact that this very close emotional relationship exists, the child has, as indicated above, more labile physiological function in almost every respect than has the adult. As a result, he may adopt as an attention-getting device any one of several disturbances of organ function which have been proven to elicit anxiety from his adult contacts. The mother who overemphasizes toilet training and the necessity of adequate bowel movement at regular intervals may, if the child feels the need for more attention from her, find that he has become constipated through the simple process of retaining his feces. Her resulting anxiety, together with the manipulation involved in the use of suppositories or an enema, may confirm in the child's mind the desirability of this somatic pattern of function. Vomiting as an attention-getting device and perhaps, in some instances, as an aggressive behavior response is a condition seen not infrequently. The conventional breath-holding temper tantrum pattern as an aggressive response to frustration or as an attention-getting device may involve fairly profound changes in somatic function such as anoxia, rise in blood pressure, increased heart rate, etc., and may be chosen by a child because he possesses a particular type of psychosomatic constitution.

R. C. is the daughter of an officer in the Air Corps and a wife to whom he had been married only two months before going overseas. The husband was killed five months before R. C. was born. She and her mother lived with her mother's step-mother for two years after she was born. The mother adjusted readily to the loss of her husband and seemed within a month or two to be concerned primarily with how she could free herself from the encumbrance of her child so that she could go on with her professional career. Her program of toilet training was distinctly casual and a source of great annoyance and concern to her step-mother, who took it over as her own responsibility. Because of her anxiety over a slight but entirely normal irregularity of R. C.'s bowel movements, she began giving the child enemas and using suppositories. Within two weeks the child was regularly withholding feces, apparently as an anxiety-producing procedure, and had no bowel movements without her enemas or suppositories. At the same time the mother, probably expressing the guilt which her step-mother was making her feel, impressed a rigid feeding program upon R. C., forcing her to eat every spoonful of every food in the diet regime furnished by the pediatrician. When she rejected a spoonful, it was forced upon her. She almost immediately adopted what appeared to be her own aggressive defence against this procedure by vomiting after every feeding.

Growth and maturation, while they are processes extending beyond infancy and young childhood, have psychosomatic implications peculiar to early childhood. It is common for various growth and maturational factors to progress independently of each other in any individual child. Any combination of acceleration and retardation in growth in height, weight, intelligence, maturation of emotional processes, or sexual maturation may occur. The child may even be relatively mature or immature in his homeostatic characteristics. This irregularity of rates of maturation and growth of various aspects of structure, function, social adaptation, and intelligence has many psychosomatic implications. A child with the body of a six-year-old may be socially unacceptable to a class of eight-year-olds, even though his chronological age is eight years. He is certain to be aware of his physical inadequacy both from the attitudes of his classmates and from the limitation his small size puts upon his ability to compete in all kinds of physical endeavor. The immaturity of his body in size, therefore, has a profound effect upon his social adjustment, his sense of personal adequacy, and very probably upon the development of drives and motivation. Similar problems of social adjustment and adaptation are presented to the child who is overlarge for his age or obese or markedly underweight. A so-called gifted child, a child whose intellectual growth is outstripping his physiological maturity, must face certain problems of group adaptation which would not be present if other aspects of his structure and function kept pace with the growth of his intelligence. The environmental forms into which the child is thrust are almost as unyielding and as disregarding of the disunities of physiological, physical, emotional growth, and maturation as are the dimensions of a man's ready-made suit, which disregards unusual stature, shoulder breadth, or other characteristics.

J. C. is an eight-year-old who, although he is mature in muscular co-ordination and intelligence, has grown very slowly in height and weight. His feeling of inferiority, probably the result of his small size, has been a potent factor in the development of an extremely aggressive pattern of behavior. He is constantly in fights, rebels against his parents, and bullies his siblings.

One aspect of somatopsychics which is peculiar to the period of childhood is the nature of the social group in which the child moves. I refer to what might be called the "unconscious cruelty" of children in their uninhibited and unthinking treatment of the physical defects and deficiencies of their fellows. "Fatty," "skinny," and "stumpy," nicknames implying physical differences, are indicative of the readiness with which children call attention even in their nicknames to the physical differences and deformities of their fellows. This emphasis on lack of physical attractiveness, or even physical deformity, comes at a time when it can be most important in terms of permanent personality delineation and emotional adjustment.

The state of bodily attractiveness of the child to the mother has considerable somatopsychic significance. As suggested above, the child may represent to the mother an accomplishment. Through him she has acquired a sense of validity for her daily existence. He may be a part of her reason for existence, and she may place in him a hope for many of the accomplishments she has failed to attain. As she becomes less attractive physically, she is apt to be increasingly concerned with the attractiveness and the beauty of her child. The effect, therefore, of an extensive facial eczema is to change greatly the relationship of child to mother. There may be conscious or unconscious disappointment on the part of the mother, often a repressed resentment. Her child is something to hide, not to display proudly, and while she cannot give in to such feelings, of course, and express them overtly, the change in her attitude is often apparent to the child. Congenital defects, such as club foot, hare lip, or hypospadias, act similarly. Eneuresis, occasionally the result of a congenital defect of the genito-urinary tract, may have a similar effect. Here again the child's physical state helps to shape his environment, which in turn involves his emotional processes.

J. S. is a boy of nine who suffered a birth injury. He is of normal intelligence but has very poor muscular co-ordination and a very slight degree of spasticity. His brother is a normal, well co-ordinated child. J. S. cannot compete in games nor can he hold his own against his brother in the tussles which they engage in. He tends to develop a speech block when he becomes at all excited. Because of his difficulties and the reaction of his schoolmates to them, he has become a problem in school. These characteristics appear to be the basis of a rather decisive rejection of him by his mother. This rejection is a partial source of his terrific insecurity and has accentuated his lack of co-ordination and speech block.

Since training to conformity with accepted social custom is an integral part of the maturation of the child, the various pressures of such a program offer many opportunities for rebellion both in overt behavior pattern and in terms of somatic function. The withholding of feces mentioned above may be, as indicated, an attention-getting device, or it may be an expression of aggression toward the training program of an overzealous parent. Resentment of this enshrouding process of teaching and enforcing social conformity may be expressed in a variety of forms of antisocial behavior, some of them, such as eneuresis, psychic vomiting, and disturbance of gastro-intestinal motor control, somatic. Socialization viewed as a process of maturation involves the application of successive pressures impinging upon the ego of the child. These pressures applied upon the child by the parents may be applied with or without skill and without insight into his need for self-approval. They may be applied with or without understanding of his drives and constitutional ease of conformity or psychosomatic constitution. As a result, they may or may not produce a significant rebellion expressed in psychosomatic behavior (refer to case of R. C.).

The energy level of a child is extremely important in determining the pattern of behavioral and personality response to an expanding environment. Every individual is endowed with certain hereditarily determined or constitutional characteristics of physical and mental vigor. These qualities, perhaps with others not so easily defined, determine his potential resistive or reactive response to environmental pressures. The strong, highly active, mentally alert child may respond to a restrictive parent with open rebellion. Furthermore, he may be able to compensate in part for what he interprets as rejection by his mother, by identifying somewhat with others. The child with a lower energy level, with less toughness and vitality, may make a passive adjustment to his mother's restrictive attitude. His pattern of behavior becomes a conforming one with a tendency to withdraw from new or anxiety-producing situations rather than to meet them aggressively.

While every individual is born with an inherited potential vitality or energy level, that energy level is, of course, subject to constant modification through nutrition, disease, and environmental pressures. Such modification of energy level, and therefore of behavior, is not limited to the period of childhood. The adult with a nutritional or thyroid deficiency reacts differently to an environmental impact than he would have in a state of excellent health; however, in childhood, the processes of growth-limited homeostasis make for much more dynamic and drastic changes in energy level in response to nutritional deficiencies or illness. Furthermore, childhood is the period of personality formation—the time when the individual is most rapidly expanding his social sphere. He is for the first time meeting and competing with fellow humans of a comparable age ınd physique. Rachitic bones and flabby muscles, with their limitation upon a child's ability to compete in a nursery school situation, have a quite different significance than would a few months of lowered vitality from a nutritional deficiency or other cause thirty years later. The state of a child's physical well-being, then, is a dynamic factor in the whole process of his social adjustment and emerging personality. The primary tubercular infection of infancy, which usually goes unrecognized, can, and often does, produce a period of poor weight gain, lessened endurance, and increased apathy. Environmental pressures elicit a different and usually less adequate and resistive response during such a period. Childhood is, more than any other segment of life, a period of fluctuating adequacy of the whole organism to resist and successfully adjust to an expanding parade of environmental encroachments.

E. H. was, up to the age of thirty months, a child who met her parents' rigidity and lack of evident acceptance and affection with some degree of aggressiveness and rebellion. She had an occasional temper tantrum and in general displayed a resistive pattern of behavior. At thirty months she developed a fairly severe pulmonary tubercular infection from contact with a temporary member of the household. While she was not hospitalized nor even kept in bed, she ran a small temperature for many weeks and retained evidence of continuing activity of the infection for nearly three years. Her energy during this period and after it was much impaired. Her adjustment was in all of its aspects much more passive and less aggressive than during early life. The presence of the infection probably changed the mother's attitude to a more solicitous and affectionate one, and may thus have modified E. H.'s adjustment to her; however, this change should not have been responsible for the very passive adjustment E. H. continued to make to all social situations.

The health of the child has another psychosomatic implication in addition to modifying his aggressive resistance to environmental impact. Severe illness can change that environment and the position of the child in it. Illness causes withdrawal of the child from his normal social situation and from contact and competition with others of his own age. Again he spends his hours in a quiescent state, often flat in bed. He is waited upon solicitously, perhaps even fed, by a mother as anxious and solicitous as during his early infancy. Illness for him means a regression or retreat to infancy, its limited family circle and extreme state of helplessness. Illness is, then, in many ways a desirable state, and he is most apt not to give up all of its advantages readily. If the episode is prolonged, he is likely to emerge with a lapse in toilet training, food fussiness or other attention-getting devices as compensations. His modified behavior patterns and the regression of his process of socialization are somatopsychic aspects of illness.

A discussion of the ways in which the psychosomatic factors of childhood differ from those during later periods of life would not be complete without mention of the so-called psychosomatic diseases. Hypertension and peptic ulcer are rarely seen during childhood. Reynaud's disease, mucous colitis, and ulcerative colitis rarely appear before adolescence. Eczema and asthma—allergies which certainly in many instances at least have an emotional component—do appear often in early infancy. Just why these allergies should occur in childhood, while organic lesions resulting from derangements of function (arteriosclerosis and peptic ulcer) should not, is not clear. Perhaps in certain organ systems the immaturity of the tissue is responsible for a resiliency of resistence to real tissue damage, while in others that factor is unimportant. At any rate there is a very real difference in the organic changes resulting from organ dysfunction of emotional origin between infancy and adulthood.

SUMMARY

Here, then, are some of the ways in which the psychosomatics of childhood are peculiar to that period of life. The most important of them are:

1. Certain aspects of behavior and body function may be modified through result of modification of fetal environment.

2. Immaturity of the child's homeostatic processes makes for a larger somatic component of response to environmental impact.

3. Many emotional factors may influence nutrition and, therefore, its dual role of providing not only for body maintenance but also for growth.

4. Limited social environment of infancy may lead to a heightened emotional relationship with parents, fluctuations of which are readily expressed as changes in behavior and somatic function.

5. Changes in somatic function may be adopted as means for heightening parental anxiety and securing more attention.

6. Irregularity in the rates of maturation and growth of various aspects of the individual may make for difficulties in adjustment.

7. The unconscious cruelty of children toward any one of them whose body does not conform to theirs in size, form, and function may be an important factor in the emotional adjustment of children lacking this conformity.

8. Lessened attractiveness of the child from severe eczema or other cause may change the mother's emotions toward him and thus his own behavior and adjustment.

9. The child's program of training to conform with social custom may, if not skillfully administered, result in rebellions sometimes expressed somatically.

10. A child's energy level or "constitutional vitality" may be very important in determining his response to what he interprets as a hostile environment.

11. Physical illness produces an abrupt change in social status, and environment may be responsible for prolonged behavior and somatic functional changes.

12. Most of the common psychosomatic diseases of the adult do not often occur in infancy, perhaps because of the resiliency of young tissue.

31

Phyllis Greenacre

INFANT REACTIONS TO RESTRAINT

PROBLEMS IN THE FATE OF INFANTILE AGGRESSION

    Dr. Greenacre takes a factor, restraint of infants, which may be situational, as in the tying up of a child in our society where this is not culturally sanctioned, or which may be due to group membership, as in the customary use of a cradle board or of swaddling in some societies. She indicates how the reaction of the developing personality to roughly the same type of objective treatment varies, for instance, with the individual's constitution as it is determined by a developmental stage. There are the further complications of the infant's primary attitude towards restraint, resultant from his previous life experiences, and of the idiosyncratic attitudes of the restrainers. However, in spite of the many and important variations, Dr. Greenacre suspects that there are some general psychological results of prolonged restraint, whether situationally or culturally imposed.

THIS paper was stimulated by clinical discussions concerning an intellectually retarded boy of four, who had been severely and frequently tied up by his mother. Questions were raised concerning the relation between the severe prolonged restraint and the intellectual impairment, and whether therapeutic unshackling of the child might also initiate a com-

NOTE: Reprinted with slight abridgment from the *American Journal of Orthopsychiatry*, Vol. 14 (1944), pp. 204–18, by permission of the author and the American Orthopsychiatric Association. (Copyright, 1944, by the American Orthopsychiatric Association, Inc.)

parable intellectual freeing. Such cases are not uncommon in out-patient clinics, and the more dramatic ones find their way into press reports. Clinical reports are in general singularly naïve and overoptimistic. It is probable that in many such cases the restraint was actually enforced partly because of the child's original impairment rather than the opposite, and that subsequently the two factors of original intellectual deficit plus severe restraint act together to the child's further detriment.

The term "restraint" is a broad one. But in all forms of restraint there is the common situation that the free response (usually partly motor) of which the subject is capable is not permitted. Restraint may be applied through physical means as in binding the child's body or shutting him up, or through psychic channels by the use of threats, warnings, and prohibitions.

It is further apparent that the problem of the effects of restraint on a young infant is indeed complex, involving at least the questions of whether the restraint approximates a passive limitation of motion, whether it involves pain and sets up marked counter-reactions, the time at which it occurs in the infant's life with reference to the special developmental stages of spontaneous growth, and finally, whether it is restraint of a single part of the body or involves most of the body. With these questions in mind, some review of the observations available was attempted from three angles: (1) clinical observations of psychiatric patients, both adults and children; (2) experimental work, chiefly in the field of psychology; (3) observations regarding customs of restraint in certain folk groups, mainly the custom of swaddling infants as found in widely diverse areas in the world.

I. CLINICAL MATERIAL

Restraint for infants varies from the extremely common tying up a child's hands to stop thumb-sucking or masturbation, to a rare case where the child may be kept in a near-embalmment state. Such cases in a clinic are generally not well studied, nor is careful study often feasible. The mother who restrains a child in this fashion is apt to be stupid, hostile to the child, and guilty. Consequently she rarely tolerates treatment of the child for a long period. The clinician must depend too much on clinical impressions gleaned over a period of years and covering a number of cases. Occasionally in psychoanalytic work with adults one glimpses through the dreams or symptoms of the patient many of the infantile situations, including those of restraint, which are seen first hand but too briefly in the children's clinic.

The problem of reaction to restraint and forcible limitation of motion was thus presented in a very complex and late form in a twenty-year-old girl who was brought to my office as a last resort several years ago, with hope of a cure.

The presenting symptom was that she had failed in her first year in college. The mother stated that the girl was bright but extraordinarily slow throughout her entire life. The mother dated the slowness in fact from the time when, soon after returning home from the hospital where the child was born, she had labored for an hour and a half to get the infant to take the breast. She explained that the nurses must somehow have known how to manage the baby, for she had been all right in the hospital. The mother had been depressed during pregnancy and later, and felt the physician should not have permitted her to nurse the baby. Instead he had only cut down the number of feedings. There had seemed to be plenty of milk, but the mother spent most of her time trying to induce the baby to nurse. The baby cried a great deal and the mother was frantic. This went on for five or six months. In telling the story the mother remarked with grim facetiousness: "First babies should be prohibited by law!" She had prided herself on keeping her temper all through the years, and now was a quivering mass of self-restraint.

The patient had been somewhat slow in learning to walk, but the facts of the situation were not entirely clear. From eighteen months to two years of age she wore braces on her legs. The mother's reason was that she thought the child's legs a little bowed and not properly developed. The physician had been averse to using braces but she had been insistent. At the same period the baby was wearing aluminum mitts. It is not clear eighteen or nineteen years later whether this had been as protection against thumb-sucking or masturbation. Mother gave no history of either. She recalled that the baby had liked the mitts and called them "toys," but she hated the braces and would bang them against her crib. The child had a nurse until she was ten or eleven. Wet the bed until about this time, but got over it the following year. She was ushered into puberty by being given a book on what girls should know.

Throughout, the child was "unbelievably stubborn." At nine or ten she would stand on the street corner looking skyward for a time before crossing the street. The mother interpreted this, probably correctly, as a negativistic response to her own anxious admonitions to look in both directions before crossing. Once when the mother purchased some blooming geraniums and set them out in a pretty border, the child broke the stems of all the plants. Again, the mother bought a lovely antique sideboard which had sixteen panes of very old glass in it. The child broke five panes in a single effort.

The mother stated spontaneously that she had wanted a boy for her first child, and then hastily explained that this was because she herself had no brothers and thought it would be easier in any family if the first child were a boy. She said she was all mixed up about sex and in a terrible state during the pregnancy. She wanted the child to be a boy for the child's own sake. In adolescence, the young girl was unattractive. Had a "large abdomen, pipe stem legs, and looked funny." Cared nothing for clothes. Her posture was bad in spite of postural exercises and dancing classes, where she was awkward and taller than even the boys. In adolescence "she had all the ills possible. If you can think of any more, she had them too." She wore special corsets to support her back which had a slight curvature. She was repeatedly treated by an orthodontist, but would not wear the dental braces consistently. In high school she was bright but seemed slow of tempo. With some tutoring she did well in examinations, even winning a special prize, and was able to enter one of the leading colleges.

The mother seemed to be in a chronic psychosis, which she covered up rather than compensated, driven by terror of husband's abandonment of her. She described her husband as "so well balanced that he has no understanding of anything else." He had been "wonderful" to her, though completely non-understanding. He was a gentleman of one of the more orderly professions, and I could vouch from my contact with him for the non-understanding his wife described. He told of his wife's nervous illness before the birth of the baby, but stated that he "did not intend to have *that* go on" and that he had cured her himself, though he could not remember exactly how. He was an ambitious, successful, ruthless, sentimental man.

The daughter's condition was really unique, best described as an angry ambulant catatonia. Treating her soon became a hopeless task. The same sort of restrictive measures, but with incredible ramifications, were applied to the girl's visiting the psychiatrist. Because of the parents' sense of disgrace, the girl was not permitted to live outside the family lest some one discover she was seeing a psychiatrist. She was instructed in minute detail how to avoid or evade detection should she meet any one in or near my office. She should ring twice before entering as a signal that it was she and that the way should be cleared. She must walk upstairs rather than ride in the elevator in order to avoid the elevator operator. The psychiatrist must never telephone her for fear the identity might be revealed to the servants. If she telephoned the psychiatrist, she must never leave her name with the secretary. She must pay the fee in cash so the psychiatrist's name appear on no checks seen by the bank; and probably also to impress on her the amount she was taxing her parents.

This account gives a circumstantially comprehensive picture of the strait-jacket in which the girl lived, at first an almost complete physical one consisting of a series of metal braces, and then an expanding psychic strait-jacket supplemented by forced propulsion along certain demanded lines. By the time she came to the psychiatrist she was practically a synthetic person, a kind of living marionette in which there were not only the external strings of control but the opposing forces from within. It seems clear that the girl was hostilely appersonated by the mother, and the major part of her energy was used in a series of blind attempts to separate herself. While there are many clinical descriptions of children and young people subjected to long periods of restraint, I know of no case where the restraint was so complete, so varied, and so subtle as this one.

2. EXPERIMENTAL WORK

The experimental work to be considered may be divided into two subdivisions—one having to do with positive restraint of motion of infants, and the other dealing with a more negative restriction, viz., the elimination of ordinary stimuli in the infant's life which would promote activity or provide situations for practice.

While there were earlier observations along these lines, Watson's work was so publicized that it influenced the experiments on infants for a decade or so afterward. About 1917, he published his observations of three types of emotional behavior responses in newborn babies which he thought corresponded to the later emotions felt as rage, love, and fear. From 1917 to 1930, he developed, restated, and popularized his ideas until they became well known in this country. The psychological literature of the twenties contains many more references to Watson's work and accounts of experiments substantiating or controverting it than has been true in the decade since. In this review we are concerned only with the responses which Watson characterized as "rage" and which he describes as follows: "Observation seemed to show that *hampering of the infant's movements* is the factor which apart from all training brings out the movements characterized as rage. If the face or head is held, crying results, quickly followed by screaming. The body stiffens and fairly well co-ordinated slashing or striking movements of the hands and arms result; the feet and arms are drawn up and down; the breath is held until the child's face is flushed. In older children the slashing movements of the arms and legs are better co-ordinated and appear as kicking, slapping, pushing, etc. Almost any child from birth can be thrown into a rage if its arms are held tightly to its sides; sometimes even if the elbow joint is clasped tightly between the fingers, the response appears; at times just the placing of the head between cotton pads will produce it" [18].

In this passage and in some others, Watson stated quite definitely that he considered the restriction or hampering of motion as producing the reaction; and the reaction which he described was certainly a vigorous one. He both stated and implied that the reaction occurred when the hampering was by slight constraint. He did not, except by implication, recognize any possible difference in reaction of the infant according to the intensity of the holding exertion, i.e., the difference between light contact and deep intense pressure. Neither did he clearly meet the question whether the infant's response is influenced by jerkiness or sudden change in the experimenter's pressure in holding. Workers who followed Watson gave various reports. The Shermans [15], in 1925, described diffuse defense reactions of the hands in a newborn baby if pressure was exerted on the chin while the head was held steadily in an assistant experimenter's hands. The "defense movements were not at all as violent or as extensive" as Watson described as rage, however. Taylor [17], at Ohio State, who somewhat attempted to follow Watson's method of arm and nose restraint, could evoke no constant patterned responses and concluded that he merely evoked a generalized activity. Other workers there [14] had earlier attempted a more quantitative method of recording observations of infant reactions, including those to holding the nose and the arms. They concluded that the reaction was general rather than specific, that the reactive activity was greatest in those bodily segments which are nearest to the region stimulated, and that a decrease in the magnitude and frequency of the activity roughly corresponds to the distance from the zone stimulated. This seems to me the nearest, in the experimental work, to any consideration of the possibly varying effects on the infant's reaction according to where the restriction of movement is applied. It would seem that there might be some difference in infantile reaction when, for example, the whole trunk was held gently and firmly from that when the hand or wrist was held and the infant was otherwise free to try, as it were, to get away from its limited captivity. The latter situation might favor strain and frustration, while the former would tend more to produce a struggleless submission. It seems possible, from the report of Pratt, Nelson, and Sun, that they may sometimes have had such a "quiet" response from the babies, but their whole report is in terms of how many movements of each type occurred, and one can scarcely keep sight of the reacting infant [7]. It appears further that they obtained results similar to but less intense than Watson's, and that they utilized less intense stimuli.

At the University of Virginia, Wayne Dennis with the aid of his wife [6] conducted experiments on a pair of twins which are more extensively reviewed later in the paper. He found that at two months the twins reacted with Watsonian rage in variable degrees and to a variety of stimulations, including holding the head between the experimenter's hands so that it could not be moved, and pressing the subject's nose with the experimenter's forefinger, as well as by using strong taste stimuli—saturated salt solution, a very bitter quinine solution, and dilute citric acid. In these latter experiments it is evident that the discomfort of too intense or unpleasant taste stimulation is added to the question of restraint. The Dennises state, however, that it was necessary to use very vigorous restraint to provoke the so-called rage response. They concluded that it was the intensity and persistence of the stimulation rather than merely the limitation of movement which was the especially provocative factor. Dennis objects to the characterization as rage; he prefers the simple descriptive terms "struggling, thrashing, and crying" or, when the reaction is present in a lesser degree, simply "restlessness and crying." It seems to me that these early reactions may be interpreted as very simple aggressive defense, whether or not we call them rage or the precursors of rage.

The Dennises found that when, in the tenth month, their twin subjects were held with pressure on the nose or with their bodies or heads immobilized, far from showing a disturbed reaction, the babies responded with a smile which sometimes lasted as long as five minutes before fussing began. They recognized that this smiling was due partly to the peculiar attachment of the subjects to the experimenters on whom they were so emotionally dependent that they tolerated and "liked" any attention given them so long as it was not severe or acutely painful. Dennis concluded rightly, I believe, that it is probably futile to attempt to test for "instincts" or native emotional reactions after the very early weeks of life, as every stimulus which is presented, even if it has never before been employed, bears some relationship to the (individual) experience of the child. In other words, there can be no isolated stimulus. He states quite definitely that restraint of movement achieved without the use of intense stimulation does not cause negative, i.e., rage-like, reactions in the newborn, but that intense and enduring stimulation, of which rough restraint in the sense of interference with customary behavior sequences is an example, gives rise to negative reactions of frustration.

The recognition that stimulation is very soon incorporated into the fabric of experience and does not remain unrelated to other experience leads to another comment or question. In most clinical situations which come to us as psychiatrists, the restraint which has been applied early has not been in an impersonal or even unemotional setting. The type of restraint most frequently encountered clinically is punishment restraint of some sort in which the effect of anger or disapproval of the person who applies it has also to be reckoned with in the total situation. The nearest we can come to impersonally applied or at least unemotionally enforced restraint would be in studying the forms of aggression shown in a group of babies who were early subjected to plaster cast treatment, as in the case of clubfeet treated by nonoperative methods. Here, of course, there would be still other factors to be considered, but such a comparative study might be worth while.

Considering further the factor of the emotional attitude of the restrainer, the question arises of the tendency of even a relatively young infant to reflect in some degree the attitude of the person who holds it [12, 5]. (This question is one I discussed in a paper on the "Predisposition to Anxiety, Part I," *Psychoanal. Quart.*, Vol. 10 (1941), pp. 66–94. One sees this empathic reflection of the person watched in the eager, stimulated, aggressive expressions, and clenched fists of the members of a prize fight audience. I believe this "absorption of emotion by reflection" may be one of the most potent factors in the heightening of emotional reactions in a crowd resulting in mob force.) My belief is that such an induction or increase of emotional pitch by empathic reflection is appreciable, and that it may indeed occur at an earlier age than would at first be thought possible. The literature on this subject is not very extensive, and the problem is not readily subjected to experimental methods. But there is some evidence of a basis for its occurring in infancy in the observations of C. Buhler [1], the Dennises, and Gesell [10], all of whom remark on the early focusing of the infant's gaze on the face of the mother or nurse.

Other factors entering into the restraint situation are discussed by Dr. Wayne Dennis in an article in which he summarizes his critique of Watson's rage theory [7]. Dennis here brought out the question of the contrast between the quiet reaction of the infant to general, slight, or moderate restraint of motion and the thrashing, active, aggressive reaction to severe or sudden restraint of motion. He thought this contrast might be due to a similarity of the first condition to the intra-uterine state of the infant, while the latter condition caused definite thwarting and, if the restraint was sufficiently intense, there might be pain. Independently I made a similar suggestion in my paper on anxiety previously noted. This has also been remarked by various other writers, discussing swaddling and other restraint customs which will be reviewed at the close of this paper.

While there is an extensive literature on the effect of practice on learning, that on the effect of deprivation of practice situations is not as well rounded. The idea of determining what infants would do if not subjected to ordinary stimulations from other human beings seems to have stirred the imaginations of philosophers and psychologists for ages. This may be an academic form of the narcissistic question of the dependence and self-sufficiency of man so well romanticized in Robinson Crusoe. Herodotus is said to have reported the case of two children reared without contact with speaking adults to determine if these children would develop speech (quoted by Dennis).

The most comprehensive report that I have located is contained in the monograph of Wayne Dennis [6]. This offers some interesting material which has both direct and tangential bearing on this discussion.

Dr. and Mrs. Dennis took a pair of twin girls five weeks old and attempted to rear them with a minimum amount of stimulation and minimal practice situations. They kept the babies through the fourteenth month. Although there are striking defects in their observations and interpretations from a medical angle, the experimenters have brought out a valuable human document. In a résumé of their work, two groups of responses may be emphasized here: the emotional responses of the babies, and the developmental locomotor behavior, i.e., sitting up, reaching, standing, walking.

For the first six months the experimenters were able to maintain quite rigid conditions. The twins were not permitted to see each other. They were handled with as complete unemotionality as the Dennises were capable of. They were taken from their cribs only for feeding and bathing or for specific experiments. If either child cried at other than feeding time, one experimenter entered and investigated, and usually changed a diaper. There was no response if the infant cried when the experimenter left. They did not speak before entering the room, did not speak to the twins, and did not speak to each other while feeding or caring for the twins. There was no demonstration of affection, petting, fondling, cuddling, or smiling. There were no rewards or punishments, and a definite effort was made in the direction of "indifference" while handling the babies. Acts which would provide examples for imitation were avoided.

The result of this emotionally sterilized environment is a little surprising, in that it was the experimenters rather than the babies who broke down and could not maintain this unnatural status. By and large, one might summarize the developing situation as follows: Since the babies had no one else to become attached to, their emotional responses became focused on the experimenter to a provocative extent. At seven weeks

the babies were observed to follow the experimenters with their eyes, and
to smile occasionally when they appeared. "From the time that the infants first showed visual regard for adults, they were most attentive to
faces rather than to other parts of the person at whom they were looking." At eight weeks their hunger-crying would stop when an adult appeared. Between nine and twelve weeks they began to laugh and coo;
but also to show something resembling impatience about feeding. Their
hunger-crying stopped on the entrance of an adult but promptly began
again if the feeding was not immediate. Between thirteen and sixteen
weeks, they began to show behavior suggestive of the beginning of disappointment reactions: they cried when the adult turned from the crib,
and if they were picked up as though for feeding and then put down
again, they cried more lustily.

During the fifth and sixth month the twins showed especially strong
fright-like reactions to certain noises. But they smiled and laughed persistently when approached unless they were hungry. Also, in the sixth
month they developed "a response which for some time seemed to serve
as an expression of interest, of wideawakeness, of excitement, and of well
being." This was called by the experimenters "extension of the extremities." In this response the arms or the legs or both would be rigidly
extended and perhaps moved slightly. This often occurred when the
experimenters bent over the crib, but might also occur when they were
not near. It was often accompanied by smiling and "loud vocalization."
From this time on the twins laughed more. One of them would laugh
almost any time she was touched. When in their eighth month, one of
them seized the experimenter's hair and face as he bent over the crib, it
seemed they had finally more or less captivated the experimenters and
begun to set themselves free from the tepidness of their handling. For the
experimenters began gradually to break down and talk and play with the
twins a little every day. The babies were then allowed to associate more
with each other, and the negative morale of the experiment dissolved
considerably.

It is unnecessary in this résumé to recount full details of the locomotor
development. Neither twin learned by herself to sit up alone by a year.
An effort was then made to test them for their ability to support their
weight on their legs, and neither baby gave any indication of doing so,
giving only a few momentary kicks. After many repetitions on that day,
both were supporting their weight for a few seconds, and after four days
of such practice were able to support their weight for several minutes
when thus balanced by the experimenters.

While the Dennises were chiefly interested in what they called the
"autogenous behavior" of the infants, of equal interest to the psychiatrist
is the degree of deferment of certain behavior responses under conditions
of prolonged deprivation of stimulus. Of additional interest is the fact
that the Dennises indicate that such delayed behavior could be brought

out readily as soon as stimuli and practice situations were amply provided.

There is a sort of sly humor of nature in the experiment, which was designed to frustrate, at least to deprive the babies. But in the emotional interreaction of the twins and the experimenters, it was the latter who in a way found themselves thwarted by the conditions of the experiment. So great was the twin's positive reaction to them that they found it almost impossible to get moving pictures of their "negative" reactions. As soon as they approached or even moved in arranging the camera, any crying or fussing would stop and smiles would supplant the negative attitude. This is certainly reassuring!

To summarize, it appears that it is not simple hampering of motion that provokes aggressive rage-like behavior in the young infant. Indeed, consistent, moderate and general restriction of motion may first quiet the infant, and intense and sudden restriction of motion with deep constriction, binding or jerking, seems to provoke negative responses. It seems possible that within the first year, though not immediately after birth, the infantile reaction may be increased by the angry or tense attitude of the person applying the restraint, and this may be of considerable clinical importance. The indication is that special types of restraint, plus an angry or tense restrainer, may provoke a strong emotional response leading to a reciprocal fear and then to anxiety. Depending on the duration and special degree of interference with expected activities, it further acts by direct limitation of practice. There is, however, no evidence that a permanent impairment of intellectual function itself is produced.

3. FOLK CUSTOMS

The folk customs which suggest themselves for study in connection with these problems of restraint are chiefly the various forms of swaddling. This practice occurs in different forms and degrees, especially according to whether part or all of the body is swaddled, as well as the time in the development of the child at which such restraint is applied. Dennis [7] has reviewed much of the literature on the subject, and the accounts here studied are taken largely from articles mentioned in his bibliography. The interest in this report, however, has a different point than that on which Dennis focused.

The practice of swaddling the infant from birth until six months or a year old has been common throughout the centuries and in widely different localities—the papoose of the American Indian, the bambino of the Italian, and Baby Bunting of our own folklore, being widely separated examples. Then there are the customs of binding parts of the body, notably the head and the feet, although there is no part, it seems, that entirely escapes this attention in some folk custom. Our interest is directed, however, only to those restraints which are applied soon after birth and continued for some time. In reviewing the literature certain general impressions stand out, viz., that the swaddling of infants has some practical utilitarian value; has commonly been used by hoi poloi in various groups; and that it is in its own way the antecedent to the cradle, the play pen, or the kiddie koop of today. In most instances the swaddling is not extremely tight, painful, or continuous. The infant is released from its confinement for several hours during the day. The swaddling, whether with or without a cradle board, serves as a kind of portable cradle in groups that do not move on wheels. In contrast to this, the head-binding and foot-binding which are painful and deforming, and the utilitarian value of which is not readily discernible, are generally marks of aristocratic distinction. Head-deforming was and still is carried on in many parts of the world, being commonest among the Indians of the northwest part of North America, especially the Columbia River region, and least common in Australia. It is said to have been practiced recently by the Haida and Chinook Indians in America, by certain tribes in Peru and on the Amazon, by the Kurds of Armenia, by certain Malay peoples in the Solomon Islands and the New Hebrides. The deformation was always done in infancy and often in both sexes. In some places it was reserved for boys, and sometimes, as in Tahiti, it was reserved for a single caste [8, 9, 11, 16]. The binding of aristocratic Chinese girls' feet continued until 1910. In this review, special attention is paid only to those reports which include some description of the infant's reaction to the swaddling or deformative binding.

Winifred de Kok [4] describes the swaddling of Italian children as an entirely happy proceeding, and concludes that the swaddling offers a comfortable support and protection to the newborn which, far from irritating, serves as a happy bridge across the abrupt birth passage and vaguely approximates the intra-uterine conditions until the baby is a little more prepared for the world. She speaks further of the ease with which the bundled baby may be handled, and of the complacency of the Italian baby in its modified strait-jacket. The papoose of the American Indian was generally "comfortably swaddled" on a cradle board with a round protruding protection for the head, for all the world like the hood on a baby carriage. The restraining jacket was laced over the baby and was comfortably padded within. (This seems undoubtedly to have a function similar to that of the restraining pack, whether wet or dry, often useful in quieting disturbed patients, as any hospital psychiatrist will recognize.) This same form of body swaddling was used by the Chinooks and other Columbia River tribes, but the special head deforming pressure was added.

A study of the swaddled infants of Albania offers a peculiarly interesting comparison with the Dennis experiment on the twins. In 1934, two Viennese psychology students, Lotte Danziger and Liselotte Frankl, published a report [3] of their painstaking study of Albanian infants in and near the little city of Kavaja. From their report, it seems that they approached their study with some of the same interests of the Dennises in setting up their experiment. Only the Viennese students wished to study the babies in their own special environment which included peculiar conditions of the infants being removed from most but not all stimulations. The Albanian children are swaddled from birth, and it is no halfway swaddling that is described either. The swaddling bands are applied tightly and the whole swaddled bundle further bound into a crude cradle. It is said that the infant generally screams when the swaddling is applied, but that the mother continually rocks the cradle until the crying gradually dies down.

The swaddling restraint is kept on almost continuously during the first year. Although the Mohammedan law prescribed that the baby should be bathed daily during the first year, the investigators did not seem to see many mothers bathing their infants. Feeding is often carried on without removing the child from its trappings. It is subjected to a further form of passive restraint in that whenever the mother is out, the cradle is put in one of the sleeping rooms which is commonly windowless, the only light coming through cracks in the roof. As soon as the mother comes home, however, she is wont to take the cradled baby out into the lighted living room where other adults and older children may be present. The investigators were not able to reach accurate conclusions as to whether the amount of time spent in the cradle varies at all with the age of the baby. Their description indicates, however, that the night and certainly the greater part of the day is spent in the dim back room. The baby is rarely taken out of doors during its first year. The babies are not played with or dandled. On the other hand, there is no absence of expressions of affection; parents and grandparents lean over the cradle, talk and smile to the baby. In this respect, the Albanian infants have the edge on the Dennis twins. It is noteworthy and unfortunate that neither the Dennises nor Danziger and Frankl say anything about the management of urination or defecation during these early months. To the psychiatrist and psychoanalyst, aware of how much these functions participate in the expression of both anxiety and aggression, and which may be affected by restraint, this is a rather glaring omission.

Danziger and Frankl compared their observations on the behavior of ten Albanian babies with similar observations on Viennese children. Their general conclusions resemble those of the Dennis experiment. All but one of their ten infants showed a distinct retardation in general development when compared with the Vienna group, but the difference was not so striking as might have been expected, especially when the amount of individual variation is taken into account. The babies showed distinctly less good motor activity, co-ordination, and general body mastery than the Vienna group. At six to ten months the babies sat up but did not crawl. When they reached for objects, the reaching was only approximate and not well directed. Even when not tightly swaddled, they showed little spontaneous movement during this first year, and then it was superficial "surface movement" which died out readily unless stimulation was continued. Three of the ten infants showed enough spontaneous movement to be worth calling it that, but at best it was weak and poorly directed. When at seven to eight months the child was offered some object, it would make movements indicating some general motion of grasping, but head movements would then occur. It would appear that the child felt drawn toward the object, but moved only diffusely and uncertainly toward it. It was noteworthy, however, that one of the babies, who at ten months showed a marked retardation in motor ability and made no attempt to slap two objects together or play with hollow blocks, seemed suddenly to "catch on" and in two hours of freedom seemed to have made up for about four months of retardation. It appeared that the latent development was there, but simply had not previously been called into definite form.

In social responsiveness and control, however, the Albanian children were better developed than the Viennese children of the same age group. Unfortunately the authors do not describe very clearly just what they mean under this heading. I presume it means indicating by facial expression or behavior the awareness of and obedience to other human beings. The authors attribute this difference to the fact that, although the babies spent so much time in the dark, when they were brought out, social-emotional contact was the one realm permitted to them. Curiously, the authors have very little to say specifically about sounds or speech. Perhaps this too is included in their social development. It may be that because they are deprived so much and so generally, the babies grow especially in the dimension in which they are permitted freedom, just as a tree seeks the sunlight. The situation again somewhat resembles the Dennis experiment, in which the concentrated spontaneous emotion of the developing babies finally dissolved the self-imposed masks of the experimenters.

Regarding the next three to four years of life of the Albanian infants, Danziger and Frankl state that the young children play little and have no toys, but are wont to stand around watching and listening to the activities of the grown-ups and are quickly drawn into participation in the work of the family. From the account of the investigators, there is no evidence whatsoever of any lasting intellectual impairment.

The head-molding carried on by the Columbia River Indians, especially the Chinooks, is also a restraint process starting at birth. Here the restraint is applied with such force as actually to deform the shape of the head. Swan [16] gives a detailed account of the process. The baby at birth is placed in a cradle made of a piece of cedar log, hollowed out and lined with softest cedar bark, then covered with soft skin and cloth. A pad is put at the baby's neck and a soft little pillow of wool and feathers, enclosing a flat stone, on the forehead. The body is laced into the soft wooden cradle with protective thongs, but the pad on the forehead is laced on very tightly and with increasing pressure, so that the head is unbelievably flattened by the end of the first year. Both Cox [2], a trader with the Indians along the Columbia River in 1814, and Paul Kane [13], an artist writing in 1925, give graphic accounts of the infant's appearance during the course of this flattening process. Kane says: "The eyes seem to start out of their sockets with the pressure"; and Cox: "The appearance of the infant while in this state of compression is frightful, and its little black eyes, forced out by the tightness of the bandages, resemble those of a mouse choked in a trap." Observers all agree that the flat-headed Indians, whose heads as a result of this treatment sometimes do not exceed an inch in thickness at the top, suffer no intellectual impairment and are no duller than those who keep their round heads. Incidentally, the latter are contemptuously called "rock-heads" (instead of blockheads), and the flat-heads are the more aristocratic.

Kane remarks that the infants seem quiet and placid when the head pressure is on, but, if for any reason it is removed, they cry loudly until the wrappings are replaced. Cox also mentions the placidity-under-pressure of the infants, but says nothing of any outcry. Kane offered the explanation that the pressure deadened the sensitivity of the child and only when it was removed did the child become aware of pain. This may be true. It may also be true that, since the pressure is gradually applied, the child gradually becomes accustomed to the discomfort and, in a sense, misses it when the pressure is suddenly released.

From a cursory review of the literature, it appears that swaddling and head binding are practiced immediately after birth, and that binding of the feet, amputation of the hands or fingers or other parts of the body seem to be done later than infancy. In one folk group, the grandmother gives up one digit for the birth of each grandson.

While the folk custom material is colorful, it might be considered too anecdotal for much specific interpretation. Still these practices, especially swaddling, have occurred seemingly independently in so many different parts of the world, that the question arises whether there are common factors in their production throughout. The impression emerges that swaddling and head binding actually do have some psychological relation to birth, as well as does the time relation—swaddling being some sort of reinstatement of intra-uterine conditions, and head binding a kind of prolonged caricature of the head molding which may occur spontaneously in a particularly difficult birth.

In the later mutilations (hand, foot binding, or amputation) a relation to castration for punishment or purification is suggested, while in the grandmother situation the mutilation would seem a combination of castration and rebirth. Throughout there is the dubious distinction of the economy of masochism which justifies itself only through the trait of endurance. However, it would take more anthropological background than is here available to understand these customs adequately. It is especially difficult to determine the exact heightening of the sado-masochistic elements of the character which would seem to be the inevitable result. One approaches the circular problem of the underlying sadistic aggression from which such frightful practices of restraint originate and then in turn contribute to the character of the new individual.

4. SUMMARY

It is apparent that the problem of the reaction to restraint in infancy is a difficult and complicated one. What the analyst sees in any stage after infancy is not only the original reaction, but the fusion of this with later experiences; quite often the interpretations belonging to these later periods reprojected back onto the early periods. Analytically, one must do a careful job indeed to dissect out what rightfully belongs to the various stages. Occasionally one has a chance to check this against the accounts of parents, early baby diaries, or hospital records, and to compare them with the direct observations on infants.

It must be recognized that *restraint* is not a simple term: that there is the positive restraint of binding and holding so as actually to limit motion; that there may be a negative sort of restraint through deprivation of situations for activity. Then there are the factors of the intensity (tight or loose swaddling), the suddenness or unevenness with which positive restraint is applied; whether it involves the whole or part of the body, and if it therefore promotes resignation or escape reactions. There is further the question of the relation of the type of restraint to the special era in the development of the child, e.g., the almost certain difference in reaction dependent on whether a baby's legs were tied down immediately after birth or at the time of the sixth month, when Dennis notes the pleasurable activity of extension of the extremities which seems to correspond with a stage in development when the baby would under normal conditions begin preliminary standing-up exercises by pushing against the mother's lap or against the floor if held in a near-standing position. Finally there is the important question not only of the infant's primary reaction to restraint, but of the augmentation of the emotional reaction by the reflection of the anxious or angry attitude of the restrainer, which must indeed be fairly frequent in the clinical cases which find their way into our clinics.

It seems, however, from the material reviewed as well as from a number of years of clinical experience, that even the reaction to severe prolonged and painful restraint does not produce any general intellectual impairment. It may temporarily retard development through limiting stimulus and practice situations, as in the case of the Dennis' twins. More important is the indication that prolonged restraint, with its accompanying frustration or submission, may be a factor in producing retardation of another kind, namely, a slowing of tempo. This in itself produces some impairment of efficiency of intellectual functioning, without destroying the intellect *per se*. Such slowing of tempo may also be a factor in chronic negativism, stubbornness, blocking, or lack of good and sustained concentration. In extremely severe cases it may produce a condition somewhat resembling the functional deterioration of chronic psychotic states. The relation of such retardation to the generation of anxiety is a problem beyond the scope of this paper. The slowing may itself be the result of the chronic anxiety associated with very strong aggressive urges, held in check and shown chiefly in the extreme ambivalence exhibited by such patients. The tendency for such early restraint to increase the sado-masochistic elements of the character is clear, but the exact pattern cannot well be established or even indicated from this material. In any event such a heightening of the sado-masochism would appear later combined with and canalized by subsequent life experiences. From a few observations on patients in the course of psychoanalytic treatment, I have thought that prolonged early restraint, producing a condition in which stimulations to the body far exceeded possible motor discharges, also resulted in an increased general body erotization and was a factor augmenting the problems generally associated with this condition.

REFERENCES

1. Bühler, C.: *The First Year of Life* (New York, John Day, 1930), p. 55.
2. Cox, R.: *Adventures on the Columbia River: A Residence of Six Years* (New York, Harper, 1832), p. 274.
3. Danziger, Lotte, and Frankl, Liselotte: "Zum Problem der Functionsreifung," *Z. für Kinderforschung*, Vol. 43 (1943), pp. 219–55.
4. De Kok, Winifred: *Guiding Your Child through Formative Years* (New York, Emerson Books, 1935), pp. 9–13.
5. Dennis, W.: "An Experimental Test of Two Theories of Social Smiling in Infants," *J. Soc. Psychol.*, Vol. 6 (1935), pp. 214–23.
6. Dennis, W.: "Infant Development under Conditions of Restricted Practice and Minimum Social Stimulation," *Gen. Psych. Monogs.*, Vol. 23 (1941), pp. 143–89.
7. Dennis, W.: "Infant Reaction to Restraint; an Evaluation of Watson's Theory," *Trans. N. Y. Acad. of Sci.*, Ser. II, Vol. 2, No. 8 (1940), pp. 202–18; this is a review of the material mentioned.
8. Dennis, W., and M. G.: "The Effect of Cradling Practices upon the Onset of Walking in Hopi Children," *J. Genet. Psychol.*, Vol. 56 (1940), pp. 77–80.
9. *Encyclopedia Britannica*, 11th and 14th editions, article "Mutilations."
10. Gesell, A., and Thompson, H.: *Infant Behavior* (New York, McGraw Hill, 1934) pp. 261, 282–3, 287–8.
11. Gunther, E.: "Klallam Ethnography," *Univ. of Wash. Pub. in Anthropology*, Vol. 1 (1920–27), p. 236.
12. Kaila, E.: "Über die Reaktionen des Sauglings auf das Menschliche Gesicht," *Psychol.*, Vol. 2 (1935), pp. 156–63.
13. Kane, Paul: *Wanderings of an Artist among the Indians of North America* (Toronto, The Radisson Society, 1925), p. 123.

14. Pratt, K. C., Nelson, A. K., and Sun, K. H.: "The Behavior of the Newborn Infant," *Ohio State Univ. Stud. Contrib. Psychol.*, No. 10 (1930), pp. 1 and 237.

15. Sherman, M., and I. C.: "Sensori-Motor Responses in Infants," *J. Comp. Psychol.*, Vol. 5 (1925), pp. 53–68.

16. Swan, J. G.: *The Northwest Coast* (New York, Harper, 1857).

17. Taylor, J. H.: "Innate Emotional Responses in Infants," *Ohio State Univ. Stud. Contrib. Psychol.*, No. 12 (1934), pp. 69–81.

18. Watson, John B.: *Psychology from the Standpoint of the Behaviorist* (Philadelphia, Lippincott, 1919), p. 200.

CHAPTER

32

Erich Fromm

INDIVIDUAL AND SOCIAL ORIGINS

OF NEUROSIS

With Dr. Fromm's paper we return to the problem of personality mal-formation that has already been discussed, especially in the papers by Kallman (p. 80) and Hunt (p. 456). Dr. Fromm points out that neu-rosis and character defects may arise equally from the situational experi-ence of the individual and from the cultural patterns of the group; he likewise recognizes the fact that these two sets of determinants influence each other. His statement of the need for more physiological and anthro-pological data, if we are to answer the problem of the origins of neurosis, implies an awareness of the constitutional determinants, though Fromm, like other neo-Freudians, tends somewhat to undervalue the biological factors in personality.

THE history of science is a history of erroneous statements. Yet these erroneous statements which mark the progress of thought have a particu-lar quality: they are productive. And they are not just *errors* either; they are statements, the truth of which is veiled by misconceptions, is clothed in erroneous and inadequate concepts. They are rational visions which contain the seed of truth, which matures and blossoms in the continuous

NOTE: Reprinted from the *American Sociological Review*, Vol. 9 (1944), pp. 380–4, by permission of the author and the American Sociological Society. (Copyright, 1944, by the American Sociological Society.)

effort of mankind to arrive at objectively valid knowledge about man
and nature. Many profound insights about man and society have first
found expression in myths and fairy tales, others in metaphysical speculations, others in scientific assumptions which have proven to be incorrect
after one or two generations.

It is not difficult to see why the evolution of human thought proceeds
in this way. The aim of any thinking human being is to arrive at the *whole*
truth, to understand the *totality* of phenomena which puzzle him. He has
only *one* short life and must want to have a vision of the truth about the
world in this short span of time. But he could only understand this totality
if his life span were identical with that of the human race. It is only in the
process of historical evolution that man develops techniques of observation, gains greater objectivity as an observer, collects new data which are
necessary to know if one is to understand the *whole*. There is a gap, then,
between what even the greatest genius can visualize as the truth, and the
limitations of knowledge which depend on the accident of the historical
phase he happens to live in. Since we cannot live in suspense, we try to
fill out this gap with the material of knowledge at hand, even if this material is lacking in the validity which the essence of the vision may have.

Every discovery which has been made and will be made has a long
history in which the truth contained in it finds a less and less veiled and
distorted expression and approaches more and more adequate formulations. The development of scientific thought is not one in which old statements are discarded as false and replaced by new and correct ones; it is
rather a process of continuous *reinterpretation* of older statements, by
which their true kernel is freed from distorting elements. The great pioneers of thought, of whom Freud is one, express ideas which determine
the progress of scientific thinking for centuries. Often the workers in the
field orient themselves in one of two ways; they fail to differentiate between the essential and the accidental, and defend rigidly the whole system of the master, thus blocking the process of reinterpretation and clarification; or they make the same mistake of failing to differentiate between
the essential and the accidental, and equally rigidly fight against the old
theories and try to replace them by new ones of their own. In both the
orthodox and the rebellious rigidity, the constructive evolution of the
vision of the master is blocked. The real task, however, is to reinterpret,
to sift out, to recognize that certain insights had to be phrased and understood in erroneous concepts because of the limitations of thought peculiar
to the historical phase in which they were first formulated. We may feel
then that we sometimes understand the author better than he understood
himself, but that we are only capable of doing so by the guiding light of
his original vision.

This general principle, that the way of scientific progress is *constructive reinterpretation of basic visions* rather than repeating or discarding
them, certainly holds true of Freud's theoretical formulations. There is
scarcely a discovery of Freud which does not contain fundamental truths

and yet which does not lend itself to an organic development beyond the concepts in which it has been clothed.

A case in point is Freud's theory on the origin of neurosis. I think we still know little of what constitutes a neurosis and less what its origins are. Many physiological, anthropological and sociological data will have to be collected before we can hope to arrive at any conclusive answer. What I shall do is to use Freud's view on the origin of neurosis as an illustration of the general principle which I have discussed, that reinterpretation is the constructive method of scientific progress.

Freud states that the *Oedipus complex* is justifiably regarded as the kernel of neurosis. I believe that this statement is the most fundamental one which can be made about the origin of neurosis, but I think it needs to be qualified and reinterpreted in a frame of reference different from the one Freud had in mind. What Freud meant in his statement was this: because of the sexual desire the little boy, let us say, has for his mother, he becomes the rival of his father, and the neurotic development consists in failure to cope with the anxiety rooted in this rivalry in a satisfactory way. I believe that Freud touched upon the most elementary root of neurosis in pointing to the conflict between the child and parental authority and the failure of the child to solve this conflict satisfactorily. But I do not think that this conflict is brought about essentially by the sexual rivalry, but that it results from the child's reaction to the pressure of parental authority, the child's fear of it and submission to it. Before I go on elaborating this point, I should like to differentiate between two kinds of authority. One is *objective*, based on the competency of the person in authority to function properly with respect to the task of guidance he has to perform. This kind of authority may be called *rational* authority. In contrast to it is what may be called *irrational* authority, which is based on the power which the authority has over those subjected to it and on the fear and awe with which the latter reciprocate.

It happens that, in most cultures, human relationships are greatly determined by irrational authority. People function in our society, as in most societies on the record of history, by becoming adjusted to their social role at the price of giving up part of their own will, their originality and spontaneity. While every human being represents the whole of mankind with all its potentialities, any functioning society is and has to be primarily interested in its self-preservation. The particular ways in which a society functions are determined by a number of *objective* economic and political factors, which are given at any point of historical development. Societies have to operate within the possibilities and limitations of their particular historical situation. In order that any society may function well, its members must acquire the kind of character which makes them *want* to act in the way they *have* to act as members of the society or of a special class within it. They have to *desire* what objectively is *necessary* for them to do. *Outer force* is to be replaced by *inner compulsion*, and by the particular kind of human energy which is channeled into character traits. As long as mankind has not attained a state of organization in which the interest of the individual and that of society are identical, the aims of society have to be attained at a greater or lesser expense of the freedom and spontaneity of the individual. This aim is performed by the process of child training and education. While education aims at the development of a child's potentialities, it has also the function of reducing his independence and freedom to the level necessary for the existence of that particular society. Although societies differ with regard to the extent to which the child must be impressed by irrational authority, it is always part of the function of child training to have this happen.

The child does not meet society directly at first; it meets it through the medium of his parents, who in their character structure and methods of education represent the social structure, who are the psychological agency of society, as it were. What, then, happens to the child in relationship to his parents? It meets through them the kind of authority which is prevailing in the particular society in which it lives, and this kind of authority tends to break his will, his spontaneity, his independence. But man is not born to be broken, so the child fights against the authority represented by his parents; he fights for his freedom not only *from* pressure but also for his freedom to be himself, a full-fledged human being, not an automaton. Some children are more successful than others; most of them are defeated to some extent in their fight for freedom. The ways in which this defeat is brought about are manifold, but whatever they are, the scars left from this defeat in the child's fight against irrational authority are to be found at the bottom of every neurosis. This scar is represented in a syndrome the most important features of which are: the weakening or paralysis of the person's originality and spontaneity; the weakening of the self and the substitution of a pseudo-self, in which the feeling of "I am" is dulled and replaced by the experience of self as the sum total of expectations others have about me; the substitution of autonomy by heteronomy; the fogginess, or, to use Dr. Sullivan's term, the "parataxic" quality of all interpersonal experiences.

My suggestion that the Oedipus complex be interpreted not as a result of the child's sexual rivalry with the parent of the same sex but as the child's fight with irrational authority represented by the parents does not imply, however, that the sexual factor does not play a significant role, but the emphasis is not on the incestuous wishes of the child and their necessarily tragic outcome, its original sin, but on the parents' prohibitive influence on the normal sexual activity of the child. The child's physical functions—first those of defecation, then his sexual desires and activities—are weighed down by moral considerations. The child is made to feel guilty with regard to these functions, and since the sexual urge is present in every person from childhood on, it becomes a constant source of the feeling of guilt. What is the function of this feeling of guilt? It serves to break the child's will and to drive it into submission. The parents use it, although unintentionally, as a means to make the child submit. There is nothing more effective in breaking any person than to give him the con· viction of wickedness. The more guilty one feels, the more easily one sub· mits because the authority has proven its own power by its right to accuse.

What appears as a feeling of guilt, then, is actually the fear of displeasing those of whom one is afraid. This feeling of guilt is the only one which most people experience as a moral problem, while the genuine moral problem, that of realizing one's potentialities, is lost from sight. Guilt is reduced to disobedience and is not felt as that which it is in a genuine moral sense, self-mutilation.

To sum up this point, it may be said that it is the defeat in the fight against authority which constitutes the kernel of the neurosis, and that not the incestuous wish of the child but the stigma connected with sex is one among the factors in breaking down his will. Freud painted a picture of the necessarily *tragic* outcome of a child's most fundamental wishes: his incestuous wishes are bound to fail and force the child into some sort of submission. Have we not reason to assume that this hypothesis expresses in a veiled way Freud's profound pessimism with regard to any basic improvement in man's fate and his belief in the indispensable nature of irrational authority? Yet this attitude is only one part of Freud. He is at the same time the man who said that "from the time of puberty onward the human individual must devote himself to the great task of freeing himself from the parents"; he is the man who devised a therapeutic method the aim of which is the independence and freedom of the individual.

However, defeat in the fight for freedom does not always lead to neurosis. As a matter of fact, if this were the case, we would have to consider the vast majority of people as neurotics. What, then, are the specific conditions which make for the neurotic outcome of this defeat? There are some conditions which I can only mention: for example, one child may be broken more thoroughly than others, and the conflict between his anxiety and his basic human desires may, therefore, be sharper and more unbearable; or the child may have developed a sense of freedom and originality which is greater than that of the average person, and the defeat may thus be more unacceptable. But instead of enumerating other conditions which make for neurosis, I prefer to reverse the question and ask what the conditions are which are responsible for the fact that so many people do *not* become neurotic in spite of the failure in their personal fight for freedom. It seems to be useful at this point to differentiate between two concepts: that of defect and that of neurosis. If a person fails to attain freedom, spontaneity, a genuine experience of self, he may be considered to have a severe defect, provided we assume that freedom and spontaneity are the objective goals to be attained by every human being. If such a goal is not attained by the majority of members of any given society, we deal with the phenomenon of *socially patterned defect*. The individual shares it with many others; he is not aware of it as a defect, and his security is not threatened by the experience of being different, of being an outcast, as it were. What he may have lost in richness and in a genuine feeling of happiness is made up by the security of fitting in with the rest of mankind—*as he knows them.* As a matter of fact, his very defect may have been raised to a virtue by his culture and thus give him an enhanced feeling of achievement. An illustration is the feeling of guilt and anxiety which Calvin's doctrines aroused in men. It may be said that the person who is overwhelmed by a feeling of his own powerlessness and unworthiness, by the unceasing doubt of whether he is saved or condemned to eternal punishment, who is hardly capable of any genuine joy and has made himself into the cog of a machine which he has to serve, has a severe defect. Yet this very defect was culturally patterned; it was looked upon as particularly valuable, and the individual was thus protected from the neurosis which he would have acquired in a culture where the defect would give him a feeling of profound inadequacy and isolation.

Spinoza has formulated the problem of the socially patterned defect very clearly. He says: "Many people are seized by one and the same affect with great consistency. All his senses are so strongly affected by one object that he believes this object to be present even if it is not. If this happens while the person is awake, the person is believed to be insane. . . . But if the *greedy* person thinks only of money and possessions, the *ambitious* one only of fame, one does not think of them as being insane, but only as annoying; generally one has contempt for them. But *factually*, greediness, ambition, and so forth are forms of insanity, although usually one does not think of them as 'illness.' " These words were written a few hundred years ago; they still hold true, although the defect has been culturally patterned to *such* an extent now that it is not generally thought any more to be annoying or contemptuous. Today we come across a person and find that he acts and feels like an automaton; that he never experiences anything which is really his; that he experiences himself entirely as the person he thinks he is supposed to be; that smiles have replaced laughter, meaningless chatter replaced communicative speech; dulled despair has taken the place of genuine pain. Two statements can be made about this person. One is that he suffers from a defect of spontaneity and individuality which may seem incurable. At the same time it may be said that he does not differ essentially from thousands of others who are in the same position. With *most* of them, the cultural pattern provided for the defect saves them from the outbreak of neurosis. With *some*, the cultural pattern does not function, and the defect appears as a severe neurosis. The fact that in these cases the cultural pattern does not suffice to prevent the outbreak of a manifest neurosis is in most cases to be explained by the particular severity and structure of the individual conflicts. I shall not go into this any further. The point I want to stress is the necessity to proceed from the problem of the *origins of neurosis* to the problem of the *origins of the culturally patterned* defect; to the problem of the *pathology of normalcy*.

This aim implies that the psychoanalyst is not only concerned with the readjustment of the neurotic individual to his given society. His task must be also to recognize that the individual's ideal of normalcy may contradict the aim of the full realization of himself as a human being. It is the belief of the progressive forces in society that such a realization is possible, that the interest of society and of the individual need not be antagonistic forever. Psychoanalysis, if it does not lose sight of the human problem, has an important contribution to make in this direction. This contribution, by which it transcends the field of a medical specialty, was part of the vision which Freud had.

CHAPTER

33

Ruth Benedict

CONTINUITIES AND DISCONTINUITIES
IN CULTURAL CONDITIONING

Dr. Benedict further develops the basic themes of Mead (p. 115) and Frank (p. 119). She reminds us that the culture of every group must have patterns for dealing with inescapable biological facts: the helplessness of infancy, the differences between the sexes, the existence in every society of persons in various age groups. These patterns have special relevance to the definition of age, sex, and other roles. This paper thus deals with the interrelations between biological, group-membership, and role determinants of personality. Dr. Benedict makes the most fundamental point of the cultural anthropologist when she insists that the anthropologist's task ". . . is not to question the facts of nature but to insist upon the inter-position of a middle term between 'nature' and human behavior. . . ." This paper, like the preceding one, is a contribution to the problem of understanding mental disorders. Dr. Benedict shows that there is no single "natural" path to maturity, that discontinuities in cultural conditioning as well as physiological factors may contribute to maladjustment and personality upheavals.

ALL cultures must deal in one way or another with the cycle of growth from infancy to adulthood. Nature has posed the situation dramatically:

NOTE: Reprinted from *Psychiatry*, Vol. 1 (1938), pp. 161–7, by permission of the author and the William Alanson White Psychiatric Foundation. (Copyright, 1938, by the William Alanson White Psychiatric Foundation, Inc.)

on the one hand, the new born baby, physiologically vulnerable, unable to fend for itself, or to participate of its own initiative in the life of the group, and, on the other, the adult man or woman. Every man who rounds out his human potentialities must have been a son first and a father later, and the two roles are physiologically in great contrast; he must first have been dependent upon others for his very existence, and later he must provide such security for others. This discontinuity in the life cycle is a fact of nature and is inescapable. Facts of nature, however, in any discussion of human problems, are ordinarily read off not at their bare minimal but surrounded by all the local accretions of behavior to which the student of human affairs has become accustomed in his own culture. For that reason, it is illuminating to examine comparative material from other societies in order to get a wider perspective on our own special accretions. The anthropologist's role is not to question the facts of nature, but to insist upon the interposition of a middle term between "nature" and "human behavior"; his role is to analyze that term, to document local man-made doctorings of nature, and to insist that these doctorings should not be read off in any one culture as nature itself. Although it is a fact of nature that the child becomes a man, the way in which this transition is effected varies from one society to another, and no one of these particular cultural bridges should be regarded as the "natural" path to maturity.

From a comparative point of view, our culture goes to great extremes in emphasizing contrasts between the child and the adult. The child is sexless, the adult estimates his virility by his sexual activities; the child must be protected from the ugly facts of life, the adult must meet them without psychic catastrophe; the child must obey, the adult must command this obedience. These are all dogmas of our culture, dogmas which, in spite of the facts of nature, other cultures commonly do not share. In spite of the physiological contrasts between child and adult, these are cultural accretions.

It will make the point clearer if we consider one habit in our own culture in regard to which there is not this discontinuity of conditioning. With the greatest clarity of purpose and economy of training, we achieve our goal of conditioning everyone to eat three meals a day. The baby's training in regular food periods begins at birth, and no crying of the child and no inconvenience to the mother is allowed to interfere. We gauge the child's physiological make-up and at first allow it food oftener than adults, but, because our goal is firmly set and our training consistent, before the child is two years old it has achieved the adult schedule. From the point of view of other cultures, this is as startling as the fact of three-year-old babies perfectly at home in deep water is to us. Modesty is another sphere in which our child training is consistent and economical; we waste no time in clothing the baby, and, in contrast to many societies where the child runs naked till it is ceremonially given its skirt or its pubic sheath at adolescence, the child's training fits it precisely for adult conventions.

In neither of these aspects of behavior is there need for an individual in our culture to embark before puberty, at puberty, or at some later date upon a course of action which all his previous training has tabooed. He is spared the unsureness inevitable in such a transition.

The illustration I have chosen may appear trivial, but, in larger and more important aspects of behavior, our methods are obviously different. Because of the great variety of child training in different families in our society, I might illustrate continuity of conditioning from individual life histories in our culture, but even these, from a comparative point of view, stop far short of consistency; and I shall, therefore, confine myself to describing arrangements in other cultures in which training which with us is idiosyncratic is accepted and traditional and does not, therefore, involve the same possibility of conflict. I shall choose childhood rather than infant and nursing situations, not because the latter do not vary strikingly in different cultures, but because they are nevertheless more circumscribed by the baby's physiological needs than is its later training. Childhood situations provide an excellent field in which to illustrate the range of cultural adjustments which are possible within a universally given, but not so drastic, set of physiological facts.

The major discontinuity in the life cycle is of course that the child who is at one point a son must later be a father. These roles in our society are strongly differentiated; a good son is tractable, and does not assume adult responsibilities; a good father provides for his children and should not allow his authority to be flouted. In addition, the child must be sexless so far as his family is concerned, whereas the father's sexual role is primary in the family. The individual in one role must revise his behavior from almost all points of view when he assumes the second role.

I shall select for discussion three such contrasts that occur in our culture between the individual's role as child and as father: (1) responsible–non-responsible status role; (2) dominance–submission; (3) contrasted sexual role. It is largely upon our cultural commitments to these three contrasts that the discontinuity in the life cycle of an individual in our culture depends.

1. RESPONSIBLE–NON-RESPONSIBLE STATUS ROLE

The techniques adopted by societies which achieve continuity during the life cycle in this sphere in no way differ from those we employ in our uniform conditioning to three meals a day. They are merely applied to other areas of life. We think of the child as wanting to play and the adult as having to work, but in many societies the mother takes the baby daily in her shawl or carrying net to the garden or to gather roots, and adult labor is seen even in infancy from the pleasant security of its position in close contact with its mother. When the child can run about, it accompanies its parents still, doing tasks which are essential and yet suited to its powers, and its dichotomy between work and play is not different from that its parents recognize, namely the distinction between the busy day and the free evening. The tasks it is asked to perform are graded to its powers, and its elders wait quietly by, not offering to do the task in the child's place. Everyone who is familiar with such societies has been struck by the contrast with our child training. Dr. Ruth Underhill tells me of sitting with a group of Papago elders in Arizona when the man of the house turned to his little three-year-old grand-daughter and asked her to close the door. The door was heavy and hard to shut. The child tried, but it did not move. Several times the grandfather repeated: "Yes, close the door." No one jumped to the child's assistance. No one took the responsibility away from her. On the other hand there was no impatience, for after all the child was small. They sat gravely waiting till the child succeeded and her grandfather gravely thanked her. It was assumed that the task would not be asked of her unless she could perform it, and, having been asked, the responsibility was hers alone just as if she were a grown woman.

The essential point of such child training is that the child is from infancy continuously conditioned to responsible social participation, while at the same time the tasks that are expected of it are adapted to its capacity. The contrast with our society is very great. A child does not make any labor contribution to our industrial society except as it competes with an adult; its work is not measured against its own strength and skill but against high-geared industrial requirements. Even when we praise a child's achievement in the home, we are outraged if such praise is interpreted as being of the same order as praise of adults. The child is praised because the parent feels well disposed, regardless of whether the task is well done by adult standards, and the child acquires no sensible standard by which to measure its achievement. The gravity of a Cheyenne Indian family ceremoniously making a feast out of the little boy's first snowbird is at the furthest remove from our behavior. At birth the little boy was presented with a toy bow, and from the time he could run about serviceable bows suited to his stature were specially made for him by the man of the family. Animals and birds were taught him in a graded series beginning with those most easily taken, and as he brought in his first of each species his family duly made a feast of it, accepting his contribution as gravely as the buffalo his father brought. When he finally killed a buffalo, it was only the final step of his childhood conditioning, not a new adult role with which his childhood experience had been at variance.

The Canadian Ojibwa show clearly what results can be achieved. This tribe gains its livelihood by winter trapping, and the small family of father, mother, and children live during the long winter alone on their great frozen hunting grounds. The boy accompanies his father and brings in his catch to his sister as his father does to his mother; the girl prepares the meat and skins for him just as his mother does for her husband. By the time the boy is twelve, he may have set his own line of traps on a hunting territory of his own and return to his parent's house only once in several months—still bringing the meat and skins to his sister. The young child is taught consistently that it has only itself to rely upon in life, and this is as true in the dealings it will have with the supernatural as in the business of getting a livelihood. This attitude he will accept as a successful adult just as he accepted it as a child.

2 . DOMINANCE—SUBMISSION

Dominance–submission is the most striking of those categories of behavior where like does not respond to like, but where one type of behavior stimulates the opposite response. It is one of the most prominent ways in which behavior is patterned in our culture. When it obtains between classes, it may be nourished by continuous experience; the difficulty in its use between children and adults lies in the fact that an individual conditioned to one set of behavior in childhood must adopt the opposite as an adult. Its opposite is a pattern of approximately identical reciprocal behavior; the societies which rely upon continuous conditioning characteristically invoke this pattern. In some primitive cultures, the very terminology of address between father and son, and, more commonly, between grandchild and grandson or uncle and nephew, reflects this attitude. In such kinship terminologies, one reciprocal expresses each of these relationships so that son and father, for instance, exchange the same term with one another, just as we exchange the same term with a cousin. The child later will exchange it with his son. "Father–son," therefore, is a continuous relationship he enjoys throughout life. The same continuity, backed up by verbal reciprocity, occurs far oftener in the grandchild–grandson relationship or that of mother's brother–sister's son. When these are "joking" relationships, as they often are, travellers report wonderingly upon the liberties and pretensions of tiny toddlers in their dealing with these family elders. In place of our dogma of respect to elders, such societies employ in these cases a reciprocity as nearly identical as may be. The teasing and practical joking the grandfather visits upon his grandchild, the grandchild returns in like coin; he would be led to believe that he failed in propriety if he did not give like for like. If the sister's son has right of access without leave to his mother's brother's possessions, the mother's brother has such rights also to the child's possessions. They share reciprocal privileges and obligations which in our society can develop only between age mates.

From the point of view of our present discussion, such kinship conventions allow the child to put in practice from infancy the same forms of behavior which it will rely upon as an adult; behavior is not polarized into a general requirement of submission for the child and dominance for the adult.

It is clear from the techniques described above, by which the child is conditioned to a responsible status role, that these depend chiefly upon arousing in the child the desire to share responsibility in adult life. To achieve this, little stress is laid upon obedience but much stress upon approval and praise. Punishment is very commonly regarded as quite outside the realm of possibility, and natives in many parts of the world have drawn the conclusion from our usual disciplinary methods that white parents do not love their children. If the child is not required to be submissive, however, many occasions for punishment melt away; a variety of situations which call for it do not occur. Many American Indian tribes are especially explicit in rejecting the ideal of a child's submissive or obedient behavior. Prince Maximilian von Wied, who visited the Crow Indians over a hundred years ago, describes a father's boasting about his young son's intractibility even when it was the father himself who was flouted; "He will be a man," his father said. He would have been baffled at the idea that his child should show behavior which would obviously make him appear a poor creature in the eyes of his fellows if he used it as an adult. Dr. George Devereux tells me of a special case of such an attitude among the Mohave at the present time. The child's mother was white and protested to its father that he must take action when the child disobeyed and struck him. "But why?" the father said, "he is little. He cannot possibly injure me." He did not know of any dichotomy according to which an adult expects obedience and a child must accord it. If his child had been docile, he would simply have judged that it would become a docile adult—an eventuality of which he would not have approved.

Child training which brings about the same result is common also in other areas of life than that of reciprocal kinship obligations between child and adult. There is a tendency in our culture to regard every situation as having in it the seeds of a dominance–submission relationship. Even where dominance–submission is patently irrelevant we read in the dichotomy, assuming that in every situation there must be one person dominating another. On the other hand some cultures, even when the situation calls for leadership do not see it in terms of dominance–submission. To do justice to this attitude, it would be necessary to describe their political and especially their economic arrangements, for such an attitude to persist must certainly be supported by economic mechanisms that are congruent with it. But it must also be supported by—or what comes to the same thing, express itself in—child training and familial situations.

3. CONTRASTED SEXUAL ROLE

Continuity of conditioning in training the child to assume responsibilty and to behave no more submissively than adults is quite possible in terms of the child's physiological endowment if his participation is suited to his strength. Because of the late development of the child's reproductive organs, continuity of conditioning in sex experience presents a difficult problem. So far as their belief that the child is anything but a sexless being is concerned, they are probably more nearly right than we are with an opposite dogma. But the great break is presented by the universally sterile unions before puberty and the presumably fertile ones after maturation. This physiological fact no amount of cultural manipulation can minimize or alter, and societies, therefore, which stress continuous conditioning most strongly sometimes do not expect children to be interested in sex experience until they have matured physically. This is striking among American Indian tribes like the Dakota; adults observe great privacy in sex acts and in no way stimulate children's sexual activity. There need be no discontinuity, in the sense in which I have used the term, in such a program if the child is taught nothing it does not have to unlearn later. In such cultures, adults view children's experimentation as in no way wicked or dangerous, but merely as innocuous play which can have no serious consequences. In some societies such play is minimal and the children manifest little interest in it. But the same attitude may be taken by adults in societies where such play is encouraged and forms a major activity among small children. This is true among most of the Melanesian cultures of Southeast New Guinea; adults go as far as to laugh off sexual affairs within the prohibited class, if the children are not mature, saying that since they cannot marry there can be no harm done.

It is this physiological fact of the difference between children's sterile unions and adults' presumably fertile sex relations which must be kept in mind in order to understand the different mores which almost always govern sex expression in children and in adults in the same culture. A great many cultures with pre-adolescent sexual license require marital fidelity, and a great many which value pre-marital virginity in either male or female arrange their marital life with great license. Continuity in sex experience is complicated by factors which it was unnecessary to consider in the problems previously discussed. The essential problem is not whether or not the child's sexuality is consistently exploited—for even where such exploitation is favored, in the majority of cases the child must seriously modify his behavior at puberty or at marriage. Continuity in sex expression means rather that the child is taught nothing it must unlearn later. If the cultural emphasis is upon sexual pleasure, the child who is continuously conditioned will be encouraged to experiment freely and pleasurably, as among the Marquesans; if emphasis is upon reproduction, as among the Zuñi of New Mexico, childish sex proclivities will not be exploited, for the only important use which sex is thought to serve in his culture is not yet possible to him. The important contrast with our child training is that, although a Zuñi child is impressed with the wickedness of premature sex experimentation, he does not run the risk as in our culture of associating this wickedness with sex itself rather than with sex at his age. The adult in our culture has often failed to unlearn the wickedness or the dangerousness of sex, a lesson which was impressed upon him strongly in his most formative years.

4. DISCONTINUITY IN CONDITIONING

Even from this very summary statement of continuous conditioning, the economy of such mores is evident. In spite of the obvious advantages, however, there are difficulties in its way. Many primitive societies expect as different behavior from an individual as child and as adult as we do, and such discontinuity involves a presumption of strain.

Many societies of this type, however, minimize strain by the techniques they employ; and some techniques are more successful than others in ensuring the individual's functioning without conflict. It is from this point of view that age-grade societies reveal their fundamental significance. Age-graded cultures characteristically demand different behavior of the individual at different times of his life and persons of a like age-grade are grouped into a society whose activities are all oriented toward the behavior desired at that age. Individuals "graduate" publicly and with honor from one of these groups to another. Where age society members are enjoined to loyalty and mutual support, and are drawn not only from the local group but from the whole tribe, as among the Arapaho, or even from other tribes as among the Wagawaga of Southeast New Guinea, such an institution has many advantages in eliminating conflicts among local groups and fostering intratribal peace. This seems to be also a factor in the tribal military solidarity of the similarly organized Masai of East Africa. The point that is of chief interest for our present discussion, however, is that by this means an individual who at any time takes on a new set of duties and virtues is supported not only by a solid phalanx of age mates but by the traditional prestige of the organized "secret" society into which he has now graduated. Fortified in this way, individuals in such cultures often swing between remarkable extremes of opposite behavior without apparent psychic threat. For example, the great majority exhibit prideful and non-conflicted behavior at each stage in the life cycle, even when a prime of life devoted to passionate and aggressive head hunting must be followed by a later life dedicated to ritual and to mild and peaceable civic virtues.

Our chief interest here, however, is in discontinuity which primarily affects the child. In many primitive societies, such discontinuity has been fostered not because of economic or political necessity or because such discontinuity provides for a socially valuable division of labor, but because of some conceptual dogma. The most striking of these are the Australian and Papuan cultures where the ceremony of the "Making of Man" flourishes. In such societies it is believed that men and women have opposite and conflicting powers, and male children, who are of undefined status, must be initiated into the male role. In Central Australia the boy child is of the woman's side, and women are taboo in the final adult stages of tribal ritual. The elaborate and protracted initiation ceremonies of the Arunta, therefore, snatch the boy from the mother, dramatize his gradual repudiation of her. In a final ceremony he is reborn as a man out of the

men's ceremonial "baby pouch." The men's ceremonies are ritual statements of a masculine solidarity, carried out by fondling one another's
churingas, the material symbol of each man's life, and by letting out over
one another blood drawn from their veins. After this warm bond among
men has been established through the ceremonies, the boy joins the men
in the men's house and participates in tribal rites. The enjoined discontinuity has been tribally bridged.

West of the Fly River in southern New Guinea, there is a striking
development of this Making of Men cult which involves a childhood
period of passive homosexuality. Among the Keraki it is thought that no
boy can grow to full stature without playing the role for some years. Men
slightly older take the active role, and the older man is a jealous partner.
The life cycle of the Keraki Indians includes, therefore, in succession,
passive homosexuality, active homosexuality, and heterosexuality. The
Keraki believe that pregnancy will result from post-pubertal passive
homosexuality and see evidences of such practices in any fat man, whom,
even as an old man, they may kill or drive out of the tribe because of their
fear. The ceremony that is of interest in connection with the present discussion takes place at the end of the period of passive homosexuality. This
ceremony consists in burning out the possibility of pregnancy from the
boy by pouring lye down his throat, after which he has no further protection if he gives way to the practice. There is no technique for ending
active homosexuality, but this is not explicitly taboo for older men;
heterosexuality and children, however, are highly valued. Unlike the
neighboring Marindanim, who share their homosexual practices, Keraki
husband and wife share the same house and work together in the gardens.

I have chosen illustrations of discontinuous conditioning where it is
not too much to say that the cultural institutions furnish adequate support
to the individual as he progresses from role to role or interdicts the
previous behavior in a summary fashion. The contrast with arrangements
in our culture is very striking, and against this background of social arrangements in other cultures the adolescent period of *Sturm und Drang*
with which we are so familiar becomes intelligible in terms of our discontinuous cultural institutions and dogmas rather than in terms of physiological necessity. It is even more pertinent to consider these comparative
facts in relation to maladjusted persons in our culture who are said to be
fixated at one or another pre-adult level. It is clear that if we were to look
at our social arrangements as an outsider, we should infer directly from
our family institutions and habits of child training that many individuals
would not "put off childish things"; we should have to say that our adult
activity demands traits that are interdicted in children, and that, far from
redoubling efforts to help children bridge this gap, adults in our culture
put all the blame on the child when he fails to manifest spontaneously the
new behavior or, overstepping the mark, manifests it with untoward bel-
ligerence. It is not surprising that in such a society many individuals fear
to use behavior which has up to that time been under a ban and trust in-

stead, though at great psychic cost, to attitudes that have been exercised with approval during their formative years. Insofar as we invoke a physiological scheme to account for these neurotic adjustments, we are led to overlook the possibility of developing social institutions which would lessen the social cost we now pay; instead, we elaborate a set of dogmas which prove inapplicable under other social conditions.

CHAPTER

34

John Dollard

THE LIFE HISTORY IN
COMMUNITY STUDIES

Professor Dollard's article deals, in the context of a case history, with the interconnections between group-membership, role, and situational determinants. One is made to see how the subject's difficulties arose out of the pressures of a particular family situation. But it is made equally clear that family attitudes and behaviors were molded by the attempts of father and mother to live up to societal and class patterns and to train their son to carry out culturally defined roles. The actual environment which pressed upon the maturing organism was a special variant of a matrix characteristic in a very general way of American society in the early twentieth century and more particularly of one class group situated in an industrial town with an ethnic population.

THE record from which these data are drawn was made on a male individual who was thirty years old at the time of the study. It represents data from the first seven years of his life, notably years four, five, six, and seven. The record then relates to a small American town between the years 1905 and 1909, since the study was made in 1931. The individual studied exhibited at the outset of the study serious deformations in his

NOTE: Reprinted, with slight abridgment, from the *American Sociological Review*, Vol. 3 (1938), pp. 724-37, by permission of the author and the American Sociological Society. (Copyright, 1938, by the American Sociological Society.)

social relations, such as acute mistrust of other people, inability to work efficiently, and a low level of satisfaction in his married life. These problems and their solution will not be reported upon, but the attempt to solve them provided adequate motivation for the acquisition of a very detailed report of the person's life along its whole length. Some of his early memories will be utilized for the following analysis.[1]

The informant remembered his town as a long, narrow one, strung along the bank of a river. On one side were the shoe factories, but no people lived over there; on his side of the town was the business nucleus which was actually on the bank of the river with the residential area flung around it. He lived at the edge of town, and he remembered that, in doing errands for his mother, it took him about fifteen minutes to walk to a store in the business section. It was, on the other hand, five minutes to open country from his house. He had a clear picture of the streaming throngs of workmen going across the bridge to the factories, of the low mill buildings and of the water tumbling over the several dams. The factory whistles blew six days a week at seven in the morning and six at night, and he remembered his parents setting the clocks by them. On Saturdays, he recalled catching rides on the backs of farmers' wagons loaded with produce for the downtown stores. The same wagons went back on Saturday afternoon, taking supplies home.

This is obviously, then, a small industrial town in a rural area. Both parents came from farms in the area surrounding the town, which would suggest the possibility that the native-born population is recruited for city life from this local area.

Though the informant was not a social analyst, his data provided immediate definition of the family form into which he came. He had the conventional recollections of being threatened with his father's authority for misdeeds at home. Father, he learned, was a very fine man but also a severe one to unruly children. He worked very hard and earned all the money that the family got. Therefore, to spend money carelessly or to throw a stone through a neighbor's window was really robbing father. It was clear to informant that mother was dependent on father, and, although she was the real boss in the house, she always pretended to defer to him and probably was at bottom afraid of him. In domestic arguments, mother might end up by crying, whereas father never cried; he only

---

[1] In the introduction to this paper Professor Dollard discusses the possibilities of using intensive life history materials for what they may reveal about group processes. The data have been extracted from a voluminous psychoanalytic case record; they are perhaps a two hundredth part of the information available on this individual. "Here, the life history is offered as an aid in picking out factors of relevance in community life. It reveals not only the processes familiar from studies on the societal level of perception but also other factors which cannot be sensed on this level. . . . A good deal of what is to come has the character of being obvious from the sociological standpoint; its only interest is in the fact that the data can be appropriately classified from the sociological standpoint; and that this can be done with the materials from a supposedly remote discipline like psychoanalysis. . . ."

glowered if his ultimate control was challenged. There was no one around to dispute father's authority—no grandfather, no uncle. Father did have a boss somewhere, but that did not mean much in the home.

The informant experienced a good deal of resentment toward both of his parents, but particularly toward his mother. She was busy with the other children, three of them, and the oldest son had to take his place in line; as more children came it seemed that the line in front of him grew steadily longer. He craved for companionship with his father and that the father would take him as a real confidant, but this never happened; the father seemed too busy and too preoccupied with his own work. When the boy brought a lunch pail to his father, as he did every day, he always hoped to be allowed to handle his father's tools and to have their use explained to him, but this again did not happen. It seemed that he only came to his father's attention when disciplinary matters were involved. Otherwise, life was too strenuous for the "old man" to be concerned about any but the youngest child during its baby period, or a little longer if it was a girl. His mother was also busy. As a usual thing, there was no help in the house; and beside childbearing there was the complete routine of family duties to be done—washing one day, baking another day, cleaning another day, and the like. The mother seemed too harried and heckled by her life to spare love for the older children. On the contrary, she was likely to stress her burdensome position and to use it as an argument in demanding absolute obedience from the informant and his brothers and sister. The informant felt, therefore, very much alone and somewhat neglected in the family circle, and, for this reason, welcomed with extreme eagerness his first play-group contacts.

A partial rejection by his parents, as described, resulted in a considerable animosity toward them which could not be expressed, but toward the brothers and sister who were in part the cause of the parental neglect he could and did feel extremely vindictive. They had the privileges of childhood which he had begun to lose. They were preferred and, so it seemed, protected by the parents. However, they were relatively defenseless against him, especially if he could find ways of abusing them which were not detectable, and he showed a devious defiance toward his father by punishing his siblings. He invented names for them, referring to their babyishness and immaturity, made up songs in which these derisive names were used, and made a number of excursions into actual physical attack. The latter was always put down by the parents but the less tangible derision was difficult to catch.

The longing for tenderness from his parents, a tenderness which would have comforted him for the many bad intentions he had toward them and the disobedient things which he actually did, accumulated on a particular Christmas in his fifth year. He had desperately expected some mark of preferment in the type of Christmas presents which he would receive. He had exploded, of course, the Santa Claus legend by this time, since he had found the presents before Christmas and overheard his parents putting labels on them and decorating the tree. This deceit on the part of his parents gave him some courage for the many deceits which he was practicing and later had to practice himself. Christmas morning came, and the presents were just ordinary presents, not much better than those of the other children; perhaps not as good in some ways. This event ended the struggle for rapport with his father and mother. All right, then, it was every man for himself, and one got affection wherever one could get it and expected nothing more for nothing.

From the foregoing, the outline of the roles of this father and mother is fairly clear. We see the isolated, small family unit of the patriarchal type, with formal authority lodged in the father. There is no assertion that this family form was typical for the community. There must have been a type of family, however, which did its "duty" toward children, was perhaps too busy to do more than this bare duty, too busy to give its children the affectionate contacts which they needed. Possibly this family type develops strong and ruthless personalities in the most deprived children just because they must turn away from family sources of security at an early age.

Clear in the informant's account of his family was the imposition of the incest taboo. He remembered that occasion when he was still sleeping with his little sister, his next sibling. He remembered waking up in the morning, and the rest is unclear. All he knew is that he felt something toward her, and that he was teasing her or that he asked her if he could do something or see something. At any rate, he crawled over on top of her, though he did not exactly know what he wanted to do. She screamed and the mother came running. The sister "told" on him, though he tried to "shush" her. The mother threatened to tell the father when he came home, and a dreadful fear settled on the informant, but he did not know of what. Nothing seemed to have happened in fact, but the fear persisted, and with it persisted an animosity toward the sister. Undoubtedly this is the psychological correlate of the incest taboo. In this connection, the informant remembered, also, that when his father went away, he used to say, "Take care of your mother," and there seemed to be a warning lurking in his voice. He was warning his son to be obedient and docile toward his mother, but the informant took the words to mean something more. Did his father know of those strange, strong feelings which the boy had when sometimes he was still caressed by his mother? What would he do, if he did know? At any rate, the informant understood that he must be very tender toward his mother, must never make any aggressive gestures toward her, and must behave like a good boy. Apart from the atmosphere of the house, which was a restraining one, and the general taboo on sexual thoughts and feelings, which was indicated by complete avoidance of reference to them, these were the only data derived directly on the question of the subject's erotic feelings toward family members.

If the informant's feelings toward mother and sister were vague and confused, he does remember strong, erotic wishes toward a maid who was in the house for a few weeks at the time of his mother's fourth pregnancy, that is, when he was six years old. He remembered wanting to go in and sleep with this maid; he wanted to see her dress and undress, and in particular he wanted to see her breasts. He was discouraged from the attempt to see her, however, by a story which his mother told about a woman who caught a man peeking through a keyhole at her and shoved a hat pin into the keyhole and into his eye. Nevertheless, on one occasion he did get up his courage to approach the maid. She was standing at the telephone (certainly a mark of high working class status for this family) and the informant reached under her dress and grabbed her by the leg. She hit him a blow on the head which sent him spinning and ended his approaches to her. She too, apparently, was within the circle of the forbidden. He felt this blow on the head, in memory, many times later when he tried to make contact with women.

The informant learned another fact about the behavior of his family. He noticed that his parents would stop a quarrel if the maid were to appear, and that he was often reproved especially for quarreling with his brothers and sister if strangers were in the house. He learned that a family puts up an appearance of solidarity before the outside world whether it actually feels it or not and that in-family aggression is to be suppressed or concealed in the presence of "others." Impertinence toward his parents was reproved with special severity if outsiders learned about it, because his parents "did not want to be made fools of by their own children." The informant was, of course, manifesting the mos which enjoins family solidarity.

In the course of our discussions, the informant enumerated all of the people that he knew in his neighborhood group. It proved to be a small group in the immediate area around his home and added up to about forty people. More important than this, these people whom the informant knew all knew one another and were related to one another dynamically in important ways. He knew the parents of all his child friends, and their parents, in turn, knew his. He noticed that there seemed to be a kind of informal policing of all the children by all of the parents, as well as informal feeding and protection. For instance, he often got cookies and candy at Aunt Ella's house next door, and at the same time Mrs. White, who lived across the street, reported to his mother that he was seen running logs on the river bank. It was a small, stable, neighborhood world into which our informant came.

The boys' group into which the informant was inducted at the age of six was a differentiated aspect of the neighborhood. For some time he was kept out of this group because he was too little, and he had to observe its interesting expeditions as an isolated outsider. If one did not belong to the group, there were only the detested younger children to play with. Finally, however, one of the group leaders approached him and made him a proposal: if our informant would agree to marry the leader's older sister when he grew up, he would be admitted to the group. The sister was very homely, but the informant felt that no price was too high to pay for this privilege, especially if payment was sufficiently deferred, so he consented to join. He was taken on the frog spearing expeditions, introduced to the hut behind the barn where you crawled in and told stories, allowed to compete with the other boys in walking the rails on the near-by railroad track and given much very specific sexual information. By age and emotional disposition our subject was always a "follower" in his play group; this has already been referred to as a chronic attitude toward authoritarian persons which was transferred to the playgroup from the family where it originated. Placation of father and playgroup leader remained for many years a distinguishing psychic characteristic of this man.

No sex practices were carried on by the gang, but the subject was told, what he did not need to be told, that his father did something to his mother at night, that something came out of his father's penis and went into the mother, that you don't get this until you are grown up, and that this "something" made a baby in mother. He was rigorously warned never to let on to his parents that he had this information on penalty of expulsion from the gang; he never did tell, and his father was so stupid when the boy was ten years older as to give him, very shyly, some of the information in a roundabout why which he had long had in such a concrete form. These data suggest the possibility that there is a definite "child level" of culture which is transmitted from one generation to the next, exclusive of the formal parental inculcation.

Another important aspect of the playgroup for the informant was that he was definitely weaned from close emotional dependency on his parents. For the first time, he was able to deceive them whole-heartedly and with the sense of group support. The thoroughgoing band-control of child behavior by a unified group of parents was met by an opposed playgroup which validated many of the behavior tendencies disapproved by the adults. Guilty deceit was changed for a more courageous deceit which was possible with the support of other and older boys. Undoubtedly, the lesson of deceit had been learned in the family itself, as the Christmas incident shows, but it was enormously strengthened by association in the playgroup. It seemed, in fact, that the playgroup performed a great service for this boy in thus mitigating the destructive effect of the high parental ideals of perfect behavior which were held up to him.

Although the neighborhood or band dominated his life quite completely at this age, he was not without knowledge that this neighborhood was limited in area. He knew, for instance, that there was a group of German boys in the same school where he went to first grade. There was fighting and stone throwing between him and his gang and this group. These boys talked a kind of gibberish at home and were said to eat nothing but sauerkraut and sausage. Their fathers worked in the mills, and they went to a separate church where they had a minister who came from a long way off. He knew also that Aunt Ella next door was a seamstress, and he used to see her working on the long veils which were used in the Polish weddings. He did not know anything about the Poles except this; they also talked a funny and different language and lived on the other side of town. They were supposed to be dirty and very much less respectable than his own family. Once his gang went down to Polack-town and exchanged a shower of stones with the Polish boys, but beat a hasty retreat when the Polacks gathered in force. Polack girls were said to be easy to play with if you could once get at them. In general, they were a low group, but were supposed to be very strong as individuals.

We see above, of course, the familiar minority group picture which has shattered the solidarity of the native white American society and which differentiates a town like this one from a similar town in England. Where social groups of contrasting customs exist side by side, the impression of crushing absoluteness is gone from primary group life.

The informant's father did not drink, except for possibly a glass of beer at a family reunion. No liquor was kept in the house, but across the street was a saloon; there was much activity there, especially on Saturday night. The subject sometimes heard shouts and sounds of fighting from his bedroom and sometimes even before he went to bed he would see a drunken man staggering out of the saloon. On one occasion, he saw two men fighting in front of the saloon in the late afternoon, and one of them picked up a stone and struck the other on the forehead. The boy was terrified at the stunned, crushed, and bleeding body of the man who had been hit. Some of the Polacks and "sausage-eaters" used to go to this saloon, as well as the younger fellows just in from the farm. Now and then one of these fellows would try to pump the informant about the maid. Did he ever see her undress? Did she drink, and whom did she go out with? He always resented these queries as invading the family privacy.

He was undoubtedly witnessing here a lower class recreational pattern which differed sharply from the conservative behavior of his parents and their immediate associates. The sight of this behavior was both appealing and terrifying, perhaps more the latter, but it did show at least that there were other modes of life than those so urgently recommended to him.

We have so far seen in our material the family form, band, and minority group phenomena; we might now note the establishment of a sharply differentiated male sex-role in our informant. He remembered, for instance, the time when his father had shamed him for playing with dolls, that his mother had hold him "not to be a calf" and cry at disappointments, that little *boys* did not do this, though they did not seem to mind it in his sister. His parents had forbidden him to attack his younger brothers and sister, but, on the other hand, he was not allowed to complain about challenges or assaults from other boys at school. His father wanted to know why he had not hit back. Less overtly expressed, but very strong also, was a wish to be like his father and to take the same confident and dominant attitude toward women which the latter did; this was one of the best reasons for growing up, for then one really could control a woman. He learned that he was not expected to be interested in cooking but that, on the other hand, mowing the grass was evidently to be his job and that running errands was more suitable for a boy. He was quite outraged once when he had to wear a coat which was made over from a garment of his mother's; he had learned that it was improper to wear women's clothes even in a disguised form. One can see from the record the persistent sex-typing of character under the influence of his parents and colleagues. Girls, of course, were excluded from the gang, as inadequate to participation in male pursuits.

In connection therewith he learned also of the dangers involved in stealing women. It came to him by a story which he heard from one of the older members of his playgroup. It was about a certain Bill Packer who came home and found another man with his wife, "doing it." Bill took out his jackknife on the spot and operated. The offender did not die, but he was not a man any more after that, and nobody did anything to Bill. Of course, this was only the case when you did something with another man's wife, but, the informant reasoned, even unmarried girls had brothers and fathers who might be possessive. His sex role was quite well consolidated by this event against pre-marital sex experience or adultery, and the taboo tended to spread to all girls, married or otherwise. It would seem an unintentional effect of our society to carry the mores against incest and adultery to objects other than the socially tabooed ones. In adult life, the image of Bill Packer was still a potent force in arousing fear in any heterosexual situation.

Due to the deprivations and dissatisfactions of childhood, there was aroused in our informant an intense desire to be grown up. Being grown up seemed the condition in which one could be independent of parental restriction and capable of building one's world on more satisfactory lines. His parents also described being grown up as a state when one had many more privileges; at a younger age, however, one had to accept the limitations of this age. "When you are a *big boy* you can . . . ." go out with mother and father, sit up late, ride a horse, have a bicycle, perhaps drive an engine or a street car, get married, have a real part in work and responsibility, be independent, and, the informant thought, be the match for these powerful adults with their unexpected prohibitions and restrictions. On the negative side, each one of the things which he wished most to do were things that could not be done by a little boy, and one had to go by the long, dreary route of school and present limitation to reach them.

Age-grading behavior is clearly delineated by the detailed life history which is adequately explored in its earlier phases. Protest against age-grading in childhood is undoubtedly one of the powerful emotional supports to social mobility in our society.

The informant's father was a plumber, and this gave the informant an unusual position in the band in which his family moved. He reported how surprised he was when one of his gang said to him, "Why, you're rich! You ought to be able to afford a bicycle because your father is a plumber." The speaker's father was a "hand" in the mills; his son was keenly aware of the difference between himself and the son of a plumber. The informant's father owned his house, had a telephone, had a bank account, and was planning to "send the children on to school." By contrast, Aunt Ella's husband was killed in the mills during the informant's childhood, and Aunt Ella was in despair; they had only paid a little on their house, there was no insurance, and the boss gave her only a very little money for her husband. The informant luxuriated in the high position of his father and tried to capitalize on it so far as his own position in the gang was concerned. He soon found out, however, that this status was likely to be a disadvantage in his playgroup, especially if he tried to utilize it outright. There was an aura about him, nevertheless, and he knew it; he valued it, while he pretended to his playmates that it did not exist.

On the other hand, there was the case of Hanky Bisworth who went to the same school as our informant. His father was rich, really rich, and they lived in the biggest house in town. It was a great, dark red brick house and the informant always walked by it with a certain amount of awe. Would he ever be able to go into a house like that, and why didn't Hanky ever ask him around to play? At home, too, there was a good deal of conversation about the Bisworths, and his parents referred to them with extreme respect. His father often said that Mr. Bisworth's career proved that it was the "Almighty Dollar" which counted in this world. It was humiliating to notice that there was someone who had a position higher than that of his own parents, those parents who had heretofore seemed so absolute in their prestige. Now he could see that merely being as good as his parents was not enough. There was something better to be than even these heroic figures.

I am sure that everyone will note in the foregoing how prestige levels are associated with wealth, because the Bisworth family made no other claim to eminence, and also how the informant perceived in a very simple sense how he stood on a status terrace with some below and others above him.

His mother was careful to tell him already at this period that she herself was an educated woman, that she had been two years in school after she finished her country school. She also told him that her father had been a teacher-farmer and stressed how well he had done with the advantages that he had. Informant's mother felt superior to his father in this respect. The father had only a grade school education, and his father, in turn, could barely read and write. The mother knew what it meant to have advanced socially in her own life from a pinched position as the daughter of a poor farm family to some position between lower and middle class status in a town. She justified many of her severe impositions on her children by stating that they were necessary to "get on" in the world. She also wholeheartedly conducted a domestic economy in the family which made it possible to save money and ultimately to realize her insistence on better opportunities for her children. The father was less concerned on this score and less willing to sacrifice for the future of the children and the security of his old age.

That the family was actually mobile is quite well shown by the fact that the informant was told about the different houses they had lived in before his memory began. These houses were all in the neighborhood, and he saw them all and had no difficulty in reaching the conclusion that the house he was living in at age six was the best of the lot. It had more conveniences and a larger lawn, was farther removed from the annoyance of the railway tracks, and finally it was actually owned in full, whereas none of the others had been. He could see, in sober fact, that his family was getting on in the world and that he was to receive a projectile push from them.

He was himself required to make renunciations for his future schooling. This came out clearly in the bicycle incident. He desperately wanted a bicycle, but it was decided by his parents that this was the type of item which the informant would have to do without in order that there might be a platform of capital for the later schooling. The father said that if he should die he would not want to leave his wife and children without some money and that he was himself giving up many things in order that this money might be accumulated. At the time, the boy was bitter over his disappointment and could have no conception of the compensating advantage of going to school, but the idea was so strong in his parents that he lived a very meager life so far as spending money was concerned during his entire childhood.

The constant atmosphere of scrimping and saving, of everlasting worry about money, undoubtedly made the parents seem heavy and depressive to the informant. The father looked like a trussed, burdened man. He could not enjoy his daily life because he was forever thinking about the uncertainties of the future. The future, in the sense of higher status, was too much present as a goal to allow for contemporary satisfactions. The mother, though also the "worrying type" of individual on other counts, was particularly eager and insistent on this score and led the campaign for immediate renunciation in favor of saving for education and old age.

In fact, all of the children of this family have had occasion to thank their parents for these sacrifices when the compensations of higher status finally began to come in; all of the children have secured college educations, and the mother and father are living an independent old age on their property. Whether the game was worth the candle or not is something no one may say with finality; what is certain is that the renunciations imposed in order to achieve higher status were in this case a real penalty and a penalty which had implications for the character formation of the subject.

The informant remembers both parents stating that they expected him to do better in life than they had done, and that they were going to aid him in getting better training so that he might thus be more secure against the hazards of life. However, the parents sometimes used this fact as a reproach and, apparently, felt their own lives to some degree as a transition point in the status march of the family. Life was not represented to the informant as something to enjoy, but rather as a mission in which he was to raise the family prestige. The hazard of Aunt Ella's lot was pointed out, how poor her husband was, how little training he had, and how badly off he had left her. Such impressive penalties for a hand-to-mouth existence were quite adequate to justify the family policy to the informant.

It would take a more intensive analysis than I have been able to make to discover to what degree the powerful mobility tendencies of this family were responsible for the neurotic discontent which the informant manifested from his childhood on and which became acute when he faced the responsibilities of adult life. Certainly we can see that from behind his memory comes discontent and rebellion and hatred against other people, both women and men. In a sense, the mobility pattern was passionately welcomed by this individual as an outlet and followed, one might almost say, with animosity. Tendencies to outdo others, to humiliate others, and to revenge oneself upon others can be displayed amply under the competition which leads to the securing of higher status. It is of greatest research interest to inquire to what extent the mobility of the family may be held accountable for the subject's neurosis. Certainly the actual adult difficulties of change in social status cannot account for it, for they were not unusual; the real disposition can only be referred to his childhood. It is possible that the total frame of the family life, which imposed such privations on this subject, was determined by the powerful mobility tendencies which were present, especially on the mother's side. In the family struggle for social advancement, it is conceivable that too much pressure was exerted on this individual and too much renunciation expected of him.

The conception of differences of intelligence became quite clear to our subject through observing Art Bovis, the feeble-minded boy. His blank, fat face, unintelligible muttering, and ineffectual social responses were early observed and derided. It gave the informant a terrible pang to hear that this condition might have been associated with masturbation, and this news acted as an effective barrier in putting out of mind the genital sensations which the informant felt at times in childhood. He wanted to have a "good mind" and was indeed told by adults that he did have one; he knew, too, that smart people got along better in the world, and this seemed an additional reason for disavowing all practices which would interfere with his growth and sanity. At this time was established the conflict between intelligence and sexual impulse which is so common among those who expect to rise or have risen in status through their own talents. Indeed, one of our informant's difficulties in adult life was that, when he was finally called upon in marriage to exercise a limited impulse-freedom, he was out of practice and carried with him many strange and unsuitable internalized taboos. This possible relationship will reward much more intensive study.

The informant early learned that there were still other forces to which his parents submitted. One of these was God and his representative, the minister. At church he heard sermons that he did not understand very well, but he did learn that most of the things you wanted to do were sins, and that death was the one final punishment for these sins. Even then, however, it was not all over. There was also Hell and the everlasting vengefulness of a rejected and outraged God. During this period of our subject's life, a male cousin died, who had been a prominent professional man but who had given up going to church. The minister was quoted as commenting on this event and pointing out how slowly the mills of the gods do grind. The informant tended to associate everything that he feared as stern and severe with religion and to feel that the minister too demanded the deprivations which he experienced. There was, he knew in a vague way, other talk about the minister, talk of admiring him, of getting comfort from him, but this aspect of the religious institution made no appeal to our subject. He could not see how the minister or God could be loving and forgiving if they did not actually let you do some of the things that you wanted to do and if they demanded that you live an unbearable life in order to be good. One of our informant's first acts of adult rebellion was to renounce the religion of his childhood—with the aid, of course, of the permissive patterns which he found in college. If this hostile attitude toward religious symbols is held by many persons from the same social group as our informant, it may be seen as a permanent force for social change. This force might exist for centuries and make itself felt only when a political or technological shift made it expedient to challenge theological authority on some other ground. Study of the life history may often show the concealed tensions which erupt in periods of social crisis.

Our discussion so far does not cover by any means the totality of societal forms which can be viewed intimately through the material of this life history, and, of course, none of the forms referred to is adequately discussed. The main point has been, however, to show that such material does submit to analysis from the societal standpoint, and that it reveals some variables in social life which cannot, perhaps, be discovered in any other way. It is, therefore, genuinely useful as an exploratory tool.

A remark may be added on the question of the type of struggle which this man had with his environment. On the one hand, there is the struggle of very early childhood, which might be phrased as one between the organism and the unified family group. In this battle, of course, the individual is always beaten, if he is socialized at all. There is a second type of conflict which this individual also exhibits and which is frequently noted by sociologists, viz., conflict in the mature person, i.e., within a developed personality, which is faced by societal segments with opposing or confusing definitions of life. In the latter case, "choice" is offered, and it is apparently this choice which results in conflict. Both types are apparent in this case. My conclusion is that without the second type of conflict this individual would still have been neurotic, but his neurosis would have been smothered by the unified front opposed to him by society. His misery would have been laid to spirit-possession, the devil, or to biological defect.

35

Bingham Dai

SOME PROBLEMS OF PERSONALITY

DEVELOPMENT AMONG NEGRO

CHILDREN

Dr. Dai's paper, utilizing the autobiographical technique previously introduced in the selection by Allport, Bruner, and Jandorf (p. 436), treats the interdependence of group-membership, role, and situational determinants.

The basic theoretical orientation of this article is very close to that of Dr. Harry Stack Sullivan. The subjects are members of a caste group, but it is shown that their personalities are affected both by participation in the culture of the larger society and by membership in the caste. In each of the case histories discussed, the individual's own conception of his role and his special family situation are seen to be the crucial factors making for differentiation of personality among Negro children, among Negro children of a particular class, indeed among children of the same family. The important point is made that a great deal of the individual's internalized normative system is internalized through role performance. Dr. Dai also stresses the importance of the sex of the child in personality formation. We know from a variety of behavioral evidence that the adjustment does differ. For example, in our society more boys than girls are thumb suckers, stammering tends to be a male disorder, stomach ulcers are at present more common in male patients.

NOTE: Thanks are due the author and the Child Research Clinic of the Woods Schools for permission to reprint this article, with abridgment, from *Sociological Foundations of the Psychiatric Disorders of Childhood*, pp. 67–100, published by the Child Research Clinic, The Woods Schools (Langhorne, Pennsylvania, May 1946).

[545

THE sociologists look at the human individual primarily as a member of society and a carrier of culture and seek to understand his behavior in terms of his social and cultural environment. The psychiatrists, on the other hand, tend to regard human personality, to use the words of a late anthropologist, as "a relatively stable system of reactivity—cognitive, affective, and conative," a pattern of life, we may say, that consistently distinguishes an individual from another. They are inclined, therefore, to interpret the behavior of a human individual in terms of his basic behavior patterns or attitude clusters that took their shape in the early years of his life, that may or may not be accessible to awareness, and that most likely will remain more or less stable in spite of the multiple roles he may later play in various social situations. In a word, while the sociological point of view focuses its attention on the individual as a member of society and a participant of culture, the psychiatrist is interested in him more as a unique personality.

For many years the sociologists and the psychiatrists worked in isolation from each other, although both groups were interested in studying practically the same phenomenon. One group would attempt to explain human behavior entirely in terms of common social situations and cultural factors that might mean very different things to different individuals, while the other would often resort to formulations that apply more aptly to lower biologic forms than to humans. In recent years, however, it has been increasingly recognized that each group has a great deal to learn from the other. In fact, through this interdiscipline collaboration, not only the social sciences have come much closer to an understanding of the unique personality and its relation to social phenomena, but psychiatry itself has shown signs of becoming more truly a science of human behavior than it has ever been before.

I. PROBLEMS AND METHODS OF THIS STUDY

In this study I propose to explore some of the problems of Negro childhood and their effects upon personality development from a point of view that looks at the Negro child both as a unique personality and as a member of his society and a participant of his culture. Data used for this exploration were collected mainly from some eighty Negro youths in the form of undirected "own stories" or autobiographies, their ages ranging from seventeen to twenty-five, except those otherwise specified. Each of these stories was supplemented by one or more interviews. In these interviews, a certain amount of free association and dream analysis was used. This initial acquaintance with Negro personality was later further supplemented by more intensive clinical contacts with a smaller number of juvenile delinquents and adult Negro psychiatric patients. Although the original purpose of most of these investigative efforts was largely therapeutic, materials on childhood experiences and family relations spontaneously came up, as they generally do in any therapeutic work. While these data are lacking in many details, they have one redeeming feature, and that is that in many cases it was possible to compare the youths' problems during childhood with their problems later in life. No attempt will be made to treat this body of material statistically or to go into the individual childhood problems in great detail. Instead, for the purpose of this presentation, I am primarily interested in pointing out the types of problem that Negro children in this country have to face in the processes of growing up, or, more accurately, problems I have come across in this exploration, and the ways these childhood problems have affected their later development as social personalities. In selecting these problems for special study, there is no intention to minimize either the importance of adolescent and adult experiences or that of therapeutic intervention in affecting the course of personality development started during childhood.

Some of the hypotheses used in this exploratory study may be briefly stated as follows:

While there are many facets to a human personality, the most important, from the point of view of this study, is an individual's conceptions of himself, for it appears that it is around the conception of the self that the many other facets of a personality are organized, and that what a person thinks of himself, consciously or unconsciously, determines his behavior to an extent not commonly recognized.

The conception an individual forms of himself usually has a social reference. It generally takes the form of some kind of relation between the self and others. In this sense the conception of self may also be thought of as a role one intends, or is expected, to play in a social situation. As a matter of fact, it is seldom possible to understand fully the meaning of any isolated trait or attitude, as any experienced clinician can testify, except in terms of the role the individual is playing in a specific social situation, either actual or imaginary.

The conceptions of the self or roles of a human individual are acquired. An individual is born with only biological needs, but acquires a self in the course of maturation and socialization. But with the growth of the self, the needs for security in self-other or inter-personal relations become as important as, and very often more than, the needs for biologic satisfaction. In fact, the self system tends to exert an over-all control over all the needs of the individual, biologic or otherwise, any serious disturbance of which control may result in varying degrees of anxiety.

It follows from the foregoing that the nature of the self system an individual acquires in the course of socialization depends largely on the kind of personalities he is associated with and the culture after which his activities are patterned, what the significant people in the environment think of him and the ways in which the socialization program is carried out.

While the number of conceptions of the self or roles an individual acquires necessarily increases with his social contacts, there seems to be a natural hierarchy among them. Those that are acquired early in the primary group are generally more important and basic than those that are acquired later in secondary group contacts. It is perhaps in this sense that Sapir once remarked that, although a person may be a black sheep in the family and may later become a professor of subversive theories, there is in his personality something that is relatively stable and consistent. That something, as I see it, is the individual's basic conception of himself or his role acquired in his early contacts with his primary group.

From these considerations, it is clear that a study of the personality problems of Negro children would amount to an exploration into the obstacles to the development of a secure and adequate self system. This statement of the problem is based on the assumption that the black child also has a self. In fact, the development of the Negro child's conception of himself or his role in society and his preferred ways to combat or to support it will be the principal point of interest in this discussion; other personality components will be considered only in so far as they may affect the basic conceptions Negro children form of themselves.

2. PERSONALITY PROBLEMS NEGRO CHILDREN SHARE WITH WHITES

Problems of, or obstacles to, the development of a secure and adequate self system among Negro children may be roughly divided into two categories: (1) problems that Negro children share with Whites; and (2) problems that are more or less peculiar to Negro children. This division, of course, is possible only for the purpose of study; actually being a Negro in an American society is apt to color practically every act and thought of the Negro child. For the convenience of this discussion, however, it is possible to examine some of the problems that Negro children share with Whites, without ignoring the fact that they are Negroes at the same time.

Psychiatric experience teaches that most of the personality disturbances of children come from their relationship with parents, or parent-substitutes, and siblings. And all indications are that it is in the process of social interaction in this primary group that a child forms his first or basic conception of himself. Negro children are no exceptions. A brief mention of a few examples will suffice.

It must be borne in mind that, while the relationship between a child and members of his primary group is essentially an inter-personal matter, referred to in this discussion as his relation to society, it is in many ways determined not only by the personalities involved but the customs of the group or culture that are embodied in these personalities. One of the problems of observing any inter-personal situation, therefore, consists of seeing how the factors designated as society and culture interplay. This point of view will be kept in mind even though our materials in most cases are too sketchy for a detailed analysis.

Thus, it may not be uncommon for the parents of a middle-class family in the South, white or colored, to slap a child of four for calling people names. Rose, the girl referred to here, is the fifth of seven children in a skilled worker's family. But the following episode would indicate something more than a matter of middle-class mores.

> When I was about thirteen years old I broke the handles off my mother's silver vase, which was on her bureau. I placed both my hands on the handles. My mother, who was ironing in the room, told me to stop putting weight on them for fear that I might break them off. Something inside of me told me to continue. This I did until one of the handles came off and my mother finally whipped me.

Something more than class mores may also be seen in the following episode.

> At the age of sixteen I received my last corporal punishment from my father. One night I was studying my lessons at home. My mother had gone out. And so I asked my father for permission to go across the street to my chum's house. He would not grant it to me. His denial made me furious, and, when he left to go to a near-by store, I left the house anyway. My father returned home before I did. When I reached the door I found him waiting with his razor strap, with which he inflicted punishment. For one week I did not speak to my father and said very little to the rest of my family.

Without going into other details, one notices from the episodes related above that, along with the love and affection that might have gone into the relationship between this child and her parents, there was an undercurrent of antagonism between them. Rose began early to think of herself as being greatly mistreated, and her habitual pattern of response to such a situation was semi-open rebellion.

As may be expected, this early conception Rose formed of herself out of such a relationship repeatedly showed itself and characterized most of her later relations with authorities. In school she became the student leader fighting against authorities for more freedom, and at work as a teacher she fought against the principal on behalf of a fellow instructor. She had similar difficulties with the school founders and the librarian. Still more interesting is her attitude towards Whites. She hates them and has fought all the White boys and girls who attempted to molest her. She has grown to be something of a race leader and has had strong desires to be a career woman.

A case showing a different pattern of reacting to parental authority may be mentioned, although the processes involved in its genesis are less clear. John is the oldest of three children in a middle-class family. One outstanding characteristic of the family is that his parents always quarrelled, with his mother forever accusing her husband of mismanaging the family finance and of squandering away income by gambling. In all these matters, the boy invariably took his mother's side. He described his father as one who loves to play the role of the "lord of the house," but who has recently become "a short grey-haired man who seems a little ridiculous with his domineering Napoleonic attitude." John's father, however, has been generous to him, and he has tried to follow his father's profession. His mother is described as a nervous person, forever complaining about her husband to her children.

As a result of this family situation, John said: "Naturally today I dislike my father more than I should. I try to treat him politely, but it is difficult." He also lacks serious ambitions. In regard to his principal interest, he said: "My only concern is how much money can I make selling my work. My point of view is thoroughly mercenary." In regard to Whites, his preferred policy is "an eye for an eye and a tooth for a tooth." In actual behavior, however, he is far from being as assertive as his words seem to indicate. He is shy and self-conscious. He bites his finger nails and prefers to be alone. In case of feeling angry, he resorts to reticence and drawing instead of direct expression. And in a piece on his philosophy of life, he has this to say about revenge:

> I find that revenge is wasteful, and that it gives the person seeking revenge no satisfaction. Nevertheless it is advisable to avoid the punishment of revenge sought by others. When conflict does occur, I find that passiveness and the passing of time dissipate the difficulty.

Now passivity or to think of self as being overpowered and unable to cope with an inter-personal situation except by appeasement may come from many sources, but in accounting for this boy's pattern of life in this culture, his relationship with his father cannot be ignored.

Another variety of response to parental authority is extreme conformity. The following excerpts are taken from the story of a fair-complexioned and nice looking girl of a northern middle-class family, the second of four daughters of a couple whose patterns of life are not compatible. Katherine's mother, following her grandmother, is extremely religious and conservative, whereas her father is more human and likes to frequent places of pleasure. The real boss of the family, however, is her grandmother, who succeeded in separating the couple, and Katherine lived most of her life with her grandmother and mother. She was forbidden to do many things, such as playing the piano on Sunday excepting hymns, and playing cards at any time. She was repeatedly whipped for receiving little love notes from boys. It seemed that one deciding factor in the shaping of Katherine's conception of herself and her pattern of achieving security was what her grandmother thought of her:

> My grandmother drilled the fact into us day and night that our father was no good. Because I looked so much like him and was stubborn like he is, I was continuously told that I would be no good like him.
> Furiously I plunged into my high school studies. I decided that I would show my mother that I could pass the subjects in which my older sister failed.

I also became rather interested in church work. . . . Soon I was regarded as a model young person by the older people.

For a while Katherine hesitatingly ventured into adolescent romance but only to come back to her familiar paths toward security, after a disappointment. As a result of this tug of war between her desire to conform to her grandmother's and mother's expectations and her own natural inclinations, or between what her grandmother and mother thought of her and what she tried to be herself, she stifled her natural desires, lost her capacity for making friends, felt "very self-conscious, and imagined that some one was always laughing at me or talking about me." She described her condition at the time of writing her story as follows:

> I am still a bewildered and restless person. I have very few intimate friends, and go out occasionally. I am still afraid of my real self. Sometimes I feel as if I would like to run away from myself. But I know this is impossible. So I still remain somewhat the type of person that my mother wishes me to be. Now and then I am really myself, and then I am really happy.

To show how two individuals may develop different conceptions of self and follow different paths toward security as a result of differences in their relations with other people and culture, it may be mentioned that Katherine's older sister is quite carefree and sociable and is far from being conforming. Apparently this sister does not have as much of the need for affection and approval as Katherine, for she is not only fairer in complexion—an asset highly valued in Negro society—but is also her mother's favorite. Hence she has no compulsive need for conformity.

Some other patterns of response to threats of insecurity in the primary group situation, especially those that come from rivalry between siblings, are forcibly demanding what one wants from the environment and straightforward destructiveness or getting rid of the obstacles to one's sense of well-being. Jean, the younger of two children in a northern middle-class family, for example, acquired what she called the "ugly duckling" complex as a result of her grandmother's and other people's preference for her elder brother because of his good looks. This state of affairs was augmented by the effects of having the rickets during her early childhood, although the results of this illness were hardly noticeable at the time she was interviewed. As a child, however, her inability to do what many other children did made her feel definitely inferior to them. At the same time, this sickness gave her a weapon whereby she could exact compliance and services from the environment. In time, this became her favorite pattern of response, not only in her relations with members of her primary group but with her boy friends. She made so many demands on a boy that he left her after three years of courtship. Jean's friends do not think that she is a bad looking girl, but, because of the conception of self as an "ugly duckling" that was forced upon her by her grandmother and others and by the psychological effects of her childhood illness, she acts as one, and her particular manner of combating the resulting anxiety apparently has not been effective. She has no respect for, or confidence in, herself and does not believe that people can genuinely like her.

The primary group situations and the problems of personality development resulting from them described above are by no means exhaustive. There are still the special problems of the only child, the only boy or girl among a group of children, the sickly child, the boy or girl brought up exclusively in the company of women or men, and many others. But space does not permit me to take up each of them. Perhaps enough has been said to show that these problems know no color line; they are mostly problems inherent in the primary group relations of any racial group, and their effects on the conceptions of the self are equally real and significant, no matter whether the children involved are black or white.

3 . PERSONALITY PROBLEMS MORE OR LESS PECULIAR TO NEGRO CHILDREN

The personality problems that are more or less peculiar to Negro children are closely associated with the peculiar social status that their elders are socially and legally compelled to occupy in this society and the peculiar evaluations of skin color, hair texture, and other physical features that are imposed upon them by the White majority. The effects of discrimination and segregation upon Negroes in the matter of making a living and in other spheres of activity are many. So far as the personality development of Negro children is concerned, the most important conditions resulting from living under caste restrictions seem to be the preponderance of lower-class families with their special codes of conduct, broken homes accompanied by the dominance of maternal authority, the special importance attached to skin color and other physical features, and the extraordinary stress on matters of social status. Each of these cultural situations is apt to leave its indelible imprint on the personality of the Negro child.

Lower-class families—Lower-class families, as understood in this country, are characterized by two things: material deprivation and low standards of conduct. Such families are more numerous among Negroes because of their limited opportunities for advancement. The culture of the lower-class families is either entirely lacking in stimulating contents or follows codes of conduct that are not acceptable in more respectable society. Cultural ennui or emptiness has been found to be associated with sleeping attacks or narcolepsy among Negro children [1]. And low standards of conduct have been found to be responsible for delinquent behavior and the unrestrained aggressive conduct of lower-class Negro youth [3]. The inter-personal relations in the family and the resulting conceptions of self are equally important in accounting for delinquent behavior just cited. How the factors designated in this study as *society* and *culture* interplay in the formation of a self system among lower-class Negro children may be illustrated by one or two examples.

Charlie, a boy of fourteen from a lower-class family, was seen at a sociological clinic for petty stealing, burglary, and other forms of misconduct. He was found not only to have been associating with delinquents but to have come from a family notorious for depending on charity and begging for things the family needed, his parents having not worked for fifteen long years. He is the youngest of six children, one elder sister having been an illegitimate mother at the age of fifteen, and the other contracting venereal diseases at the age of thirteen. The family professes the Christian religion, but of such a kind that forbids card playing on Sundays but does not prohibit demanding aid from people or even taking things from others when it can be done without being caught. At the same time, Charlie was also found to have been associating with children of higher social classes. One of his predominant desires was to dress well, which desire was responsible for most of his stealing episodes.

Our knowledge of Charlie's lower-class background and his associates or cliques, however, is not sufficient for an adequate understanding of his delinquent behavior, for all of his brothers were working and were not known to have delinquent records. Upon further inquiry, it was found that Charlie had been more or less his parents' favorite. His father, however, enjoyed whipping children, including Charlie, and often boasted of his sexual potency in spite of his age. His mother remarked that Charlie had been very "girlish" since early childhood and that he was "as dainty as the girls." He had been unusually concerned about his appearance, especially his clothing. He had enuresis till six, has had a bad temper, and is always greedy in eating. He often dreamed of being grabbed by some man in the dark, and one of his noticeable desires was to impress on his parents that he was no longer a child. Quite early he began to steal clothing from his brothers, and later this activity was extended to people outside the family. In other words, Charlie's delinquent behavior may be said to have served some definite function in his conception of himself, which may be roughly described as that of a mother's baby, who feels acutely insecure in the presence of his father and definitely inferior to his siblings and who has found a path toward security in his "girlish" looks and the attitudes that go with it. The need of self is the end, and the patterns of his social class or clique are the means.

Another example is a typical slum boy, who is more articulate than the boy just referred to and whose story will make it even clearer to the reader what it is like to be the child of a broken home, poor, and a Negro all at the same time. This is what Peter says about his parentage:

> I was born before the marriage of my mother and father had received religious and lay sanction. My father had cut a man and was hiding. . . . I had been around "raising hell" six months before my father saw me. "Pop" was a mulatto. . . . "Mama" was typically negroid in features.

At the age of four, Peter followed his parents to a northern city, and they made their home in a slum area. Then a light-skinned sister was born, whom he "often threatened to throw out the window." Soon a divorce followed, and Peter was kidnapped by his father and taken to his paternal grandmother. Since then he has seldom seen his mother, and his grandmother to a certain extent took her place in his affections. Some of the factors that went into the conception of himself as a child are described as follows:

> Since I had walked at seven months and talked fluently from a year old, the family in general regarded me as an unusual child. I was not particular about showing off my talents and often drew unfavorable comments on my skin color and defective eyes. School work was easy, though I hated arithmetic and the time limits imposed. I liked drawing, music, and most of my subjects. Teachers I did not like so well. My early school years were uneventful, except that, even though I played all the childhood games well, I would rather read and be alone.
>
> I was never terribly anxious about winning. I did my best, win or lose, and did not care which. This was always my biggest fault. I did a thing just to prove to myself that I could do it, and when I had done it I would lose interest. My grandmother prodded me all the time.

As Peter grew, he began to participate in most of the things a slum boy does:

> Money was always a problem for my grandmother and her four grandchildren. Her three sons, including my father, were spendthrifts and shiftless; and in order to offset this grandmother took in washing, and we stole wood and coal and food. I worked from the time I was six at odd jobs in the afternoons and all day during vacation months. My conscience never bothered me about this stealing, since I was never caught. I knew hunger and homelessness intimately, and the other children and I were adept at securing sympathy and a stay of ejection proceedings from judges. . . . Hunger, cold, insufficient clothing, frequent illness, ill-housing, this was my lot up to my second year in college.
>
> As a slum child I had frequent clandestine sex experiences. No attempt was made to hide the facts from me. People laughed at small children's acts toward sex expression.

Something of Peter's own personality can be seen in his relation with the gang:

> I was a part of the group, and yet not a part of it. I was tolerated, but not actually accepted. I was the best and brainiest thief, the fastest runner, and the best little all-around athlete in my neighborhood, but I wasn't an indispensable part of the gang, and I resented their non-acceptance.

To compensate for his sense of lack of belongingness, both at home and with his gang, he became an insatiable reader and lived more and more in a fantasy world. Seeing this, some members of his family predicted that he was going to be a drunken shiftless being like his father, and called him a "black bastard" or a "big-headed fool." And before he graduated from junior high school, his grandmother, his only support, died, and he then realized fully what it was to be lonely and insecure, poor and a Negro at the same time.

> The bottom fell out of everything for me, and I was more disturbed than I had been before or since. My main prop was gone. I had not realized how much I had depended on her for support. I didn't cry at the funeral because my father and uncle cried. I was mad, not just angry, mad with hatred. My grandmother had died of acute alcoholism. They taught her to crave bootleg whiskey. We had to bury her on money borrowed from friends. We were too poor to carry insurance. My sister died in infancy, and she was buried on borrowed money, too. A feeling of hopelessness came over me. Nothing mattered—high school, nothing. It was then that the whole rotten business of being poor and a Negro at the same time caught in my throat. I became an articulate rebel. I became argumentative and critical. Why? Why? Why? that word became an obsession!

In Peter's delinquent behavior we find the workings of his class or clique mores. But we also find a conception of himself as a unique personality that is characterized by being a member of his grandmother's family and yet not actually feeling that he belonged to it, desiring to be a member of his gang and yet feeling apart from them, and being recognized by his family as an unusual child and yet constantly referred to as a "black bastard." To these we must add his conception of being a poor Negro boy. The acute sense of insecurity and uncertainty about himself as a person is self-evident. How these feelings of a growing self and the behavior patterns of his group might be interwoven into adolescent or adult conduct, therefore, becomes an interesting problem. But we must suspend our curiosity about Peter's development till we come to a later section.

Space does not permit me to go into the full implications of being a child of the lower-class family and a Negro and having problems of the self all at the same time. Perhaps enough has been said to show that while the means of satisfying the needs of the self as resorted to by the lower-class Negro children may be condemnable, the needs themselves are essentially the same as those of any child. And the lower-class means are used often because they are the only ones available.

It is also evident from these illustrations that such cultural factors as class mores alone are not sufficient in explaining the behavior of a Negro child; his biologic wants and especially his needs as a social self are equally important, if not more.

Broken homes—Broken homes, impermanent family life, absence of the father, and dominance of maternal authority can almost be said to be as much a cultural pattern among Negroes as the preponderance of the lower-class families. We have some information not only on how a child in a broken home situation feels about himself, but how the effects of the situation on his self system show themselves in later situations. Something of the broken home situation in a lower-class family has already been shown in the case just cited. In the following we shall take up some cases from the middle-class families.

Sam is the son of a romantic match, his father being a skilled laborer in a southern town. Unfortunately, when he was two months old his father died in an accident at his plant. His mother soon lost interest in Sam and left him with her parents before he was two years old. He was given good care by his grandparents, especially his grandmother and later his aunt and uncle, who is a very successful business man. He was taught to read early and later was found to have superior intelligence. At the age of nine he followed his grandparents to the North and stayed with his aunt and uncle. When he was about thirteen his grandmother died. What happened to his conception of himself at this stage can be seen from the following.

> It was the hardest blow I have ever had in life. I again assumed residence with my aunt. Here I had plenty of all material things, but I never felt at ease, or that the affection which my grandmother had lavished on me was present. My uncle-in-law tried to make me feel that I was his own son, but I never felt that it was sincere.

Although he seldom saw his mother, he thought of her and would not allow his aunt to make any disparaging remarks about her. It happened that his aunt was jealous of him because his uncle liked and respected him for his superior intelligence, and life around his aunt turned out to be "a round of bitter quarrels." He read extensively, but had no interest in school work and recurrent attacks of bronchitis followed. "It was during this period," he said, "that the slovenly study habits I possess now were formed," referring to the time when this story was written. He summed up his conception of himself and his pattern of obtaining security at this time in these words:

> The feeling that I had no especial place in any home created an inferiority complex, which I defeated, when I thought it necessary, by thinking of my intellectual powers.

He further described himself as follows:

> It seems now, as I look back, that I have spent perhaps half my life building a shell around my emotions. Rationality is my god, and yet I wonder if I am rational. Emotions and sentiments have no place in my world; even my affections are cold and calculating to the greatest extent to which I can control them.

The hungry and yet cold and suspicious self showed itself unmistakingly in Sam's relations with his girl friend. After quarrels and reconciliations he said:

> I accepted her with mental reservations. All the old wariness returned. I gave nothing unless I received twice. I became proficient at

mental torture, both for myself and her. Distrust, suspicion—those are my watch-words!

Sam is a highly intelligent boy but has shown no sign of making use of his superior endowment except for purposes of self inflation. Knowingly or unknowlingly, most of his energy has been consumed by the problems that come from an insecure self system acquired during childhood and unabated during adolescence.

Another pattern of combating insecurity popular in this culture is for a man to identify himself with the Christian religion in some manner. The tenets of this religion seem to have a particular appeal to those individuals who have the need for suppressing their natural impulses and who suffer from gnawing feelings of inadequacy, for within a prescribed environment they are often enabled to obtain a certain measure of social approval through renunciation and the sense of superiority through humility, though frequently at the expense of his adjustment to the larger world. Such a security-seeking pattern was adopted by Fred, who was seen as a psychiatric patient. Fred, a farmer's son, was left with his grandmother at the age of five, when his parents were separated and when his mother had to go to the city to work. Although he was well taken care of, he often heard his grandmother remark that she was sick and tired of taking care of other people's children. He felt somehow that he did not belong there, and at the age of fourteen he ran away from his grandmother's place and went to stay with his mother. In his relations with women he was morbidly jealous and sexually impotent. As a child he had enuresis from five to twelve, was afraid of darkness, and often dreamed of somebody chasing him and of falling and flying. His conception of himself may be roughly described as one of a frightened child, who is not sure of his place in his immediate society. His favorite means of acquiring security has been to be good, and his present profession is an old-fashioned but consecrated minister. And at the time he was seen, Fred had not been to a motion picture for over ten years.

This saintly role that Fred acquired through his secondary group contacts, however, has not been able to answer the needs of a more basic conception of himself that was the result of his earlier interaction with his primary group environment. His acute sense of not belonging and his urgent need for reassurance, as well as his mistrust of human affections, manifested itself repeatedly in morbid jealousy and inability to have satisfactory sex relations with women, except those who have shown unconditional loyalty to him. Spontaneity in impulse life as a whole has been greatly curtailed, and normal desires for pleasure and self-assertion appeared only in dreams. And relations with associates and plans for the future were permeated with ambivalence and uncertainty accompanied by gastro-intestinal psychosomatic symptoms.

While the paths toward security may differ with different individuals, the conceptions children of broken homes form of themselves seem to have something in common; it is the feeling that they do not belong and that they are not really wanted. Hence, the insatiable need for reassurance and the psychiatric problems it entails.

Dominance of maternal authority—One interesting feature of the broken home situation among Negroes is the dominance of the mother or mother-substitutes, such as grandmothers, aunts, and sisters. This phenomenon may also be found in homes that are not broken, but homes where the fathers are no longer important; they are, therefore, about as good as being absent. Another related feature of the situation is the preference for girls shown by many Negro mothers and grandmothers. Owing to the limitations of space, only the briefest mention of some cases is permissible.

In one middle-class family, in which the father is an invalid, weak and mild, and the mother a striver, ambitious and demanding and assuming most of the responsibility as well as the authority, and in which the older sister is preferred by both the grandmother and the mother, Joe, the second boy, developed typical compulsive and obsessive symptoms during early adolescence, with the mother as the chief object of concern and the sister second. Most of these symptoms centered around the thought that unless certain rituals were followed something would happen to either his mother or sister or both. Interestingly enough, Joe's youngest brother later developed practically the same symptoms, revolving also around the mother as the chief object of concern. This phenomenon is quite different from what is often found in families where the father is more of an authority; there the object of substitutive aggression is often the father and not the mother. The self system of a person with obsessive and compulsive behavior may be briefly described as one dominated by ambivalence toward people and uncertainty about self. Hostile impulses are generally repressed because of the desire to conform and the need for affection, and substitutive magical power operations, therefore, very often become the only means of self expression.

Another effect of a predominately matriarchal family situation, where premium is put on being a girl, is for a boy to identify himself with the maternal authority and unwittingly become effeminate. Harris is a case of this kind. He was seen as a psychiatric patient. No clear evidence of any physiological basis for his homosexual tendencies could be found. His developmental history shows, however, that ever since early childhood he has wished to be a girl as his mother desired, and in the course of time has acquired many feminine interests, such as playing with dolls, doing house work, and being his mother's help-maid.

The case of Katherine discussed in a previous section, showing extreme conformity to maternal expectations, may be reviewed in this connection. Other types of self-conceptions and patterns of achieving security resulting from the dominance of maternal authority will be taken up in a later context.

Preoccupation with skin color and other racial features—The most obvious, but none the less detrimental, obstacle to the growth of a secure self system among Negro children is the blind acceptance of White racial prejudices and measuring one's personal worth by the degrees of proximity to white complexion or other Caucasian features. These evaluations of skin color and other physical features, however, do not affect the Negro child directly before he comes in close contact with White children; they are mostly mediated by the child's parents and other significant people in his primary group. How these evaluations affect the conceptions a Negro child forms of himself may be seen in the following cases.

Tom, the youngest of six children in a lower middle-class southern family, was reared together with other siblings by a mother who has a strong urge to get ahead. He is dark, but tall and strong, and "in the schoolroom," he said, "I was easily superior to almost all of the children." When he first moved to a northern city with his mother, he was given three promotions in the school within a short time. "These experiences," he recalled, "made me feel superior to other children in the school." But somewhere around eight years of age, as a result of what he had heard and what people called him, he became conscious of his complexion. About this experience he says:

> I have mentioned the fact that I was in doubt about what I was until I was quite large. It is difficult for me to say what I mean. I can remember looking in the mirror time after time only to leave it and wonder what I looked like. I heard my people laughing about other people who were "black." When I looked into the mirror I was dark. Since no one made direct fun of me, I was confused about the difference between me and the other people who were ridiculed as "black." When Mamma often refused to let me play with other little children, I wondered why. She seemed to feel that I was better than they. But then she beat me and often told me that I talked too much. With this sort of bewilderment, I found myself very lonely by the time I was eight years old. People outside of the family had informed me in many ways that I was dark, often in the form of insult. Never had I had any way of expressing these things. My relations with Mamma were always concerned with her giving me something or her disciplining me. Mamma never seemed to talk to me. So at eight years of age most of my thoughts and feelings went on in my head.

This problem became more acute as he grew, especially when time came for him to associate with girls.

> My conflicts about status included shame of home, family, and paternal origin, and perhaps shame of my color. *Color* is a word with emotion for Negroes. At this time in my life—between eight years and twelve years—and even later, I took many subjective jolts about being dark. Mamma in her middle-class striving always managed to live among middle-class Negroes. Among this group . . . color is the sign of their mobility. My little friends were often light brown skinned and had curly hair which

was nearer the texture of Whites than my own. That became a problem whenever girls came into the question. As I entered high school, that question became most important to me.

One of the ways he found to be useful in counteracting this inner sense of unworthiness was to befriend someone who was definitely inferior to him and whom he could dominate. He also picked up the habit of reading. "I read everything I could get," he said of his experience of this period. "Soon I had the notion that the fellows who stood around with girls were really inferior to me mentally."

As to how he appeared to his teachers and how near he was to doing something desperate, one can get some idea from the remarks made by one of them when Tom went back for a visit some years later. This teacher said: "I am glad you are doing so fine educationally, Tom. We were worried about you around here. We felt that you were dangerous. You were morose and unco-operative and yet you had a good mind. We were worried about whether that good mind would be eventually turned to crime and violence rather than to constructive living." When these words were said, Tom was well on his way in his educational career, but experienced disturbing problems in interpersonal relations that can not be gone into here. Among these problems, sensitivity about his dark complexion remained a very important one.

In the case of Tom, therefore, we have a chance to observe how a child took over the White majority's evaluations of skin color from his own group, and how these evaluations created a sense of inferiority or unworthiness that became increasingly unbearable as time went on. In other words, the color of one's skin, which does not occupy the consciousness of children of other cultures, is here made an issue of primary importance, and the personality problems thus created are almost as difficult to get rid of as the dark skin itself.

Another familiar solution of the same problem is to adopt the attitude of indifference. This is what Jane seemed to have done. She is the daughter of a northern upper-middle-class family. When she was born, the doctor taking care of her declared her to be the most perfect and the darkest baby he had ever delivered. She was brought up in favorable family conditions and possesses superior native intelligence. In spite of these, Jane has shown no serious interest in anything, and her philosophy of life is summed up in these words:

> I have never given very much thought to my philosophy of life or whether or not I even have one, but, now that I think of it, I believe it can be summed up in the slang expression "so what?" In other words, my attitude toward life, its joys and perplexities, is one of indifference. I seldom worry or become demonstratively elated. I accept things as they are without much criticism, adverse or otherwise.

While such an attitude of indifference toward life, or low level of aspiration, may be due to many factors, in the case of Jane we must take into account the girl's conception of herself that was given to her the moment she was born and that was continuously brought to her attention in her subsequent contacts with a group of people among whom "being the darkest baby" is such a clear-cut liability that it is able to offset almost any amount of assets an individual may have. There's a sense of uselessness that poisons many youths of this kind, in spite of their superior native endowments.

A rather unusual way of combating the sense of unworthiness that comes from accepting White people's evaluation of skin color is to identify one's self with Whites in attitude and thereby become a White in one's own estimation in spite of the fact that one's complexion is quite the opposite. How this is done can be seen in the following case. Alice is a dark-skinned southern girl, who at the age of ten went with her mother to the North after her father's death. It happened that her mother worked for a couple of very wealthy old White people, who had no children of their own. She describes her relations with this White couple as follows:

> Everybody in the household seemed very interested in me . . . and, because they liked me and considered me "cute," they adopted me after a fashion. Clothes from the best and most expensive stores were mine; gifts of every description were given me. Expensive children's books were bought for me, and I was given the best educational and cultural opportunities; but the "madame" insisted that I was not to play with other colored children. . . . For the most part I played alone and shared my lovely possessions with no one.
>
> In this house I had to be quiet, lady-like, and not "act colored," and because of this reason I developed an over-reticent nature, which has followed me, regretfully, until today.

The method of bolstering her deflated self system that she finally arrived at, and the conflicts it gave rise to, are described in these words:

> My behavior is patterned after the behavior of wealthy white people. I cannot wholly appreciate this attitude in myself, because I have become too dissatisfied in and too critical of my own group of people, with whom my future years are to be spent.

As examples of some of the things she found fault with in her own people, Alice mentioned their lack of table manners and modesty! This transformation from black into white through the process of identification is certainly an ingenious psychological maneuvering, but it only enhances Alice's lack of the sense of belonging; she strives to be acceptable to both white and colored groups, but ends by belonging to neither. She remains a lonely soul.

When the problem of skin color is tied up with such other problems as sibling jealousy, the situation becomes much more complicated than has been described. Lily is a brilliant girl and not bad looking, judging by what her friends knew about her. But she was very unhappy and often wished that she were dead. She constantly worried about her looks, especially her complexion and her hair. She bitterly resented the fact that ever since childhood people always told her mother how cute her younger sister was but never said anything about her. The fact is that her sister is lighter complexioned. She has become very much devoted to her mother, and up to the time when she was interviewed she still let her mother choose her friends and make decisions for her. She hesitated to make friends with boys and was thereby unpopular, for she was afraid that once she got intimate with a boy he might hurt her. She looked forward to marriage and childbirth with the greatest apprehension. She blamed the whole unbearable situation on her mother's having children other than her.

In fact, Lily's sense of unworthiness or unattractiveness, as measured by the White evaluations of skin color, had gone so far that she imagined that people were talking about her, and found it difficult to believe it whenever her friends made any favorable comment on her clothing or her appearance.

Equally disquieting is the unusual concern about hair color and texture, prevalent among many Negro families. In the following example, this preoccupation is also complicated by sibling and, in addition, parental rivalry. Nellie, the oldest of three children in an upper middle-class family, is light complexioned, but worried about her hair. Her mother has Indian complexion and "beautiful black hair." Her younger brother and younger sister, too, have good looking black hair, whereas she and her father have brown hair. In this respect, she and her father are in one camp, while the rest of the family is in the other. So much importance was attached to the looks of their hair that her mother often told her: "When you are ready to get married, choose a good looking man with good hair; you have to think of your children." As a result of such a special stress on the color and the kind of hair, Nellie has developed a great interest in people from non-Caucasian countries and has flatly refused to associate with people of her own group who are very dark and who have bad hair. As may be expected, Nellie's relationship with her associates was far from being satisfactory. Details, however, can not be entered into here.

Extraordinary stress on social status—One other variety of problems of the development of a secure self system that are most noticeable among Negro children, although they may also be found among the descendants of many immigrants and children of some other minority groups, is the tremendous importance their elders attach to matters of social status. This seems rather natural to a group whose self-respect is constantly put on the defensive. Just as a drowning man will grab almost anything in sight that will give him a temporary lift, so a group of people being constantly downtrodden may be expected to hold on to almost any label or sign of deference that will give them a momentary sense of well-being. Only in this sense, perhaps, can one understand the insistence of some Negro parents, for example, that their boy must get all A's in his school work, or that their girl must join such and such a sorority. And there are Negro men who cannot afford to support a family, but who often pay the greatest attention to the style of their clothes and the looks of their car. Perhaps the most significant of all these endeavors, from the point of view of character formation among children, is the persistent struggle for a higher class status among many Negro parents. How this unusual stress on class status may affect the personality development of a child may be seen in the following sample cases.

David, a dark and physically strong boy of about twenty, came from a family of very humble origins. His mother severed relations with her lower-class family with its loose sexual codes, only after a hard struggle, and worked her way through college. Her father, too, had some years of college education. But when David was born, he said, his family was one "that has no status, no money, and no established place in society." By hard work and thrift, however, his parents saved, bought a house and gradually began to move among the middle-class families.

David is the oldest of three children and has always been something of a model boy to his mother. He began to work along with his school work when he was eleven, and helped his mother wash clothes and clean house. In spite of this, his mother is said to have used rather harsh measures in disciplining him and urging him to get ahead. His mother's method of discipline and his reaction to it are described in these words:

> Although my mother loved me very much, in a fit of anger she would whip me severely for things which seemed to me trivial. For example, if she said for me to come home at 1:00 o'clock and if I got there at 1:15, she would have me undress so that she would not wear out the clothes whipping me. And then she would have me get down on my knees, and what followed seemed to me to be very brutal. I used to think of reporting her to the authorities in charge of inhuman treatment of children. I began to feel that I would rather be dead than alive.

What seems to irk David most is his mother's saying to him all the time that he was nothing but "an old fool." This saying has persisted in his mind and apparently has helped shape his conception of himself to a marked extent. For one thing, he wanted to be independent and get away from home. At the same time, he also wanted to prove to his mother that he was still a good boy. Since the age of thirteen he began to have his own income and occasionally ate away from home. Since sixteen, when he entered college, he has not been home for a number of years, but what he earned from working in the summer months he sent to his mother. On the one hand, he seems to miss his mother, but, on the other hand, every letter he receives from her is apt to upset him, for it is usually full of advice, and even the English of his letters to her is subject to the severest criticism.

In his relationship with other people, David is extremely sensitive but competitive and seems always prepared for a fight, boxing being his serious hobby. In his relations with girls, he is almost always treated like a brother and not as a boy friend; his own attitude toward them is uniformly formal.

It would seem, therefore, that by using high pressure methods in exacting complete obedience and perfect performances from David, in her striving for a higher social status, his mother has inadvertently cultivated in him a self system that is plagued by ambivalence toward his mother, but uncertainty as to his own worthiness, and by suspiciousness and aggressiveness toward people at large. David's familiar paths toward security are being good and working hard, but they are accompanied in his case by perpetual anxiety, the lack of the sense of humor, and the inability to enjoy the pleasures of life.

Not only the families that are striving for a higher social status, but those which have already arrived at a certain social level, on account of the special value attached to class position, may also create situations threatening to a growing self.

Ann, a light brown skinned girl of slight build, is the oldest of three children in an upper-middle-class family in a northern city, where there are a lot of other Negroes. "Most of the Negroes there," says Ann, "are jealous and narrow and hate to see any member of their race progress." She thinks that "perhaps this is due to the fact that the majority, or ninety-nine per cent of them, are employed in domestic services, act in the capacity of red caps, and work on the railroad." The relation between her family and these other Negroes in the community is described as follows:

> Although my father is in the public's eye, his family has always held themselves aloof from the other Negroes in the city. The Negroes there have done so much to keep my father from making a living that we have cut off all associations with them completely.

As a result of this situation, Ann says, "as long as I have been there I have never had contact with other girls and boys of my own age. Even now, when I go home in the summer, I give up all relations with girls and boys."

Another result of this situation is that Ann has come to form a strong attachment to her family, especially her mother. She describes this situation as follows:

> To keep a happy state of mind, I find my happiness in my immediate family. . . . My mother, knowing how I felt, did all she could to supply the companionship that I lacked. She was both a mother, friend, and confidant. All my little hurts, desires, and happiness were confided in her.

The mother-daughter tie was so strong that once, when Ann suffered a setback in her relations with her schoolmates, she called her mother on the phone four times in the course of a week. And, while in school, she just would not have anything to do with schoolmates who are very dark or whose parents are in domestic service. She had the desire to study sociology but would hate to work with poor people.

As further indices of her personality, it may be stated that she expressed the greatest hesitation about getting married or having children; she was afraid of pain of any kind. In school she was generally considered a "stuck up," and would not speak to a lot of the boys because they did not have what she desired. And the relationship with the boy she did finally befriend was a succession of quarrels and reconciliations. In other words, her adjustment to her companions was far from being satisfactory.

It would seem that this girl has almost a compulsive need for being superior, and her discrimination against Negroes of darker color and lower class level was a means to that end; and it would not be too far wrong to hazard a guess that this need for superiority was a defense against her own profound sense of insecurity and immaturity which had its roots in the family situation.

Probably it is personalities of the type just described that make a religion out of class distinctions and that use social status relentlessly as a weapon for intra-group aggression.

As a matter of fact, the effects of this intra-group aggression, on the basis of differences in social status or in skin color or hair texture, upon the growing self are more far-reaching than they are ordinarily realized to be. Besides inflicting more or less permanent injuries on the self systems of those children thus discriminated against, such fratricidal warfare, as the cases cited above show, definitely reduces the opportunities for cultivating tenderness and affectional warmth even among those relatively fortunate ones. Intimate acquaintance with such phenomena among Negro students once led me to the conclusion that "in the handling or direction of the hostile impulses among students . . . lies perhaps the most important and urgent task of Negro education in this country" [2]. Subsequent studies only reaffirm this observation.

4. SUMMARY AND DISCUSSION

In this study the problems of personality development among Negro children are approached from the socio-psychiatric point of view. More specifically, they are discussed in terms of the difficulties in the growth of a secure and adequate self system or a conception of self that is conducive to mental health. This conception of self that a child forms early in life is considered as basic in the total organization of his personality, though not easily accessible to consciousness in its full implications.

It is found that the basic conception of self that a Negro child forms early in life depends a great deal upon how his needs are satisfied and how he is thought of by his parents and other significant people in his primary group environment. The patterns of behavior current in this primary group environment are in turn determined to a great extent by the larger social and cultural conditions pertaining to the peculiar social position Negroes in this country occupy.

Under these social and cultural conditions, the problems of personality development among Negro children are found to be of two major kinds. One kind consists of problems that seem to be inherent in the primary group situation in this culture, and, therefore, they are shared in common by both Negroes and Whites. They generally result from inter-personal relations between the child and his parents, or parent-substitutes, and siblings. According as a child is made to feel wanted or unwanted, for example, his conception of himself will be favorable or unfavorable; and on the basis of this emerging conception of himself, and within the permissible limits of his culture, a child will find ways and means to obtain the sense of security and worthiness that are as natural and necessary to the social self as physical well being is to the biological self. These ways and means, in the forms of habits and attitudes, may lead to healthy development or increasing difficulties with himself and society. Some of the major types of the primary group situation and the problems they may bring to the growing self are illustrated with brief case material.

The other kind consists of problems that are more or less peculiar to Negro children because of the peculiar social position their elders occupy in this society and the special cultural emphasis this social position entails. These problems are discussed in relation to the preponderance of lower-class families, broken homes, the dominance of maternal authority, the special concern about skin color, hair texture, and other physical features, and the extraordinary importance attached to matters of social status. Although some of these situations can also be found among other social groups in this country, they are especially true of the Negroes; and each of these situations is found to be capable of producing effects on the growing self of a Negro child that will be his liabilities instead of assets.

While the primary group situation and the larger social and cultural conditions are separated above for the purpose of discussion, they almost always function together in actual behavior. The inter-relationship between social (or inter-personal) and cultural factors in the formation of the basic conception of self is brought out most clearly in the discussion of the problems of the lower-class Negro children and those involving sibling jealousy and the evaluations of skin color their elders adopt from the Whites.

REFERENCES

1. Bender, Lauretta: "Behavior Problems in Negro Children," *Psychiatry*, Vol. 2 (1939), pp. 213–28.
2. Dai, Bingham: "Personality and the Learning Process," *Harvard Educational Review*, Vol. 16 (1946), pp. 173–93.
3. Davis, Allison, and Dollard, John: *Children of Bondage* (Washington, D. C., American Council on Education, 1940).

36

Allison Davis

AMERICAN STATUS SYSTEMS AND THE
SOCIALIZATION OF THE CHILD

The next selection continues with the effects of class and caste membership upon personality formation.[1] The present paper begins with the molding of biological drives under cultural pressures, a subject previously treated by Mr. Frank (p. 119). Roles and the symbols associated with them are discussed. Variations in the family situation are considered. In short, in Dr. Allison Davis' paper we find explicit treatment of all four classes of determinants. The biological animal is clearly recognized, though no attention is paid to the fact that different organisms have different biological equipment. There is considerable integration of learning theory and of psychoanalytic theory. In his conclusion, Dr. Davis makes the transition to the application of "culture and personality" knowledge, which will be the focus of Part Three.

IN MOST human societies, the instigating symbols and socially defined goals of children, as well as of adults, are ordered to systems of appropriate age, sex, and kin behaviors. These roles usually exist within a

NOTE: Reprinted from the *American Sociological Review*, Vol. 6 (1941), pp. 345–54, by permission of the author and the American Sociological Society. (Copyright, 1941, by the American Sociological Society.)

[1] See also, Kingsley Davis, "Mental Hygiene and the Class Structure," *Psychiatry*, Vol. 1 (1938), pp. 55–65.

hierarchy of privileges. There is great variation between societies as to the complexity and rigidity of age and sex stratification. Usually, however, male adults in their prime have the highest rank; in religious or political councils, aged males frequently are accorded precedence.

Along with the definition of roles and ranking of privileges by age and sex, our Western society includes a third type of hierarchical relational system, which limits and defines the approved responses and goals of the child. This is a type of hierarchy which ranks people in defined subordinate-superordinate relationships, without regard for their age, sex, or kinship roles. Listed in order of increasing degrees of in-marriage, the status groups of this third type include (1) social classes, (2) minority ethnic groups, and (3) castes.

The aim of this paper is to call attention to the fundamental importance in America of age, sex, and class instigations and goals in the socialization of the human organism. It must be admitted immediately that the psychological reinforcements of appropriate status behavior, that is, the nature of the striving instigations and prestige responses which apparently motivate the child's internalization of social controls, is not understood. What "prestige," "approval," "acceptance," and "mastery" are in terms of bio-psychological dynamics we do not yet know. Indeed, the physiologist has not yet satisfied himself concerning the nature of the biological processes underlying the "primary" eating, sexual, and pain responses of man. It seems clear, however, that even these so-called "biological drives" reach their psychological threshold only in socially determined form. Although they are less complex functions of social training than is competition for status, nevertheless, eating, or sexual behavior, or even reaction to post-operative shock certainly includes psychosocial determinants. These cultural formants differentiate between variously socialized men with regard to the bio-psychological instigations of "hunger," "sexual drive" or "anxiety."

In the present state of psychological research, one may simply accept the repeated testimony of observation and experiment that social ("secondary") reinforcements of human behavior exist in great number and complexity. Although this type of motivation is not clear, it certainly involves a seeking for, and responsiveness to, social stimulus and reward —a "drive" which may be observed in children under one year. Whether or not the infant's smiles and vocalizations, his demands for the mother's presence (even when he is not hungry or in pain), and his rejection of food in response to the social stimulus of visitors are responses to stimuli which in turn are conditioned to food, it is apparent that they constitute social intercourse. At the infantile level, the psychological meaning of such social responses is already related to the age-hierarchy in which the infant's exploration of the world must be pursued. In general, then, it appears that *social* instigations and goals (vocalization, smile, caress) are integrated into the motivational pattern of the child as early as the infantile level. If the physiological "drives" are more apparent at this age, they become increasingly obscured as weaning, cleanliness, and genital training are internalized.

A child, therefore, learns (acquires discrimination) not simply by being denied or allowed to achieve biologically pleasant states, but also by experiencing acceptance, approval, or disapproval as expressed in social symbols of age prestige. If the mother appears omnipotent to the child at the age of one, and is therefore clung to tenaciously by him, it is not to be assumed that he values her only, or chiefly, as his avenue to food or rest. She very likely also appears as an adult who can still his anxiety concerning his refusal to eat solid food, or his failure to achieve cleanliness. To the child, in other words, the mother is the adult at the top of his age hierarchy consisting of parents, adult visitors, and children; to the extent that he can control her, he not only quiets his hunger but also stills his anxiety as a relatively helpless being in a household of physically and socially older and more powerful individuals.

If the situation is of this nature, it seems rather useless to speak of a hierarchy of "primary" and "secondary" goal-responses. It seems more promising to face at the outset the problems of the emotional and social processes in learning. The aim of such a study of human socialization would be to develop one construct which would integrate findings dealing with both the affective and the central nervous systems into one conceptual scheme of socialization. The ultimate accomplishment of such an integrated construct of learning, such as is now being explored at several centers, including the Institute of Human Relations, would be to abolish the conceptual dichotomy between biological and psychosocial behavior. It would no longer be necessary to think of affection, anxiety, or the prestige goal responses as being of a different *order* from those which are presumably controlled by the central nervous system. Failure to learn the required social controls, as well as emotional blocks to learning, would also be dealt with under such a theory of learning, which must face the problems of both the socially adaptive and maladaptive roles of anxiety in learning.

Whatever the psychological nature of the prestige responses may be, it seems clear that all forms of status and rank in our society are maintained by the enforcement of biological, emotional, and social privileges. Any physical, geographical, or social pattern may be used at times to symbolize prestige relationships. A child's hair is cut differently from an adult's, and a male's from a female's; in Mississippi, a lower-class Negro's hair is cut and dressed in a style quite unlike that chosen by upper- and upper-middle-class Negroes. An adult sits at the head of the dinner table, a child on the side; the parent's rank is likewise symbolized by his use of the front or master bedroom in middle- and upper-class homes. Lower-class status is expressed geographically by residence "across the tracks," or in the hollows, flats, and most dilapidated housing areas in cities, and on impoverished or tenant lands in the country. In most low-status churches, women are kept from the pulpit and chancel in regular religious services. Language is also employed to symbolize rank; certain phrases and tones are forbidden to children by adults, to Negroes by whites in the South, and to lower-class persons by those of high-class positions. Clothes symbolize not only age and sex rank, but also class position; houses and automobiles express in part class status. Likewise, occupational, financial, associational, and recreational behavior are partly ordered by age, class, and caste in America, and to a lesser degree by sexual status.

Like other social hierarchies, age, sex, and class position give rise to socially controlled antagonisms between the levels of the hierarchy. Patronage, protection, and mastery are the socially prescribed behavior of the superordinate to the subordinate. Deference, compliance, and minor types of sabotage are the behaviors permitted to inferiors in their relationships with superiors in each of these hierarchies. Biological, emotional, and social rewards and punishments maintain the relative privileges of the ranked groups. As socialization proceeds, these controls are internalized as adaptive forms of anxiety. This socially approved anxiety of the child may be an instigation either to strive for the appropriate behavior, or to flee from the unrewarded or punished behavior with regard to age, sex, or class roles.

The child's learning of that behavior which is appropriate to his age and sexual status is motivated not only by social instigations, but also by the emotional interaction between him and his parents and siblings. The history of these affective identifications and hostilities determines the ease with which the age-sex behavior and evaluations are acquired.

For example, the internalization by the child of the feeding, exploratory, property-respecting, cleanliness, and genital controls is a process which leaves both cultural and affective marks upon the organism. In this process, the social realism and psychic ease with which the age-sex roles are learned seems to depend upon the degree of adaptive or maladaptive striving for parental and sibling roles which the child experiences. The strength of the child's emotional identifications in the age-sex hierarchy of the family is as critical as the biological and social reinforcements in his learning of the cultural requirements.

Age-subordination, which most individuals in our society must learn to accept in economic and social relationships until relatively late in life, theoretically would appear an especially difficult adaptation for the first, or only, child to make. During the first eighteen months of life, a child in any birth-position appears to accept the mother as omnipotent and to depend upon her almost entirely for his biological and emotional nurture. When feeding, cleanliness, and other types of training are required by the infant of this age, they must necessarily be instigated and maintained without the aid of prestige motivation. At the very early age in our society when most infants are required to attempt such learning, the age-prestige motivation of "acting like a big boy or girl" or "acting like Daddy or Mother" cannot yet be effective. As they grow older, however, children with siblings near them in age have constantly before them the goal of the older siblings' behavior to pace them in learning the appropriate age-sex behavior. The only child, the first child, or a child separated by about six years from his nearest sibling, on the other hand, has to face a tremendously steep age-barrier. In some instances, the only or first child is stimulated constantly by his parents to strive for adult privileges. Since the only child has no siblings to break the impact of parental stimulation, nor to serve as therapeutic targets for aggression displaced from the parents, his goals are often set too near the adult level.

A child in this position and family-type also faces added pressure to achieve, for he is the class banner-carrier of the family in the next generation. If, in addition, the family is attempting to increase its rank in the class system, the first or only child is most likely to be overstimulated. At the opposite pole is the child who, owing either to "spoiling" or to excessive parental dominance, does not strive for the increasingly complex age-type behaviors. One result of such extreme parental love-demands or of over-protection is to intensify the age-subordination of the child, and to fix lifelong marks of child-status upon his behavior. This consequence seems more likely if the child in question has no siblings near his own age from whom he obtains prestige responses.

Competition for the parents' favor and care (which is the essential factor in rivalry between children in the same family) likewise is expressed through the system of age-privileges. Each sibling has a stake to defend in this struggle and therefore has a real source of anxiety. To "beat the game" of age-privileges with one's siblings is to win evidence of parental favor, acceptance or protection, greater than one's age-role permits. In lower-class families, such competition appears to be moderated by the custom of entrusting the parental role to an older child. Usually each child thus receives his turn to act as parent-surrogate to a younger child. A weak form of this hostility-reducing relation exists as a device in middle and upper-class families, by which the older sibling is allowed to assume minor care and supervision of a younger sibling, or is told: "This is your baby."

The problem of securing to the child in our society an adequate level of aggression for later adult life appears to center in the early management of age roles and privileges in his relationship to parents and siblings. As the child moves upward in the age-hierarchy, his social clique and school, as well as his familial relationships, define the appropriate age behaviors and evaluations. These family, clique ("gang," "peer," or "bunch"), and school roles are psychologically maintained by sanctions which instigate the child toward the allowed prestige goals, and penalize inappropriate age participation. Within the family, bio-social privileges such as food and sweets, clothes, a room to one's self, an allowance, lunch money, courting and sexual exploration, use of the automobile, etc., are age-typed. Forms of play are likewise age-graded by both the family and the child's clique.

The school is our most thoroughly age-graded institution. With com-

pulsory promotion now operating in most public school systems, we have
a form of automatic, involuntary age-grading which has had few parallels
in primitive societies. In the social life of the elementary or secondary
school pupil, great differentiation in rank and clique behavior exists between groups separated by only one or two age-grades. Caroline Tryon
has described by quantitative methods the very marked variations in social
personality roles and recreational patterns among adolescent school children in Berkeley, according to narrow age-gradations.

Probably as a result of the history of their identification with, and
competition for, the parents, children differ with respect to their adaptation to the age-hierarchy of the family and larger society. Some accept
the required role, others strive vigorously for the privileges of a higher
age-group, and still others flee downward from the appropriate age-
demands to an earlier level. The child who pushes hard against the system
of age-rank, fighting by cunning and aggression for the privileges of
older siblings or parents, is likely to meet especial difficulty in the adolescent period. At this time, when he is maturing very rapidly in sexual and
physical status, he excessively resists the sub-adult status which society
fixes upon him by economic, sexual, and educational subordination. When
such upward age–striving children are also basically aggressive toward a
parent, they would be expected to rebel fiercely against sub-adult status in
adolescence. Individuals of this type are probably more likely than are
those who either accept or flee downward from their age-status, to strive
also for participation with a social class higher than that of their parents.
In this upward class striving, they can subordinate the parent by acquiring
etiquette, or educational and occupational symbols which the society
recognizes as superior to those of the parents.

With regard to age-roles, it is also interesting to notice that in relationships between the color castes in the South, or between lower-class and
upper-class individuals of the same color group, the individual of subordinate rank is treated as if he had child status. White servants, for
example, as well as almost all Negroes, are called by their first names by
the high-status whites; a Negro man, furthermore, is called "boy" and a
Negro woman "girl" by most whites.

Sex-typing of behavior and privileges is even more rigid and lasting
in our society than is age-typing. Indeed, sexual status and color-caste
status are the only lifelong forms of rank. In our society, one can escape
them in approved fashion only by death. Whereas sexual mobility is
somewhat less rare today than formerly, sex-inappropriate behavior,
social or physical, is still one of the most severely punished infractions of
our social code.

In most familial, occupational, and political structures the male is
trained for the superordinate roles while the female is restricted largely
to subordinate positions. As in other types of inferior social rankings,
however, the female position allows a certain degree of chronic aggression, sabotage, and cleverness against the superior rank. The modes of ex-

pressing personality traits such as fear, aggression, and affection are also socially typed for sex. The sexual role and personality are trained by the family and school, through their insistence upon sex-appropriate language, clothes, hairdress, gait, pitch and intonation of voice, play, recreation, and work. For most of these sex-appropriate behaviors there is certainly no biological basis of sex-linked traits.

Our understanding of the processes by which sex-appropriate or inappropriate behavior is established is made especially difficult by the early presence of bio-social sexual instigations in the child. There is evidence in clinical and exploratory studies of children, however, to suggest that the child's imitation of a sex-role is functionally related to (1) his early genital training, (2) his learning of the out-marriage rule of the family, and (3) the relative strength of his cross-sex and same-sex identifications with parents. It is certain, moreover, that in middle-class child training the sexual impulses are still heavily tabooed. The degree of severity or abruptness in parental controlling of the sexual impulses seems partly to determine the child's adjustment to the sex-role.

This penalizing of the sex drive itself is intensified in those families where the child is pressed to assume the appropriate sex-role too early or too completely. Parents who are anxious concerning their own sex-typing are likely to overemphasize those controls upon the child. The learning of sex-appropriate behavior also depends upon whether the child's imitation of the same-sex parent is motivated by the effort to escape constant punishment and disfavor, or by positive reinforcements of acceptance and prestige. If he fails to imitate more fully the same-sex parent, owing either (1) to that parent's failure to reward him, or (2) to the cross-sex parent's greater power, or fuller acceptance, the child is likely to reveal the marks of inappropriate sex-typing.

The sexual role, which is first defined by the early family training of the child, is in successful cases greatly strengthened by the sex-typing controls to which he is subjected later. In his social clique, his school, and his formal organizations, the child gains prestige if he learns the sex-appropriate code, rituals, and goals, but meets extreme social—and at times physical—punishment if he does not. Sexual segregation not only is maintained in most kinds of play until full adolescence, but also in the school and church, especially in lower-class environments. In the white or Negro lower-class in the South, men and women seventy years of age, as well as children, are segregated spatially in church or lodge meetings.

In America, as in most societies, the crucial definition of the sex-appropriate role is made at adolescence. The study of Berkeley adolescents by Tryon, previously referred to, indicates that the earlier physical and social maturation of adolescent girls leads to rather serious adjustment problems for the boys of the same age-grade. For a time, sex roles appear to be reversed, but in a year or two the normal adult patterns usually are established. This research leads one to question whether the actual roles of males and females at different periods in childhood and adolescence may not be quite different from the unchanging dichotomy which we popularly assume.

Within broad limits, the expected behavior of age and sex groups in our society differs according to their social class. Within the lower class in Mississippi, for example, white or Negro pre-adolescents and women usually smoke, drink, or curse in public without meeting punishment from their family or clique. The corresponding age and sex groups in the lower-middle class, on the other hand, are forbidden such behavior. Age-sex roles in farm and house work, in family discipline and child-rearing, in school, and in church likewise differ according to class.

A social class system restricts intimate participation to a limited group within a society, above and beyond the age-sex restrictions. Social class relationships are extensions of intimate clique and family relationships; they limit participation where the basis of a *pattern* of traits (such as family rank, plus occupation, plus education, plus manners, plus clothes, plus language, etc.), all of which are differentiated according to rank in the class hierarchy. By defining the group with which an individual may have intimate clique relationships, our social class system narrows his training environment. His social instigations and goals, his symbolic world and its evaluation are largely selected from the narrow culture of that class with which alone he can associate freely.

A child's social learning takes place chiefly in the environment of his family, his family's social clique, and his own social clique. The instigations, goals, and sanctions of both the family and of the intimate clique are a function principally of their class ways, that is, of *the status demands in their part of the society*. The number of class controls and dogmas which a child must learn and struggle continually to maintain, in order to meet his family's status demands as a class unit, is great. Class training of the child ranges all the way from the control of the manner and ritual by which he eats his food to the control of his choice of playmates and of his educational and occupational goals. The times and places for his recreation, the chores required of him by his family, the rooms and articles in the house which he may use, the wearing of certain clothes at certain times, the amount of studying required of him, the economic controls to which he is subjected by his parents, indeed his very conceptions of right and wrong, all vary according to the social class of the child in question.

Our knowledge of social class training and of the biological and psychological differentials in child development as between class environments is now sufficient to enable us to say that no studies can henceforth generalize about "*the* child." We shall always have to ask, "A child of what class, in what environment?" Very few of the statements which one might make concerning the physical growth, the socialization, or the motivation of lower-class children, for example, would hold for upper-middle-class children, even in the same city.

Class ways in child training, as well as the class-motivating factors in the child's social learning, differ sharply even when the observer considers only the classes having low status. The social instigations and goals of the lower-middle class, for example, are fundamentally unlike those of the lower class. In education, the ineffectiveness of middle-class sanctions upon the great masses of lower-class children probably is the crucial dilemma of our thoroughly middle-class teachers and school systems. The processes underlying this failure are not yet clear, but it seems probable from life histories that lower-class children remain "unsocialized" and "unmotivated" (from the viewpoint of middle-class culture) because (1) they are humiliated and punished too severely in the school *for having the lower-class culture* which their own mothers, fathers, and siblings approve, and (2) because the most powerful reinforcements in learning, namely, those of emotional and social reward, are systematically denied to the lower-class child by the systems of privilege existing in the school and in the larger society.

The culture which the child brings to school with him has been instilled by the class environment of his family and his intimate associates. Except in the case of class-striving families and children, this culture is maintained by this same status-bound class world, undergoing relatively slight modifications from classroom instruction. In the middle-class family and environment, it is true, the teacher meets support for her methods and goals in child training. If she is to pit herself against the lower- and upper-class child and family, who are from a quite different culture, however, she and the school administration have need for socially more skillful methods and less ethnocentric, middle-class bias with regard to manners, aggression, and recreation than they now reveal.

For in all these last named patterns of behavior, child training in the lower class and lower-middle class which have been selected here for illustrative purposes, differ markedly. In the lower-middle class, parents exert a strenuous and unrelenting push to motivate their children to study their lessons, to repress aggression at school, to inhibit sexual impulses, to avoid lower-class playmates, to attend Sunday School regularly, to avoid cabarets, beer parlors, pool parlors, and gambling houses. They keep steadily before the child, often in the face of economic disaster, the status goals of a "nice" play group and social clique, a high-school education, skilled or white-collar occupation, and a "good" middle-class marriage.

In lower-class white or Negro society, on the other hand, a child lives in a different cultural environment; he is surrounded by people who have habits quite different from those of the lower-middle class, and who make other demands and set different goals before him. Among lower-class urban *whites* in the South, for example, extramarital partnerships are common for both husband and wife; separations are the rule; fighting, shooting, cutting, gambling, and frequently magic are accepted classways; church and lodge participations scarcely exist. With regard to sex, education, occupation, recreation, and marriage, the goals which the lower-class family, white or Negro, sets before the child are basically unlike those in the lower-middle-class family. This difference is greatest in those areas of behavior which middle-class society most strongly controls, i.e., aggression, sex responses, and property rights.

As the middle-class child grows older, the effective rewards in maintaining learning are increasingly those of status; they are associated with the prestige of middle- or upper-class rank and culture. The class goals in education, occupation, and status are made to appear real, valuable, and certain to him because he actually begins to experience in his school, clique, and family life some of the prestige responses. The lower-class child, however, *learns* by *not* being rewarded in these prestige relationships that the middle-class goals and gains are neither likely nor desirable for one in his position. He discovers by trial-and-error learning that he is not going to be rewarded in terms of these long-range, status goals, if he is a "good little boy," if he avoids the sexual and recreational exploration available to him in his lower-class environment, or if he studies his lessons. In this learning, he is often more realistic than his teacher, if one judges by the actual cultural role which the society affords him.

In order to motivate the great masses of lower-class children who crowd our elementary and secondary schools so that they will learn the educational and technical skills, the sexual and aggressive controls, and the manners which will enable them to gain higher privileges and greater social and economic efficiency, educators must first know lower-class culture and understand the instigations and goals of the lower-class child. If these old habits and reinforcements of the lower-class child are to be replaced by new learning which will enable the school to recruit the child into the middle-class way of life (with an attendant increase in the social and economic efficiency of our society), the school must (1) remove the class punishments from the lower-class child within the school society and (2) concretely reward his tentative striving for prestige in the school community. The striving which the middle-class pupil exhibits is driven by socially adaptive forms of anxiety, learned in his class world. As yet, it seems, our society must depend upon this process for maintaining the long-range instigations which effective socialization in the high-skill roles demands. In order to reinforce the lower-class child in such striving, the teacher and social worker must learn to reward him. To be capable of this type of education, they must be able to view their own middle-class status and culture with a wholesome degree of objectivity.

37

Alex Inkeles

SOME SOCIOLOGICAL OBSERVATIONS ON CULTURE AND PERSONALITY STUDIES

Dr. Inkeles's paper brings both a word of caution and a chart of little-explored fields, closing this section on the Interrelations among the Determinants with an indication of how much more remains to be done.

STUDY of the interrelations of personality and society represents a major focus for the articulation of theory and research in anthropology, sociology, and psychology. The bulk of the investigations commonly referred to as studies in "culture and personality" have, however, been concerned with only one major dimension of the field, largely neglecting, as their title indicates, the interrelations between personality and social structure [17, 22]. This paper will, therefore, address itself primarily to an examination of some sociological considerations which have special relevance for the study of personality in society.

To set this discussion in its general setting, four major foci in the literature on personality in society should be noted. These are: 1) the study of social and cultural influences reflected in the individual personality; 2) the delineation of group regularities in the personalities of populations in given societies; 3) the specification of casual factors accounting

NOTE: This paper was written especially for this volume and represents its first publication.

[577

for these observed regularities; and 4) the exploration of interrelations between social systems and observed group regularities in personality. Although the first three foci will be briefly commented on, our attention will be mainly concentrated on the fourth.

1. *The study of socio-cultural influences reflected in the individual personality*. An understanding of the adult personality may, of course, be approached predominantly in terms of biologically given characteristics such as the prolonged dependency of the human infant, or biological potentialities such as the human maturational cycle on which Freudian and neo-Freudian personality theory rests so heavily. The adult personality may also be examined largely in terms of learning theories, which view the individual primarily as the end product of a process of conditioning of behavioral response in a matrix of reward and punishment patterns. In recent decades, however, more effort has been made to understand the individual's behavior as expressive of value orientations and as representative of patterned or statistically normative ways of acting which reflect his development in a given cultural milieu and a specific institutional environment.

The center of interest in such explorations is primarily in the individual as such, and socio-cultural influences are considered primarily as an aid to understanding the dynamics of the unique personality. The development of effective methods for studying the socio-cultural determinants of the individual personality is, however, clearly important as a necessary pre-condition for adequate investigation of socio-cultural influences on the personality patterns of larger numbers of individuals. In addition, it should be noted that some effort has recently been made to use the individual personality as a source of data on strains and tensions in the social system [1].

2. *The delineation of group regularities in personality in the populations of given societies*. The basic problem facing researchers in this area is that of selecting from among the enormous range of personality patterns a manageable number of strategically important personality variables to be investigated, and then of developing reliable and valid criteria for the measurement of those variables in a total population or an adequate sample thereof. Adequate performance of this operation is clearly crucial if further steps are to be taken in relating group regularities in personality to cultural and social structural variables. Unfortunately, in the existing studies the selection of the components of personality to be studied has frequently been haphazard, impressionistic, and arbitrary. Little has been done to insure comparability from one study to another. In addition, the existing studies frequently give so little specification of the methods whereby regularities were delineated, that it is difficult if not impossible to replicate and thereby verify findings. Although this is in part a product of the lack of adequate instruments for recording and measuring personality data, it is no less a product of failure to apply rigorous thought to the problem.

The term "national character," for example, has been used in the broadest possible sense as being more or less synonomous with all learned cultural behavior [4, 27, 28]. If this definition applies, then to relate character to culture is largely to relate character to itself. Clearly so global a concept does not permit the precise gathering of data. The suggestion of Bateson [3] that character differences be measured in terms of standardized behavior revolving around such themes as dominance-submission and succoring-dependence relations is much more concrete and amenable to precise measurement [19]. But in this instance, as in the case of many other fruitful suggestions of criteria for measuring regularities in personality, little has yet been done to convert the idea into an instrument for recording and measurement which can actually be placed in use in studying given populations. Even insofar as there exist clinical instruments such as the Rorschach and the Thematic Apperception Test, which appear to have potentialities for measuring regularities in personality in a given population, there remains the question of their proper use. Until such time as data are collected with such instruments from adequate numbers of individuals selected according to appropriate sampling procedures, the findings must be regarded as little more than suggestive. The importance of such reservations is, of course, greatly increased when, as in the case of modern industrial society, the populations under study are large, heterogeneous, and characterized by widely variant life experiences.

In much of the available literature, furthermore, the effort to delineate group regularities in personality has largely been equated with the search for uniformities or elements common to the majority of the total population [15]. We leave for later discussion the theoretical relevance for the functioning of social systems of such shared characteristics, and speak here only of the problems involved in delineating such regularities. Many, if not the great majority, of the enduring national societies have populations composed of highly heterogeneous subgroups with distinctive sub-cultures, to a degree which makes the assumption of common or uniform *personality* patterns highly dubious. This was most forthrightly stated by Ruth Benedict [4], who declared that "no anthropologist would attempt to study 'the' character structure of such a welter of cultures as were included within the national boundaries of Yugoslavia." Similar, although lesser difficulties, she continued, are posed by the different regions of England, France and the United States. To handle these areas, according to Benedict, it is clearly necessary "to multiply the number of investigations up to any desired point . . . ," and there is the likelihood that the number of character structures which will be discovered may be as large as the total number of investigations. It is suggested here that in modern national states the "welter" of class and occupational groups, to name but two additional variables, requires the same multiplication of investigations.

Although these facts have been explicitly acknowledged by most students of personality regularities in national populations, they show an unfortunate propensity for slipping inadvertently into a search for and description of "the" national character of a given population such as that of the United States [14, 27]. Although this is understandable in terms of the strain placed on the researcher who is concerned with uniformities, nevertheless the demands of methodological rigor require that it be persistently called to attention. It may be suggested that many of these difficulties would be obviated if there were a shift in theoretical emphasis away from the search for "national character" towards the delineation of *modal* personality structures.

This carries the implication that in any population there may be several or many modal personality structures, which is clearly more compatible with expectations created by the size and diversity of population in many contemporary large scale industrial societies. Since the modal personality structure types, or their distribution, could be different in different national populations, this approach does not deny the basic premise on which national character studies rest. At the same time, by facilitating comparisons between groups which are from different national populations but which occupy comparable positions in their respective social systems, it would permit important explorations into the relation of personality and society on a comparative structural basis cutting across national and cultural lines.

3. *Specification of the causal factors accounting for observed group regularities in personality.* The most intense controversy which has revolved around culture and personality studies has concerned their more or less common emphasis on infant care disciplines as the most crucial factor in shaping adult personality. This assumption, along with the proposition that most children raised in the same culture undergo essentially the same crucial early infant and childhood experiences, constitute the two major theoretical foundations for the greater part of the so-called culture and personality studies.

The adequacy of the first assumption has recently been the subject of very extensive critical review [12, 25, 33] and will therefore be commented on only briefly. Two major difficulties have been manifested in many of the efforts to relate specific infant care disciplines to adult personality. The first of these derives from the prevalence of a marked tendency to analogical reasoning. Gorer, for example, sees the scheduled feeding to which American infants are frequently exposed as producing in American adults the symptomatic fear "that America will be reduced to want, perhaps to actual starvation, if it lets its food or resources or money outside of the country [14]." The direction of the analogy may be reversed, as in the case of Erikson's alleged expectation of tensions surrounding toilet training in the Yurok on the basis of "anal character" traits found in the adult [13]. There is a widespread feeling on the part of many critics that an adequate and necessary connection has not been established between such patterns of infant training and specific characteristics of adult behavior. This is not to deny the existence of any connection between these two phenomena, but rather to emphasize the need for a fuller recognition of the fact that the adult personality is the end product of a long developmental process in which the infant training disciplines are simply one important starting point.

This leads directly to a second major difficulty. Many explorations of the relationship between infant care disciplines and adult character, while acknowledging the importance of other forces in shaping personality, have a tendency in practice relatively to neglect the variables intervening between infancy and adulthood. Even if one accepts the proposition that the major "core" components of the personality such as patterns of aggression and attitudes towards authority are laid down in infancy, it hardly necessarily follows that experience in childhood associations, adolescent groupings, and early adult activity will be so congruent with infant training as to support and reinforce these early tendencies. The anthropologist's concern with the unifying themes and patterns of culture manifested in various aspects of the social system naturally leads him to seek and perhaps to "see" such support and reinforcement. Legitimate and important as this search for unifying themes is, it should not obscure the important and universal fact of discontinuities in the cultural conditioning of the child to which Ruth Benedict has called attention so persuasively [5]. And in the institutional diversity and heterogeneity of modern industrial society, the possibilities of such discontinuities in conditioning and experience are multiplied. The emphasis on infant training disciplines, premised on the intimacy of their connection with the formation of core personality components, has led in culture and personality studies to extreme concentration on the family and other related primary social groupings. This has meant an unfortunate neglect of the major economic, political, and (nonprimary) social institutions in which the maturing individual becomes progressively more involved and in interaction with which he may undergo many significant changes or at least adaptations in the core components of his personality.

The proposition that most children raised in the same culture undergo essentially the same crucial early infant and childhood experiences has perhaps been less subject to critical comment. It is, however, one with which the student of social structure is centrally concerned. It may be assumed that accidental factors in the development of particular individuals, chance variation in family patterns, and similar influences will not exert a statistically significant effect on the regularities in personality manifested in any population. Patterned institutionalized differences in the infant and childhood experiences of major subgroups in any population, however, can be expected to yield important subgroup differences in modal personality structure.

It seems appropriate, therefore, to reiterate here the point made in the preceding section. In dealing with complex societies adequate account must be taken of the fact that the same social system may include a variety of "cultures," each with its own distinctive patterns for infant care. Indeed on this very point there is considerable evidence that there are crucial major differences in child rearing patterns along regional, class, and ethnic lines in the United States [6, 7, 8, 21]. Such differences, furthermore, are not limited to instances in which there are important subcultures along the lines indicated, but may under conditions of rapid social change be manifested along generational lines within groups which have the same broad cultural heritage [29]. Such variations, particularly those across class, regional, and generational lines, need not be unrelated, and indeed they are likely to be variations in a given area such as feeding or even on a central theme such as dominance-submission. But insofar as one holds to the position that patterns of infant care are crucially related to personality, it follows that such group variants hold at least the potentiality of producing variations in personality and related action patterns of strategic importance in their consequences for social structure.

4. *Specification of the interrelations between the social system and observed group regularities in personality.* The patterns of infant care and socialization discussed in Section 3 may legitimately be considered part of the social system. The decision to treat them separately was therefore an arbitrary one made primarily in recognition of the importance attached to that area in the existing literature. We turn now to the more general problem of which the relation between infant care disciplines and adult personality is but one facet.

Any social system may be viewed as a complex of functionally interrelated institutions. Each institution, in turn, involves a complex of status positions, in each of which inhere certain specific role patterns. Continued functioning of the system depends upon the regular and adequate performance of the social acts which constitute the given role pattern. These social acts are carried out by individuals who must meet certain specific requirements before assignment to the given status positions. Assignment to the role, however, does not guarantee adequate performance. In the case of all achieved as against ascribed statuses [26], and even in the case of ascribed statuses beyond those biologically determined such as age and sex, the society must develop in the requisite number of individuals qualities which make possible their assignment to any given status. And following this assignment, the society must insure the required level of role performance.

Clearly many factors which are not of central interest in this paper enter into this situation. In the case of occupational status, for example, there are institutionalized means for training the specific requisite technical skills which make assignment possible. To secure continued adequate performance of the role there is the host of sanctions and other social pressures, which are generally systematically treated in the study of any society. It is clear, however, that there is an additional element which operates to insure the availability of people who meet the specifications for given status positions, and which secures the effective performance of roles once they are assumed by the individual. That element is, in its broadest sense, human motivation. More specifically, it is the fact that individuals have values, attitudes, and sentiments which predispose them to act in certain ways under specified social conditions. The precise content of these values, attitudes, and sentiments determine the social personality of the individual, and group variations in such content may serve to differentiate the populations of different societies and sub-populations in the same society.

Before proceeding to examine the implications of this position, it is essential to clarify and amplify the meaning of the term "social personality" as used here. All statements about personality must be recognized as abstractions from individual behavior, for personality is not directly accessible to the senses as is the physical person. Inevitably, therefore, there is a great deal of variation among students of personality as to which aspects of behavior they emphasize in assessing personality, as well as in the assessment they make on the basis of that behavior. For our purposes it is not necessary to pursue the merits of any particular scheme, but simply to recognize that generally personality is treated at one or more of three different levels.

The first of these levels involves statements about personality "in depth," which deal with basic characteristics or mechanisms of the personality and utilize such concepts as aggressiveness, anality, depressive tendencies, etc. There is a second or intermediate level of statement about personality, which describes some of the more specific yet highly generalized orientations of the personality to itself and to significant others. An example of this level of description may be found in Geoffrey Gorer's assertion that Great Russians tend to view all authority as "evil, suspicious, and pervasive [16]." Finally, at the third level, there are statements which are primarily descriptions of some aspect of the more or less conscious idea systems of individuals. These idea systems are composed of the concrete, specific, and interrelated beliefs, attitudes, values, and sentiments which individuals directly express in speech and related behavior. As an example one might take the definition of good government which is expressed in the statement (frequently made by former Soviet citizens): "A government should have the same relation to its citizens as a parent to his child—it should support, aid, and nurture its people."

It must be admitted, of course, that the levels of personality distinguished here are intimately linked, and that any one of the levels may be used as an avenue of approach to the others. At the same time, it is not firmly established that there is a simple one-to-one correspondence between the content of the various levels. Indeed, common characteristics at the depth level may be manifested in relatively diverse orientations at the level of idea systems and in consequent behavior. Thus, deep seated dependency needs may underlie an ideology which extols powerful, omniscient leaders, or one which involves over-assertive demands for individual "independence" and for the elimination of "restraints"; distrust of authority may be manifested in glorification of evasive and rebellious behavior against all authority, or in the compulsive insistence on the need for a "good" authority. For the student of social structure such idea systems imbedded in attitudes, values, and sentiments are of prime importance. For these idea systems have crucial relevance to the individual's social action, and consequently for the social system.

Since psychoanalytic psychology has been almost the sole major psychological school to influence American anthropologists [20], it is to be expected that culture and personality studies would concentrate on the depth level in their descriptions of personality regularities in given populations. Although some attention has been given to the intermediate level of personality, there has been relatively little done to elaborate the specific content of the attitudes, values, and sentiments which have here been described as making up the social personality. And where such components have been dealt with, the areas concerned have been largely those of primary relationships—love, friendship, marriage, etc.—rather than the areas of major economic and socio-political institutional involvement.

Yet the relative neglect of major economic and socio-political institutions in the existing literature on personality in society may be charged only in part to the fact that culture and personality studies have focused on culture rather than social structure, and on basic personality rather than what has here been called social personality. It is a product, as well, of sociological concentration on delineating "structure" rather than on analyzing "functioning," and of general sociological neglect of the role of psychological factors in the operation of social systems. Just as it is important for the anthropologist to give more attention to the relations of group regularities in personality to social structure, so is it imperative that the sociologist focus attention on the individuals whose social action makes possible the functioning of the structures he delineates.

To illustrate the point being made here it is appropriate to turn to some concrete examples. Although there has been no systematic effort to explore the question of whether or not there are personality patterns requisite to the functioning of *any* society [2], there have been attempts to describe traits or characteristics of personality which are generally important to the functioning of certain specific social systems. In the case of Pueblo society, for example, Esther Goldfrank [11] has noted that the dependence upon water control as a means to survival under extremely adverse agricultural conditions requires extensive co-operation among adult males in the fields, in the preparation and maintenance of irrigation facilities, in the construction of houses, etc. Under these conditions it is the "deeply disciplined" man—either not characterized by, or able to suppress initiative, ambition, and intense personal loyalties—who is essential to the effective functioning of Hopi and Zuñi society. As is to be expected, this personality type represents both the social ideal and apparently the actual statistical norm in the Pueblo populations. Although the means by which this personality type is developed is not a central concern here, it is relevant to indicate Goldfrank's view that the infant care and training disciplines would have led one to anticipate a rather different pattern in adult personality. In her view the explanation is to be found in experiences in later childhood and adolescence, reinforced by the cultural milieu of adult life.

Another example is provided in Goldschmidt's analysis of the Yurok-Hupa societies of Northwest California [13], which is of particular interest because of the close correspondence between the socio-economic structures and the ethico-religious systems of those societies and that of Western Protestant nations. According to Goldschmidt, Yurok-Hupa society is characterized by concern with property, money, and wealth; an open-class system in which mobility is emphasized and power and prestige are intimately linked with the possession of goods; and an ethical system emphasizing a moral demand for work and the pursuit of wealth, for self-denial, and for individual responsibility. He sees these structural features as rewarding certain personality configurations—aggressiveness, hostility, competitiveness, loneliness, and penuriousness—to such an extent that they come to dominate the social scene and to be transmitted from parents to children. And he concludes that "indeed it is hard to see how a Hupa or Yurok could operate effectively in his society without these traits."

In dealing with more complex large-scale societies, however, it does not appear that one can anticipate finding traits or personality complexes having such pervasive importance for the effective functioning of the society. This is, of course, an empirical question which should not be prejudged. The exposition here, however, will be restricted to delineating personality patterns which appear crucial to the effective functioning of specific segments of the social system in modern industrial society.

In the case of the American occupational structure, for example, Talcott Parsons [34] has stressed that it is characterized by the relative predominance of patterns of rationality, universalism, and functional specificity of roles. Clearly the ability to function in roles requiring such action patterns depends upon the presence of certain appropriate constellations in the personalities of the participants in the occupational structure. It appears that our society develops adequate numbers of individuals whose personality is such as to enable them to operate in such roles, at least to the extent necessary to maintain both their individual personality integration and the effective functioning of the occupational structure. It also appears that different cultural and social milieus may produce individuals who, by and large, cannot function effectively in such contexts. Kardiner [18] has noted, for example, that the Comanche could not perform effectively in a situation demanding the patterns of behavior Parsons has delineated, and this may be assumed to apply quite apart from the differences in the formal institutions of the societies concerned. It is, of course, not necessary to go so far afield. There is evidence to indicate, for example, that the Spanish-American population in the American Southwest is able to function within the framework of our occupational structure only with severe strain for the individuals concerned, to leave aside for the moment the strain on the occupational structure itself.

As another example of a problem of interest to the student of social structure which also relates directly to the study of personality in society, we may consider certain aspects of the American kinship system [35]. One of the central patterns in that system involves the establishment of the independent conjugal home upon marriage, with relatively abrupt and complete separation from the family of origin of both husband and wife. Although it has taken its most complete development in the United States, modifications of this pattern appear in a variety of contexts in Western industrial society. There are, of course, many structural factors which account for the pattern. In particular, it appears to be associated with industrialization, with its emphasis on assignment to occupational roles on the basis of technical rather than traditional considerations, its requirement of labor mobility, the development of large urban centers, etc. The intensification of the pattern in the United States is probably a result of our extreme emphasis on individual social mobility based on achieved status in occupational roles, relatively independently of the status of the family of origin. The nature of the break from the family of origin is socially and psychologically considerable, however, and seems therefore to be associated with distinctive courtship and marriage patterns which have been summed up in the expression "romantic love complex."

It appears that romantic love is a behavior pattern of which very large numbers of Americans are not only capable, but which they manifest with a high degree of regularity. This manifestation of personality, when combined with the appropriate social pressures, insures the continued operation of our courtship and marriage patterns. At the same time, it must be recognized that "romantic love" is relatively unusual in the known cultures of the world and to the participants in a great many cultures is essentially incomprehensible. This certainly need not be considered a matter of accident, since in societies where the family structure demands extensive restrictions on the choice of marriage partners, and in many instances precisely prescribes those partners, the appearance of romantic love patterns as we know them would be exceedingly disruptive to the existing system. Clearly it is important to investigate the process by which such personality orientations are developed in American and other societies, and by which they are prevented from appearing in a significant number of individuals in certain other social milieus.

The pattern of analysis suggested here can of course be extended to almost any institutional complex in any social system. Essentially the same pattern could well be applied with varying degrees of refinement. In the case of the occupational structure dealt with above, for example, fruitful exploration could be made of particular occupational realms such as bureaucracy, the professions, industrial labor, etc., on down to the level of particular professions. Fortunately some highly suggestive work along these lines has already been undertaken, particularly that by C. W. Mills on the competitive "sales" personality [32]. In the case of the family patterns mentioned, it would be fruitful to explore further the social conditioning of some of the widespread emotional reactions, such as sexual jealousy, following the model of analysis provided by Kingsley Davis [9].

The variety of roles and role clusters made available in institutionally diversified large-scale social systems permits effective functioning for so wide a range of personality patterns that only the most general congruence of "basic" personality with the over-all social structure is required as a minimum condition for individual participation in the system. In such societies, furthermore, the growth of formal organization, with its attendant elaborate division of labor, permits a surprising degree of participation in major social roles whose chief requirement is formal adherence to impersonal rules and routine performance of mechanical tasks without necessary extensive involvement of the deeper lying elements of personality. The research orientation proposed here, therefore, is one that emphasizes isolating statistically relevant modes in the manifestation of carefully limited and selected components of social personality assumed to be strategically related to the functioning of specific institutional structures. The basic assumption underlying this more limited and partial approach to the study of personality in society is that the functioning of a given institution requires only that a specific number of individuals participating in the relevant positions manifest the underlying personality patterns relevant to the performance of their particular roles. It is not necessary for other members of the society to have the same general personality constellations, nor is it necessary for the individuals to do more than manifest specific forms of behavior without commitment of their total personality.

This is not meant to prejudge the question of the limits on the extent to which individuals can function in diverse institutional roles so constituted that the relevant personality characteristics most adaptive to the given roles are non-congruent. Nor does this approach render less legitimate the search for uniformities, for traits manifested by all or most members of a society, and the attempt to treat the total personality. Yet it must be recognized that the concentration in the existing culture and personality literature on such general uniformities, and on the total personality, has served to deflect attention away from those crucial aspects of social personality which most directly relate to the functioning of specific social institutions. This may in large degree account for the dissatisfaction which representatives of other disciplines, notably economics and political science, have expressed in relation to the existing studies of national character. It has been pointed out [23], of course, that frequently the fault lies with the user of these materials, whose lack of training in dealing with psycho-cultural data does not enable him to make adequate use of the hypotheses offered. Perhaps with equal frequency, however, the difficulty arises from the more or less exclusive emphasis on the relation of personality to culture as against its relation to the functioning of major political and socio-economic institutions.

*   *   *   *   *

The pattern of analysis suggested here will undoubtedly strike many readers as indicating belief in some kind of social teleology, that is, the assumption that the society is able to "determine" for itself the kinds of personality constellations necessary for its effective operation, and then sets about producing those personalities in the requisite numbers. This is a very important critical point which demands careful consideration.

Taken at a single point in time, any social system can be seen as a more or less integrated institutional structure in which the balance of cultural influences and institutional pressures tends on the whole, through its influence on significant parental figures, to produce a socialization process for the child, and to create a socio-cultural milieu for the adolescent and adult, which yield personality constellations generally adaptive for life in the given society. Clearly this is not the time or place to explore this integration of personality and social system. It is essential, however, to give due weight to the fact that at the time it is analyzed any social system represents simply a point in a continuous historical process, during which the relative integration of the elements of the society has been effected. That integration between the personality patterns developed in the population of any society and the requirements of the roles provided and required by the ongoing social system may be highly imperfect. And since personality resides in individuals and not in social systems, that personality, once developed, becomes a more or less independent entity which is in interaction with the social system. Consequently, group regularities in the personality structure of given populations, however much initially determined by the social system, may play a major role in determining the functioning of the social system and the institutional components thereof, and may result in major changes therein.

The most general exploration of the effects of personality constellations on social institutions is to be found in the work of Abram Kardiner and his associates [18]. Working, however, primarily with limited aspects of personality, in particular depth components which he calls basic personality structure, Kardiner has found that of necessity he has been limited to statements about the impact of personality on projective systems, notably religion and folklore. In the case of non-literate societies, Kardiner affirms that these projective systems can be shown to be very largely determined by the basic personality structure common to a given population and created by common infant-care and child-rearing practices. We need not concern ourselves here with the merits of Kardiner's formulation. From the point of view of the student of social structure, however, it is necessary to indicate that Kardiner found that basic personality was not adequate to account for institutional systems of a non-projective nature, and particularly for those he described as the "empirical-reality systems," including instrumental processes, technical knowledge, science, political and economic institutions, etc.

It is obvious, of course, that it is precisely with such institutional components that the student of social structure is concerned. In addition, it must be recognized that as the social system grows in complexity to the level of large-scale industrialized societies the number, scope, and diversity of these empirical-reality systems grows enormously. As a consequence of this fact, that portion of all socially relevant behavior which falls outside the realm of possible explanation in terms of the basic personality concept as used by Kardiner also increases enormously. This is painfully clear in the difficulties, of which Kardiner was not unaware, which he experienced in the attempt to analyze the materials on Plainville, U.S.A. It is my opinion that the difficulty inheres in the severely restricted approach to personality represented by the basic personality structure concept. A conception which takes in more and broader dimensions of personality, approximating the notion of social personality mentioned earlier in this article, would certainly greatly facilitate exploration of the impact of socially conditioned group regularities in personality on the functioning of social institutions.

An example of the type of study needed may be found in Merton's paper on bureaucratic structure and personality [31]. Merton notes that the operation of a bureaucracy depends upon rigorous discipline manifested in high reliability of behavior and adherence to prescribed rules. As a particular example of the general proposition presented earlier in this paper, he notes that this discipline is insured and buttressed by strong appropriate attitudes and sentiments in the bureaucratic personnel. Bureaucratic structure includes elements for inculcating and reinforcing such sentiments, and, as Merton recognizes, there is also the strong likelihood that bureaucratic organization both selects and attracts the appropriate personality types. The factors which account for these personality orientations are not at issue here. The central fact to be noted for our purposes is that as a result of an apparent displacement of the original goals of the bureaucracy, which results from identification with discipline and conformity to rules as an immediate value for the bureaucratic personnel, rigidities are introduced into the bureaucracy's operation which may in significant degree serve to prevent it from most efficiently performing the social purposes for which it is created.

This pattern of analysis could fruitfully be extended to other aspects of the social system including the occupational structure, the kinship system, political institutions, etc. It should be noted, in this connection, that if such studies were based on a concept of social personality which included basic attitudes, values, and sentiments, important opportunities would be provided for the utilization of the valuable data accumulated in public opinion polls and surveys which to date have been inadequately utilized in national character studies. At the same time, such an orientation could serve as a challenge and stimulus to those engaged in making opinion and attitude surveys to collect data with greater relevance to the problems of students of the structure and functioning of social systems.

Despite such promising leads, however, the effort to explain the functioning of specific institutions, and particularly of complex social systems, on the basis of observed group regularities in personality, must be approached with extreme caution. The complexity of the forces at work and the relatively uncertain state of our knowledge and methodology in this area set marked limits on the conclusions which may safely be ventured.

Indeed it is most unfortunate for the development of future work in this direction that individuals, not adequately equipped or predisposed to handle the complexities of the analysis of large-scale institutional functioning and change, have made such sweeping generalizations in their studies of national character. The most striking manifestation of this tendency is perhaps to be found in the work of Geoffrey Gorer, particularly in his work on the Russian people [16], but he certainly is not alone in this respect. This is not the place to attempt a systematic analysis of the defects of Gorer's work, and indeed this has been effectively done elsewhere [12]. The essential point which must be recognized is that the main features of large-scale institutional components of modern society, such as a system of factory production, are determined by forces intimately related to the specific functions which those institutions must perform. They are, therefore, relatively uniform from society to society, and in their main structural features are relatively independent of the influence of group regularities in personality.

This is not to say, of course, that particular patterns within the framework of such an institutional structure may not be significantly affected by such regularities of a cultural or characterological nature. Neither is it to say that characterological forces may not play a major role in determining the nature and direction of large-scale social change. Such characterological forces must be seen, however, as but one element in a larger field in which a great many other semi-independent forces are simultaneously operating. The diversity of the forces operating in the field places serious limitations on any effort to explain major social process predominantly in terms of group character structure. By contrast, as has been well illustrated in some of the work on Germany [10] and notably that of Kurt Lewin [24] and Talcott Parsons [36], emphasis on the social impact of characterological regularities in a given population can serve a tremendously valuable function in enriching the analysis of major institutional process when integrated with other established modes of social analysis.

REFERENCES

1. Aberle, David F.: *The Psychosocial Analysis of a Hopi Life History, Comparative Psychology Monographs,* Vol. 21, No. 1 (Berkeley, University of California, December, 1951).
2. Aberle, D., Cohen, A., Davis, A., Levy, M., and Sutton, F.: "The Functional Prerequisites of a Society," *Ethics,* Vol. 60 (1950), pp. 100–11.
3. Bateson, Gregory: "Some Systematic Approaches to the Study of Culture and Personality," *Character and Personality,* Vol. 11 (1942), pp. 76–82.
4. Benedict, Ruth: "The Study of Cultural Patterns in European Nations," *Trans. N. Y. Acad. of Sci.,* Ser. II, Vol. 8 (1946), pp. 274–9.
5. ——: "Continuities and Discontinuities in Cultural Conditioning," *Psychiatry,* Vol. I (1938), pp. 161–7.
6. Dai, Bingham: "Some Problems of Personality Development among Negro Children," reprinted in Kluckhohn and Murray: *Personality in Nature, Society, and Culture,* pp. 437–58.
7. Davis, Allison: "American Status Systems and the Socialization of the Child," *Am. Soc. Rev.,* Vol. 6 (1941), pp. 345–54.
8. Davis, Allison and Havighurst, R. J.: "Social Class and Color Differences in Child-Rearing," *Am. Soc. Rev.,* Vol. 11 (1946), pp. 698–710.
9. Davis, Kingsley: "Jealousy and Sexual Property," *Social Forces,* Vol. 14 (1936), pp. 395–405.
10. "Germany After the War," *Am. J. of Orthopsych.,* Vol. 15 (1945), pp. 381–441.
11. Goldfrank, Esther: "Socialization, Personality, and the Structure of Pueblo Society," *American Anthropologist,* Vol. 47 (1945), pp. 516–39.
12. Goldman, Irving: "Psychiatric Interpretation of Russian History, A Reply to Geoffrey Gorer," *The American Slavic and East European Review,* Vol. 9 (1950), pp. 151–61.
13. Goldschmidt, Walter: "Ethics and the Structure of Society: An Ethnological Contribution to the Sociology of Knowledge," *American Anthropologist,* Vol. 53, No. 4 (Oct.–Dec., 1951).
14. Gorer, Geoffrey: *The American People: A Study in National Character* (New York, Norton, 1948).
15. ——: "The Concept of National Character," *Science News* No. 18, pp. 105–22 (Penguin Books, Hammondworth, Middlesex, 1950).
16. ——: *The People of Great Russia* (London, Cresset Press, 1949).
17. Haring, Douglas G.: *Personal Character and Cultural Milieu,* revised edition (Syracuse, Syracuse University Press, 1949).
18. Kardiner, Abram (with Ralph Linton, Cora Du Bois, James West): *The Psychological Frontiers of Society* (New York, Columbia University Press, 1945).
19. Klineberg, Otto: "A Science of National Character," Bulletin of the Society for the Psychological Study of Social Issues, *Journal of Social Psychology,* Vol. 19 (1944), pp. 147–63.
20. Kluckhohn, Cylde: "The Influence of Psychiatry on Anthropology in America During the Past One Hundred Years," *One Hundred Years of*

American Psychiatry (New York, Columbia University Press, 1944), pp. 589–617.

21. Kluckhohn, Clyde and Florence R.: "American Culture: Generalized Orientations and Class Patterns," *Conference on Philosophy, Science and Religion* (New York, Harper, 1946), pp. 106–28.

22. Kluckhohn, Clyde and Murray, Henry A.: *Personality in Nature, Society and Culture* (New York, Knopf, 1948).

23. Leites, Nathan: "Psycho-Cultural Hypotheses about Political Acts," *World Politics*, Vol. 1 (1948), pp. 102–19.

24. Lewin, Kurt: *Resolving Social Conflicts* (New York, Harpers, 1948).

25. Lindesmith, A. R. and Strauss, A. L.: "A Critique of Culture-Personality Writings," *American Sociological Review*, Vol. 15 (1950), pp. 587–600.

26. Linton, Ralph: *The Study of Man* (New York, Appleton-Century, 1936).

27. Mead, Margaret: *And Keep Your Powder Dry* (New York, William Morrow, 1943).

28. ——: "The Application of Anthropological Techniques to Cross National Communication," *Trans. N. Y. Acad. of Sci.*, Ser. II, Vol. 9 (1947), pp. 133–52.

29. ——: "The Implications of Culture Change for Personality Development," *American Journal of Orthopsychiatry*, Vol. 17 (1947), pp. 633–46.

30. Merton, Robert K.: "Social Structure and Anomie," *American Sociological Review*, Vol. 3 (1938), pp. 672–82.

31. ——: "Bureaucratic Structure and Personality," *Social Forces*, Vol. 18 (1940), pp. 560–8.

32. Mills, C. Wright: "The Competitive Personality," *Partisan Review*, Vol. 13 (1946), pp. 433–41.

33. Orlansky, Harold: "Infant Care and Personality," *Psychological Bulletin*, Vol. 46 (1949), pp. 1–48.

34. Parsons, Talcott: "The Professions and Social Structure," *Social Forces*, Vol. 17 (1939), pp. 457–67.

35. ——: "The Kinship System of the Contemporary United States," *American Anthropologist*, Vol. 45 (1943), pp. 22–38.

36. ——: "The Problem of Controlled Institutional Change," *Psychiatry*, Vol. 8 (1945), pp. 79–101.

Part Three

SOME APPLICATIONS TO

MODERN PROBLEMS

INTRODUCTION

THE SELECTIONS of Part Three all state implications of knowledge of personality in nature, society, and culture for practical issues. In the broadest sense, we may say that studies in this field contribute to the understanding of ourselves both as individuals and as members of groups. The perspective of viewing other individuals, in our own society and in diverse societies, and the clues gained as to the origins of various ways of reacting, aid in attaining that measure of detachment essential to self-comprehension. More specifically, the papers that follow offer useful orientations for dealing with problems of education, administration, race prejudice, maladjustment (alcoholism and mental disorder), social work, international relations, and political movements. Space does not allow illustration of approaches to other questions, e.g., speech disorders.[1]

These selections are concerned with predictions as to the modal personalities of groups rather than with specific individuals. When one is dealing with a single individual, one tries to say which of the many courses of behavior within his physical and mental capacity he will characteristically take. To do this, one must have intimate knowledge of that individual's constitution and of his particular life experience. As previous readings have shown, one must not assume that the organic motors of action in different persons are alike either quantitatively or qualitatively. There is variation in the autonomic and central nervous systems. Each man has his own threshold for each drive, activity level, frustration tolerance, and capacity to repress or inhibit specific drives. Moreover, even if two individuals could be matched as to anxiety level at birth, the threshold at maturity would differ in accord with the intensity and frequency to which each has been subjected to fear stimulation in the course of his life experience.

A group can no more have a "common character" than they can have a common pair of legs. No individual is exactly like any other individual

---

[1] See Jules and Zunia Henry, "Speech Disturbances in Pilagá Indian Children," *American Journal of Orthopsychiatry*, Vol. 10 (1940), pp. 362–9; and Stanley Newman and Vera G. Mather, "Analysis of the Spoken Language of Patients with Affective Disorders," *American Journal of Psychiatry*, Vol. 94 (1938), pp. 913–42.

either in his separable personality attributes or in his constellation of these traits. The variations from family to family within each society tend to be greater—so far as significance for the more detailed differences in personality are concerned—than the patterned variations between cultures. On the other hand, neither does any individual fail to show some features of personality highly similar to those of other members of this group. And, for purposes of gross prediction, the individual differences within a group tend to a considerable degree to cancel each other out. Behavior results from forces that have direction. Therefore, all behavior depends to a large extent on the cognitive structure which people have acquired from their culture or sub-culture. The goals of individuals within a group have much in common, although the range in intensity and the deviations from the abstract pattern are highly significant when one's eye is on the specific individual rather than on the central tendencies within the group. The typical conflicts of infancy and early childhood transcend cultural lines, but each culture has its selective defences against these conflicts and its preferred forms of resolution.

The details of causation are still obscure, even if certain interdependences are reasonably certain. So far as applications are concerned, the most profitable approach at present is to ask: What antecedents have what consequences? What associations appear to be relatively consistent? If children are socialized according to a given cultural pattern and a given group environment, what statistical predictions can be made as to how most of that group will react to a particular stimulus-situation in ways differentiated from the modal responses of other groups?

Just as the revolutionary advances in physics in recent years have been due less to new facts than to new ways of thinking about facts, so also in contemporary problems of group life and personal adjustment the need is for examining the familiar in terms of fresh categories and especially of a field approach.

CHAPTER

38

Hortense Powdermaker

THE CHANNELING OF NEGRO

AGGRESSION BY THE CULTURAL

PROCESS

The relations between groups of different race, language, religion, and culture constitute one of the crucial practical problems of our time. Studies in culture and personality illuminate both facets of the problem. On the one hand, what kinds of personalities are prone to group prejudice? On the other hand, how do persons in the victimized group handle the tensions created in them? Frenkel-Brunswick and Sanford have ably dealt with the first facet.[1] The second facet, which is the subject of Dr. Powdermaker's paper, has less often been treated in a scientific manner. Well-meaning persons, including scientists, tend too frequently to consider only the aggression manifested on the side of those guilty of "race prejudice."

NOTE: Reprinted, with slight abridgment, from the *American Journal of Sociology*, Vol. 48 (1943), pp. 750–8, by permission of the author and the University of Chicago Press. (Copyright, 1943, by the University of Chicago Press.)
[1] Else Frenkel-Brunswick and R. Nevitt Sanford, "Some Personality Factors in Anti-Semitism," *Journal of Psychology*, Vol. 20 (1945), pp. 271–91; see also, Else Frenkel-Brunswick, Daniel J. Levinson, and R. Nevitt Sanford, "The Antidemocratic Personality," in *Readings in Social Psychology* (T. M. Newcomb and E. L. Hartley, ed. 1947), pp. 531–41.

WE SHALL attempt in this article to look at one small segment of our cultural process—namely, a changing pattern of aggressive behavior—caused by the interracial situation. We limit ourselves to considering, at this time, only the Negro side of this complex of interpersonal relations; and we shall do no more than offer a few rather broad hypotheses on the relation between the forms aggression has taken during different historical periods and changes in the cultural processes at these times. For our hypotheses we are indebted to history, anthropology, sociology, and psychoanalysis; to the first three for understanding how social patterns come into being at a given point in time and how they are related to each other; and to the fourth, psychoanalysis, for a clue to the mechanisms by which individuals adopt particular social patterns. We shall concentrate on an analysis of two forms of adaptation where the aggression seems to have been concealed and, therefore, less understood. The two forms are that of the faithful slave and that of the meek, humble, unaggressive Negro who followed him after the Civil War. Since there is much more data on the latter role, this is the one we shall discuss in detail.

Education includes learning to play certain roles, roles which are advantageous to the individual in adapting himself to his particular culture. As the culture changes, so does the role. Adaptation to society begins at birth and ends at death. Culture is not a neatly tied package given to the child in school. It is an ever-changing process, gropingly and gradually discovered. The family, church, movies, newspapers, radio programs, books, trade-unions, chambers of commerce, and all other organized and unorganized interpersonal relations are part of education. All these are part of the cultural process, which determines how behavior and attitudes are channeled.

The cultural milieu of the Negro in the United States has run the gamut from slavery to that of a free but underprivileged group who are slowly but continuously raising their status. From the time slaves were first brought to this country until today, there have been barriers and restrictions which have prevented the Negro from satisfying social needs and attaining those values prized most highly by our society. How the resentment against these deprivations is channeled depends largely on cultural factors. Each historical period has produced certain types of adaptation.

Much has been written as to whether slaves emotionally accepted their status or whether they rebelled against it, with the consequent aggressive impulses turned against their masters. There is no categorical answer. Aggression can be channeled in many ways, and some of these are not discernible except to the trained psychiatrist. But others are quite obvious. The fact that thousands of slaves ran away clearly indicates dissatisfaction with their status. Crimes committed by the slaves are another evidence of lack of acceptance of status and of aggressive feelings toward the whites. The fact that these crimes were committed in the face of the most severe deterrents—cutting-off of ears, whipping, castration, death by mutilation —bears witness to the strength of the underlying aggression. Equally cruel was the punishment of those slaves who broke the laws against carrying firearms, assembling, and conspiring to rebel. The Gabriel conspiracy in Richmond, the Vesey conspiracy in Charleston, the Nat Turner rebellion, and others resulted in the massacre of whites and in the burning, shooting, and hanging of the Negroes. These attempts were undertaken despite the fact that the superior power of the whites made it virtually impossible for a slave revolt to be successful.

But the overt aggression was very probably only a small part of the total hostility. The punishments imposed by the culture for failure were too severe, and the chances for success too slight, to encourage the majority of slaves to rebel to any considerable extent. There were large numbers of loyal and faithful slaves, loyal to the system and to the masters. It is this loyalty that we try to understand.

Psychologically, slavery is a dependency situation. The slave was completely dependent upon the white master for food, clothing, shelter, protection—in other words, for security. If he could gain the good will or affection of the master, his security was increased. In return for this security, the Negro gave obedience, loyalty, and sometimes love or affection. With certain limitations, the situation of slave and master corresponds to that of child and parent. The young child is completely dependent on his parents for food, shelter, love, and everything affecting his well-being and security. The child learns to be obedient because he is taught that disobedience brings punishment and the withdrawal of something he needs for security. Basic infantile and childhood disciplines relating to sex are imposed on this level. In our culture, parents forbid and punish deviations by a child, who in turn renounces his gratification to gain the parent's approval. "The parent is needed and feared, and must therefore be obeyed; but the hatred to the frustrating parent, though suppressed, must be present somewhere." [2]

We mentioned above that there are certain limitations to our analogy. Obviously, the bondage is greater for the slave than for the child. Equally obviously, while there was love in some master-slave relationships, it was certainly not so prevalent as between parents and children. Again, the child always has a weak and undeveloped ego while the adult slave may have a strong, developed one. But most important is the difference in the reasons for the dependency attitude. The limited strength and resources of the child and his resulting helplessness and anxiety are due to biological causes. But the slave's dependency is imposed on him by culture and has nothing to do with biological factors. The structure of the two dependency situations is, therefore, very different. Nevertheless, func-

---

[2] Kardiner, Abram, and Linton, R.: *The Individual and His Society* (New York, Columbia University Press, 1939), p. 24.

tionally they have something in common. To attain the only security available to them, both the slave and the child repress, consciously or unconsciously, their hatred for the object which restricts their desires and freedom. At this late date, it is impossible to determine to what degree aggression occurred in slaves' fantasies or in minor overt acts. It probably varied from one slave to another, as it does for children. Neither all children nor all slaves repress their aggression all the time. Running away is a pattern for both groups. Disobedience is followed by punishment for both. Another alternative for both is open rebellion. Finally, children and slaves may accept their dependency and repress their aggression when compensations are adequate. They may even identify with the frustrating object. The picture of the faithful slave who helped the white mistress run the plantation while the master was away fighting, fighting the men who would liberate the slave, is only superficially paradoxical.

Data from psychoanalysis indicate that those children who do not permit their aggressive impulses to break through even in fantasy, not to mention overt behavior, have great difficulty as adults in entering into any personal relationship which does not duplicate the dependency pattern of parent and child. A legal edict of freedom did not immediately change the security system for the slave, conditioned over years to depend on the white man for all security. Time was needed for the compensations of freedom to become part of the ex-slave's security system. The process of growing up, or becoming less dependent, is a long and difficult one.

With emancipation the slave, from being a piece of property with no rights at all, attained the status of a human being—but an underprivileged one. Psychological dependency did not vanish with the proclamation of freedom. In the period following the Civil War, the slave's illiteracy, his complete lack of capital and property, the habituation to the past, and the continuous forces wielded by the whites in power created new conditions for the continuance of the old dependency. The recently freed Negro was dependent on the whites for jobs, for favors, for grants of money to set up schools, and for much of his security. In the South, following the Reconstruction Period, it was by obtaining favors from whites rather than by insisting on his rights that the Negro was able to make any progress or attain any security. The set of mores which insured the colored man's status being lower than that of the whites was and is still firmly intrenched. The denial of the courtesy titles (Mr., Mrs., Miss); the Jim Crowism in schools, buses, and trains, in places of residence; the denial of legal rights; the threat of lynching—these are among the more obvious ways of "keeping the Negro in his place." He is deprived of what are considered legal, social, and human rights, without any of the compensations for his deprivation which he had under slavery.

The same questions we asked about the slave occur again. Did the Negro really accept his position? Or was aggression aroused, and, if so, how did the culture channel it? This is an easier situation to study than the slavery of the past; for varied ways of reacting or adapting to this situation became stereotyped and still persist today. They are therefore susceptible of direct study.

First, there is direct aggression against its true object. Since the whites had, and still have, superior power, and since Negroes are highly realistic, they rarely use this method on any large scale except in times of crisis, and then as a climax to a long series of more indirect aggressive behavior patterns. The knocking-down of a white overseer, the direct attack on other whites, has occurred, but only occasionally. One of the reasons advanced by many southern white planters for their preference for colored share croppers to white ones is that the former do not fight back like the latter.

A second method consists in substituting a colored object for the white object of aggression. This was, and still is, done very frequently. The high degree of intra-Negro quarreling, crime, and homicide, revealed by statistics and observation, can be directly correlated with the Negro's frustration in being unable to vent his hostility on the whites. The mechanism of the substitution of one object of aggression for another is well known to the scientist and to the layman. The substitution of Negro for white is encouraged by the culture pattern of white official and unofficial leniency toward intra-Negro crime. Courts, more particularly southern ones, are mild in their view of intra-Negro offenses, and the prevailing white attitude is one of indulgence toward those intra-Negro crimes which do not infringe on white privileges.

A third possibility is for the Negro to retreat to an "ivory tower" and attempt to remain unaffected by the interracial situation. But this type of adjustment is very difficult and consequently a rare one.

Another form of adaptation consists in the Negro's identification with his white employer, particularly if the latter has great prestige. Some of the slaves also identified themselves with the great families whom they served. This pattern may likewise be observed in white servants. Still another adaptation is the diversion of aggression into wit, which has been and still is a much-used mechanism. We have not sufficient data on these two mechanisms to discuss them in detail.

But we do want to analyze in some detail a very frequent type of adjustment which occurred after the Civil War and which has persisted. We mean the behavior of the meek, humble, and unaggressive Negro, who is always deferential to whites no matter what the provocation may be. The psychological mechanism for this form of adaptation is less obvious than some of the other types, and a more detailed analysis is therefore needed. We have called this Negro "unaggressive," and that is the way his overt behavior could be correctly described. All our data, however, indicate that he does have aggressive impulses against whites, springing from the interracial situation. He would be abnormal if he did not have them. Over and over again, field studies reveal that this type of Negro is conscious of these resentments. But he conceals his true attitude from the whites who have power. How has he been able to conceal his aggression so successfully? His success here is patent. What is the psychological mechanism which enables the Negro to play this meek, deferential role?

A clue appears in certain similarities of this kind of behavior to that of the masochist, particularly through the detailed analysis of masochism by Dr. Theodor Reik in his recent book on that subject.[3] The seeming paradox of the masochist enjoying his suffering has been well known to psychoanalysts. He derives pleasure, because, first, it satisfies unconscious guilt feeling. Second (and here is where Dr. Reik has gone beyond the other psychoanalysts in his interpretation), the masochist derives another kind of pleasure, because his suffering is a prelude to his reward and eventual triumph over his adversary. In other words, he gets power through his suffering. We must not be misunderstood at this point. The meek Negro is neither neurotic nor masochistic any more than the slave was biologically a child. But the unaggressive behavior has some elements in common with (and some different from) the behavior of the masochist; and a comparison of the two gives a clue to the understanding of the strength behind the meek, humble role played by so many Negroes.

First, there are essential differences between the Negroes we are describing and the masochists analyzed by Dr. Reik and others. The Negro's sufferings and sacrifices are not unconsciously self-inflicted (as are those of the masochist) but are inflicted on him by the culture. The Negro plays his social masochistic role consciously, while the psychologically compulsive masochist does it unconsciously. These two important differences should be kept in mind while the similarities are discussed.

Our hypothesis is that the meek, unaggressive Negro, who persists today as a type and whom we have opportunity to study, feels guilty about his conscious and unconscious feelings of hostility and aggression toward the white people. These Negroes are believing Christians who have taken very literally the Christian doctrine that it is sinful to hate. Yet on every hand they are faced with situations which must inevitably produce hatred in any normal human being. These situations run the scale from seeing an innocent person lynched to having to accept the inferior accommodations on a Jim Crow train. The feeling of sin and guilt is frequently and openly expressed. In a Sunday-school class in a southern rural colored church a teacher tells the tale of a share cropper who had worked all season for a white planter, only to be cheated out of half his earnings. The teacher's lesson is that it is wrong to hate this planter, because Christ told us to love our enemies. The members of the class say how hard it is not to hate but that since it is a sin they will change their hate to love. They regard this as possible, although difficult.

One woman in the same community, who plays the deferential role to perfection and who, whites say, never steps out of "her place," tells me she feels guilty because she hates the whites, who do not seem to

---

[3] Reik, Theodor: *Masochism in Modern Man* (New York, Farrar & Rinehart, 1941).

distinguish between her, a very moral, respectable, and law-abiding person, and the immoral, disreputable colored prostitutes of the community. She says that God and Jesus have told her not to hate but to love—and so she must drive the hatred and bitterness away. Almost every human being in our culture carries a load of guilt (heavy or light as the case may be) over his conscious and unconscious aggressive impulses. It is easy, then, to imagine how heavy is the load of guilt for the believing Christian Negro who lives in an interracial situation which is a constant stimulus to aggressive thoughts and fantasies. By acting in exactly the opposite manner—that is, meekly and unaggressively—he can appease his guilt feelings consciously and unconsciously. It is this appeasement which accounts, in part, for his pleasure in the unaggressive role he plays with the whites.

But only in part. The unaggressive Negro enjoys his role also because through it he feels superior to the whites. Like the masochist, he thinks of his present sufferings as a contrasting background for his future glory. His is the final victory, and so he can afford to feel superior to his white opponent who is enjoying a temporary victory over him. My own field work and the work of others give many examples. Dr. Charles S. Johnson, in his recent book on rural colored youth in the South, discusses the dissimulation of many of the young people studied.[4] Any expression of antagonism would be dangerous, but this is not the whole story. It is not just that this boy and others avoid danger by meek negative behavior. There is a positive element in that he and others are insuring eventual victory. This was expressed by a colored servant who is a model of deferential behavior when with the whites. However, to me she says, partly scornfully and partly jokingly, that she considers it ridiculous that having cleaned the front porch and entrance she has to use the back entrance. She hates having to walk in the back door, which in this case is not only the symbol of status for a servant but the symbol that a whole race has a servant status. She adds that she expects to go to Heaven and there she will find rest—and no back doors.

The Christian doctrines, "the last shall be first, and the first shall be last" and "the meek shall inherit the earth," and all the promises of future reward for suffering give strong homiletic sanction to the feeling that the Negroes' present status and suffering is a prelude to their future triumph. Colored ministers give very concise expression to this attitude. A sermon heard in a colored church in rural Mississippi related

the story of a rich woman who lived in a big house and had no time for God. When she went to Heaven she was given an old shanty in which to live and she exclaimed: "Why that's the shanty my cook used to live in!" The cook, who on earth had given all her time to God, was now living in a big house in Heaven, very much like the one in which her former mistress used to live.

[4] Johnson, Charles S.: *Growing Up in the Black Belt* (Washington, D. C., American Council on Education, 1941), pp. 296-7.

The Christian missionaries of the pre-Civil War period emphasized the reward for the meek and their contrasting glories in the future, partly because it was an important part of Christian doctrine and partly because it was only by negating the present and emphasizing the future that the evangelists could get permission from the planters to preach to the slaves. The general theme of many of these sermons was that the greater the suffering here, the greater would be the reward in the world to come. Sermons, past and current, quite frequently picture Heaven as a place where whites and Negroes are not just equal but where their respective status is the opposite of what it is here.

This fantasy of turning the tables on the oppressor is not always confined to the other world; sometimes the setting is our own world. An example of this is the fantasy of a young colored girl in a northern town who had publicly taken quite meekly a decision that the colored people could not use the "Y" swimming pool at the same time white people were using it. Privately she shows her anger and says that she wishes the colored people would build a great big, magnificent "Y," a hundred times better than the white one, and make that one look like nothing. Her fantasy of triumph over the whites obviously gives her real pleasure and allows her to carry the present situation less onerously. Another example of the same type of fantasying occurs in the joking between two colored teachers who obey a disliked white official with deferential meekness. The joking consists of one of them boasting in some detail about how he has fired the white official; and the other one, in the same tone, describing how he "cussed out" the white official over the telephone.

Another aspect of the unaggressive Negro's pleasure is his feeling of superiority because he thinks he is so much finer a Christian than his white opponent. He, the Negro, is following Christ's precepts, while the white man does the opposite. The white man oppresses the poor and is unjust; in other words, he sins. He, the Negro, is virtuous and will be rewarded. One Negro, referring to a white man's un-Christian behavior, says: "It reflects back on him."

This feeling of superiority is a third characteristic of the unaggressive Negro's pleasure and is not limited to the feeling of Christian virtue. He feels superior to the whites because he is fooling them. His triumph is not completely limited to the distant future, but he enjoys at least a small part of it now. One of my informants in Mississippi, who plays this role to perfection, told me how he has the laugh on the whites because they never know his real thoughts. He quite consciously feels that he and the other Negroes like him have the upper hand through their dissimulation. He says very openly that it makes him feel superior. One woman, who presents an appearance of perfect meekness, laughs with a kind of gleeful irony when she tells me how she really feels, and her meekness drops away from her as if she were discarding a cloak. Another chuckles when she relates how much she has been able to extract from white people, who would never give her a thing if they knew how she really felt about them.

A Negro official who holds a fairly important position in his community knows that he is constantly being watched to see that he does not overstep his place, that his position and contact with whites has not made him "uppity." As he goes around humbly saying, "Yes, ma'am" and "Yes, sir," waiting his turn long after it is due, appearing not to heed insulting remarks, he is buoyed up with a feeling of superiority because he is really fooling all these whites. He is quite aware of his mask and knows it is such and not his real self. This mask characteristic comes out particularly when one of these individuals is seen with the whites and then later with his own group. One woman who has been particularly successful in the deferential, humble role with the whites gives a clear impression of meekness and humility. Her eyes downcast, her voice low, she patiently waits to be spoken to before she speaks, and then her tone is completely deferential. An hour later she is in the midst of her own group. No longer are her eyes downcast. They sparkle! Her laugh flashes out readily. Instead of patiently waiting, she is energetically leading. Her personality emerges, vibrant and strong, a complete contrast to the picture she gives the whites. These people enjoy wearing their mask because they do it so successfully and because its success makes them feel superior to the whites whom they deceive.

The deferential, unaggressive role just described and well known to students of Negro life has a very real function besides the obvious one of avoiding trouble. As Dr. Reik says in his book on masochism, "The supremacy of the will is not only expressed in open fights." It is, as he says, likewise expressed "in the determination to yield only exteriorly and yet to cling to life, nourishing such fantasies anticipating final victory." [5] Our unaggressive Negro, like the masochist, imagines a future where his fine qualities are acknowledged by the people who had formerly disdained him. This, in good Christian manner, will be brought about through suffering. This philosophy and its resulting behavior obviously make the Negroes (or any minority group who have them) very adaptable to any circumstance in which they find themselves, no matter how painful. They continue to cling to life, in the assurance of ultimate victory. They cannot be hurt in the way that people without this faith are hurt. The adaptability of the Negro has often been noted. This hypothesis may give some further clue to understanding it.

A special combination of cultural factors—namely, oppression of a minority group and a religion which promises that through suffering power will be gained over the oppressors—has channeled one type of adaptive behavior similar to that of the masochist. This behavior pattern has given the Negro a way of appeasing his guilt over his aggressive impulses and a method of adapting to a very difficult cultural situation. Because of the understanding given us by psychoanalysis of the pleasure derived through suffering, of the near and distant aims of the masochist,

---

[5] Reik, Theodor: *op. cit.,* p. 322.

we are given a clue to the psychological mechanism underlying the so-called "unaggressive" Negro's behavior. This Negro is not a masochist, in that his sufferings are not self-inflicted and he plays his role consciously. He knows he is acting, while the masochist behavior springs from inner compulsion. Again, there is a real difference in structure, as there was in the dependency situations of the child and the slave; and again there is a real similarity in function. The masochist and the meek, unaggressive Negro derive a similar kind of pleasure from their suffering. For the Negro as well as for the masochist, there is pleasure in appeasing the guilt feeling; for each there is the pleasure derived from the belief that through his suffering he becomes superior to his oppressors; and, finally, for each the suffering is a prelude to final victory.

Neither the slave nor the obsequious, unaggressive Negro, whom we have described, learned to play his role in any school. They learned by observation and imitation; they were taught by their parents; they observed what role brought rewards. Since the Civil War, the Negro has likewise seen the meek, humble type presented over and over again with approval in sermons, in literature, in movies, and, more recently, through radio sketches. By participating in the cultural processes, the Negro has learned his role. This was his education, far more powerful than anything restricted to schools; for the kind of education we are discussing is continuous during the entire life of the individual. It is subtle as well as direct. One part of the cultural process strengthens another part, and reinforcement for the role we described comes from every side.

But the cultural process continues to change with resulting changes in behavior. Just as the completely loyal and faithful slave disappeared, so the meek, unaggressive, and humble Negro, the "good nigger" type, is declining in numbers. In the rural South, and elsewhere too, the tendency of Negro young people (in their teens and twenties) is to refuse to assume the unaggressive role. The passing of the "good nigger" from the scene does not entail a civil war, as did the passing of the faithful slave. But it does indicate a psychological revolution. For the slave the Civil War altered the scope of the dependency situation. Today, without a civil war, equally significant cultural changes are taking place. The Negro is participating now in a very different kind of cultural process from that which he underwent fifty years ago.

Some of the differences occurring today are here briefly indicated. There is a decline in religious faith. The vivid "getting-religion" experience prevalent in the past has become increasingly rare for young people. Today they use the church as a social center. Gone is the intensity of religious belief that their parents knew. The young people are not atheists, but they do not have the fervor and sincerity of belief in a future world. They are much more hurt by slights and minor insults than are their parents, because they do not put their faith in the promise of a heavenly victory.

Along with changes in the form of religious participation have come many other changes. The illiteracy of the past has disappeared. A lengthening of schooling and a steady improvement in educational standards tend to give the Negro the same knowledge and the same tools enjoyed by the white man and to minimize cultural differences between the two. A more independent and rebellious Negro type is making its appearance in literature, as, for instance, the character of Bigger in the best seller, *Native Son.*

The steady trek of the rural Negroes to cities, North and South, has changed the milieu of masses of Negroes from the rural peasant life to the industrial urban one. Here they come under the influence of the trade-union movement, which slowly but gradually is shifting its attitude from one of jealous exclusion to one of inclusion, sometimes cordial and sometimes resigned. The shift is not anywhere near completion yet, but the trend is there. In the city the Negro is influenced by the same advertisements, the same radio sketches, the same political bosses, the same parties (left or right), and all the other urban forces which influence the white man.

The Negro's goals for success are thus becoming increasingly the same as those of the white person; and these goals are primarily in the economic field, although those in other fields, such as art and athletics, are not to be minimized either. The securing of these goals is in this world rather than in a future one. They are attained through the competition and aggressive struggle so characteristic of our culture rather than through meekness and subservience. The compensations available to the loyal slave and the humble, unaggressive, free Negro no longer exist or, at least, are steadily diminishing. The white man can no longer offer security in return for devotion, because he himself no longer has security. The whites of all classes have known a mounting social insecurity over the past decade, and they obviously cannot give away something which they do not possess. Thus the material rewards for obsequiousness and unaggressiveness are fading away. Gone, too, is the religious emphasis on rewards in Heaven. When the cultural process takes away rewards for a certain type of behavior, dissatisfaction with that behavior appears and there is a gradual change to another form which is more likely to bring new compensations. Obviously, one can expect, and one finds, a growing restlessness and uncertainty which occur in any transition period, when old goals have been lost. The new goals are the standard American ones. But the means for attaining these goals are not yet as available to the Negro as they are to the white. Economic and social discriminations still exist. Unless some other form of adaptation takes place and unless discriminations are lessened, we may expect a trend toward greater overt aggression.

However, there are no sudden revolutions in behavior patterns, and this holds for the patterns of aggression. They change slowly; the old ones persist while new ones are being formed, and opposing patterns exist side by side. But change occurs. The cultural process in which the Negro has participated from the time when he was first brought to this country until today has involved a constant denial of privileges. The denial has taken various forms, from the overt one involved in slavery to the more subtle ones of today. The compensations for the denial have varied from different degrees of material security to promises of future blessings in Heaven, and from the feeling of being more virtuous than the white to the feeling of fooling him. Today these compensations are fading away. Equally important, ideological fetters of the past have been broken by the Negro's increasing participation in the current urban industrial processes.

The Negro's education, formal and informal, has consisted of his participation in this ever changing cultural process, one small part of which we have briefly examined. Slavery, religion, economic and other social factors, have channeled his activities, offering him alternatives within a certain cultural range. We have examined only two of the alternatives in any detail—namely, the roles of the faithful slave and of the humble, meek Negro who was a fairly common stereotype following the Civil War; we have concentrated on the latter because he still exists and we therefore have more data on him. Both appear unaggressive. A functional comparison with the psychoanalytical analysis of the dependency situation of the child and of the problem of masochism has indicated how the aggression may have been present, although concealed, in these two roles.

CHAPTER

39

Talcott Parsons

ILLNESS AND THE ROLE OF THE PHYSI-
CIAN: A SOCIOLOGICAL PERSPECTIVE

Professor Parsons subjects illness and the role of the physician in our society to the kind of searching analysis made possible only by the co-ordinate use of the disciplines of anthropology, sociology, and clinical psychology. One of his conclusions, that mental illness should be viewed as an alternative to other forms of deviance, and that if there were less mental illness there might perhaps be more crime, is one which may come as a surprise but it stands, apparently, on reasonably firm ground.

THE present paper will attempt to discuss certain features of the phenomena of illness, and of the processes of therapy and the role of the therapist, as aspects of the general social equilibrium of modern Western society. This is what is meant by the use of the term "a sociological perspective" in the title. It is naturally a somewhat different perspective from that usually taken for granted by physicians and others, like clinical psychologists and social workers, who are directly concerned with the care of sick people. They are naturally more likely to think in terms of the simple application of technical knowledge of the etiological factors in ill health and of their own manipulation of the situation in the attempt to

NOTE: Reprinted by permission of the author and The American Orthopsychiatric Association, Inc. from the *American Journal of Orthopsychiatry*, Vol. 21 (1951), pp. 452–60. (Copyright 1951.)

control these factors. What the present paper can do is to add something with reference to the social setting in which this more "technological" point of view fits.

Undoubtedly the biological processes of the organism constitute one crucial aspect of the determinants of ill health, and their manipulation one primary focus of the therapeutic process. With this aspect of "organic medicine" we are here only indirectly concerned. However, as the development of psychosomatic medicine has so clearly shown, even where most of the symptomatology is organic, very frequently a critically important psychogenic component is involved. In addition, there are the neuroses and psychoses where the condition itself is defined primarily in "psychological" terms, that is, in terms of the motivated adjustment of the individual in terms of his own personality, and of his relations to others in the social world. It is with this motivated aspect of illness, whether its symptoms be organic or behavioral, that we are concerned. Our fundamental thesis will be that illness to this degree must be considered to be an integral part of what may be called the "motivational economy" of the social system and that, correspondingly, the therapeutic process must also be treated as part of that same motivational balance.

Seen in this perspective illness is to be treated as a special type of what sociologists call "deviant" behavior. By this is meant behavior which is defined in sociological terms as failing in some way to fulfill the institutionally defined expectations of one or more of the roles in which the individual is implicated in the society. Whatever the complexities of the motivational factors which may be involved, the dimension of conformity with, versus deviance or alienation from, the fulfillment of role expectations is always one crucial dimension of the process. The sick person is, by definition, in some respect disabled from fulfilling normal social obligations, and the motivation of the sick person in being or staying sick has some reference to this fact. Conversely, since being a normally satisfactory member of social groups is always one aspect of health, mental or physical, the therapeutic process must always have as one dimension the restoration of capacity to play social roles in a normal way.

We will deal with these problems under four headings. First, something will have to be said about the processes of genesis of illness insofar as it is motivated and thus can be classed as deviant behavior. Secondly, we will say something about the role of the sick person precisely as a social role, and not only a "condition"; third, we will analyze briefly certain aspects of the role of the physician and show their relation to the therapeutic process, and finally, fourth, we will say something about the way in which both roles fit into the general equilibrium of the social system.

Insofar as illness is a motivated phenomenon, the sociologist is particularly concerned with the ways in which certain features of the individual's relations to others have played a part in the process of its genesis. These factors are never isolated; there are, of course, the constitutional and organically significant environmental factors (e.g., bacterial agents), and undoubtedly also psychological factors internal to the individual personality. But evidence is overwhelming as to the enormous importance of relations to others in the development and functioning of personality. The sociologist's emphasis, then, is on the factors responsible for "something's going wrong" in a person's relationships to others during the processes of social interaction. Probably the most significant of these processes are those of childhood, and centering in relations to family members, especially, of course, the parents. But the essential phenomena are involved throughout the life cycle.

Something going wrong in this sense may be said in general to consist in the imposition of a strain on the individual, a strain with which, given his resources, he is unable successfully to cope. A combination of contributions from psychopathology, learning theory, and sociology makes it possible for us to say a good deal, both about what kinds of circumstances in interpersonal relations are most likely to impose potentially pathogenic strains, and about what the nature of the reactions to such strains is likely to be.

Very briefly we may say that the pathogenic strains center at two main points. The first concerns what psychiatrists often call the "support" a person receives from those surrounding him. Essentially this may be defined as his acceptance as a full-fledged member of the group, in the appropriate role. For the child this means, first of all, acceptance by the family. The individual is emotionally "wanted" and within considerable limits this attitude is not conditional on the details of his behavior. The second aspect concerns the upholding of the value patterns which are constitutive of the group, which may be only a dyadic relationship of two persons, but is usually a more extensive group. Thus rejection, the seducibility of the other, particularly the more responsible, members of the group in contravention of the group norms, the evasion by these members of responsibility for enforcement of norms, and, finally, the compulsive "legalistic" enforcement of them are the primary sources of strain in social relationships. It is unfortunately not possible to take space here to elaborate further on these very important problems.

Reactions to such strains are, in their main outline, relatively familiar to students of mental pathology. The most important may be enumerated as anxiety, production of fantasies, hostile impulses, and the resort to special mechanisms of defense. In general we may say that the most serious problem with reference to social relationships concerns the handling of hostile impulses. If the strain is not adequately coped with in such ways as to reduce anxiety to manageable levels, the result will, we believe, be the generating of ambivalent motivational structures. Here, because intrinsically incompatible motivations are involved, there must be resort to special mechanisms of defense and adjustment. Attitudes toward others thereby acquire the special property of compulsiveness because of the need to defend against the repressed element of the motivational structure. The ambivalent structure may work out in either of two main directions:

first, by the repression of the hostile side, there develops a compulsive need to conform with expectations and retain the favorable attitudes of the object; second, by dominance of the hostile side, compulsive alienation from expectations of conformity and from the object results.

The presence of such compulsive motivation inevitably distorts the attitudes of an individual in his social relationships. This means that it imposes strains upon those with whom he interacts. In general it may be suggested that most pathological motivation arises out of vicious circles of deepening ambivalence. An individual, say a child, is subjected to such strain by the compulsive motivation of adults. As a defense against this he himself develops a complementary pattern of compulsive motivation, and the two continue, unless the process is checked, to "work on each other." In this connection, it may be especially noted that some patterns of what has been called compulsive conformity are not readily defined as deviant in the larger social group. Such people may in a sense be often regarded as "carriers" of mental pathology in that, though themselves not explicitly deviant, either in the form of illness or otherwise, by their effects on others they contribute to the genesis of the kinds of personality structure which are likely to break down into illness or other forms of deviance.

Two important conclusions seem to be justified from these considerations. The first is that the types of strain on persons which we have discussed are disorganizing both to personalities and to social relationships. Personal disorganization and social disorganization are, in a considerable part, two sides of the same concrete process. This obviously has very important implications both for psychiatry and for social science. Secondly, illness as a form of deviant behavior is not a unique phenomenon, but one type in a wider category. It is one of a set of alternatives which are open to the individual. There are, of course, reasons why some persons will have a psychological make-up which is more predisposed toward illness, and others toward one or another of the alternatives; but there is a considerable element of fluidity, and the selection among such alternatives may be a function of a number of variables. This fact is of the greatest importance when it is seen that the role of the sick person is a socially structured and in a sense institutionalized role.

The alternatives to illness may be such as to be open only to the isolated individual, as in the case of the individual criminal or the hobo. They may also involve the formation of deviant groups as in the case of the delinquent gang. Or, finally, they may involve a group formation which includes asserting a claim to legitimacy in terms of the value system of the society, as in joining an exotic religious sect. Thus to be a criminal is in general to be a social outcast, but in general we define religious devoutness as "a good thing" so that the same order of conflict with society that is involved in the criminal case may not be involved in the religious case. There are many complex and important problems concerning the genesis and significance of these various deviant patterns, and their relations to each other, which cannot be gone into here. The most essential point is to see that illness is one pattern among a family of such alternatives, and that the fundamental motiv tional ingredients of illness are not peculiar to it, but are of more general significance.

We may now turn to our second main topic, that of the sense in which illness is not merely a "condition" but also a social role. The essential criteria of a social role concern the attitudes both of the incumbent and of others with whom he interacts, in relation to a set of social norms defining expectations of appropriate or proper behavior for persons in that role. In this respect we may distinguish four main features of the "sick role" in our society.

The first of these is the exemption of the sick person from the performance of certain of his normal social obligations. Thus, to take a very simple case, "Johnny has a fever, he ought not to go to school today." This exemption and the decision as to when it does and does not apply should not be taken for granted. Psychiatrists are sufficiently familiar with the motivational significance of the "secondary gain" of the mentally ill to realize that conscious malingering is not the only problem of the abuse of the privileges of being sick. In short, the sick person's claim to exemption must be socially defined and validated. Not every case of "just not feeling like working" can be accepted as such a valid claim.

Secondly, the sick person is, in a very specific sense, also exempted from a certain type of responsibility for his own state. This is what is ordinarily meant by saying that he is in a "condition." He will either have to get well spontaneously or to "be cured" by having something done to him. He cannot reasonably be expected to "pull himself together" by a mere act of will, and thus decide to be all right. He may have been responsible for getting himself into such a state, as by careless exposure to accident or infection, but even then he is not responsible for the process of getting well, except in a peripheral sense.

This exemption from obligations and from a certain kind of responsibility, however, is given at a price. The third aspect of the sick role is the partial character of its legitimation, hence the deprivation of a claim to full legitimacy. To be sick, that is, is to be in a state which is socially defined as undesirable, to be gotten out of as expeditiously as possible. No one is given the privileges of being sick any longer than necessary but only so long as he "can't help it." The sick person is thereby isolated, and by his deviant pattern is deprived of a claim to appeal to others.

Finally, fourth, being sick is also defined, except for the mildest cases, as being "in need of help." Moreover, the type of help which is needed is presumptively defined; it is that of persons specially qualified to care for illness, above all, of physicians. Thus from being defined as the incumbent of a role relative to people who are not sick, the sick person makes the transition to the additional role of patient. He thereby, as in all social roles, incurs certain obligations, especially that of "co-operating" with his physician—or other therapist—in the process of trying to get well. This obviously constitutes an affirmation of the admission of being sick, and therefore in an undesirable state, and also exposes the individual to specific reintegrative influences.

It is important to realize that in all these four respects, the phenomena of mental pathology have been assimilated to a role pattern which was already well established in our society before the development of modern psychopathology. In some respects it is peculiar to modern Western society, particularly perhaps with respect to the kinds of help which a patient is felt to need; in many societies magical manipulations have been the most prominent elements in treatment.

In our society, with reference to the severer cases at any rate, the definition of the mental "case" as sick has had to compete with a somewhat different role definition, namely, that as "insane." The primary difference would seem to center on the concept of responsibility and the mode and extent of its application. The insane person is, we may say, defined as being in a state where not only can he not be held responsible for getting out of his condition by an act of will, but where he is held not responsible in his usual dealings with others and therefore not responsible for recognition of his own condition, its disabilities, and his need for help. This conception of lack of responsibility leads to the justification of coercion of the insane, as by commitment to a hospital. The relations between the two role definitions raise important problems which cannot be gone into here.

It may be worth while just to mention another complication which is of special interest to members of the Orthopsychiatric Association, namely, the situation involved when the sick person is a child. Here, because of the role of child, certain features of the role of sick adult must be altered, particularly with respect to the levels of responsibility which can be imputed to the child. This brings the role of the mentally sick child in certain respects closer to that of the insane than, particularly, of the neurotic adult. Above all it means that third parties, notably parents, must play a particularly important part in the situation. It is common for pediatricians, when they refer to "my patient," often to mean the mother rather than the sick child. There is a very real sense in which the child psychiatrist must actively treat the parents and not merely the child himself.

We may now turn to our third major problem area, that of the social role of the therapist and its relation to the motivational processes involved in reversing the pathogenic processes. These processes are, it is widely recognized, in a certain sense definable as the obverse of those involved in pathogenesis, with due allowance for certain complicating factors. There seem to be four main conditions of successful psychotherapy which can be briefly discussed.

The first of these is what psychiatrists generally refer to as "support." By this is here meant essentially that acceptance as a member of a social group the lack of which we argued above played a crucial part in pathogenesis. In this instance it is, above all, the solidary group formed by the therapist and his patient, in which the therapist assumes the obligation to do everything he can within reason to "help" his patient. The strong emphasis in the "ideology" of the medical profession on the "welfare of the patient" as the first obligation of the physician is closely related to this factor. The insistence that the professional role must be immune from "commercialism," with its suggestion that maximizing profits is a legitimate goal, symbolizes the attitude. Support in this sense is, so long as the relationship subsists, to be interpreted as essentially unconditional, in that within wide limits it will not be shaken by what the patient does. As we shall see, this does not, however, mean that it is unlimited, in the sense that the therapist is obligated to "do anything the patient wants."

The second element is a special permissiveness to express wishes and fantasies which would ordinarily not be permitted expression in normal social relationships, as within the family. This permissiveness must mean that the normal sanctions for such expression in the form of disapproval and the like are suspended. There are of course definite limits on "acting out." In general the permissiveness is confined to verbal and gestural levels, but this is nonetheless an essential feature of the therapeutic process.

The obverse of permissiveness, however, is a very important restriction on the therapist's reaction to it. In general, that is, the therapist does not reciprocate the expectations which are expressed, explicitly or implicitly, in the patient's deviant wishes and fantasies. The most fundamental wishes, we may presume, involve reciprocal interaction between the individual and others. The expression of a wish is in fact an invitation to the other to reciprocate in the complementary role, if it is a deviant wish, an attempt to "seduce" him into reciprocation. This is true of negative as well as positive attitudes. The expression of hostility to the therapist in transference is only a partial gratification of the wish; full gratification would require reciprocation by the therapist's becoming angry in return. Sometimes this occurs; it is what is called "countertransference"; but it is quite clear that the therapist is expected to control his countertransference impulses and that such control is in general a condition of successful therapy. By showing the patient the projective character of this transference reaction, this refusal to reciprocate plays an essential part in facilitating the attainment of insight by the patient.

Finally, fourth, over against the unconditional element of support, there is the conditional manipulation of sanctions by the therapist. The therapist's giving and withholding of approval is of critical importance to the patient. This seems to be an essential condition of the effectiveness of interpretations. The acceptance of an interpretation by the patient demonstrates his capacity, to the relevant extent, to discuss matters on a mature plane with the therapist, who shows his approval of this performance. It is probably significant that overt disapproval is seldom used in therapy, but certainly the withholding of positive approval is very significant.

The above four conditions of successful psychotherapy, it is important to observe, are all to some degree "built into" the role which the therapist in our society typically assumes, that of the physician, and all to some degree are aspects of behavior in that role which are at least partially inde-

pendent of any conscious or explicit theory or technique of psychotherapy.

The relation of support to the definition of the physician's role as primarily oriented to the welfare of the patient has already been noted. The
element of permissiveness has its roots in the general social acceptance
that "allowances" should be made for sick people, not only in that they
may have physical disabilities, but that they are in various ways "emotionally" disturbed. The physician, by virtue of his special responsibility
for the care of the sick, has a special obligation to make such allowances.
Third, the physician is, however, by the definition of his role, positively
enjoined not to enter into certain reciprocities with his patients, or he is
protected against the pressures which they exert upon him. Thus giving
of confidential information is, in ordinary relationships, a symbol of
reciprocal intimacy, but the physician does not tell about his own private
affairs. Many features of the physician-patient relationship, such as the
physician's access to the body, might arouse erotic reactions, but the role
is defined so as to inhibit such developments even if they are initiated by
the patient. In general the definition of the physician's role as specifically
limited to concern with matters of health, and the injunction to observe
an "impersonal," matter-of-fact attitude without personal involvement,
serve to justify and legitimize his refusal to reciprocate his patient's deviant
expectations. Finally, the prestige of the physician's scientific training, his
reputation for technical competence, gives authority to his approval, a
basis for the acceptance of his interpretations.

All of these fundamental features of the role of the physician are given
independently of the technical operations of psychotherapy; indeed they
were institutionalized long before the days of Freud or of psychiatry as
an important branch of the medical profession. This fact is of the very
first importance.

First, it strongly suggests that in fact deliberate, conscious psychotherapy is only part of the process. Indeed, the effective utilization of these
aspects of the physician's role is a prominent part of what has long been
called the "art of medicine." It is highly probable that, whether or not the
physician knows it or wishes it, in practicing medicine skillfully he is always in fact exerting a psychotherapeutic effect on his patients. Furthermore, there is every reason to believe that, even though the cases are not
explicitly "mental" cases, this is necessary. This is, first, because a "psychic
factor" is present in a very large proportion of ostensibly somatic cases
and, secondly, apart from any psychic factor in the etiology, because illness is always to some degree a situation of strain to the patient, and
mechanisms for coping with his reactions to the strain are hence necessary, if the strain is not to have pyschopathological consequences. The
essential continuity between the art of medicine and deliberate psychotherapy is, therefore, deeply rooted in the nature of the physician's function generally. Modern psychotherapy has been built upon the role of the
physician as this was already established in the social structure of Western

society. It has utilized the existing role pattern and extended and refined certain of its features, but the roles of the physician and of the sick person were not created as an application of the theories of psychiatrists.

The second major implication is that, if these features of the role of the physician and of illness are built into the structure of society independent of the application of theories of psychopathology, it would be very strange indeed if they turned out to be isolated phenomena, confined in their significance to this one context. This is particularly true if, as we have given reason to believe, illness is not an isolated phenomenon, but one of a set of alternative modes of expression for a common fund of motivational reaction to strain in the social system. But with proper allowances for very important differences we can show that certain of these features can also be found in other roles in the social system. Thus to take an example of special interest to you, there are many resemblances between the psychotherapeutic process and that of the normal socialization of the child. The differences are, however, great. They are partly related to the fact that a child apparently needs two parents while a neurotic person can get along with only one psychiatrist. But also in the institutions of leadership, of the settlement of conflicts in society and of many others, many of the same factors are operative.

We therefore suggest that the processes which are visible in the actual technical work of psychotherapy resemble, in their relation to the total balance of forces operating within and upon the individual, the part of the iceberg which protrudes above the surface of the water; what is below the surface is the larger, and, in certain respects, probably still the more important part.

It also shows that the phenomena of physical and mental illness and the counteraction are more intimately connected with the general equilibrium of the social system than is generally supposed. We may close with one rather general inference from this generalization. It is rather generally supposed that there has been a considerable increase in the incidence of mental illness within the last generation or so. This is difficult to prove since statistics are notably fragmentary and fashions of diagnosis and treatment have greatly changed. But granting the fact, what would be its meaning? It could be that it was simply an index of generally increasing social disorganization. But this is not necessarily the case. There are certain positive functions in the role of illness from the social point of view. The sick person is isolated from influence upon others. His condition is declared to be undesirable and he is placed in the way of re-equilibrating influences. It is altogether possible that an increase in mental illness may constitute a diversion of tendencies to deviance from other channels of expression into the role of illness, with consequences less dangerous to the stability of society than certain alternatives might be. In any case the physician is not merely the person responsible for the care of a special class of "problem cases." He stands at a strategic point in the general balance of forces in the society of which he is a part.

CHAPTER

40

Nathan Leites

* TRENDS IN AFFECTLESSNESS

As the clinical psychologist reads a Rorschach protocol or the stories of a Thematic Apperception Test, so has Dr. Leites read Albert Camus' The Stranger. But Dr. Leites has taken a further step, treating this novel as a protocol of the ethos of our time.

ALBERT CAMUS' *The Stranger* [1] is one of the most prominent post-1939 French novels, and one which has been widely regarded as conveying a new "philosophical" message. This paper is not concerned with the connections between the novel and the philosophic writings of the author (in particular *Le mythe de Sisyphe*, 1942). It attempts to trace relationships between the novel and certain trends in the temper of the age. In doing so, it offers a psychological interpretation of the content of the novel.

* * * * *

The outline of the plot is as follows: Meursault, a small French clerk in Algiers, receives news of his mother's death; he attends her funeral; he begins an affair and becomes engaged; he kills, under a blazing sun, without any adequate conscious motivation the brother of an Arab prostitute

NOTE: Thanks are due to the author and the *American Imago* for permission to reprint this paper with abridgment of page references to the novel and some quotations in the original French. The reader is referred to the original article in the American Imago, Vol. 4 (1947). Pp. 3–26. Copyright, 1947.

whose pimp he vaguely knows; he is condemned to death and will presently be executed. He is the narrator.

Three attitudes may be taken toward the hero. Firstly, that he is the
incarnation of a metaphysical affirmation which is known to be held by
the author.

Secondly, the hero may be considered unintelligible.

The third position, which shall be set forth here, is that Meursault's
behavior is largely intelligible.

*   *   *   *   *

The novel contains only a few explicit indications—almost entirely
overlooked even by the sophisticated critics—about the hero's past. But
these few are quite significant. As to his father, "I never set eyes on him."
As to his mother, "For years she'd never had a word to say to me and I
could see she was moping with no one to talk to." ". . . neither Mother
nor I expected much from one another." "As a student I'd had plenty of
ambition. . . . But, when I had to drop my studies, I very soon realized
all that was pretty futile." The child and the adolescent is thus shown as
reacting with withdrawal of conscious affect in intrapersonal relations
(that is, the relations between the various components of the self) and in
interpersonal relations. He is thus reacting to the guilty rage induced by
the severe deprivations which were imposed by an absent father, an indifferent mother, and a withholding wider environment.

It is this characteristic defense which the hero perpetuates and elaborates in his adult life, and which gives his personality—conveyed in a
style appropriate to this dominant trait—its particular aura. I shall now
discuss the various major manifestations of the hero's affectlessness.

Firstly, and most obviously, the hero is usually rather clearly aware
of the *absence or weakness of affects in response to intrapersonal and interpersonal stimuli*. This is in sharp contrast to the intensity of his reactions to external nonpersonal stimuli—to the colors, smells, tactile values,
and sounds of cityscape and landscape. These he knows to be his "surest
humblest pleasures." The hero is also presented as feeling a persistently
strong and unbrokenly euphoric sexual attraction towards his girl friend—
almost the only point in which I would question his plausibility. Perhaps
the author, so free from many illusions, is here still presenting a derivative
of the Western myth on the transcendent position of "love" in human nature. "I could truthfully say I'd been quite fond of mother—but really,
that didn't mean much." His affects appear to him as *questionable* rather
than as inevitable and valid: "I came to feel that this aversion [against
talking about certain things——N.L.] had no real substance." He is much
aware of the almost total dependence of his affective on his somatic state
(cf. p. 80)—conforming (though in extreme fashion) to what is probably
a contemporary trend.

While the hero is acutely aware of his atypicality as a "stranger" to
the world, he *spontaneously subsumes most of his few near-affective ex-*

periences in interpersonal relations under general categories. When his lawyer asks him whether he had loved his mother, he replies, "yes, like everybody else." When his girl friend asks him, "Suppose another girl had asked you to marry her—I mean a girl you liked in the same way as you like me—would you have said 'Yes' to her, too?" the hero does not find such a hypothesis inconceivable and his emotions towards Marie unique. He answers, apparently effortlessly: "Naturally." In this, he presumably manifests a widely diffused trend in the quality of Western "love" experiences in this century.

It may be surmised that such "generalizing" procedures are in part a defense against the unconscious threat of overwhelming affect. When the hero learns of his mother's death, he arranges for keeping "the *usual* vigil" beside the body. The owner of his habitual restaurant affirms "there's no one like *a* mother" and lends him a black tie and mourning band procured for the occasion of an uncle's death.

The hero shows *a high degree of detachment towards decisive impacts of his environment on him.* During most of his trial he feels as if somebody else is about to be condemned to death. ". . . He ["one of my policemen" ——N.L.] asked me if I was feeling nervous. I said 'No,' and that the prospect of witnessing a trial rather interested me." When danger mounts, "the *futility* of what was happening seemed to take me by the throat."

All value judgments have ceased to be self-evident, as they have in some variants of contemporary empiricist epistemology. There is a *tabula rasa* where the traditional ethical postulates stood. But the hero attains at the end a state of exaltation in contemplating the *certain facts of his present aliveness and impending annihilation:* "It might look as if my hands were empty. Actually, I was sure of myself, sure about everything, far surer than he [the prison chaplain——N.L.]; sure of my present life and of the death that was coming. That, no doubt, was all I had; but at least that certainty was something I could get my teeth into—just as it had got its teeth into me." The cathexis withdrawn from norms is in part displaced to very general aspects of facts.

Choices made, then, appear as quite inevitable and correct (for they have been made) and as quite arbitrary (for they were choices). In his final exaltation the hero comes to feel that "I'd been right, I was still right, I was always right. I'd passed my life in a certain way and I might have passed it in a different way. . . . I'd acted thus and I hadn't acted otherwise. . . . And what did that mean? That, all the time, I had been waiting for this present moment, for that dawn [of execution——N.L.] . . . which was to justify me."

The hero *abstains from morally reacting to others* as much as to himself. His incapacity for moral indignation is again related to certain contemporary trends [3].

What are the behavioral counterparts to the hero's valuelessness? His tendency is to *minimize overt action,* symbolic as well as motor. He tends to react with silence to communications of others, perpetuating the wordlessness of his relations with his mother. He shows a preference for the *maintenance of his personal status quo*, at any given moment and with reference to his overall mode of life. When an evening conversation imposed on him is prolonged, he feels that "I wanted to be in bed, only it was such an effort making a move." When his employer offers him a Paris job, "I saw no reason for 'changing my life.'" Getting out of bed requires an intense effort.

Whenever he contemplates *alternative courses of action*, he becomes convinced that they lead to an *identical result*. Thus nothing is a "serious matter." When his boss offers him a Paris job, "really I didn't care much one way or the other." When he is present at a tense underworld encounter which may instantly develop into shooting, "it crossed my mind that one might fire or not fire—and it would come to absolutely the same thing."

Correspondingly, the hero tends to feel a situation as *invariant* which according to the judgment of non-"strangers" has varied radically. When his boss, offering him a Paris job, stresses the advantages of a "change of life," "I answered that one never changed his way of life."

"The same thing" to which all conceivable courses of action lead is a negative thing. When the hero accompanies the coffin of his mother to the cemetery in the Algerian heat, he is told: "If you go too slowly there's the risk of a heat-stroke. But, if you go too fast, you perspire and the cold air in the church gives you a chill." He adds: "I saw . . . [the] point: *either way one was in for it*." There is no conceivable intermediate optimal point between too much and too little. Similarly—and again with the heat in the role of the great depriver—at a certain point during the morning which ends with the crime, the hero stands before a house he is expected to enter: ". . . I couldn't face the effort needed to go up the steps and make myself amiable to the women. But the heat was so great that it was just as bad staying where I was. . . . To stay or to make a move—it came to much the same. After a moment I returned to the beach"—a move which leads to the murder.

"One is cooked both ways" because *death is the terminal state of any sequence of acts—which is therefore equivalent to any other sequence.* "Nothing, nothing had the least importance, and I knew quite well why. . . . From the dark horizon of my future a sort of slow, persistent breeze had been blowing towards me, all my life long, from the years that were to come. And on its way that breeze had levelled out all the ideas that people tried to foist on me. . . . What difference could . . . [it] make . . . the way a man decides to live, the fate . . . he chooses, since one and the same fate was bound to 'choose' not only me but thousands of millions of privileged people. . . . Every man alive was privileged; there was only one class of men, the privileged class. And all alike would be condemned to die one day. . . . And what difference would it make if . . . he [the prison chaplain instead of the hero——N.L.] were executed . . . since it all came to the same thing in the end?" This indifferentism with the horizon of death acts manifestly as a defense against distress. Despairing of his girl friend's faithfulness, the hero asks in the context of the passage just quoted: "What did it matter if at this very moment Marie was kissing a new boy friend?" The fatherless and motherless (in more than one sense) murderer asks: "What difference could they make to me, the deaths of others, or a mother's love or . . . God?" Awaiting execution and faintly hoping for the success of his appeal, the hero attempts with conscious and only partly successful effort to make himself see that "it makes little difference whether one dies at the age of thirty or three score and ten. . . . Whether I die now or forty years hence, this business of dying had to be got through, inevitably. . . ."

Presumably the belief in death as annihilation has been, and is, spreading and deepening in Western culture. Reactions to this major trend seem to be (as one would expect) polarized. On the one hand there is an increasing tendency to "scotomize" death, i.e. to minimize its role in conscious awareness. On the other hand, a breakthrough of this awareness may dominate the consciousness of the individual in a somewhat new fashion: The massive fact of death may appear as establishing the pointlessness of life beyond any possibility of mitigation. The *tabula rasa* as to previous conceptions of Good and Evil may thus be accompanied by a *tabula rasa* as to previous conceptions of the Meaningful Life.

But the residual certainty of the very fact of life which is accentuated when valuelessness has been established may become the new meaning-creating factor: If death is annihilation, life—all we have—is infinitely precious (a resurgence of the *carpe diem* theme in "highbrow" speculation which has often accompanied the breakdown of civilizations). This is the metaphysical and axiological aspect of the passage on "certainty" quoted above. The sentence quoted, "every man alive was privileged," may thus become one denying or affirming the meaningfulness of life according to whether the temporariness or the availability of the "privilege" is stressed. If the positive accent is chosen, the desired though unobtainable "life after the grave" appears as "a life in which I can remember this life on earth. That's all I want of it!" The hero attempts to get rid of the intruding prison chaplain with "I'd very little time left and I wasn't going to waste it on God." But there is a price to this reversal: The affects which the negative view intended to ward off by the depreciation of life reappear. The hero awaiting the coming of the executioners every dawn knows every morning that "I might just as well have heard footsteps and felt *my heart shattered into bits*." On the other hand intermittent phantasies of survival give him the task "to calm down that sudden *rush of joy* racing through my body and even bringing *tears* to my eyes." Thus belief in the meaninglessness of life appears in the hero related to the successful repression of affect, and belief in its meaningfulness to the return of the repressed from repression.

What typical conscious motivation remains available to the hero in his predominant negative phases? The hero tends to choose a certain action "*for want of anything better to do,*" or as "I had nothing to do," or "as a last resource against the unspecified displeasure of more complete inaction." He tends to comply with demands made of him in a "*why not?*" fashion. When his pimp acquaintance asks him for a favor "I wrote the letter [requested——N.L.] . . . I wanted to satisfy Raymond as I'd no reason not to satisfy him." When the pimp thereupon says, "so now we are pals, ain't we," "I kept silence and he said it again. I didn't care one way or the other, but he seemed so set on it, I nodded and said, 'yes.' " When his boss offers him a Paris job, "I told him I was quite prepared to go; but really I didn't care much one way or the other." When his girl friend asks him if he would marry her, "I said I didn't mind; if she was keen on it, we'd get married." When she objects to this reaction, "I pointed out that . . . the suggestion came from her; as for me, I'd merely said 'yes.' "

Presumably the predominance of such "negative motivations" is overdetermined. Affectlessness is here not only a defense against the various phantasied dangers of involvement but also an instrument of aggression against (and contempt for) those persons who expect a fuller response from the hero. The aggression proceeds by spiteful obedience: Demands are complied with in the letter but not in the spirit.

A major "positive" motivation of the hero is *sleep*, functioning as defense against the overwhelming impact of dangerous stimuli. It makes him feel good to look forward to sleep in a short while. At a certain "evening hour . . . I always felt so well content with life. Then, what awaited me was a night of easy dreamless sleep." "I can remember . . . my little thrill of pleasure when we entered the first brightly lit streets of Algiers [returning from his mother's funeral——N.L.] and I pictured myself going straight to bed and sleeping twelve hours at a stretch."

Where other psychic structures would react with intense affect, the hero tends to react with *fatigue and somnolence*—which he often attributes to the external physical rather than to an internal psychic "heat."

To the high tendency towards sleep corresponds the *vagueness and poverty of internal and external psychic perceptions* during the hero's waking hours. (As was implied above, his perceptions of nonhuman aspects of his environment are rich and acute.) Low awareness of self is shown in extreme fashion when the hero after a long time in prison suddenly "heard something that I hadn't heard for months. It was the sound of a voice, my own voice, there was no mistaking it. And I recognized it as the voice that for many days of late had been sounding in my ears. So I know that all this time I'd been talking to myself." Similarly, the hero at important occasions—where again, he is apt, projectively to hold the heat of the day responsible—fails to hear or to understand correctly what others are saying to him or about him. When his lawyer pleads for him in court, "I found that my mind had gone blurred; everything was dissolving with a grayish, watery haze."

More particularly, the hero—having interfered so severely with his own affects—is *highly inhibited in the perception of affects of others, especially of those having himself as their target.* This lack of empathy facilitates the unconsciously aggressive and self-punitive candidness of the hero which will be discussed below. The same trait also induces a *poverty of the prognostic horizon.* After the hero has witnessed an altercation between the pimp and a policeman, the pimp tells him "he'd like to know if I'd expected him to return the blow when the policeman hit him. I told him I hadn't expected anything whatsoever."

Affects of others, if perceived, appear *"embarrassing."* Explicit or implicit demands of others to express empathy (and sympathy) by words responding to their words, or to express other nuances of affect, tend to elicit a "I have nothing to say" reaction. When the hero's girl friend answers his indifferent marriage consent by declaring that she "loves" him because he's a "queer fellow," but that she might "hate" him some day, the hero reports: "to which I had nothing to say, so I said nothing." When the magistrate asks the hero about his "reputation of being a taciturn, rather self-centered person," the hero answers: "Well, I rarely have anything much to say. So, naturally I keep my mouth shut." His habitual silence is thus not based on conscious restraint against verbalizing a rich subjectivity, as one recognizes the limitations of words and values privacy. It is a silence expressing a subjective void. Only in rare and fugitive instances does this void appear as a disguise of plentitude. At one moment during his trial the hero feels like "cutting them all short and saying: '. . . I've something really important to tell you.' However, on second thought, I found I had nothing to say."

In many instances the hero has nothing to say because *he experiences what others say as meaningless.* (One may recall the central importance, in contemporary empiricist epistemology, of the designation of certain types of sentences as "meaningless"—and hence neither "true" nor "false" —which had been regarded as "meaningful" before.) This is particularly the case when *others talk about the hero's subjective experiences.* When the examining magistrate, using an "intimate" technique, tells the hero "what really interests me is—you," "I wasn't quite clear what he meant, so I made no comment." When his lawyer asks him whether he had felt grief when his mother died, "I answered that of recent years I'd rather lost the habit of noting my feelings and hardly knew what to answer." When Marie "asked me if I loved her, I said that sort of question had no meaning really, but I supposed I didn't." When Marie asks again some time later, "I replied much as before that her question meant nothing or next to nothing—but I supposed I didn't." Her sentences on one's own affects appear either as meaningless or as difficult to test—but never as evidently true or false, a quality which earlier epistemology usually attributed to introspective statements.

The syndrome of affectlessness which I have sketched in the preceding passages is a largely *ego-syntonic* one. The hero does not share in the historically typical despair about, and revolt against, psychic impotence—

a change which is probably, again, related to contemporary trends. When his boss offers him a Paris job, the hero declares that "my present . . . [life] suited me quite well. . . . I saw no reason for 'changing my life.' By and large it wasn't an unpleasant one." While he is incapable of "explaining" his subjective state, it isn't a problem for him either. Affectlessness is reacted to affectlessly.

Such is the syndrome which dominates the usual life atmosphere of the hero. But, besides the affect inhibitions hitherto described, the hero shows a set of *affect substitutes*. Some of them are *somatic*. When the hero leaves a consciously entirely flat conversation with the pimp, "I could hear nothing but the blood throbbing in my ears and for a while I stood still, listening to it." Other affect substitutes are non-somatic. Certain *perceptions of detail*, closely associated with the central matters to which the hero does not react consciously, show a *high intensity* and at the same time *unreality*. When he is holding a wake at the coffin of his mother without any conscious grief, ten of her friends are with him: "Never in my life had I seen anyone so clearly as I saw these people. . . . And yet . . . it was hard to believe they really existed." At the funeral itself he perceives in a father figure—the "boy friend" of his mother in the home for the aged—the somatic image of his affect inhibition: "His eyes were streaming with tears. . . . But because of the wrinkles [in his face——N.L.] they couldn't flow down. They spread out, crisscrossed and formed a smooth gloss on the old, worn face." The hero is bored by the prosecutor's speech at his trial: "The only things that really caught my attention were occasional phrases, his gestures, and some elaborate tirades —but these were isolated patches."

Related to such affect substitutes is *doubt about details* closely associated with central matters. The novel begins with a doubt about a detail of the mother's affectlessly experienced death—reminiscent of Leonardo's slip in his diary notation on his father's death [2]. "Mother died today. Or, maybe, yesterday. I can't be sure. The telegram from the Home says: YOUR MOTHER PASSED AWAY. FUNERAL TOMORROW. DEEP SYMPATHY. Which leaves the matter doubtful; it could have been yesterday."

* * * * *

I have up to now discussed the hero's defenses against affect. What are the particular affects which have been repressed and whose repression is secured by the defenses described? I shall suggest that a major affect involved is murderous rage originally directed against the depriving parents.

The hero presumably experiences intense *guilt about the death of his mother* towards whom he has felt conscious, though consciously feeble, death wishes. In accordance with his overall techniques of defense against affect, he has largely repressed guilt feelings. When the magistrate asks him whether he regrets his murder, he answers that "what I felt was less regret than a kind of vexation." When the public prosecutor accuses him of his lack of guilt feelings about the murder: "I'd have liked to have a chance of explaining to him in a quite friendly, almost affectionate way but I have never been able really to regret anything in all my life."

Having repressed his guilt feelings, the hero additionally projects the accuser into the outer world. While he consciously feels innocent of his mother's death, he believes exaggeratedly or at least prematurely that others accuse him of it, or of his behavior in connection with it. (As will be seen below, he behaves presumably in part with the unconscious intent of provoking accusations.) He tends spontaneously to react to such ac·cusations apologetically rather than counterassertively. When he fixes up a two days' leave with his employer to attend his mother's funeral, "I had an idea he looked annoyed, and I said, without thinking: 'Sorry, sir, but it isn't my fault, you know.'" When the warden of the Home, in which his mother died, briefly recapitulates her history in the Home, "I had a feeling that he was blaming me for something [i.e. sending his mother to a Home——N.L.] and started to explain." At the wake near his mother's coffin "for a moment I had an absurd impression that they [his mother's friends who are present——N.L.] had come to sit in judgment on me." When he for the first time makes love to Marie the day after his mother's death and when Marie learns about this death, "I was just going to explain to her that it [his mother's death——N.L.] wasn't my fault, but I checked myself as I remembered having said the same thing to my employer, and realizing that it sounded rather foolish. Still, foolish or not, somehow one can't help feeling a bit guilty, I suppose." Finally, the public prosecutor at his trial affirms emphatically that "this man . . . is morally guilty of his mother's death," and adds—referring to a parricide which is on the court's agenda—that "the prisoner . . . is also guilty of the murder to be tried tomorrow in this court." The projected tends to return from projection: when the hero's "callousness" about his mother's death is shown in court, "I felt a sort of wave of indignation spreading through the court-room, and for the first time I understood that I was guilty."

Another major manifestation of the hero's intense unconscious rage consists in the commission of acts which aggress his environment and provoke it into aggressing him, thus alleviating his guilt. The self-destructive aspect of these acts is only little conscious; and the same is true for the aggressive aspect of some of them. For example, the hero adopts a "free association" policy in his verbal utterances, however grave the consequences of this may be. When he is asked to speak at his trial, "I said the first thing that crossed my mind," "as I felt in the mood to speak." The hero is consistently and consciously *frank, in words and acts in expressing the nonconformism corresponding to his affectlessness.* (For many of the hero's fictional predecessors the refusal to conform to conventional modes of expressing affect had been related to the awareness of a unique intense nuance of affect which insisted on its own channels.) He indicates clearly his lack of grief about the death of his mother and refuses to go through the paces of conventional mourning behavior, his atheism, his lack of response to others as conveyed by silence.

The effect, of course, is to stimulate hostilities directed against himself. But the hero scarcely—or only belatedly—recognizes this. He has, indeed, little empathy for his environment's aggressive reactions to acts of his own which are interpreted as aggressive. As he represses destructiveness in himself, he denies it in others. When he communicates to his lawyer his unfavorable attitudes toward his mother, the lawyer makes him promise "not to say anything of that sort at the trial." Thereupon the hero attempts to satisfy the lawyer: "Anyhow I could assure him of one thing: that I'd rather Mother hadn't died." He has no awareness of the unfavorable impact of such a communication. On the contrary, a recurrent conscious motivation of his is "to keep out of trouble," for example, by complying with demands made by others. Similarly, at the trial (where he behaves with resigned passivity and his usual provocative candor) he has great difficulties in realizing emotionally the seriousness of the public intent to murder him for his murder. He exaggerates the mildness of the world, and this is the counterpart to the presentation of the world as deprivational by withholding: the world isn't sufficiently interested in me to be out for my skin. (This belief also serves as a defense against panic.) Thus the imprisoned hero lacks up to a late moment the conviction that he is in danger: "At first I couldn't take him [the examining magistrate—— N.L.] quite seriously. The room in which he interviewed me was much like an ordinary sitting-room . . ."; "it all seemed like a game"; "When leaving, I very nearly held out my hand and said 'Good bye.' " When a routine of conversations between the magistrate, the hero, and his lawyer has become established, "I began to breathe more freely. Neither of the two men at these times showed the least hostility toward me, and everything went so smoothly, so amiably that I had an absurd impression of being 'one of the family.' " When he first sees his jury, "I felt as you do just after boarding a street-car and you're conscious of all the people on the opposite seat staring at you in the hope of finding something in your appearance to amuse them. Of course, I knew this was an absurd comparison. . . ."

Besides chronic covert self-destructive aggressiveness stand major explosions of overt aggressiveness: the murder of the Arab and an assault on the prison chaplain. A closer analysis of these acts may show how they fit into the character structure hitherto described. How are the defenses against completing rage overcome?

(1) The aggressions are felt as *inexplicable explosions originating outside the self and overwhelming it.* "Then [at a certain point in the protracted exhortation addressed by the prison chaplain to the hero awaiting execution——N.L.] I don't know how it was, but something seemed to break [*crever*] inside me, and I started yelling at the top of my voice. I hurled insults at him. . . . I'd taken him by the neckband of his cassock, and, in a sort of ecstasy of joy and rage, I poured out on him all the thoughts that had been simmering in my brain."

The murder of the Arab is presented as forced upon the hero by the primarily somatic and only secondarily psychic impact of the *heat* (a projection of the hero's impulses, we may surmise, onto emanations of the sun, a frequent paternal symbol): "As I slowly walked towards the boulders at the end of the beach [where the hero will find his victim—— N.L.] I could feel my temples swelling under the impact of the light." When he is near the Arab, whose posture is then entirely defensive, "it struck me that all I had to do was to turn, walk away, and think no more about it. But the whole beach, pulsing with heat, was pressing on my back. All the sweat that had accumulated in my eyebrows splashed down on my eyelids, covering them with a warm film of moisture. Beneath a veil of brine and tears my eyes were blinded; I was conscious only of the cymbals of the sun clashing on my skull." (The sun as inducer of "daze," while a bout of intense motor or psychic activity is performed, recurs throughout the novel.)

The murder itself appears—with a projection of destructiveness onto a target yet more remote from the self—as a cataclysmic release of violence in nature: "Then everything began to reel before my eyes, a fiery gust came from the sea, while the sky cracked in two, from end to end, and a great sheet of flame poured down through the rift." When violence begins, the projective impulsion by heat ceases: "The trigger gave. . . . I shook off my sweat and the clinging veil of light."

The hero then goes on to fire four more shots. This the magistrate and the public prosecutor take as a conclusive indicator of "deliberateness." The novel gives no clues on the immediate context of these "loud, fateful rap(s) on the door of my undoing."

(2) The hero consciously attempts to *resist* the explosion of aggression. First, he prevents his pimp acquaintance from shooting the Arab whom he later kills. When he walks on the beach in the direction of his victim, "each time I felt a hot blast strike my forehead, I gritted my teeth, I clenched my fists in my trouser pockets and keyed up every nerve to fend off the sun and the dark befuddlement it was pouring into me . . . my jaws set hard. I wasn't going to be beaten." But the very attempt to resist the consummation brings it nearer: The sun appears at this moment as "trying to check my progress" and thereby leading the resisting hero unknowingly towards his victim. Once confronted with this victim, the sun bars retreat (cf. above), although the hero, taking another step forward "knew it was a fool thing to do."

(3) The hero encounters his victim *without knowing* that he will do so: "I was rather taken aback. . . ."

(4) The hero appears to himself as *acting in self-defense*. His fantasies of being destroyed (suggestive of castration anxieties) go far beyond the real threat. When "the Arab [at a safe distance and with an apparently defensive purpose——N.L.] drew his knife and held it up towards me, athwart the sunlight," "a shaft of light shot upward from the steel, and I felt as if a long, thin blade transfixed my forehead. . . . I was conscious

. . . of the keen blade of light flashing up from the knife, scarring my eyelashes and gouging into my eyeballs."

(5) The hero has a *conscious instrumental motivation* in performing the acts leading up to his crime: this motivation is *superego-syntonic*. He approaches, unknowingly, his victim, who is in the shadow and near water, in search of relief from the sun. Walking on the beach "the small black hump of rock [behind which the Arab is lying——N.L.] came into view. . . . Anything . . . to retrieve the pool of shadow by the rock and its cool silence . . . I couldn't stand it [the heat——N.L.] any longer, and took another step forward. I knew it was a fool thing to do; I wouldn't get out of the sun by moving on a yard or so. But I took that step, just one step forward. And then the Arab drew his knife. . . ."

(6) Guilt-alleviating and anxiety-enhancing factors such as the ones mentioned facilitate the temporary but total return of the repressed rage from repression. The *self punitive* significance of such a return has the same effect. The heat driving the hero to murder and hence to his execution "was just the same sort of heat as at my mother's funeral." The hero atones for that funeral by arranging his own. He feels at the end that he is going to be "executed because he didn't weep at his mother's funeral."

In addition, "perhaps the only things I knew about him [the hero's father——N.L.] were what Mother had told me. One of these was that he'd gone to see a murderer executed." Presumably, the hero's own execution is in part for the benefit of this fantasied spectator who "had seen it through," although "the mere thought of it turned his stomach" and he (the spectator) afterwards was violently sick. (Also "when we lived together, mother was *always watching* me.") One may surmise that inducing the father to be sick has the significance of having a *sexual relation* with him. One may also surmise that being executed in front of the scoptophilic father has an *exhibitionist* meaning. One may further assume that this act is an *expiation* for aggressive tendencies towards the father in general and his scoptophilia in particular: "at the time [when the hero's mother told him the story——N.L.] I found my father's conduct rather disgusting. But now I understood: It was so natural." Furthermore, the hero identifies himself with a fantasied spectator of executions (his own?) and thus induces an oral, scoptophilic, and sado-masochistic ecstasy bordering on panic: "The mere thought of being an onlooker who comes to see the show and can go home and vomit afterward, flooded my mind with a wild, absurd exultation . . . a moment later I had a shivering fit . . . my teeth went on chattering. . . ."

These gratifications induce rationalizing elaborations of the fantasy of attending executions: "Often and often [awaiting his own execution—— N.L.] I blame myself for not having given more attention to accounts of public executions. One should always take an interest in such matters. There is never any knowing what one may come to. . . . How had I failed to recognize that nothing was more important than an execution:

that, viewed from one angle, it's the only thing that can genuinely interest a man. And I decided that, if I ever got out of jail I'd attend every execution that took place.

Some *speculative* points may be ventured here. The hero's interference with his destructive tendencies towards his father is accompanied by an identification with the (projectively) destructive father. In his pre-crisis equilibrium the hero is the consciously unintentional spectator of a number of beatings and the equally unintentional audience of stories about beatings: He sees or hears an old man living in his house maltreating his dog, the pimp beating his unfaithful girl, a policeman beating the pimp, the pimp beating the Arab and being counterattacked by him. The pimp tells him about previous beatings of the girl and of the Arab. (While these habitual "affectionate-like" beatings of the girl "ended as per usual," the hero's love-making to Marie is free from overt destructive admixtures.)

In the murder of the Arab, however, he makes the decisive transition towards an act of violence of his own. On behalf of what may be a degraded father-figure (the pimp) he aggresses (by his complicity in the pimp's revenge scheme) against what may be a degraded mother-figure (the pimp's Arab girl) and destroys her brother-defender-himself. If so, his murder would be a suicide not only in its unconscious provocative intent but also in its immediate significance.

(7) The hero's extreme aggressions are probably unconsciously intended to *extort concern*—be it in the form of indulgences or deprivations —from a world whose real or fantasied neglect had induced chronic suicide by affectlessness. This syndrome breaks down when the hero, in the extreme situation of impending execution, is exposed to affect indubitably directed towards him. He is then able to drop the image of the world as essentially uninterested in him and to react with felt affect to perceived affect. Having gotten a rise out of the world, he can get one out of himself. When at a certain moment of his trial the hero "for the first time realized how all these people loathed me," "I felt as I hadn't felt for ages. I had a foolish desire to burst into tears." When the owner of his habitual restaurant testifies in his favor with moist eyes and trembling lips, "for the first time in my life I wanted to kiss a man."

(8) Presumably, *punishment* means for the hero—among other things —*love*. In this context the story of Salamano, a degraded old man living in the hero's apartment house, and his degraded dog is relevant. (Thinking of Salamano "for some reason, I don't know what, I began thinking of mother.") The spaniel—of whom the hero's mother had been very fond— is "an ugly brute, afflicted with some skin disease . . . it has lost all its hair and its body is covered with brown scabs." The dog's relation to his master is presented as an extreme sado-masochistic one, with most of the overt sadism on the side of the master. Salamano repeatedly utters death wishes towards the dog. But when the maltreated spaniel finally escapes (to his death), Salamano is in despair and expresses unconditional love for, as well as total dependence on, the lost object.

(9) Presumably the "you'll be sorry afterwards" theme just alluded to enters into the hero's self-destructiveness. In prison he finds only one scrap of an old newspaper reporting a murder case: A son is murdered by a mother who ignores his identity. When she learns about it she commits suicide. "I must have read that story thousands of times. In one way it sounded most unlikely, in another it was plausible enough." If this were a "real life" case, the exposure of the subject to the story would be regarded as accidental and his reaction to it as significant. As this is a fantasy case, both are significant. This point is implicit in certain preceding passages of this paper, e.g., in the discussion of the hero as a frequent spectator of beatings.

(10) The hero's acts of violence discharge to some extent his destructive tendencies; hence a *diminution of the counter-cathexis* which is expended on interfering with them. Therefore, the degree of the hero's affectlessness decreases during his sojourn in prison. He discovers memory; time has been more difficult to kill in freedom. Now he can *express affectionate tendencies towards the punishing world:* "I can honestly say that during the eleven months these examinations [by the Magistrate—— N.L.] lasted I got so used to them that I was almost surprised at having ever enjoyed anything better than those rare moments when the Magistrate after escorting me to the door of the office would pat my shoulder and say in a friendly tone: 'Well, Mr. Antichrist, that's all for the present.' " After his second act of violence—directed, this time, against an obvious father figure, the prison chaplain, and rendering his execution doubly certain—his affectionate tendencies are more fully released and attain the level of serene happiness: "It was as if that great rush of anger had washed me clean." After the assault the hero falls asleep by exhaustion. Awakening, he experiences a reconciliation with the indifferent world and its prototypes, the indifferent parents; he can love them though he knows they do not love him: ". . . for the first time, the first, I laid my heart open to the benign indifference of the universe. To feel it so like myself, indeed so brotherly. . . ." "Almost for the first time in many months I thought of my mother." He "understands"—and identifies with—her having played at making a fresh start just before her death by taking on a "fiancé" in the Home for the Aged where she lived. The novel ends with the re-evocation of the father's execution scoptophilia and with a lead from the acceptance of the indifferent world to the acceptance of the punishing world: "For all to be accomplished, for me to feel less lonely, all that remained to hope was that on the day of my execution there should be a huge crowd of spectators and that they should greet me with howls of execration."

*   *   *   *   *

A satisfactory language of esthetics scarcely exists. If it did, I would attempt to formulate in it a favorable judgment on *The Stranger*. Assuming such a judgment, it may be surmised that the psychological plausibility (in this case, the "common sense" implausibility) of the content contributes to the esthetic value of the novel—which, on the other hand, depends on "formal" characteristics. Among these there is one of general importance which can be particularly well shown in this novel with it's unusual terseness and concreteness of style: The high degree of *implicitness* of the unconscious content layers. That is, the reader is *not* (as he is in numerous "psychologically oriented" contemporary productions) told in *so many words*, or *obtrusively* led to see, connections of the order of those discussed in this paper. Conceivably the author is less than fully aware of these connections. The conditions and consequences of the antagonism between explicitness and esthetic impact—an antagonism which has, of course, already received psychoanalytic attention—seem to warrant further speculation and research.

REFERENCES

1. Camus, Albert: *L'Etranger* (Paris, Librairie Gallimard, 1942); English trans. by Stuart Gilbert (New York, Knopf, 1946). All italics in quotations are supplied.
2. Freud, Sigmund: *Eine Kindheitserinnerung des Leonardo da Vinci, Gesammelte Werke*, chronologisch geordnet, Vol. 8 (London, 1940), pp. 190–1.
3. Kris, Ernst and Leites, Nathan: *Trends in 20th Century Propaganda*, in Geza Roheim (ed.): *Psychoanalysis and the Social Sciences* (New York, International Universities Press, 1946).

41

Raymond A. Bauer

THE PSYCHOLOGY OF THE SOVIET
MIDDLE ELITE: TWO CASE HISTORIES

It is only a half truth that "It takes all kinds of people to make a world," if by "world" is meant a particular society. Although it takes many kinds of people, they must be in the right proportions and in the right positions, and when a society finds that although it has many people, there are not enough of the right kinds in the right places, something will give. In this analysis of two case histories Dr. Bauer describes the personality type which the Soviets are trying to develop and why.

EVERY social order favors to a certain extent some particular orientation of personality and finds other personality types less compatible with its needs. Soviet Communism in developing into a modern industrial society and in assuming its present totalitarian structure has reacted against many of the passive, undisciplined characteristics which typified the Russian population, and has favored a more highly disciplined, energetic, and responsible type of person. The population of every large state is sufficiently heterogeneous to include the full spectrum of psychological types. The distribution of such types, however, may be such that persons with certain personalities are not available in sufficient numbers to meet the varied demands of the social order. This is essentially what happened in the Soviet Union

NOTE: This paper was prepared especially for this volume and represents its first appearance in print.

during the period of rapid industrialization and collectivization of agriculture. The result has been an extensive program of character training in the Soviet Union to develop "The New Soviet Man," a person who is supposed to incorporate in himself the qualities of discipline, will, vision, boundless energy, flexibility, and allegiance to the Soviet state [1].

As was said above, the emphasis on developing this new type of conscious, purposive individual is to a great extent a reaction against the stereotype which the Bolsheviks held of the deficiencies of the older Russian character, which did not seem to be suited to the disciplined routine of a modern industrial order. There was not, of course, a total absence of the desired type or person. In fact, changes in Russian social structure in the half century preceding the Revolution seem to have facilitated the emergence of a class of highly motivated and disciplined entrepreneurial tradesmen and peasants. The latter group, known as the *kulaks*, were numerically the more important and might have served as a reservoir of manpower for the Soviet elite, but because of the conflict of their economic interests with those of the regime, they became a focus of resistance to Bolshevism. Rather than an asset, this group became one of the most important obstacles to overcome.

The specifications of the "New Soviet Man" are drawn from the characteristics of the idealized Bolshevik leader. They reflect the activist orientation toward man and toward society that Lenin and Stalin represent in Russian politics, and that seems to have characterized many of the men who flocked to their banner. The problem of the Soviet order has been to add to their number a sufficient body of new recruits adequate to fill responsible administrative positions, and to condition in the general population certain traits of this character which will enable it to function more effectively in the new social and political order.

The two case studies presented here, one a colonel in the Red Army and the other a former idealist Communist, represent two variations of the type of person whom the Bolsheviks are attempting to develop [2]. Both, of course, although products of the Soviet regime, are not the results of the program of character training of recent years. They are rather persons who, having the proper personality characteristics, succeeded in the system. The similarity between them, it will be found, is superficial. Beneath the surface they are distinctly different persons, and these differences in basic personality have, as we shall see, significant implications for the functioning of Soviet society, and for the allegiance of such persons to the Soviet system.

Kamen: Bolshevik Militarist

Kamen was born in 1914 into a military family. His grandfather, his father, and his older brother had been officers in the Czarist Army. His father died shortly after the Revolution. The brother joined the Red Army during the Revolution; he was a colonel from the beginning of the Bolshevik regime until his death early in World War II. At the father's death, about the time of the Revolution, this brother "became the commander of the family." Almost a generation older than Kamen, he served as his conscious role model. However, Kamen resolved to advance further than his brother and become a general officer. He describes his mother as "non-political." His social and political attitudes developed under the influence of Soviet schools and of the family's military tradition.

Kamen joined the Red Army in 1932 and had risen by regular promotions to the rank of major when the war broke out. By 1943 he had become a full colonel. In 1944 he was wounded and captured by the Germans. He chose not to be repatriated after the war, presumably for fear of being treated as a traitor.

Our concern is with the developmental history of this man's attitudes toward the regime; yet the truth of the matter is that there is little "development" in the area of these attitudes. But for the accident of war, he would still be an officer in the Red Army, and even several years of exposure to Western ideology have made no perceptible dent on his original ideology. Except for the fact that he occasionally identifies with the Western position in the present power struggle, he remains basically loyal to the Soviet power and seems to have precisely the same attitudes which he had under that regime. As we shall see, his loyalty to the Soviet power stems not from any ideological affinity to Bolshevism, but from the fact that the Soviet regime gave him his status as an officer.

He gives a much less detailed history of how he acquired his attitudes than most informants. On the other hand, he is extremely vocal in stating his opinions. Furthermore, his personality expresses itself very clearly, so that the relationship between personality, his attitudes, and his role as an officer in the Red Army stand out very clearly. Hence, the discussion of this case is somewhat more static and functional than developmental.

Kamen is an extreme elitist and has strongly authoritarian attitudes. His underlying impulses are repressed and subject to very rigid and presumably brittle external controls. He lacks a unified sense of values and needs to find a definition to himself in his relationship to the external world. He is a man with little flexibility and adaptability, but one who had found a niche in Soviet society in which he could operate very effectively.

Some former partisans of the regime once favored, or still favor, its avowed social goals and certain of its pronounced humanitarian values, but disapprove of the regime's methods and its authoritarianism. But it can be said of this respondent that he approves of the regime's methods and the authoritarian nature of the Soviet order, is indifferent to its social goals, and might even be characterized as antagonistic to its humanitarian pronouncements.

He considers himself first and foremost a soldier. "I myself was not active in politics. I'm really a soldier. I would serve Nicholas or Ivan, no matter." That regime is best which conforms to his ideals of efficiency and good organization. "Someday, in the end, the world must be commanded by engineers and professors." His favorite example of the laxness of the

West is the International Refugee Organization, which he abuses on every possible occasion.

> There is so much bureaucracy here. Take the IRO. That is the most horrible example of the West that I have ever run across. In the Soviet Union such things would not be tolerated. Just think of what would happen to complaints there. If you complained in a restaurant or in a hotel that there was a mouse in your room or at your table, you could get the director fired. There are specific terms set. You have to submit complaints through a certain channel within forty-eight hours and nobody will dare delay forwarding them. Measures will be taken promptly. After all, they try to satisfy your demands. Not so the IRO.

He is a complete elitist and sees both the natural and the supernatural world as hierarchically arranged, with the people on top running things to their own taste. "You know," he says in one of his rare moments of identification with the West, "I have my own recipe for solving the world's crisis without a war. Give me a regiment to command and drop us over the Kremlin. We'll take care of the very top of the whole structure, and the rest will be no problem. Then afterwards, we can ask the people, 'Did we do the right thing or not?' "
Even the relations of man to God are authoritarian:

> There is probably some higher world. I once didn't think so. You know, when the illiterate village priests tell you so you don't believe it. But when somebody like Einstein says so, perhaps there is something to it. But if there is some higher world above ours, I have a hunch it doesn't give a damn about us. It doesn't care what we do.

Kamen regards secret informers in the Red Army as generally contemptible, but he justifies them on the grounds that they were essential for the maintenance of Soviet power.

> We called them Judases. But of course, all armies have such units. The regular rank-and-file Red Army did not even know about their existence. Like other armies, the Red Army has the system of Seksoty (secret "co-workers") among the troops. They play on your patriotism and while the Red Army and the people did not like them, they do a job essential for the authorities. They are a pillar of Soviet power.

The basic image of society which is reflected in these statements is mirrored in the organization of Kamen's personality. He takes the same attitude toward his own impulses; they must be disciplined, kept under tight control. He has succeeded in developing a rigid external personality that is almost a complete denial of what lies beneath.
His hyperactivity and militant masculinity seem quite clearly, on the basis of Rorschach and other data, to be reaction formations against an underlying passivity and feminine identification. Throughout the psychological tests, he shows a basic confusion of sexual identity. He reacts to

this by adopting the surface façade of extreme masculinity and by denigrating everything female. Women are regarded as objects of immediate sexual pleasure. "I often daydream of . . . *a beautiful woman and a glass of vodka*," is his way of finishing an incomplete sentence. He cites his brother's advice to enjoy women but not get married:

> . . . but if you married them, he said, the next day you find the bag with messy hair next to you in bed. The second day, she'll dress sloppily, the third day, you find her sitting on a chamber pot. That's what he used to say.

The manner in which he protects himself against the fear of being feminine is illustrated by his tendency to assimilate women into masculine patterns. On the sentence completion test, he says: "Semion thought that women . . . *mean to be men*." Speaking of his wife: "My wife is a German girl, a real soldier, she is a doctor, but she should be a sergeant." In the Rorschach test, he places soldiers' helmets on two female figures.

Substantially the same problem is posed for him in his relationship with other persons. Apparently in reaction against underlying tendencies toward dependency he vigorously asserts his independence. He did the unprecedented thing of refusing American cigarettes from the interviewer, ostentatiously smoking his own brand. This one act was so unusual in the situation as to be a topic of conversation among the interviewers for several days. In the sentence completion test, he affirms this need for independence: "He was willing to do anything . . . *to become independent*." This attempt to be independent of others, and to keep his own emotions from pushing him into close ties which might endanger this independence resulted in his being unable to establish successful warm relationships with other persons. Again his completion of an unfinished sentence is symptomatic: "When he was told that his best friend spoke against him . . . *he sent him to the devil* [told him to go to hell]." Thus, he rejects the idea that this event could be emotionally disturbing to him.

He is like a considerable number of other respondents who were interviewed in the extent to which they tended to evaluate themselves on the basis of external manifestations rather than in terms of some central concept of themselves. With such people, usually military personnel, there was a preoccupation with clothing, and particularly with the uniform. One young man looked down at his drab clothes, shook his head sadly, and then said to the interviewer: "But you should have seen me when I was in my uniform!" Kamen displays this tendency dramatically when on being asked about his first impression of foreigners, he replies: "The first foreigner I met was a German . . . the impression I got—his suit was shabby." Similarly, "Gregory was ashamed . . . *of his shabby suit*." With him, it is almost as if the suit, the uniform, the status of an officer were more real than the person.

He continues to identify with the Red Army, and makes it quite clear to the interviewer that it is not only an exceedingly good military outfit, but a highly desirable social institution. He repeatedly refers to the officer group as one of the most privileged sectors of the Soviet population, one which had the support of the regime and the respect and love of the people.

A blind reading of his Rorschach protocol led the analyst, Mortimer Slaiman, to predict that the respondent had a paranoid trend. The present writer had been struck, entirely independently, by the paranoid tone of the respondent's comments on one topic—the betrayal of the Red Army by spies. He is the only person interviewed, to my knowledge, who took seriously the trials of the Soviet marshals for collaboration with the Nazis in 1938. Furthermore, he elaborates this "betrayal" of the Red Army with a series of other "betrayals" in 1941. These stories are supplemented by the fantastic rumor of an eccentric beggar named Yasha who hung about the city of Gorky for fifteen years. Yasha, he says, indulged in such behavior as molesting girls in public so as to give the appearance of being "harmless." "Suddenly in 1937, Yasha was declared a spy. Now we heard that Tukhachevsky was Yasha's best friend, that Yasha had in reality a villa near Moscow, where Tukhachevsky visited him. I don't know how true it is but that story sounded credible." All the early set-backs of the Red Army are explained on the basis of betrayal. It is as though he would go to any extremes to distort reality in order to preserve in his mind the image of the Red Army as invincible, because his own importance, his own strength, his own security, lay in his identity with the Red Army.

His attitudes are by no means logically consistent. He is not guided by any central unifying set of values to which his beliefs and behavior are referred. Opinions have an instrumental value of relating him to his external environment, of defining that environment in terms which are acceptable to him, and of enabling him to influence that environment by the statements he makes. Consistency is a secondary consideration. His opinions are compartmentalized so that the contradictions of their implications never trouble him.

An example of this compartmentalization of attitudes is the fact that he has been closely associated with the most anti-Soviet of the émigré groups, because he finds their hierarchical principles compatible, and because they, as he does, favor "one, united Russia."

One of the major functions of his opinions is to reaffirm his unity with the system with which he identifies. Thus, he makes semi-contradictory statements about the Soviet treatment of returned prisoners of war. He says on a number of occasions that all cases were treated "on an individual basis," that if there were no evidence of intent to go over to the enemy, the returned prisoner was treated well. On the other hand, when asked about the practice of executing prisoners returned during the first Finnish war, he defends this practice vigorously: ". . . this was entirely justified. Those bastards had no business deserting to the enemy. If you want to desert, do so before or after the war. But during military operations, it is unfair to the other soldiers." Similar formal inconsistencies in his attitudes could be documented. The point of consistency is always the intent which lies behind his saying what he did; in this case, his intent was to defend the Soviet power.

Not all of these ideas, however, fall into this category. The officer's code to which he subscribes is indeed a set of guiding principles for his conduct; and violation or seeming violation of that code, as, for example, being captured by the Nazis, produces a feeling of guilt in him. If he can be said to have anything approaching internalized values, it would be this officer's code. But it is really a set of values of only limited applicability, which sets the norms for proper conduct in certain restricted settings. His officer's code has a quality of externality. This code does not define his "self," an entity which remains essentially without definition. It is not so much a part of himself as a set of imperatives for conduct to which he must conform rigidly regardless of consequences.

This basic type of personality is one frequently encountered among the younger men in the Soviet emigration [3]. In some instances these persons have deteriorated badly under the moral strain in which they found themselves. Having no essential self-conception, they are dependent entirely on the cues from their environment: "Do people like me and/or respect me?" "Am I succeeding?" Needing constant reaffirmation of their own adequacy and worth, they strive constantly for these tokens of success. Many persons of this type get into a situation where they have to compromise what principles they have. Futhermore, they become propelled progressively into situations of ever-increasing tension. They take refuge in some form of narcosis, usually excessive drinking. They are characterized by the same problems of impulse controls that mark this respondent, and employ the same primitive, rigid device of repression that he does. Being as a rule less successful in controlling their behavior, they commit indiscretions and are subject to involuntary expressions of the feelings they try to hide. They get into progressively greater difficulties and involvements.

For this type of personality, Kamen is the optimum of success. He functioned well within the system and gave the appearance of competence and self-possession. His success is somewhat a function of his very considerable ability, of the strength of the controls he has imposed on his own impulses, and on the compatibility of his personality with his life situation in the Red Army. His place in the military hierarchy and the respect which came to him as an officer were the external assurances of his success and adequacy. He was entirely capable of meeting the technical requirements of his work; he found the disciplined military life entirely congenial. External controls are, in fact, a source of comfort to such persons who have difficulties in controlling their own behavior. Dicks speaks of such persons as requiring a tight "moral corset [4]."

The allegiance of such a man to the Soviet regime is not based on political conviction, but on his need to find some satisfactory adjustments to his external milieu. The Soviet regime offers him a framework within which to advance. It particularly rewards him for his façade of unquestioning loyalty. Also, being a strongly repressive regime, it offers him some measure of security *vis à vis* his own impulses. Disaffection from the regime on the part of such persons will come only when the regime loses power, or if the individual is threatened with rejection by the regime. This respondent suffered doubt concerning the regime only because he feared that *it* would reject him.

> From 1937 on, I think, I thought that I myself might get into trouble. I began to feel that, true enough, defense demanded purges, but that I was insecure myself—although this would not have been a cause for me to desert.

For such a person as this, ideology and belief are symptoms rather than causes of allegiance. The ideology of the system of which he is a member is adopted both as an adaptive device within the system and as a means of reassuring himself of his unity with the system. Given a commission in a European army, his admiration of the Soviet system would diminish quickly, but he would remain as authoritarian as ever.

Chestny: The Idealist Communist

Chestny was born in 1914, the fourth son of a minor supervisor in a coal mine. His father died in 1918, leaving the mother and four sons behind. His mother was forced to work as a cook and laundress to support the two younger boys. His two older brothers fought in the Civil War, one with the White Guardists, the other with the Red Army. His mother was alternately beaten up by troops of both sides as they came in search of the son fighting on the other side. The eldest son, fighting with the Whites, was killed. The second son returned after the Civil War and changed his name to hide the fact of his "bourgeois" origin and of his connection with the White Guardists. This brother left the family so as better to hide his background. Chestny was not apprised of the politically questionable background of his family and successively entered the various Party organizations—Pioneers, the Komsomol (Young Communists), and finally the Party itself at the age of seventeen, one year younger than the legal limit.

> Thus, until 1937, I did not know my real origin. Mother did not tell me the truth. I was a Komsomol, then a member of the Party, and always said that I was the son of a worker and that I had no repressed relatives. . . . I was devoted to Communism and believed in it. When I entered the Pioneers, my mother was very angry. When she learned I was getting into the Komsomol, she said: "Go and resign, there is nothing there for you." I said, "Mother, you do not understand anything in these matters. You are an old-fashioned person."

Membership in the Party was an honor to him:

> I took this as a reward and as an honor and filled out (the questionnaire). I and other people being received into the Party were praised in an

assembly as "worthy individuals." They did not say that we wanted to join, but that the Party was ready to accept us.

Because of his mother's marginal employment, they had been continually in difficult material circumstances. In 1929, at fourteen, he went to work as a carpenter. About 1931, he was made a rate-setter, in the same organization. From 1929 to 1932, he went to a preparatory school in the evening. In 1932, he entered a technical institute and graduated as an engineer in 1936. Because of the vacancies created by the purges, he rose by 1938 to be chief of a construction office, and, at the age of twenty-two, he was boss over more than one thousand workers.

At this time his faith in the regime began to be shaken by the large-scale arrests. At first he accepted the official view that the arrested persons were enemies of the regime and saboteurs. Punitive action against groups which opposed progress toward the goal of building socialism was not inherently bad in his mind so long as he thought they were enemies of the new order in which he believed—a new order which would improve the welfare of the Soviet people. But gradually, he came to realize that the regime was punishing the people who had been loyal to it and who should be benefiting from it. He told of the arrest of an engineer with whom he worked. This man, he says was a capable and conscientious person whom the respondent did not like personally, but who did excellent work. One morning this man did not show up at his office, and his secretary told Chestny that the man had been arrested the night before.

> At first I thought that if they had taken him away, there was something there, but I knew that his work was irreproachable. I knew his work like the palm of my hand and regardless of an antipathy to him, I began to think, would he be a saboteur? . . . Two months later, they put someone in his place, someone who was not as capable, but was really stupid. I began to think the capable man was in jail and an incapable man had taken his place. I could not answer this question. I knew this fellow's origin. He was not of a class which was hostile to the regime and the arrest made a big impression upon me, because I could not explain to myself the reason for this arrest.

He reacted to these doubts by beginning an individual plan of self-education and investigation to explore the reasons for those events which he did not understand.

> As a member of the Party, I asked myself the question, "Why are there so many saboteurs?" "Why are there so many enemies of the people?" I could not understand it. I could not understand why one part of the population must accuse the other, and I wanted very much to know how these questions were solved in other countries. I began to go to libraries and I began to read newspapers very carefully, but I found an answer nowhere. I wanted to find an objective answer to my doubts. Then in the fall of 1938, I accidentally met in the library a representative from a school of journalism. We talked together.

As a result of these conversations, the respondent enrolled in the school of journalism. His personal reasons for enrolling were primarily to continue his education in order that he might understand better the events around him.

About this time, it should be noted, his mother gave him the full story of his own family background. Up to this point, it must be remembered, he thought it proper that there should be discrimination against people who came from certain social classes.

> I . . . felt that it was fully correct, and that all who were not poor peasants or workers should be considered as enemies of the people who wished to return to the old order. (What happened when you learned about your real origin?) I was very deeply upset and started asking myself these questions: "I am an honest Communist. I obey all orders of the Party. I do all the work voluntarily that I must do. I give all my strength to the regime. I want to be a good Communist. I believed in the Party. I felt no opposition, and I wondered whether it would pay to announce that my father and my brother were not the people I wrote about. . . . I asked myself this question, I asked myself: 'where is the truth, where is the justice?'" In characteristic fashion, he continues immediately: "I found that I had not enough education in the humanities and I wanted to add to my knowledge with further study."

His experience as chief of the construction unit raised other doubts. Whereas he had initially considered it plausible that men in responsible positions were sabotaging the efforts of the regime, he found when he achieved such a position that the regime put intolerable demands on him and that he was subject to interminable investigations.

> In 1938, in the space of one and a half to two months, there were twenty-eight investigations. For example, there was a commission that investigated norms, one commission was there to check and see whether I was wasting funds for salaries, there was a commission from the Union which told me I did not build houses correctly, and I was accused of favoring my own workers against the miners. There was a commission on construction control, and I was told by them that I used wrong materials. This was true, I had done it, but on the orders of higher authorities.

The injustice of his own position, of the excessive demands put on him, and of the unjustified suspicion with which he was regarded spurred the doubts which had been growing in his mind.

During the years 1938 to 1940, he continued his education in the school of journalism. By 1940, his defection from the regime crystallized: "Suddenly and unexpectedly I felt in the depth of my heart that I was no longer a Bolshevik." He resisted the conclusion, until he met another much older Party member who worked in responsible Party posts. He discovered that this man was also anti-Bolshevik. This strengthened his conviction that he had been wrong in believing in the Communist system. Whereas many respondents systematically suppressed their doubts in order not to disturb their accommodation to Soviet society, he admitted them to himself, but resisted them seemingly out of a feeling of guilt for renouncing a set of values which were part of him.

When the war began, he had already decided that he was opposed to the regime, but he could not bring himself to surrender to the Germans, even though his comrades did so. He fought through the entire war with the Soviet Army and left only at the very end. To a large extent his decision not to surrender earlier was the result of his observation of the behavior of German soldiers. If it were not for the fact that, to use his own words, he is "slow to make decisions" it seems possible that he might have surrendered at the very beginning of the war, before having had extensive contact with the German troops. The issue for him was not patriotism in the abstract, but the way in which the Nazis treated people. He deserted only when the Nazis were defeated. His leaving was prompted by the movement of the Red Army to take Prague "before it fell to the capitalists."

The flight to the Allied side was marked by unexpected drama. Chestny drove off in a truck only to discover a group of Red Army soldiers sleeping in the body of the truck. En route, he encountered a group of Russian women who had been conscripted by the Germans. These women greeted him as a liberator, and plied him with questions about life in the Soviet Union. Under terrific emotional strain, he left the soldiers in charge of the group of women and crossed over to Allied territory. One of the women, whom he "instinctively trusted," he took with him. She is now his wife.

This constitutes in outline the history of this man's relation to the Soviet regime. With Chestny, it is possible to reconstruct quite faithfully the attitudes which he held as a Party member, partially because of the extreme objectivity and the intellectual manner in which he recounted his past history, and partially because to a large extent many of the traits of character which made him a devoted and faithful member of the Party are still an integral part of his personality. The salient features of his personality are a pronounced idealism characterized by a high value for human beings and a strong capacity and disposition toward productive work. In fact, in his attitude toward work he is somewhat compulsive. In addition to these traits he values "objectivity" very highly and is much inclined to seek an intellectual formula for that which happens about him.

It is in his humanitarian idealism and his value of productive achievement that we find the source of his early affiliation to the regime. The conflict between the avowed idealist goals of the regime and its poor treatment of its subjects explains his defection from the regime. His description of the development of the Russian child dramatizes the conflict which occurred in his own person:

> The education of the Russian child includes Russian psychology which in itself means love for men and some sort of an invisible, undeveloped

Christianity. It is invisible under its cloak of Marxism, but it is there. It means to share what you have. The Bolsheviks cannot change this Russian character; they must swim alongside it.

The stated purposes of collectivization of agriculture and of industrialization were consistent with his values. His complaint is that the regime had other goals than those announced and that it led the peasants and workers into the socialization of the economy by false promises.

The real reason for collectivization was that the Bolsheviks needed a labor force for the industrialization and mechanization of the country. At the same time as they announced collectivization, they announced industrialization. You must understand that the slogan of industrialization among all the strata of the population was popular. . . . I felt that had any other people been in the place of the Russian people, and from whom such efforts for a better life had been demanded, any people would have agreed and made these sacrifices.

But, Chestny says, the population was deceived into becoming slaves of the Bolsheviks. They created not Communism, but state capitalism, which "Marx said was the worst form of capitalism."

As a loyal supporter of the regime, he saw himself as one of a devoted core of shock workers, leading the fight for a better life. He saw membership in the Party as bringing not advantages, but responsibility. He warned the interviewer not to be misled by members of the emigration who stressed the privileges and advantages of Party membership. Frequently he enumerated the duties and responsibilities of Party members.

I had to be very disciplined and behave well; I could not be a drinking man or engage in hooliganism and I could not have unbecoming behavior. . . . You have to attend meetings, pay dues, go to the Party school, study the short course of the history of the Party. . . .

Particularly at the present time, the Bolshevik Party takes into its ranks the most honest, the most work-loving and modest, and people who in the depth of their soul are not sympathetic to the regime.

Despite the fact that his income and status as a Party engineer would have afforded him a comfortable living, he lived quite frugally, being primarily concerned with the job he was doing. When asked to compare the general conditions of his work with those which obtained throughout the U.S.S.R., he commented laconically: "I was not concerned about my job satisfaction!"

Chestny is outstanding among our respondents in his positive evaluation of hard work, and his belief in the possibility of the Soviet citizen's improving his lot through the use of initiative. On several occasions, he attributed the poor plight of individuals to their own indolence.

I do not believe those workers who said that they starved in the Soviet Union and I will explain it in this way. I lived in a town of 17,000 people and in the name of my enterprise, the Promkombinat, I had to distribute 300 hectares of land to the workers. . . . I really had to force people to

accept this land because the people did not want to take it willingly re-gardless of the fact that a tractor would work on this land free of charge.
. . . Bad food is not the result of the regime, but the result of the psychol-ogy of the past that drags the people back; they do not strive for im-provement.

The esteem which he has for hard work and initiative, and his own capacity for sustained disciplined work, made him an effectively func-tioning Party member and spurred his enthusiasm for the long-range pro-gram of building and expansion that the Party initiated in the late twenties.

Chestny is typical of many people raised under the Soviet system. He was influenced by the Soviet system of education to accept the Bolshevik regime, and then repulsed by the realities of Soviet life that ran counter to the very values by which he had become a partisan of the regime. But some of the more detailed aspects of his relationship to the regime can be understood only on the basis of certain of his deeper psychological char-acteristics.

To some extent the firmness of his early allegiance to the regime must be related to the absence of any stable male figure in his own family. The dilemma of a young boy raised in an entirely feminine environment is reflected in the various psychological tests. In these tests, he identifies with both male and female figures and, on one occasion, mistakenly uses the female pronoun for a male character. It would seem that in the absence of father or other older man after whom to model himself, he adopted an ab-stract, idealized role model, the "ideal Party member." He says of himself: "I was trying to find out what the ideal Party member should do, or should be like, but I could not find this ideal."

Whether by coincidence or shrewd insight, his mother broke to him the news of his actual social background at the precise time when Chestny's doubts about the regime were budding. The extent of his reac-tion to this news cannot be understood entirely in terms of the situa-tional dangers which this knowledge spelled for him; by this time, he was aware of the fact that millions of Soviet citizens had successfully disguised their social origin. This news came at the time when he was beginning to doubt the ideal image in which he would mould himself, and it meant that certain essential elements in his self-conception were false, that he was living a lie. Thus the idealized picture of himself as a model Bolshevik crumbled simultaneously from within and without. The discovery of his true social origin probably facilitated this defection.

Under the influence of the ideal Bolshevik role model, Chestny had re-jected his mother's religious beliefs, even making her pray in secrecy and hide her ikons.

. . . Under the influence of my views, she had to pray secretly. She had to hide religious books. (Did she have ikons?) She hid them also.

With the failure of his attempt to find an ideal after which to guide him-self in the Bolshevik regime, he reverted to the once rejected value system of his mother and is today a strong religious idealist, as fervent in his acceptance of religion as he once was in attacking it.

The extent of his early identification with the Party, however, was such that he is still not entirely able to divest himself of it. More than almost any other respondent interviewed, Chestny continues to identify with the rank and file Party members; he sees them as the best elements in the population, imbued with favorable characteristics, and he attributes to them those very traits which most distinguish himself. Despite his rejection of the regime, he cannot reject the Party member, because in the Party member he sees his male identity. He reconciles these two factors by stressing the extent to which Party members are coerced into joining and thereby dissociates them from the regime. This appears to be related to the history of how the discovery that an old Party member was opposed to the regime helped him in breaking from allegiance to the Party. This event seems to have suggested to him for the first time that the Party member could be dissociated from the regime.

Tentatively we may pose the hypothesis that the presence of an adequate male role model in his family would have lessened his early tendency to identify himself with the regime and would have made it easier for him to break with the regime once doubts had risen in his mind. As a hypothesis of general significance this suggestion is strengthened by a variety of bits of evidence that the *bezprizorniki* [homeless waifs] of the early period of the regime selected real or ideal political figures as role models. It is a proposition that can be investigated in other of our cases.

The fact of his close association with his mother, and the fact that he identified himself to a large extent with her as well as with the father-surrogate, the "ideal Party man," made the shift of ideology from Bolshevism to religious idealism easier for him.

This man must be differentiated sharply from certain other "easy shifters" among the one-time adherents to the regime (such persons as Kamen above). The most common type of "easy shifter" seems to be a person who is without any specified value system, but rather is a congenial accommodator to any milieu and system, apart from the content of its ideology. These are people who have essentially no self-image. They seek for identity not in any system of values, but in terms of the approval of the social milieu in which they find themselves at the time. In the present case, it is possible to specify ideological systems to which he could not give allegiance. In the instance of these other persons such specification is impossible since the content of ideology is virtually irrelevant.

In one other way he is in essential contrast to this other group. His compulsive work habits and his tendency to intellectualize the problems with which he is confronted, to phrase problems in abstract, detached terms, rather than react to them emotionally, do seem to be defenses against a considerable amount of inner conflict which is revealed on psychological tests. However, these traits, whatever their genesis, are adaptive traits, making him an effective and competent member of society. The members of this contrasting group are not always capable of, or disposed toward, sustained creative effort, and often succeed merely on the basis of their ability to conform, to exhibit loyalty and reliability. His competence permits him to steer a more independent path.

Chestny's relationship to his mother should be commented on in some greater detail. While he seems to have been constantly at odds with her throughout childhood and youth over his allegiance to Soviet ideology, nevertheless their emotional relationship was a warm and stable one. From her, he seems to have gained the human values, the inclination to cherish the welfare of the individual human being, that made his continued allegiance to Bolshevism impossible. From her, also, he seems to have received the emotional security which makes him an integrated and effective person.

While he has turned against the Bolshevik regime, it must be strongly underscored that he has not turned against all aspects of the existing order of events in the U.S.S.R. Consistent with his concern for human values, he favors the retention of the social security aspects of the regime. In all respects an essentially paternalistic pattern pervades. His search for better comprehension of the world about him, however, has given a somewhat original twist to many of his ideas about the type of order he would like to see there. Of social insurance, for example, he says: "Social insurance is a drug for a man or an organism which is already sick. I am in favor of prophylatic methods to prevent sickness. The government . . . must take preventive measures." In the economic sphere, he says he favors "private ownership," but "private ownership" it turns out means for him a form of syndicalism, in which the factories would be owned by the workers. He favors free speech, even for Communists if they will forsake illegal means of gaining power—a stand which few of our respondents took. More than the average respondent, he favors, also, a respect for the form of law and an adherence to general principles even at the sacrifice of immediately desirable objectives. Thus, his present ideology continues to reflect the human values that led him to Communism, the general ethical principles which are part of his religious orientation, and his highly intellectual bent of mind.

Such idealistic Communist Party members carry in themselves a strong potential for disaffection in the disparity they experience between their own value system and the actual conditions of Soviet life. Being persons directed by an inner system of values, they are potentially capable of resistance to the regime to an extent that is unthinkable in the more pliant conformers. Furthermore, being highly capable and highly motivated, they rise disproportionately fast in the system. Therefore, their defection becomes a serious matter for the regime, since it takes place in the most responsible sector of the population.

SUMMARY AND DISCUSSION

The Soviet elite, if defined to include the full range of professional persons, scholars, and artists, embraces an extremely wide variety of personality types. If one restricts himself, however, to a consideration of individuals in positions of responsibility, with administrative or executive functions, the range becomes narrower. The cases discussed above represent two of the major types of "responsible" Soviet elite.

Chestny is the value-oriented idealist. His loyalties are primarily to a set of objectives and only secondarily to the system within which he works for those objectives. Kamen is the system-oriented conformist. The recognition and support that he needs could come from any system that can use his personality and skills. His loyalty is primarily to a system which gives him stability.

The latter type falls within the Soviet category of "careerist," but it does not by any means exhaust this category. Another type of careerist, not represented here, is one who dresses himself in the trappings of allegiance to the regime in order to gain the material advantages and prestige that come from advancing in the system. In recent times, this has come to be a more or less accepted course of action for capable and ambitious young people. Such a person is primarily loyal to his own interests, but in practice this is usually identical with primary loyalty to the system, as long as the system remains stable, since it is only through the system that his goals are served.

A person like Chestny is an "ideal" Communist in both senses of the word "ideal": he is motivated to personal ideals, and his is the ideal type toward which the regime professes to aspire. This ideal type, however, poses problems for the regime. The ideal Soviet citizen is supposed to be one whose goals and ideals are identical with those of the Soviet order. He is supposed to have the ability to see the interests of the state and then pursue them energetically, without deviating from the path to the goal. While he himself should have this understanding, he should also recognize that the Party, and specifically the highest echelons, has the highest level of comprehension of the goals of the state; and he should be prepared to follow quickly any change in the line defined by the Party. It is this simultaneous insistence on individual responsibility and complete subservience, on being both autonomous and an automation, that create problems with the idealist Communist. Having internalized the values and goals of Communism, he becomes an authority for deciding when they are being implemented and when perverted. He is not content to exercise his reason only to understand the wisdom of the course laid down by the Party. The difficulties which the Bolsheviks have had with such idealistic Communists are documented by the political purges of 1936–8 in the U.S.S.R., and by post-war purges in the satellite countries. Under the German occupation, these idealists were the first of all the groups of collaborators to break with the Nazis. Under both regimes their loyalty remained with their ideals [5].

The Soviet system of education and propaganda has been focused in recent years on restricting this idealism and on gearing the individual citizen's ideals into the needs of the state. Properly speaking "idealism" in the sense of a single life goal is not encouraged. "Ideals," however, are encouraged. Such ideals include love of work, Soviet patriotism, and goal-orientation of subordinating the interest of the individual to the group and of disciplining one's emotions. These are not unitary life goals but principles of restricted application that make the individual a more effective Soviet citizen. Nevertheless, "idealism" in the broader sense mentioned above does seem to infect a portion of the youth, and these youngsters become imbued with a burning devotion to a set of ideas. In many instances they become disillusioned at an early age and revert to cynical careerism. To the extent that they become disillusioned only with the Soviet regime but retain their faith in these ideals they pose a problem for the system. Such an idealist presents a dilemma for the regime. It is from this type of person that the regime could find its strongest and most effective support; yet he is the greatest source of potential spontaneous disaffection.

A conformist like Kamen presents no such threat of spontaneous action counter to the regime's policy. Kamen is unquestioning in his loyalty, and his obedience is immediate and unthinking. Nevertheless, he is relatively rigid and limited in his behavior. His brittle psychological defenses hamper creative imagination and capacity to review critically his own behavior and beliefs. In well-defined situations, he functions extremely well, but many posts in Soviet society demand a flexibility and ability to improvise that would run counter to the rigidity of this man.

The purely opportunistic careerist is not necessarily hampered by the handicaps of either of these men. However, he will never identify himself as thoroughly with the goals of the regime as do either of the foregoing types. He is likely to experience moral conflict only to the extent that in order to succeed he is forced to support principles or methods of which he does not approve. In practice, however, most such careerists are essentially apolitical in their values and are not faced with a moral conflict.

There is presently little evidence on the relative distribution of these types in contemporary Soviet society. Observation of members of the recent migration suggests a preponderance of the rigid, system-oriented conformist [6]. But there are reasons to believe that the Soviet emigration is selectively weighted in favor of this type, particularly because of the over-representation of the military in the emigration.

Regardless of the relative frequency of such types at present, each type exists and presents certain advantages and disadvantages for the regime. As the social order becomes more routinized, it may be anticipated that there will develop a selective preference for the pure careerist and the system-oriented conformist who are less likely to disturb the smooth functioning of an orderly society.

REFERENCES

1. The evolution of this concept and its implications are treated in considerable detail in Raymond A. Bauer: *The New Man in Soviet Psychology* (Cambridge, Harvard University Press, 1952); and in Margaret Mead: *Soviet Attitudes Toward Authority* (New York, McGraw-Hill, 1951).

2. These cases are selected from interviews collected by the Harvard Project on the Soviet Social System. They were selected because of their relevance to the manpower problems of the Soviet elite. The Harvard Project, under the sponsorship of the Russian Research Center and the Human Resources Research Institute of the U. S. Air Forces, gathered an extensive series of such interviews with refugees from the U.S.S.R. in Europe and the United States during the years 1950–1.

   I am indebted to Helen Beier and Eugenia Hanfmann for access to the clinical materials they gathered on these cases, and to Helen Beier and Mortimer Slaiman for sharing with me their analyses and understanding of this material. Naturally, I alone bear the responsibility for the form in which these interpretations appear here.

3. Henry Dicks: "Observations on Contemporary Russian Behavior," to be published in *Human Relations*, Vol. 15, No. 2 (May, 1952), (seen in manuscript form) considers the general type of which this is a variant to be the dominant personality pattern among the present generation.

4. Dicks: "Observations on Contemporary Russian Behavior."

5. These conclusions have been reached by Alexander Dallin, who had done extensive work on Soviet popular reactions to wartime occupation. Cf. Alexander Dallin: Popular Attitude and Behavior under the German Occupation, an interim report, Cambridge, 1952, unpublished.

6. Dicks: "Observations on Contemporary Russian Behavior," and the interviews of the Harvard Project.

CHAPTER

42

Margaret Mead

SOCIAL CHANGE AND CULTURAL

SURROGATES

A change in culture ought, if our theory is correct, to produce a change in the modal personalities of the group. The reverse is also true, of course. Constitutional determinants or situational pressures, or a combination of the two, produce unusual personalities. Under certain circumstances, these personalities create social inventions that are generally enough accepted to bring about major social changes. Sometimes, as Dr. Mead suggests is the case in our society today, there is a widespread "state of readiness" to innovate and accept innovation. It is this ceaseless interaction between cultural forms and cultural agencies [1] that is the theme of the following selection.

Dr. Homer Barnett has also contributed importantly to our knowledge of the relationship between personality and culture change.[2] Dr. Barnett has adduced evidence for the view that, in general, it is the deviant or maladjusted individuals who are innovators themselves or who are disposed to follow the innovating leads of others.

FIFTEEN years ago, students of education talked about the educational process or the educational experience with relatively little attention to the

NOTE: Reprinted with slight abridgment from the *Journal of Educational Sociology*, Vol. 14 (1940), pp. 92–110, by permission of the author and the *Journal of Educational Sociology*. (Copyright, 1940, by the Journal of Educational Sociology, Inc.)

[1] See also John W. Bennett, "Culture Change and Personality in a Rural Society," *Social Forces*, Vol. 23 (1944), pp. 123–32.

[2] Homer G. Barnett, "Personal Conflicts and Cultural Change," *Social Forces*, Vol. 20 (1941), pp. 160–71; and "Cultural Growth by Substitution," *Research Studies of the State College of Washington*, Vol. 10 (1942), pp. 26–30.

way in which that which was taught was mediated to the growing child. Students of culture—especially students of primitive society—recognized that the most diverse sets of cultural behavior could be transmitted to the growing child with equal success—that a newborn child among the Eskimos became an adult Eskimo, a complete version of Eskimo culture, with the same inevitability that a newborn Hawaiian became a Hawaiian. Among these students of primitive society there was, therefore, a tendency to emphasize the inevitability and complete effectiveness of the transmission of culture to the new generation on the one hand, and on the other, the extreme flexibility of the human organism which was capable of taking on such diverse behavior patterns. There were occasional discussions also of methods of education in which it was pointed out, by a collection of random examples, that children could be hurried and delayed, cuffed or bribed, into becoming adults. And, meanwhile, in Europe and America, the progressive-education movement was growing up, which insisted that our former techniques of formal teaching were all wrong, and that, if the individual child were permitted to unfold amid rich and stimulating surroundings, a new sort of human being, and ultimately a new world, would develop.

Among these approaches to a discussion of education, there was a sort of discontinuity. If methods were so all-important, why did we find that in different primitive cultures one method seemed to work as well as another, and any method produced a fully acculturated adult? There was a missing term in the discussion. It was necessary to shift the emphasis from how does culture A, faced with its newborn, induct them into being full-fledged members of the society characterized by culture A, to the question of what kind of character structure do these members of culture A have, as adults. Was there not a systematic relationship between the culture forms, the method by which the newborn were inducted into them, and the kind of character structure that individuals so educated would have? Granted that any educational technique would work—in the sense that a group of adults sharing a homogeneous culture would always succeed in imparting it to their children—if we shifted our attention from the accuracy with which the child repeated the behavior of his forebears to the internal mechanisms within the child that permitted him to act, in one case like an Eskimo, in another like a Hottentot, would we not find a significant difference there?

This new emphasis developed out of the interaction of Freudian psychology and the study of the socialization of the child. The Freudian psychologist, concentrating on the mechanisms within the individual, was the first to give a working account of the way in which children, born into Euro-American bourgeois culture of the last fifty years, took on that culture. This was not, of course, the way in which they phrased their accomplishment. They were very slightly conscious of cultural differences and assumed that the mechanisms of cultural transmission which they identified among their well-to-do patients were the mechanisms by

which Man became a socialized human being. (It was not until psychoanalysis reached into the clinics, and individuals from contrasting social
backgrounds were analyzed, that the first recognition of the different
mechanisms characteristic of different cultures—here seen as class mores
—began.) "Superego formation" was the name that they gave to this
process of cultural transmission. In its simplest terms, this theory stated
that, in the socialization of the child, the parent, or some individual who
stood *in loco parentis*, played a paramount role, and that only as the child
took unto itself its conception of the adult behavior as *the* right behavior
did the child undergo socialization. The psychoanalysts then proceeded to
explore the mechanisms by which this process was expressed in the character structure of individuals who had been, with various degrees of success, made to undergo it.

The next step in the understanding of the socialization process was not
so difficult to take. It was necessary to take over and generalize the psychoanalytic findings so that they would have cross-cultural validity; to
say, in effect, that culture is transmitted to the young organism through
the mediation of persons who become the surrogates of the culture into
which the child is being inducted. It would then be possible to investigate,
as one dimension of the cultural process, the question of who (parent,
grandparent, nurse, elder sibling, member of age group, masked dancer,
etc.), in the sense of individual in what status relationship to the child,
became the mediator of any given aspect of the culture. The recognition
of this next step was considerably delayed by two tendencies. (1) The
psychoanalyst, assuming that his description of a process characteristic of
Euro-American middle-class society in the late nineteenth century must
be universal, tended to force other cultures into the same mold—and
found in every culture a superego which to all intents and purposes was
identical with, although perhaps not as rigid or as exacting as, our own.
(2) The anthropologists, on the other hand, impressed by the Freudian
formulation, tended to identify the Superego with Culture. Both these
approaches suffered from insufficient generalization of the premises.

If we, however, do make the necessary generalization, it becomes possible to compare cultures in these terms and ask: How is the culture
transmitted, by whom, when, and with what sanctions? In this paper I
shall confine myself to a brief cross-cultural discussion of the surrogates
of the culture, the individuals who, in the various cultural settings, mediate
the culture to the child. This will provide a background for considering
what is happening to the process of cultural transmission in our own society today.

In a "pure case" of the functioning of our own system of cultural
transmission, the child at a very early age, soon after learning to talk, begins to take as its model the parent of the same sex. The child accepts the
standards of that parent, as stated *to* the child, as its own, until, in the absence of the parent, the child learns to act *as if* the parent were still there,
and to choose and reject courses of action as it conceives the parent

would have done. Failure to conform to these standards induces in the individual a retrospective discomfort, which is technically called "guilt" (traditionally referred to by the idea of "conscience"), which is independent of actual discovery by the parent, or any other member of the society. There are certain conditions which are essential to the occurrence of this "pure" type. The adults in the society must think of the child as qualitatively different from themselves, in that the child has not yet attained their moral stature, but is subject to innate impulses which, if permitted unchecked expression, would eventuate in an adult character different from and morally inferior to that of the parent. Furthermore, the parent, or a surrogate of the parent who acts as if he or she were the parent, must administer the culture prohibitions and admonitions, and must accompany this administration with various sanctions—the threat of punishment or of withdrawal of love and support, etc. The child's fear of the invocation of these sanctions is played upon to enforce the parental standard of behavior. Within such a society, teachers, the clergy, judicial officials, etc., partake of the character of the judging parent who metes out reward to the individual child (or, later, adult) who makes a satisfactory approximation to the desired behavior, and punishment to him who does not. A conception of the Deity which sees Him as primarily concerned, like the parent, with moral behavior, and as backing up the parent in dealing with the potentially immoral child, completes the classical picture.

If we examine other cultures, we find very different conditions. The Samoans think of children as quantitatively, not qualitatively, different from themselves. The child has not yet developed the ability to behave according to cultural standards—an ability which is conceived of as unfolding in the same way that an ability to carry a tune might develop. Individuals differ in their capacity to develop these abilities, for some, for instance, obedience to those of superior status, a desirable trait, comes hard. They are said to "listen with difficulty," as we might speak of some one who "has no ear for music." During the years before children have attained to full socialization, they are regarded as somewhat of a nuisance, as likely to shatter the austere dignity of a headman's council with a shrill cry, or outrage the etiquette of some stately function by standing upright when convention prescribes that every one should sit. It is necessary to watch them continually, and for this rather unimportant task small girls are selected. It is not their duty to train the infants entrusted to their care, but merely to see that they do not trespass upon adult attention by outraging the rules of etiquette, or by getting hurt. If the small nurses fail, and the baby does cry within earshot of adults gravely occupied with their own affairs, it is the nurse and not the baby who is punished. A baby who misbehaves, i.e., is noisy, does not conform to the social rules for excretion, grabs promiscuously at the possessions of people of rank, etc., is dragged out of earshot. Children learn, not as our children learn: "If I am to receive reward and escape punishment I must be good," but "if I am to be let alone and allowed to stay where I like, I must keep quiet, sit still, and conform to the rules."

The administrative forms of the society are also congruent with this picture; recalcitrant individuals are expelled from the household, the village, or the status which they have attained, and gods are conceived, on the pattern of the formally occupied adults, as concerned about their own affairs and presiding graciously over the affairs of men as long as men keep quiet and conform to the rules. The whole community is committed to following a way of life, and to the extent to which the child learns to follow it, he is permitted to participate in it. Obviously in such a setting, there is no room for guilt. Transgression and non-transgression are matters of expediency. The ties between parents and children are attenuated by the existence of large households, by the use of child nurses, by the system of household government through which it is the head of the household, whether he be father, uncle, grandfather, or cousin, who has the authority over the children, and by the extension of dependence behavior to a wide group of kin. Adults are regarded as persons who, with varying degrees of facility, follow the cultural forms better than children, and children offer as the alibi for any piece of disapproved behavior the simple statement: "*Lai titi a'u*" (I am young).

Among the Balinese, a related but different system of socialization is found. Children are regarded as partaking a little more of heaven than do adults and at the same time as not quite ready to become fully participating members of the community. Whereas in Samoa they are socially ignored and hurried off the scene, in Bali the steps by which the baby, in a series of ceremonies, is inducted into full citizenship in this world—and concomitantly sheds his membership in the other world—are of interest to everyone. A six-month-old baby, dressed in silk patterned in gold, put through all the paces of religious ritual, made to pray, to receive holy water, and to waft toward itself the virtue of the offerings, is the center of attention. Before there is any hope of an infant's learning cultural acts, it is put through them again and again, not necessarily by its parents, but by any one, adult or child, who happens to be carrying it. A three-month-old baby is made to act out the rule that nothing must ever be accepted from another with the left hand. Children are carried on the left hip, so that the right hand is pinioned and spontaneous gestures toward an object held out by another are all made with the free left hand. Monotonously, the person carrying the child pulls back the left hand, extracts the right hand, and extends it.

The whole teaching process is symbolized by the method of teaching dancing; the teacher stands or sits behind the pupil and pulls the body into the prescribed postures. At every turn a premium is put upon passive acceptance of others shaping one's behavior into the desired form. If, however, a child departs from this passive acquiescence and begins to act in unacceptable ways, to wander off when its guardians want it to stay near, to wander out into the midst of the actors, or stray toward the altar where the priest is praying, to reach for the property of another, or destroy some fragile object, the guardian—parent, relative, child nurse, or comparative stranger, for children are passed from hand to hand among girls and women, and every one assumes responsibility for keeping them from unacceptable activities—will first pull the child back, casually, without affect, without punishment, merely lift it up, or pull it down from the place where it does not belong to the place where it does. If it has an object which it should not have, she will replace this object, which she removes, with a substitute. If the child wanders again, or reaches again, the same process is repeated. If, however, the child gets out of hand, runs too far to be pulled back without changing her position, or grabs for the same forbidden object over and over again, she has another resource. She may mimic fear, exclaiming with an exclamation of tense horror: "Caterpillar!" "Centipede!" "Wild cat!" "Feces!" etc., at the same time imitating all the motor behavior appropriate to fear, so that the child comes running to her, to find reassurance in a terror that is shared with a bigger person. Or, if the object is not to bring the child back from some wandering, but to still its wails or whines in a setting in which such sounds are out of place, the mentor will attempt to call upon a pleasanter mood in the child by the recital of pleasurable experience; suggestions about listening to the orchestra or buying cakes or watching a shadow play will be recited with emphasis. These are not necessarily promises that if the child is quiet it will be taken to hear an orchestra or a shadow play, although to the Western ear that is the way they sound at first. They are a diverting recital of the pleasures of life, attention to the thought of which may be expected to supersede the mood out of which the child has been wailing. And for display of skill or cultural sophistication, passing the betel box to a visitor, or executing a ritual gesture correctly, the child is given extravagant admiration, "Beh, you are clever, how clever, how very clever!"

From all this the child learns that a pleasant mood and cultural conformity to fixed patterns occur together, and meets any possibility of deviation from that pattern with vague, uncertain distrust. This relaxed dependence upon a known way of life and distrust of all unknown ways is expressed in definite, spacial terms; a Balinese is only relaxed and able to function with easy rhythmic skill when he knows what day it is, where the religious cardinal points are, and the caste and status, phrased as whether they "sit" above or below him, of every one present. He is deterred from attempting the new or the strange by an unspecific dread which is not placed in any time sequence of rewards and punishments.

Balinese social structure is congruent with this type of character structure. It is a static formal society, in which behavior is prescribed and specific, within which individuals function, as it seems to us, motivelessly, tirelessly, in following out endless ritual forms, dedicated to gods who have no personalities, whose names are meaningless abstractions.

Among the Iatmul of New Guinea, a still different system of socialization is pursued. The Iatmul are an upstanding gay people, putting a premium on self-assertion and vigor. Mothers take more care of their babies than they do in Samoa and Bali. The child nurse, although present, plays less of a role; children, as soon as they become somewhat independent, tag along after elder siblings of the same sex more as appendages than charges. The mother assumes, from the time the infant is able to assert its wants definitely, that the child is possessed of a will and determination similar in kind to her own, and she acts as if the child were as strong as she is. She does not nurse the infant when she thinks it is hungry, any more than an Iatmul adult ever does anything for another adult merely because he thinks the other needs it. Instead she waits until the child has cried and cried hard and made her come across to it before she satisfies its demands. If the child does something to annoy her, she slaps it, an instantaneous quick slap, without any warning. But she is not punishing a young individual who has not yet reached moral stature for being "bad," she is simply slapping another person who has happened to displease her. If children continue to annoy, the mother will withhold food from them, and as a routine procedure she does withhold it each day, for an hour or so after she has cooked it. The parents are in control of the necessities of life, and the children are not allowed to forget this. So, the small boy who has annoyed his mother by eating up some food she had been saving for another purpose, or who has annoyed his father by setting fire to a neighbor's sago patch, and so involving his father in a first-rate quarrel, may run away unscathed in the first instance. But hunger will bring him home to his mother, and mosquitoes will bring him home within reach of his father, who has only to lie in wait beside his mosquito bag. Iatmul children are, on the whole, pretty well behaved, and they are always well behaved in the immediate proximity of a potential slap.

The Iatmul child learns from this treatment: "If I do not assert myself, I will get nothing; and if I anger other people, I will get slapped; and if I temporarily escape from being slapped, hunger and mosquitoes will drive me back again within range of retribution." In such a system there is again no room for guilt. The child merely learns to accommodate himself to a world in which every one is assertive and protectively high-tempered. His safety does not lie, as in Samoa, in looking carefully to the formal setting before he acts, nor, as in Bali, in never venturing outside carefully prescribed paths, nor as in twentieth-century America, in being good and so placating the parent ideal which he has accepted. His safety lies in getting angry first, in asserting his will over against others, before they assert theirs against his.

Iatmul social organization relies on these mechanisms embodied in political forms. The community is made up of a series of patrilineal clans, grouped into moieties. This division is crosscut by a division into initiatory age-grades. None of these groups has any means of disciplining its own members, but if an own member offends, it is an outside group, the other moiety, another age-grade, in extreme cases another village, which is called in to take punitive measures against an offender. Even in the disciplining of a wife, a man calls in members of another age-grade to reduce her to compliance. The individual and the group are both oriented toward and controlled by external sanctions.

Many forms of cultural transmission as contrasting as these have been described by social anthropologists, and there are undoubtedly many more forms, some actually existing at present, others that have existed in the past or are still potentialities of human beings acting within our historically defined cultural forms. Geoffrey Gorer has described a type of socialization among the Lepchas of the Himalayas, where the child is continually rewarded by food and casual affection for complete passivity. The North American Indian culture area is distinguished by its reliance upon shame as a principal sanction, and the fact that the parents do not set themselves up as punishing and rewarding surrogates of the culture but instead continually refer the child to "what people will say," in which fear of social disapproval becomes a principal mechanism, social disapproval enforced by scare dancers, or the licensed joking of relatives, etc. Another widespread type of character formation is that in which a child of noble birth is told that "people of rank don't do that sort of thing," in which expulsion from the social category of his parents is the sanction, and pride in his own class or caste membership is fostered and used as the sanction, as opposed to the fear of the opinion of others.

Seen within this cross-cultural perspective, our own superego system of character formation appears as a special and rather complicated development. Very few cultures have attempted the kind of explicit internalization of parental standards upon which ours depends. Our system might work quite smoothly in a stable culture which was changing very slowly. For an essential element in the system is that the child is expected to become like the parent, is expected to take the parent as a model for his own life style. In periods of rapid change, and especially when these are accompanied by migrations and political revolutions, this requirement of the system is unattainable. The child will never be, as an adult, a member of the same culture of which his father stands as the representative during his early years. This introduces a flaw into the working of the system in the ideal form in which we described it above. It is possible that many of the difficulties which are now attributed to too rigid or too strong a superego structure would not result in psychoneurotic manifestations in individuals who lived in a stable society where they had not been faced with such a great discrepancy between the content of the parental ideal and the possibility of living this ideal out in detail in their own lives in the same terms.

Now it is a recognized feature of our society that children soon after starting school begin to substitute the standards of other children for the standards set by their parents. This tendency becomes steadily aggravated until at adolescence it often results in a crisis in parent-child relations. Although this substitution of age-group standards for home standards is often regarded as a phenomenon rooted so deeply in the psychology of maturation as to be inevitable, cross-cultural investigations show that this is not so. In Samoa, the young boys and girls are given increasing status in the community as they reach and pass adolescence, but there is no period when they rebel against the authority of the head of the household and substitute instead a set of counter and antagonistic standards. (This is so in spite of the fact that Samoa tends to educate, during the later years of childhood in age-grade categories, so that children are very much influenced by the way in which group standards reflect the average household standard of the community, and children from deviant households become, through this mechanism, more like the average Samoan than like their deviant households.) Because we, like the Samoans, tend to educate children in age-grade terms, to classify behavior, that is, as appropriate for individuals at one stage of development, and inappropriate for those above and below that age, our society has in readiness the age-grade standard to apply to other ends than perpetuation of the *status quo*. And the role we give to parents, that they must pose as better and more complete representatives of their culture than they really are, also exposes growing children to almost inevitable disillusion. Furthermore, the notion that children are different in kind from adults fosters attitudes in children's and adolescents' groups which are qualitatively different from the attitudes of the adults. If young people bring home expressions of a point of view antagonistic to that of the adults, the adults have a category in which to place it; this is again a display of original sin, not yet eradicated, and contains a threat that the adolescents will turn out badly; that is, express in their adult activities the impulses which the parents have been at such pains to identify and brand during the process of education.

All of these aspects of our method of socialization would not, however, in a stable or very slowly changing society, have the drastic effects that they have under the present conditions of American society; but when the world in which the children already live is a different world from that in which their parents grew up, and when the world in which they will be adults will have a still different pattern, the socializing function of the age group becomes very much intensified. The children who continue to adhere to the standards set up by their parents carry the stigma of being "old-fashioned," "out of date," "prigs," "prudes," or "lacking in social consciousness," depending upon the slant of the group of their contemporaries among whom their lot is thrown. And the majority of young people respond eagerly to the approval and shrink away from the disapproval of their age mates. When they do so, what are they doing in terms of our original formulation?

The surrogates who carry the cultural standards have changed. They are no longer the parents, omnipotent and belonging to another order of being, but one's everyday companions, with the same strengths and weaknesses as oneself. Their power is transitory; by moving to another town one could escape them altogether, whereas to parents children are bound by inalienable ties. Now by examining the process of character formation among American Indians, and especially among the Zuni where we have particularly good material, we can judge something of the effect in devaluing the self which is developed in individuals who are taught to accept as cultural surrogates individuals who are not highly respected. There is a difference in the positive self-valuation of the individual who is attempting to meet standards represented by remote and highly respected persons, which in a mild way is what happens in Samoa, and the individual who is striving hard to meet the standards of persons who inspire no great respect. Furthermore, in our society age mates are always being adversely criticized by those whom the adolescents still, in spite of themselves, respect enormously, parents and parent surrogates. The rejection of parental standards in favor of late-recognized and antagonistic age-grade standards results, therefore, in an attenuation of self-respect and a weakening of the internalized standards of behavior upon which the operation of our culture is still postulated. Shame, the agony of being found wanting and exposed to the disapproval of others, becomes a more prominent sanction behind conduct than guilt, the fear of not measuring up to the high standard which was represented by the parents.

This shift in the process of socialization has several theoretical implications. It is illustrative of the way in which the process of socialization may alter in response to changes in the culture, in this case to changes in the rate and quality of social change. All adults are to some extent out of touch with the newest patterns of behavior as they most particularly affect the behavior of adolescents. (For example, children seldom use in school now the textbooks their parents used; if they do so, it is regarded as a sign that the school is an inferior and out-of-date school.) This condition is partly due to rapid change, and partly associated with the circumstance that in America many adults were reared in a different national, and often also a different class, pattern from that in which their children are growing up. Both circumstances are important contributing factors to this substitution of age-grade standards for parental standards. This shift to a greater dependence upon age-mate surrogates is also profoundly relevant to any plans for guided social change. There has been a great deal of emphasis upon the care that the totalitarian states have taken to bring up a generation in accord with the new standards, but there has been less emphasis upon the way in which a disjunction between parental standards and adolescent standards, already reflecting cultural change and particularly reflecting the break between those who reached maturity before and after the War of 1914, opened the way and made easy such a mass approach to youth. In cultures which, like those of Western Europe, had relied upon the parental standard as the major socializing mechanism, the use of age-grade standards as an element of cultural stability was not highly developed. Furthermore, the effect of a subordination of the self to age-grade standards in the devaluation of the self can give a strong support to mass movements; the quantity of the surrogates replaces their quality, for a single individual or pair of individuals, who are highly respected as different in kind and better than the child, is substituted the *number* of age mates who approve and follow a certain course of behavior.

Our historical form of character structure, the character which is sometimes referred to in current shorthand as the "Puritan" character, had potentialities for the kind of change which those who are involved within it regard as progress, because each fully socialized individual was striving to avoid the self-reproach of failing to realize an unobtainable ideal, the picture of the parent which he had conceived in childhood; and this culturally engendered discontent was an effective stimulus for "progress." But historically new aims have had to be integrated with the old ideals, if these aims were to be pursued with the energy derived from the recurrent guilt at falling short. Under such circumstances, a continuity of cultural trend was, if not guaranteed, at least made probable. It is this type of character that is invoked by those who believe that because Americans have always stood for democracy, or religious freedom, or some other social ideal, they will always do so. But those who trust to this increasingly rare type of character, with its moral emphases, are neglecting the enormous and growing prevalence of age-grade standards which are replacing the older sanctions. Parental standards are still, however—because grounded deep in such cultural forms as the biological family, the one-family abode, the responsibility of parents before the law, the habits of training children very young to achieve cleanliness, order, respect for property, and the banishment of the impulse life—part of our early socialization process. When the developing child repudiates them for the age-group standard he has in some degree surrendered his sense of moral autonomy for the comfort of a crowd.

Under these circumstances, the standard of the crowd becomes binding upon the individual, and the content of that standard is without significance, for in repudiating the early moral standard he repudiated the unconscious insistence on the intrinsic difference between good and bad. Any demagogue can sway crowds so constituted in any direction, provided he can capture the allegiance of enough of them at once.

A demagogue has no such easy task in societies which rely upon sanctions other than a superego of the parent-ideal type. Samoa, Bali, Zuni have all been very resistant to and resilient from culture contact with its potentialities for rapid change. It is the anomaly, the discontinuity in our present system of socialization which presents the element of instability. The young people who stand listening to a demagogue expound doctrines that are in the deepest conflict with all that they have been taught stand there because the others stand there, but they stand there *guiltily*, because they have deserted, for the sake of being in accord with the unrespected crowd, the moral allegiance they gave their parents. And it is this element in their character structure which leaves room for the leader, a parent surrogate who will lift their conformity to the mob onto a higher level again and make them feel less guilty of apostasy toward their own infantile acceptance of their parents' dictated systems of morality. There are no leaders in Samoa, in Bali, or in Zuñi, and among the Iatmul a man is acclaimed only so long as he can personally outshout and out-threaten the rest. Those societies are adjusted to different sets of sanctions. Our type of character structure has always left open the way for the leader, because he could become a parent surrogate; but only because of this hiatus in allegiance to the parental standards, which is the concomitant of such rapid and disjunctive social change, is the way open for *any* leader who advocates *any* doctrine which is accepted by a *large* enough number of people.

The existence of this state of readiness for leadership in the young people of this country represents a potentiality for desirable social change. Some organization will undoubtedly step in and fill the role which the home has been forced to abandon as a standard-setting agency. The school could be that organization. If educational leaders became sufficiently aware of the possibilities of using this mass willingness to follow a congenial solution, a willingness which is so characteristic of young people today, and if they were able to enlist young people in the task of creating new patterns of living congruent with the aims of a democratic society, this readiness for any new path might be used in building a more democratic state rather than a less democratic one.

CHAPTER

43

Margaret Mead

ADMINISTRATIVE CONTRIBUTIONS TO
DEMOCRATIC CHARACTER FORMATION
AT THE ADOLESCENT LEVEL

The following paper by Dr. Mead is clear and cogent. It will perhaps be useful to amplify her discussion of one point. She maintains that the type of personality integration easily achieved in many relatively homogeneous primitive or folk cultures is impossible in a stratified, mobile modern society. Only those who continue to live in a narrow sector of our society, say an isolated rural village, can attain to the organization of personality around a fairly stable set of values and constant interpersonal relationships. Others achieve adjustment by pseudo-integration—by compartmentalizing their personalities to fit different segments of their lives. The lesson of primitive society for us is not a magical formula for personality integration; rather, the comparative study of personality in culture offers us invaluable clues as to what kinds of personality fit with what kinds of institutions, what kinds of personality are characteristic products of child training in a given institutional milieu.

I HAVE been asked to discuss with you the influence of culture upon personal development, to discuss what social anthropology, and particularly that branch of social anthropology concerned with problems of person-

NOTE: Reprinted from *Journal of the National Association of Deans of Women*, January 1941, by permission of the author and the editor of the *Journal*.

ality and culture, can give to a group whose major responsibility is the relationship between the structure of the school and college and the personality of the individual students who live within that structure. I want to narrow this very wide problem down to a particular point and to assume that you have asked me a more precise question: *What has social anthropology to say about ways in which administrative and guidance officers in the schools and colleges of New York State now can contribute to the formation of a type of personality which will be effective in a democratic society?*

The first question is whether there is a genuine connection between the administrative forms and the personalities of individuals who must function under these forms. During the last ten years, a great deal of work has been done to bring together the two divergent approaches, best exemplified by the classical economic interpretation of history (which put the entire onus upon institutions and insisted that, if we altered the institutions, personality would take care of itself), and the psychological interpretation of history (which put the onus upon the organization of each individual personality, and insisted that, if we altered the personalities, the institutions would take care of themselves).

But every administrative or guidance official knows that actually the two aspects are inextricably combined, that there is a continuous and dynamic interrelationship between social forms and personality, that an autocratic system administered by a democratic personality is not the same system as an autocratic system administered by a convinced autocrat, and that either type of administrator is limited in his results by the forms within which he must work. There have been many recent studies to show that modern economy not only produces a certain type of character but is in turn dependent upon that type of character to perpetuate its peculiar institutions. The combination, then, that deans present, of work with structure and work with individuals, is a particularly happy one, for deans are in a position to attack both ends of the problem at once, altering the structures which are in their charge, and modifying the personalities of administrators and students so as to reify those alterations.

I have further limited the question so that it would apply to the present local American scene. You are primarily concerned with producing a more democratic personality here in the United States this year. This concern has several implications. It means that those individuals in whom this type of personality is to be fostered and encouraged have all been born within the last twenty-five years and reared for the most part in the United States. It means that they present to you a given type of character structure, which has been developed by the circumstances under which they grew up, and which presents you with definite potentialities and definite limitations. You are not working with raw human nature. You are working with American adolescent character, and this character is set in two ways: by the past and by the nature of the world into which these young people know they are going It is, therefore, of primary importance to take into account what this character is, within the limitations of which all work with adolescents in New York must be done.

Here there are two significant peculiarities of American character structure: its emphasis upon moral choice and its dependence upon achievement measured against the achievement of near equals. The first, the emphasis upon the importance of moral choice, Americans share with all those brought up within the Northern European tradition, and we are accustomed to think of this type of character as human. When the psychoanalyst discusses the "formation of the superego" or the moralist discusses "conscience," these terms are equated with "man's natural development within society." Comparative studies, however, demonstrate that this type of character—in which the individual is reared to ask first not "Do I want it?" or "Am I afraid?" or "Is this the custom?" but "Is this right or wrong?"—is a very special development, characteristic of our own culture, and of a very few other societies. It is dependent upon the parents personally administering the culture in moral terms, standing to the child as responsible representatives of right choices, and punishing or rewarding the child in the name of the *right*. Under this system the child grows up to be able to act in the parent's absence *as if* the parent were present, and later to substitute for his parent's actual conduct, which he finds wanting, an ideal standard towards which his conduct, usually wanting, is oriented. From the discrepancy between this ideal and his actual achievement, the individual derives a feeling of guilt, which provides the motivation for another special development of our civilization, an urge toward progress. From the emotional experience with parental reward and punishment, given in moral terms while he was a child, the individual derives the ability to bulwark a moral choice with his full emotional allegiance, a type of behavior upon which our form of democracy is postulated.

The second significant peculiarity of the adolescent personalities with which you work is especially characteristic of the present epoch in American culture; that is, the emphasis upon comparative achievement. Because of the extreme rapidity of culture change in America, and because of the divergence in national backgrounds, American parents do not depend upon appeal to the motivation of "being like father" or "being like mother" after a child enters school, but substitute instead the motivation of "doing as well as" a brother or sister, which means keeping up with the record set by older siblings and keeping ahead of young siblings. The process of American character development thus can be divided into (1) the pre-school period, in which the child is primarily preoccupied with the relationship to parents and parent surrogates, seeking to win love and approval and avoid punishment, in direct child-parent contacts; (2) the school period, in which the motivation of comparative achievement with members of the age group is substituted for child-parent relationships, and in which the child is swayed by age-group standards, always interpreted by him, however, in terms learned in the preceding period; i.e., as "right" and "wrong." In the earlier period he was preoccupied by conflicts between his own desires and his need and dependence upon his parents; and now to the precipitate from that earlier period—the fear of not doing right—is added the fear of not keeping up, plus the strain of discrepancies between the age-group standards and his own home. His principal armor against guilt, when his age-group standards contravene his parental standards, is that his parents themselves have told him to go out and excel, that they have told him to stand upon his own feet and measure himself, not against them, but against his age-mates.

With this complex set of psychodynamic mechanisms, the adolescent enters high school, and the administrators have the task of devising a set-up in which this sort of personality can reach a mature integration. Here it is necessary to turn from the student's developmental history to the future role which we anticipate for him. What sorts of integration, as seen against the background of the socio-economic institutions, could we envisage for him? Here studies of comparative culture provide us with a wealth of material. There is the sort of integration which is given by living all one's life in the same social group, dealing always with the same individuals, in which one derives one's integration from the knowledge of one's relative strength, wit, persistency, etc., in a known constellation of human beings, in a single geographical setting. The integration which we find in peasants is of this sort. We cannot aim towards it for present day adolescents, who must move from group to group, who must be members of constantly shifting and overlapping groups, who must act now with one set of persons and now with another.

Among the Iatmul of New Guinea we find a primitive tribe, democratic in the sense that what government there is is by the people, with no system of chieftainship or permanent leadership, a government, in fact, based on a large number of overlapping pressure groups. The integration of the society is accomplished by this very overlapping; each group is arrayed against one or more other groups, but, since the membership overlaps, the community, though almost always involved in a brawl, is unable to split. Individuals are called upon to act sometimes in terms of membership in a clan, sometimes an age-grade, sometimes a moiety, etc.; and two men who fight side by side today will be on opposite sides tomorrow. It is to be noted that the Iatmul native lacks the integration which comes to the member of the very small community with no overlapping groups. He is required to act emotionally, in association with or against the same persons, according to the dictates of various group memberships. He maintains his integration by acting in the same way in all situations; if he is noisy and tempestuous in one alignment, he is noisy and tempestuous in all his contacts; he yells at his wife's relatives, at another clan, or at another age-grade in an identical tone of voice. If he is a milder and more indirect type, he is consistently mild and indirect. In a society without a hierarchy of any sort, with no status which must be given emotional recognition, this consistency of behavior pattern is possible. Without enduring personal ties, with continual demands upon him to react emotionally because a group membership involves him, the Iatmul, nevertheless, is able to maintain a sort of integration. A somewhat similar sort of integration is attained among us by the perennial protestor, the individual who has found a consistent and satisfying type of behavior to apply to every sort of situation. But this sort of consistency, this fixity of behavior regardless of context, is not the sort of integration upon which our democratic system is postulated. The Iatmul never makes a choice except to choose, where there is overlapping, with which group he will shout and stamp. Issues are always irrelevant.

The Balinese mountain community is also a democracy, although of a very different order. Within the community every full citizen has a fixed place determined by the order in which he came of political age. The oldest citizen who has not been retired because of some social or physical defect is the titular head of the community. There is an enormous code of rules governing every conceivable aspect of community relationships, and, in the rare cases where there is no rule, resort may be had to communication with the village gods through divination and oracles. It is the duty of the council of citizens to interpret and enforce the rules, and this is done without any emotional participation by the members of the community, either the culprits or the executants. In deliberations it is never asked: "Is this right?" "Is this advisable?" but only: "Is this the correct interpretation of the rules?" And in such deliberations one voice which suggests "There is no precedent and so we do not dare to act" always will overrule the rest. Each individual moves through this minutely controlled pattern, without emotion, speaking in words of respect to those above him in the hierarchy and words of familiarity to those below, completely uninvolved. The degree of detachment which is requisite for this type of smooth, effortless, and unenterprising citizenship is attained by the Balinese during a type of childhood experience very different from that experienced by our children. At the present time, individuals reared and forced to live under American institutions attain this particular emotional detachment at very great psychic cost. Unless possessed of special artistic gifts, such individuals are more likely to inhabit mental hospitals than legislative halls, to be an actual drain upon the working of our form of democracy. The solution of our problem does not lie in teaching individuals not to feel emotional involvement in social issues.

If we come nearer home for a moment, we have a certain amount of data as to the ideal of integration which has been fostered in recent years in Germany. It is especially instructive to us, because the German culture shared, historically, the type of child-training which has been discussed above, in which the parent teaches the child, by reward and punishment, to make moral choices. Now this capacity to choose includes the capacity to blame oneself for a wrong choice, and this sort of self-blame is only tolerable if it does not have to be done too often. It is necessary for the individual who values right and wrong to be right at least some of the time; the alternatives are the rearing of a new standard of right and wrong, or more often the formation of a compensatory type of personality, in which the intolerable amount of responsibility is projected outward; someone else is to blame, the individual is perfect. It is significant that in the present German system this perfect individual whose enemies are always to blame is no longer asked to shoulder the type of moral responsibility which we expect of citizens of a democracy. For the democratic principle of self-determination is substituted the leadership principle; for emotional allegiance to issues is substituted emotional allegiance to a leader; for internal self-criticism, and the psychic strain which goes with it, has been substituted another type of internal balance, emotional submission to those above, emotional bullying of those below.

We find a conspicuous case of an analogous personality split among some South African tribes, in which individuals are expected to react differently to different relatives; a man is stern and harsh to his own children, lenient and indulgent to his sister's children; a woman reverses the process, as mother she is indulgent, as father's sister she shares the paternal role of harsh disciplinarian. Children spend part of their childhood at home, under the father's rule and part in the village of their mother's brothers, and they learn as children both to submit and to bully, and are prepared for this dual role when they grow up. There obviously is no room in such a personality for moral choice; emotional attitudes are determined by status, and in the adult political world, one is submissive to the king, autocratic to political inferiors.

Our system is not so defined and simple as any of the three primitive systems which I have discussed. Furthermore, what has happened in Germany is earnest of what profound alterations can be wrought upon the basis of our Western European type of character structure under too great social pressure.

The young adolescent American is already carrying a heavy psychic burden. The kind of character structure with which we endow him provides him with little immediate satisfaction. He has been made to care specifically about issues; to ask "Is it right?" or "Is it wrong?" He has been taught to orient his behavior to that of others of his own age and sex, to measure himself, not against adults but against age-mates. He has been taught further that to be grown up is to be independent of parents and parent surrogates, whose immediate standards he already has discarded for age-grade standards.

This is the human material with which a dean has to deal. If you wish to avoid the Iatmul type of integration, avoid turning out habitual "yes men" or habitual "protestors," who only seek to maintain a consistent type of emotional behavior regardless of issues, then it is necessary to avoid trying to run a school or college in terms of overlapping groups— pitting class loyalty against house loyalty, against club loyalty, regardless of the issues involved. Such a set-up, although it may keep the school community from splitting wide open, does not produce the type of democratic citizen who is capable of choosing among issues; it simply produces pressure group members.

If you wish to avoid the Balinese passive, emotionless enactment of routines, then it is necessary to have school routines which are amenable to individual choice and manipulation, not something imposed by impersonal gods whom the individual is afraid to displease for fear of some indefinite and unknown penalty. All the disadvantages of the Balinese system are manifest in institutions like orphanages, in which the children move through a meaningless routine of minute observances upon which what security they know depends. A sufficient amount of sufficiently irrelevant red tape may beget in the individual student the sort of paralysis of initiative and emotional participation which in Bali makes a good citizen, and which in America makes a candidate for a mental hospital. The essence of the Balinese system is that it is meaningless, and one's very life depends upon following the rules. The essence of our democratic system is that every part of it should have meaning for those who actively participate in it—whether it is the structure of a school or of the wider community.

Finally, there is the very real danger that we may impose upon our already psychically burdened adolescents, pressure towards another personality split, by demanding that they behave not as free, morally choosing, participating individuals, but with submissiveness, dependency, placation, to ingratiate, to follow unquestioningly those in authority. The recent experiments of Kurt Lewin at Iowa have demonstrated how an autocratic atmosphere, imposed upon children with typical American character structures, has certain very definite effects. The children lose initiative and power to choose, follow the leader with a passive dependency, and adopt a contrasting emotional attitude towards other children, becoming bullying, aggressive, destructive in social relationships. In his experiments, Lewin has to operate, just as you do, with already formed American characters. His experiments are not so much a demonstration of how people behave under autocracy, as of how people with our traditional type of character structure behave under autocracy. An individual who has been taught that he must stand up and choose, or else his parents will withdraw their love and punish him, cannot stand a situation in which the representatives of parents, teachers, and others in authority, suddenly deny all of his training and start giving blind orders and expecting him to give blind obedience. With his delicately balanced personality structure, there is a shift in emphases, a rearrangement of affects; in proportion as he submits and toadies, he also will dominate and bully, to restore some sort of integration within his outraged personality.

These considerations have very definite implications, then, for administrators of schools and colleges. To the extent that they provide a framework within which the student is given initiative, within which he is asked to participate in terms of moral choices, emotionally bulwarked, rather than in terms of mere group allegiance or the slavish following of a meaningless routine or blind allegiance to authority—to that extent he will develop an integrated personality. He will be made capable of functioning in our democratic system, capable of acting in terms of issues, not pressure groups; in terms of meaning, not fear; in terms of a whole moral approach to life in which he is able to treat everyone with whom he comes in contact as a co-partner in a democratic enterprise.

CHAPTER

44

Robert Waelder

THE SCIENTIFIC APPROACH
TO CASE WORK

WITH SPECIAL EMPHASIS ON PSYCHOANALYSIS

Dr. Waelder's paper has implications for the problem of social change treated by Dr. Mead (p. 651), but its main purpose is to summarize for the professional social worker, as well as lay persons, some of the insights and orientations of psychoanalysis. Applications of the cultural point of view in social work may be found in studies by Adland and by Cohen and Witmer.[1]

THE last two centuries have witnessed tremendous triumphs in the development of the natural sciences—physics, chemistry, and biology—and their application in technology and medicine. There has been no comparable development in the science of man or in the technique of influencing human affairs to keep pace with the rapid progress in the

NOTE: Reprinted from *The Family* (October 1941), by permission of the author and the Family Welfare Association of America. (Copyright, 1941, by the Family Welfare Association of America.)

[1] Charlotte Adland, "Attitudes of Eastern European Jews toward Mental Illness," *Smith College Studies in Social Work*, Vol. 8 (1937), pp. 87–116; and Eva Cohen and Helen Witmer, "The Diagnostic Significance of Russian Jewish Clients' Attitudes toward Relief," *Smith College Studies in Social Work*, Vol. 10 (1940), pp. 285–315.

[671

understanding and the command of external nature. Many of the problems of our time have their roots in this fact. A world that has become rich in gadgets is not very much better equipped with adequate methods of social engineering than was the simple world of our forefathers.

Many efforts are now under way to increase scientific insight into man's mind and society and to develop methods of deliberately influencing them. Social work is one of these efforts. In some instances, as in case work, one approaches individual cases; in others, one aims at a scientific technique of social planning.

Social case work, with its attempt at remedying unsettled conditions of individuals, to some extent takes the place occupied centuries ago by the priest. The priest, to be sure, did not offer any financial support, but he acted as adviser to his clients in difficult life situations, offered comfort to the suffering, moral support to the wavering, and relief from guilt to all.

The theoretical background for his social and therapeutical activity was not in doubt; it was the dogmatic philosophy of the church. According to this philosophy, the soul was immortal and earthly life but a period of trial and preparation. To expect happiness on earth was as futile as it was pretentious. Suffering had its rightful place in this necessarily imperfect world of creation. Happiness, derived from the fulfilment of earthly appetites, was as deceptive as it was transient, and no happiness counted in the long run other than that which came from peace of mind and the hope of divine grace.

In two points, no doubt, the priest was in a position superior to that of his present-day successors: he was accredited with supernatural authority, and he had the power of absolution. He could therefore relieve his clients from guilt feelings far more effectively than any present psychotherapist can.

The Protestant minister takes his place midway between the priest of the Roman Church and the modern case worker. He no longer claims divine authority. In the Protestant conception, the individual conscience is autonomous, and the minister merely tries to help the individual find his own way to his God.

One step further in the development of individualism leads to the conceptions of modern case work and modern psychotherapy in which, without supernatural implications, people can be helped to work out their own destiny.

Within the methods of present-day psychotherapy, we find the same difference between guidance and autonomy of the individual that prevails between the Catholic and the Protestant approach, and between the latter and secular psychotherapy. This is the difference between non-analytic psychotherapies and psychoanalysis.

The non-analytic psychotherapies offer encouragement and moral support, persuasion and suggestion, training and advice, comfort and reassurance; they educate and re-educate, and give some sort of guidance. The psychoanalytic therapy, on the other hand, works with interpretations. It tries to bring to consciousness all the forces operative in the individual, thus affording the integrative efforts of the ego a chance of working out a solution. Thus, in the non-analytic methods of psychotherapy, leadership is still offered; whereas, in the psychoanalytic methods, such leadership has been reduced to the possible minimum—it is merely help in the process of gaining insight into oneself, without any active interference with the final solution.

We may say that the roles of Catholic priest, Protestant minister, non-analytic therapist, and analytic therapist present a series in which the degree of autonomy of the individual increases while the amount of external guidance decreases. After this glance at the place of case work in the realm of time, let us now face the problem of its methodological place in the realm of contemporary science.

The social world has a highly complicated structure. Man, with his own complicated mental life, faces other human beings, the conditions of physical nature, and the culture in which he lives. Out of this structure, social case work singles out a particular problem presented by a particular individual in his environment. Society, with its political, economic, cultural conditions, forms the background of the picture. Society is beyond the reach of case work and has to be accepted as it is. The variables of case work are merely the individual himself and the conditions of his personal environment. The social conditions, many of which are bound to put strain on the individual, are mere constants in the problem. This is different from other attempts at influencing human affairs; in social planning, for example, the institutions of the culture are the variables of the problem.

Thus, though the whole of the social sciences will have some bearing on the problems of social case work, two chapters of knowledge are of particular importance: the psychology of the individual and social conditions as they present themselves in his everyday life.

There are two problems. One is diagnostic and one is therapeutic; one asks for the facts and the other for the possibilities of change and the ways of influence.

I. DIAGNOSTIC PROBLEMS

First, there is the familiar distinction between external and internal pressures. To what extent do external and internal pressures share in the responsibility for a particular state of unsettlement?

Whenever, in human affairs, we try to determine what share various factors have in bringing about a certain result, we meet with a difficulty: we cannot make an experiment. The natural scientist goes into his laboratory, isolates the various factors, and can thus determine their respective roles in causation. But we cannot make such an experiment with man. Sometimes a quasi-experiment will be made in thinking: we may ask ourselves how other people respond to the same pressures, and how the same individual responded to challenges of a different nature but of equal severity. This may help both to prevent us from giving too much weight to the external conditions and to convince the client that internal difficulties contribute to a problem which he himself so often feels to be of a purely external character.

Psychoanalysts, in general, will be inclined to believe that only very seldom does external pressure bear the full responsibility for a serious unsettlement of life or for a lasting condition of unhappiness. For example, economic frustrations, even of a severe variety, are accepted far more graciously when they are a common destiny than when they befall just "me." Does this not indicate that it is not so much the economic frustration itself to which we react as the fact that we feel discriminated against? Is it not the injustice of fate, the lack of love to which we feel subjected, which upsets us, rather than the actual want?

Moreover, psychoanalysts had the opportunity of observing time and again that people who were near the breaking point under a particular strain knew how to cope successfully with other no less difficult situations which did not touch their own internal problems, their own psychological weaknesses. And if inner difficulties have been removed, the external situation, though unchanged, often proves to be no longer unbearable.

Within the environmental conditions, however, only one group is sociological. To it belong such factors as possibilities of unemployment, financial responsibilities, and the like. Other environmental factors are psychological; they consist of the attitudes of people in his environment towards the client, and of the character or neuroses of the persons near to him—his parents, marital partner, children, or employer.

However, in distinguishing between the psychology of the environment as an external factor and the psychology of the client as an internal one, a complication is brought about by the psychological interplay between a person and his environment. Man has an almost uncanny ability of getting into contact with those who are the nearest complement of himself. The sadist will sense the masochist at a distance; he who desires being supported will instinctively, as it were, spot those who are desirous to support others; and he who is out to deceive others will be attracted by those who are willing to be deceived. Furthermore, though people do not consciously understand the unconscious tendencies of others, they frequently behave as if they had understood them. The neurotic reactions of various members of a family, for example, often form some sort of a dialogue between their unconscious fantasies. This factor must be considered when we try to change the environment, as when we place a child in a foster home. The child cannot help but try to establish with the new objects the same relationship he had with the old ones, and the change will be successful only if these new objects are so different from the old ones that they do not lend themselves to an attempt at transferring the old relationship, or if they are wise enough to refuse to co-operate in the game.

This last point leads us to the psychology of the individual himself. Of course, the whole of psychiatric and psychoanalytic knowledge can be brought to bear on the diagnostic and the therapeutical problems. Three aspects, however, seem to me to deserve our special attention. These are the role of internal conflicts, the infantile roots, and the sexual aspect.

Among the internal pressures, inner conflicts between various tendencies in our mind are especially important in neurotic maladjustments. If a country is in danger from without, internal disunion or co-operation of some citizens with the enemy will increase the danger. This is not to say that internal union is in itself a guarantee that the enemy outside will be successfully dealt with; but internal disunion will no doubt considerably diminish whatever chances of survival or victory may exist.

Likewise, the individual is tremendously hampered in his attempts at coping with his external situation if he is divided in himself. A person looking for employment will be far less efficient in dealing with his difficult task if there is something in himself that desires to be dependent on others; he has to fight an internal enemy together with the external one.

Human life is full of internal conflicts, and even among the normal some of these conflicts never get settled. Unsettled conflicts between instinctual desires and fears are the core of neurotic difficulties. Usually there is one main conflict characteristic for an individual which permeates his life. If we succeed in spotting this main conflict, we have made an important step towards understanding him.

The second important aspect of the individual is his life history. Psychoanalysis maintains that neurotic difficulties have their background in the life history of the individual, especially in his childhood. This theory has been challenged in many quarters. Neurotic difficulties, it is claimed, can be adequately explained as reactions to the strain of current reality. What can it avail us, we are asked, to go back to the past, as this would merely push back the problem but not solve it?

Experience shows that every neurotic individual fails only in dealing with specific challenges to which he is allergic and deals adequately with other situations that other people are not able to handle. Some neurotics fail in every competitive situation, others in the relation to their superiors, but either group may prove equal to situations that the other one cannot stand. Everyone responds neurotically to situations that touch a sensitive spot in his personality, usually situations similar to those he was unable to solve in his childhood. A man who has once broken his leg is likely to break it in the same place when he falls again.

The ego of an adult person is, to a considerable extent, capable of facing facts, bearing frustrations, and dealing with difficulties. For the most part, it will only fail wherever a weak spot has been left from un-solved conflicts of the past. It is not surprising that the child failed to solve the problem. His ego was weak, he had little knowledge of reality, was hardly able to influence events, and his capacity of enduring pain was small; he was likely to fail in a difficult situation. Every neurotic mal-adjustment is, therefore, to some extent a carry-over from the past.

Furthermore, the sexual aspect of neuroses must be kept in mind. Psychoanalytic experience has shown that in every problem that leads to neurosis a conflict over a sexual desire is involved. The term "sexual," of course, does not merely refer to impulses related to propagation; it is meant in the broader sense of desires for pleasure that go hand in hand with some sort of physical excitement.

Passivity and the desire to be supported, which so often form a stumbling-block in the way of treatment, have usually a libidinal nucleus. They have frequently been described as a shrinking back from responsi-bilities, or as a desire for protection and for a sheltered existence, as a child-adult relationship. Such descriptions are correct but incomplete. There is also an instinctual, libidinal desire of a passive or feminine or masochistic character, and it is precisely this instinctual basis that makes it so difficult to modify such attitudes.

We can find such libidinal implications in all neurotic symptoms and neurotic habits. In recent years, observation, especially of children, has shown that even anxiety, when it proves irritating, does so because of a sexual contribution. Fear as an expectation of an anticipated evil is one thing, but it has nothing to do with sexual tendencies and in itself alone usually proves bearable though disagreeable. But anxiety with continuous excitement is another thing; it is mostly not merely the expectation of an evil but some sort of indulgence in fear or flirtation with the danger. Fear is then sexualized, as it were; a secret pleasure is attached to it, and it then becomes exciting, irritating, and unbearable. Incidentally, this is one more example of how external problems are aggravated by internal conflict.

But how can we explain that just sex, in the broader sense of the term, is the main trouble-maker of life? The answer seems to be this: The non-sexual tendencies of man are easily adjustable but the libidinal impulses are very difficult to adjust. There is a certain element of insatiability in them; however often rejected, they always tend to come back. Neuroses are maladjustments; thus, the ill-adjustable parts of our make-up bear the brunt of responsibility for them.

An attempt at understanding as much as possible of the psychological problems of a neurotic individual, on the basis of the material available from short-term observation, may therefore be centered around the fol-lowing questions: What are the main internal conflicts of the person? What were his childhood problems, and what was his infantile neurosis? How is the present difficulty related to these? In what libidinal pleasures does he indulge in his neurotic habits?

2. THE THERAPEUTIC PROBLEM

The accessibility to influence of the environmental factors, such as destitution or unemployment, will be judged according to the knowledge of the social conditions and of the remedies available; no specific criteria for this can be suggested. But we can give some criteria for the possibilities of changing the psychological condition of the individual. What are the possibilities and the limits for such a short-term psychotherapy as social case work can offer?

There are, first of all, the limitations presented by the nature of the mental or psychological disfunctioning. Psychoses for the most part present an almost absolute limit; feeblemindedness leaves room merely for restricted possibilities of training; the possibility of psychotherapeutically influencing delinquency, drug addiction, or other psychopathies is still largely a field of experimentation. For neurotic maladjustments, however, a few general criteria can be worked out.

One is, of course, the severity of the disturbance and the length of its duration; another one is the client's age. The older he is, the less are we entitled to expect profound changes because of the increasing petrification of the neurotic patterns, the diminishing elasticity of the personality, and the ever-increasing amount of lasting residuals of his own neurotic activities with which the person is surrounded and which he cannot undo.

In addition to these well known factors, three criteria for the chances of psychotherapy may be considered: the degree to which the present disturbances seem to have their root in the infantile rather than in the current conditions; the period of life as being one of relative strength or relative weakness of the ego in comparison with the instincts; and the potentiality or carrying capacity of the healthy parts of the personality.

1. There is an infantile background for every psychological disturbance or maladjustment. Nevertheless, the share that the carry-overs of the past and the challenges of the present have in building up or keeping alive the psychological disturbance may vary considerably. An old wound may still be bleeding or else be cicatrized, as it were. In the first instance, no psychotherapeutic influence is likely to help, unless it succeeds in revealing and working out the infantile problem. Unless this is done, every treatment of a particular difficulty, though temporarily successful, would still leave the breeding-place of such difficulties in operation. In the second instance, however, the old problems, though not really settled, may yet not create difficulties out of their own initiative. Once the current problem has been solved, the neurosis may be in abeyance again for a long time.

Whether a particular neurotic difficulty belongs to the first or to the second category may be judged from the following considerations: Has it persisted for most of the life, or did it intrude only recently after a long period of comparative health? Have the symptoms themselves to a marked degree an infantile character inconsistent with an adult mind?

2. Another criterion refers to the particular period of life in which the maladjustment makes its appearance. There are periods of life in which normally the instincts are stronger than the ego and others in which the ego has a chance to prevail. That is to say, the curves that show the development of instinctual strength and the development of the strength of the ego from the cradle to the grave are not parallel. The instincts climb to a first peak in childhood, approximately at the age of three to six, the time of the prime of infantile sex. After this, in the period of latency, which is approximately the time of elementary education, they somewhat quiet down. In adolescence, there is a sudden influx from biological sources, and the instincts reach a new and probably all-time high; and when adolescence is over they tend to settle down to the level of the mature age. Once more, they increase markedly in the female sex, and to a much less degree in the male, in the years before the climacteric, as if the anticipation of forthcoming decline would stimulate a new but last climax. They then definitely sink down in old age.

The curve of the ego development, on the other hand, in sharp contrast to the irregular curve of instinctual development, has a simpler shape. There is a gradual and rather regular increase of the strength of the ego up to the time of maturity; it then remains on the level thus achieved for most of the life and may, but need not, decline a little in old age.

Thus, there are three periods in life in which the strength of the instincts is superior to that of the ego and the ego has a hard job in dealing with them and in keeping balance. These are the height of infantile sex from about three to six, adolescence, and the years preceding the climacteric. They are the periods of unsettlement in mental life and most productive of neuroses. There are three other periods of comparative strength of the ego—latency, maturity, and old age—which, therefore, offer best chances for settled psychological conditions. They are also the periods in which psychotherapeutic influence has more chance to succeed than in the periods of turmoil. Psychotherapy can find as its ally in the ascending part of the curve of life the increasing possibilities of substitute gratifications which the expansion of life offers, and in the declining part the older person's increased capacity for resignation.

3. Finally, there is the criterion of the potentiality of the healthy parts of the personality. We may distinguish between the disease on one side and the other aspects of the personality that are not afflicted on the other. In some instances, the borderline is distinct: a few symptoms are insulated, as it were, within an otherwise normal personality. In other instances the neurosis seems to have permeated almost everything. We may say that the person either may have a disease or may be diseased.

The chance of psychotherapy is dependent on the capacities of the healthy parts of the personality. When Archimedes discovered the principle of the lever, he is said to have exclaimed: "Give me a solid joint in the universe and I shall move the earth out of its joints."

In combatting neurotic symptoms or behavior patterns we need some solid ground within the mind of the individual, something that can be used as an ally, and this can be offered only by the healthy parts of the personality. The carrying capacity of the unafflicted parts of the personality may be judged from the following: the strength of character as revealed by the ability to stand pain and frustration, to take blows, and, above all, by the morale that has been shown in emergencies; the record of past achievements; the ability to sublimate, that is, the capacity of finding substitute gratifications in socially evaluated activities; intellect and talents on which such ability to sublimate partly depends; the strength of conscience; in a man, the degree of his masculinity as against feminine tendencies and, in a woman, the degree to which feminine attitudes prevail in her relationship to men and children.

These three criteria may be used, in addition to those of diagnosis and age, to get some fair estimate of the chances of psychotherapeutic influence, especially of short-term psychotherapy as an internal treatment of emotional needs in case work is bound to be.

Finally, one point which has no counterpart in mechanical engineering may be considered when we deal with human affairs and try to do some human engineering—what seems to be the most intelligent solution of a problem is hardly ever the wisest one. There is all the difference in the world between intelligence and wisdom. The seemingly most intelligent solution would be the best one only if man were a totally rational being. Wisdom will take into account his fallibility, and the less perfect solution may easily be the more workable one.

Another consideration is the time that must be allowed for the assimilation of stimuli, both external and internal. It takes time until unpleasant realities can be faced, transitions of attitudes can be worked out, and adjustments can be made.

The problem of case work is always that of some personal or environmental unsettlement; its aim a new settlement. We cannot forget, however, that such settlement can seldom be more than relative and will hardly long endure. This is due not only to the gravity of situations and to the limited devices of help available, but to a considerable degree also to the character of our present-day life. Highly institutionalized cultures like the medieval Christian culture confronted the individual with conditions which he could hardly hope to change. Most people had no reasonable chance of changing their place of living, their kind of work, their living conditions. Many desires met with unsurmountable institutional barriers. The individual had to yield or else to break. In our modern society, however, limitations are far less stringent, and considerable changes of life conditions are possible. Such a state of affairs keeps appetites and aspirations alive and thus intensifies inner conflicts. In a sense, it is with a great amount of personal unsettlement and neurosis that our culture pays the price for the possibility of constant change which it cherishes.

CHAPTER

45

Donald Horton

THE FUNCTIONS OF ALCOHOL IN
PRIMITIVE SOCIETIES

A CROSS-CULTURAL STUDY

*Dr. Horton examines the connections between culture and one par-
ticular form of personal maladjustment—alcoholism. His paper, of which
only the concluding section is reprinted here, is of special interest as an
exemplification of the methods developed by the Cross-Cultural Survey
at Yale University. In the body of the paper, data from a large sample of
societies are analyzed statistically and compared.*

I . SUMMARY AND CONCLUSIONS

IT HAS been the aim of this research to develop and test a general psy-
chological-cultural theory of the use of alcoholic beverages. We assumed
that a correct general theory would be valid for any culture, and
could be submitted to the test of cross-cultural comparisons. Before a
theory broad enough to satisfy this requirement could be constructed, it

NOTE: Thanks are due the author and the *Journal of Studies on Alcohol* for per-
mission to reprint a portion of this paper from the *Quarterly Journal of Studies on
Alcohol*, Vol. 4 (1943), pp. 292–303. (Copyright, 1943, by the Journal of Studies on
Alcohol, Inc.) The chapter reproduced here has been revised by the author.

was necessary to adopt a consistent position on the relationship between culture and the behavior of individuals. Our position has been, in brief, that social behavior consists of learned habitual responses to (ultimately) external stimulation. The external stimuli most significant in social behavior are culturally conditioned, and in anthropological shorthand are generalized and described collectively as "the culture." In developing a general theory of drinking behavior, we therefore assumed that our theory must describe the interrelations of two sets of variables: the independent variables are the psycho-physiological processes of response and habit formation (and the directly physiological effects of alcohol on these processes), and the dependent variables are the relevant aspects of culture. From such a theory it should be possible to predict the concomitant variation of cultural conditions and psychological response, and to test these predictions by a simple correlation technique applied to the available data from a representative sample of the primitive and folk societies of the world.

In practice, some further modifications in method were necessary. Physiological and psychological processes are for the most part unobservable (especially under conditions of field observation) and must therefore be inferred from overt behavior. The reports of drinking behavior found in most of the anthropological literature available for this study are highly generalized, and usually describe only the customary degree and frequency of insobriety. The only measurable variables were, therefore, insobriety (which could be taken as an indicator of the strength of the drinking response) and cultural conditions (including the nature and strength of the available alcoholic beverages). In order to proceed from the general theory to these specific variables, it was necessary to deduce a set of theorems appropriate to the data, and from them to develop specific predictions of concomitant variation which could be tested by the cross-cultural statistical technique.

After an examination of the relevant literature on the physiological and psychological effects of alcohol, psychoanalytic theory, stimulus-response psychology and anthropology, we constructed a general theory on the assumption that the primary function of alcohol is reduction of anxiety. The greater the amount of alcohol consumed, other conditions being equal, the more completely anxiety is reduced; and conversely, the greater the initial anxiety, the greater the amount of alcohol required to reduce it. Anxiety-reducing acts are inherently rewarding and, therefore, tend to be habit-forming. Since anxiety is a universal reaction to certain conditions of social life, all peoples to whom alcoholic beverages are available are potential habitual drinkers. But since anxiety is the agent of inhibition, anxiety reduction tends to reduce inhibition and release previously inhibited responses. Inhibitions are themselves the result of punishments imposed by society, according to its cultural tradition, for certain proscribed forms of action (especially sexual and aggressive acts). Release of such behavior tends to re-invoke the original punishments, which then produce responses in opposition to the rewarding act of drinking. To these punishments may be added others, self-inflicted, or socially administered, resulting from alcoholic impairment of physiological functions. Actual drinking behavior, therefore, represents a dynamic psychological equilibrium, in which initial anticipation of anxiety reduction is the chief motive for drinking, and anticipation of punishments for the personally or socially unacceptable consequences of drinking is the chief counter-acting response.

From this very general theory we derived the following set of theorems, from which concrete prediction of anthropologically observed behavior could be formulated.

1. The drinking of alcohol tends to be accompanied by release of sexual and aggressive impulses.

2. The strength of the drinking response in any society tends to vary directly with the level of anxiety in the society.

3. The strength of the drinking response tends to vary inversely with the strength of the counter-anxiety elicited by painful experience during and after drinking.

Corollaries—The sources of such painful experiences are:

 a. Actualization of real dangers as a result of impairment of physiological functions.

 b. Social punishments for impairment of functions.

 c. Social punishments invoked by release of sexual impulses.

 d. Social punishments invoked by the release of aggressive impulses.

Predictions of specific concomitant variations of insobriety and cultural conditions were then formulated and tested by means of the Chi-square correlation technique. Data from fifty-six primitive and folk societies, distributed throughout the world, were available for this purpose in the files of the Cross-Cultural Survey in the Institute of Human Relations at Yale University. The results of this statistical analysis are summarized below.

Theorem 1—The first theorem is rather in the nature of a basic postulate than a true theorem, since it is not expressed as a quantitative relationship between two variables. In the review of the data it was reported that the literature furnished strong evidence of the release of aggressive impulses as a result of drinking, and less strong but significant evidence of sexual responses following drinking. In a later section, in which drunken aggression was discussed, it was found possible to make some quantitative predictions with respect to the release of aggression as a consequence of drinking. Although these predictions, depending as they do on the presence of sorcery as an index of the level of inhibited aggression, have not been verified for all the societies in the sample, they have been verified for half of the societies for which there are data on drunken aggression. A reliable negative association between sexual anxiety and insobriety was obtained. This association inferentially supports the theorem that the drinking of alcohol tends to be accompanied by release of sexual responses. The inverse character of the relationship between sexual anxiety and insobriety explains in part the fact that overt sexual responses are not often observed. We conclude that directly and indirectly this theorem has been tentatively verified.

Theorem 2—The second theorem was tested by predicting that certain relevant variables would be found to be positively associated.

1. The customary degree of insobriety (of men) in any society is positively associated with the type of subsistence economy, i.e., the more primitive the subsistence activity, the greater the degree of insobriety.

2. The customary degree of insobriety is positively associated with subsistence hazards.

3. The customary degree of insobriety is positively associated with subsistence hazards, including hazards due to acculturation.

On the basis of the observed associations we concluded that the theorem is tentatively verified. Insobriety varies directly with anxiety as measured indirectly in terms of the anxiety-provoking conditions of subsistence insecurity and acculturation.

It was demonstrated that acculturation, when the acculturation process involves damage to the subsistence economy, is perfectly associated with strong insobriety.

An attempt was made to determine the relationship between warfare and insobriety, but the attempt was unsuccessful. It was concluded that without further analysis of the actual psychological concomitants of warfare in primitive society, no relationship with insobriety could be verified.

The relationship between insobriety and sorcery was tested. These contingencies were exploratory rather than tests of a specific prediction. There was thought to be a possibility that the belief in sorcery might be a manifestation of an anxiety capable of motivating drinking or of a counter-anxiety capable of reducing the strength of the drinking response. Neither of these alternative expectations was supported by the statistical test.

Theorem 3, corollaries a and b—Corollaries *a* and *b* of Theorem 3 were discussed but were not tested by statistical means.

Corollary *a* states that actualization of real dangers as a result of drinking will tend to elicit counter-anxiety to drinking, and drinking will tend to vary inversely with this anxiety. No statistical test of this theorem could be obtained. An examination of the data led to the conclusion that the predicted counter-anxiety occurs but does not effect a weakening of the drinking habit in most societies. Instead, it motivates a cultural solution to the problem presented by such actualization of objective dangers. The solution takes the form of special customary precautions against dangers, and the arrangement of drinking occasions to minimize interference with normal social and economic activities. This interpretation cannot be verified statistically, since the postulated counter-anxiety which motivates the adoption of such arrangements must become overt only temporarily, during a period of crisis or dilemma, in which new habits are tried out. Once these have obtained the status of custom, the motives remain hidden. Anxieties with respect to such dangers of drinking are then kept at a low level and can seldom be detected by the casual observer.

The second corollary could not be tested because of the absence of data. There is, however, considerable evidence that the reaction to impairment of function has been, in many societies, to change the drinking situation in such a way that impairment of functions need not be punished, but merely controlled.

It is concluded, with respect to corollaries *a* and *b*, that the painful experiences postulated do play an important role in determining the nature of drinking customs; they produce adjustments in the conditions of drinking behavior, not in the amount of drinking. These corollaries of Theorem 3, therefore, are incorrect as stated, but there is no evidence that the theorem itself is invalidated by this fact.

Theorem 3, corollary c—The third corollary states that drinking will tend to vary inversely with the counter-anxiety elicited by punishment of the sexual responses released by drinking.

The prediction was made that drinking will be inversely associated with the strength of punishments for pre-marital sexual behavior. This prediction is based on the assumption that the punishments for pre-marital sexual behavior can be taken as an index of the probable (but not reported) punishments for sexual responses released by alcohol.

Two predictions were made with respect to this theorem:

1. Drinking will be inversely associated with the strength of the punishments for pre-marital sexual behavior.

2. Drinking will be inversely associated with the strength of punishments for pre-marital sexual behavior in those societies in which the level of subsistence anxiety is not high.

The expected inverse relationship was observed, but the association is reliable only in the second case where the anxiety motivating drinking is not especially strong. The counter-drive elicited by punishment of sexual responses is not strong enough to compete with high subsistence anxiety.

The association is believed to verify corollary *c*. There is a tendency for sexual counter-anxiety to be inversely associated with insobriety.

Theorem 3, corollary d—The slight evidence to support the fourth corollary has been presented. Five societies were found in which a high level of aggression could be deduced from the presence of sorcery; in these societies, sorcery coexists with superordinate social control. It was assumed that in these societies drunken aggression would be punished and drinking would therefore be moderate. This prediction is borne out, but there are so few cases that no statistical value can be claimed for this evidence. In roughly half of the societies in the present sample, drunken aggression is relatively unrestrained. In the others, the condition postulated in the third theorem may exist, but the data required to test this possibility were not available. It is concluded that corollary *d* remains un-verified.

When the variables isolated in this research are viewed not as pairs of related variables but as systems of interrelated variables, several distinctive patterns of drinking behavior become evident. Each of these occurs in societies in different parts of the world. The common features in each case are expressions of the similarity in the crucial psycho-cultural variables. A detailed study of each of these patterns would no doubt reveal further common elements—attitudes, customs, beliefs related to the central psychological and cultural elements. Such a study is beyond the scope of the present research. But we consider it worth while to devote a few paragraphs to these patterns because they illustrate the effectiveness of the cross-cultural method of research in revealing new aspects of culture.

The first pattern occurs in societies that have a high level of subsistence anxiety and a strong and active belief in sorcery. Sexual restraints are usually weak, but even where they are strong they are ineffective against the powerful anxieties that motivate excessive drinking. Insobriety is always excessive and is almost invariably accompanied by extreme aggression. Social control is co-ordinate, and there is no power in the society capable of suppressing drunken aggression. The cultural adaptation to this condition takes the form of restraints and precautions, usually entrusted to the women. The women drink less than the men, if at all. Catharsis of aggression occurs, and normal life seems to be relatively unaffected by the orgy of aggression at the drinking bout.

In the present sample there are eight societies which manifest the basic elements in this configuration (i.e., level of subsistence anxiety, sorcery, customary degree of insobriety, and strength of drunken aggression). These are Chukchee, Wolof, Murngin, Menominee, Huichol, Tehuelche, Abipone, and Jivaro. Three other societies come close to this pattern: Papago, Choroti, and Taulipang.

The second pattern is marked by high subsistence anxiety but an absence of sorcery, or the belief in sorcery is relatively unimportant. Insobriety is invariably excessive but is accompanied by only moderate aggression in most cases. The anxiety motive is too strong to be countered by sexual anxieties. No attempt has been made to obtain data on control of aggression in these societies; but in view of the unimportance of sorcery, one might expect relatively few cases of strong aggression and no necessity for powerful counter-anxieties on this score. Drinking can be unrestrained, and aggression does not become excessive. There are, however, in several societies, ritual and practical precautions, and provision is made for the care of drunken men.

Societies manifesting this second pattern include Lakher, Samoyed, Ainu, Naskapi, Cahita, Toba, Apinaye, and Tupinamba. There are three other societies in which the same conditions of anxiety and lack of sorcery prevail, but excessive aggression occurs. The reported insignificance of sorcery in these cases is evidently not a measure of the level of inhibited aggression. These societies are: Mongol, Omaha, and Goajiro. One other society is exceptional, in that the customary level of insobriety is moderate, rather than strong, while aggression is also moderate (Macusi).

The third pattern has already been roughly described in a preceding section. This is the pattern of fear and restraint in drinking. Hopi, Zuñi, Azande, and several others manifest this pattern; their restraint may extend to a taboo on distilled liquor (Hill Maria Gond) or to a complete taboo on all alcoholic beverages (Hopi).

In a recent paper, Hallowell [1] offers some very interesting comments on what is essentially the first of the drinking patterns outlined above. The paper reports an attempt to recover data on the psychological characteristics of the Northeastern Woodlands Indians of North America from the literature of the seventeenth and eighteenth centuries. The literature proved to be unexpectedly rich in psychologic materials; a convincing picture of the "structure of the emotional life" of these Indians emerges from the analysis of the comments of travelers and missionaries. Emotional restraint, stoicism, fortitude under torture, the inhibition of all expression of aggression in interpersonal relations, a culturally demanded amiability and mildness in the face of provocation to anger, and suppression of all open criticism of one's fellows, are typical characteristics. Hallowell comments that "the whole psychological picture is one that suggests a suffusion in anxiety—anxiety lest one fail to maintain the standard of fortitude required no matter what the hardship one must endure; anxiety lest one provoke resentment or anger in others." This pattern of inhibition was coexistent with an absence of superordinate authority; despite the minimal power of the chiefs, open conflicts were rare. But covert slander was a constant expression of the inhibited aggressive impulses. The Indians also had at their disposal a highly institutionalized means of covert aggression, namely, witchcraft or sorcery. The picture is that of a society with co-ordinate control, social control through sorcery, and inhibition of aggressive impulses.

When liquor was introduced to these Indians by Europeans, the consequences were disastrous. Alcohol "permitted the discharge of suppressed hostility in the form of overt physical aggression which in the sober state was inhibited and overlaid by an effective façade of amiability." Hallowell quotes a statement by Le Clercq which could hardly put the matter more succinctly:

Injuries, quarrels, homicides, murders, parricides are to this day the sad consequences of the trade in brandy; and one sees with grief Indians dying in their drunkenness: strangling themselves: the brother cutting the throat of the sister: the husband breaking the head of his wife: a mother throwing her child into the fire or the river: and fathers cruelly choking little innocent children whom they cherish and love as much as, and more than, themselves when they are not deprived of their reason. They consider it sport to break and shatter everything in the wigwams, and to brawl for hours together, repeating always the same word. They beat themselves and tear themselves to pieces, something which happens never, or at least very rarely, when they are sober.

Despite these extravagances of aggression, no authority existed among the Indians which could maintain restraints. Subsistence conditions were precarious, and, in terms of our analysis, the anxiety drive was a summation of hunger anxiety and aggression anxiety. No institution of the native society could oppose this drive by eliciting powerful counter-drives.

Many other penetrating observations made by Hallowell have not been given in this résumé, but enough has been said to indicate how well this description coincides with the first pattern of drinking behavior isolated by the present analysis. One striking aspect of this pattern of inhibition and covert hostility, pointed out by Hallowell, is the contrast between the aggressive fury released by alcohol and the normal appearance of amiability. In our literature this same contrast is widely reported of precisely those societies in which sorcery is important and drunken aggression is strongest. It is our impression that behavioral patterns similar to those described by Hallowell are of world-wide occurrence.

Hallowell's paper has been emphasized because it provides a confirmation, independently reached, of some of the conclusions offered in this paper, and also because it emphasizes the nature of the psychological insights that the anthropologist can obtain through a study of drinking habits. As Hallowell puts it: "The conduct of the Indian when drunk is, in a sense, a natural experiment, a cue to his character. If his primary emotional balance was one that led to the suppression of a great deal of affect, in particular aggressive impulses, then we would expect that these might be released in a notably violent form under the influence of alcohol." It is this revealing character of drunken behavior that makes its study a convenient key to covert aspects of culture. The present research is believed to have demonstrated some of the crucial psychological and cultural variables involved in this behavior.

The one anthropological field-research that has effectively utilized the insights afforded by a study of drinking habits is reported in a paper by Ruth Bunzel [2]. Bunzel studied and compared the role of alcohol in two Central American communities, Chichicastenango, in Guatemala, and Chamula, in Mexico. The unique feature of her study, as compared with the usual ethnographic account, is its analysis of the drinking customs and drinking behavior of these two contrasting Indian communities in relation to the full context of their distinctive cultures; the patterns of drinking are viewed in relation to the economic, social, and religious institutions, and in relation to the distinctively structured character of the people whose life is shaped by these institutions.

The broad pattern of drinking behavior in Chichicastenango has many elements in common with the first pattern described above. Drinking is excessive and is accompanied by the release of sexual and aggressive responses which are normally inhibited. Sorcery is institutionalized and widely feared; the hostilities manifested in the sorcery phantasy are engendered by frustrations which derive from the unequal system of land tenure and inheritance and from the autocratic authority of the father as landowner and lineage representative. The paternal despotism is supported by supernatural lineage-ancestors. Blind obedience to the father is demanded; all aggression must be inhibited. Pre-marital chastity and post-marital fidelity are mandatory, and lapses are punished by the supernaturals, from whom no act is hidden. Fear of the ancestors, fear of sorcery, and anxiety about sin are the typical responses to these conditions.

Although the formal institutions of social control are not described, it is evident that sorcery is here coexistent with some degree of superordinate control; but the police institutions seem to be relatively weak, and at the markets and fiestas which give occasion to heavy drinking the municipal officers themselves become too drunk to maintain any degree of control over others. Drinking begins ceremonially but ends in unorganized drinking, sometimes in "colossal sprees in which men stay drunk for days on end." Excessive drinking is more common among men than among women. Women are permitted to drink as freely as men, but when a man is on a spree his wife "follows him around, waiting to pick him up and take him home when his violence has spent itself and yielded to exhaustion." The responses released by drunkenness—sexual and aggressive—are not formally punished, but they are punished retrospectively by a deepening of the sense of guilt. Drunkenness is feared and associated with sin. The anxieties built up by drinking and its consequences contribute to the succeeding phase of the cycle in which the general anxiety level rises sharply to the next catharsis in drinking.

A major difference between Chichicastenango and the societies in our sample which manifest a somewhat similar basic pattern of drinking behavior is the relative economic prosperity of this community. There can be no question of subsistence anxiety at anything like the level of anxiety required to motivate excessive drinking. Drinking is motivated by, and reduces, aggressive impulse anxieties and sexual anxieties. This observation supports the conclusion that where there is effective superordinate authority, capable of imposing effective punishments for sexual and aggressive acts, the impulse anxieties will operate as counter-anxieties; but where authority is ineffective, the impulse anxieties may operate directly as drinking motives.

Another aspect of this drinking pattern has been emphasized—the guilt resulting from drunken transgressions of the moral code. It may be offered as a hypothesis that in many of the societies in which excessive drinking and aggression have been reported, and in which there are apparently no significant social and psychological consequences of this catharsis, feelings of guilt and sinfulness are elicited and manifest themselves, in a covert way, in an ambivalent attitude toward drinking as well as in other phases of the culture. Their effects might be found ramifying widely through the culture, and especially in the religious institutions.

Indian communities in Mexico are generally addicted to alcohol, and Chamula, the second community reported by Bunzel, is recognized as one of the worst in this respect. Excessive drinking is learned in childhood as a matter of custom, even infants at the breast being given distilled liquor. Excessive drinking is the only important pastime in Chamula. "At fiestas the whole town is in varying degrees of intoxication for a day or a week." Drinking is always ceremonial. Fights occur, but drunken aggression is exceptional. The usual course of a spree is peaceful; it ends in stupor. Its course is marked neither by aggression nor by sexual behavior. Certain paranoid individuals become violent when intoxicated, but these are deviant, not typical of the community. The attitude toward drinking is free of guilt; there is some anxiety about drinking, but it is related to the heavy economic penalties for indulgence, the cost of liquor, the adverse effect on the standard of living of the community, and the inevitable consequence of debt and wage-slavery on the plantations.

The cultural conditions of this drinking pattern include family life free of the tensions characteristic of Chichicastenango. Bunzel points out that, although children are taught to drink at an early age, they generally reject alcohol. Excessive drinking only begins when the boys are sent to the plantations to work and there, in a hostile environment, where they are encouraged by the plantation managers to drink themselves into debt, begins a vicious cycle of drinking, debt, and anxiety. But excessive drinking continues even after men have returned to their native villages and have settled on their own farms. The character of the anxieties which continue to motivate this drinking is not made too clear in Bunzel's account, but she interprets the character of Chamula drunkenness as manifesting a regression to the warmth and helplessness of infancy. Due to conditions of character-formation in Chamula, its people are unable to cope successfully with the responsibilities of adult life. Their culture offers them no adequate substitutes for the lost gratifications of infancy. Although this may be an adequate description of the pattern of drinking behavior, it leaves unspecified the sources of the anxieties which can only be successfully reduced by this alcoholic regression. Presumably this question will be answered when Bunzel's data on Chamula are published.

The analyses of drinking behavior made by Hallowell and by Bunzel have been reviewed at some length because, in a sense, the present study is complementary to them. Some of the psycho-cultural relationships tentatively verified in this paper on a level of broad generalization are, in them, shown in the deeper context of specific cultures. Their studies, on the other hand, reveal some of the great variety of particular cultural conditions and patterns of response which have yet to be assimilated into a general theory of drinking behavior.

The present study was undertaken in the belief that progress in the interpretation of culture depends on a developing and reciprocal relationship between intensive analyses of particular cultures and extensive generalizations about the processes of culture. Anthropology has, to date, produced many descriptive monographs, but few verified principles. The cross-cultural research reported here is submitted as a demonstration of a possible method for the verification of psycho-cultural generalizations. At the same time it is hoped that the hypotheses formulated and tested in this research can be considered a contribution to the general body of anthropological theory.

REFERENCES

1. Hallowell, A. Irving: "Some Psychological Characteristics of Northeastern Indians," in *Man in Northeastern North America*, Frederick Johnson, ed. (Papers of the R. S. Peabody Foundation for Archaeology, Vol. 3, Andover, 1946), pp. 195–225.
2. Bunzel, Ruth: "The Role of Alcoholism in Two Central American Cultures," *Psychiatry*, Vol. 3 (1940), pp. 361–87.

CHAPTER

46

Henry J. Wegrocki

A CRITIQUE OF CULTURAL AND
STATISTICAL CONCEPTS OF
ABNORMALITY

It is clearly established that the differing incidence of ailments, both "physical" and "mental," in different societies is not simply a matter of variations in group biology. For example, malignant cases of essential hypertension are quite rare among unacculturated African Negroes but quite common among American Negroes. In the fifteenth and sixteenth centuries there seem to have been an enormous number of persons suffering from a certain form of hysteria that is exceedingly rare in the same groups today.

Anthropological research has greatly widened and deepened the psychiatrist's conception of the normal. Some types of personality fail to find fulfillment in one culture, whereas there is some reason to suppose they might have flourished in another. Some cultures allow latitude for a variety of personal adjustments; in others the individual who does not conform to the single pattern is punished so cruelly as to become neurotic or, perhaps in the case of constitutional predisposition, psychotic. Behavior regarded as abnormal in one culture is socially acceptable in another.

Not so long ago, standards for normality seemed on the point of disappearing in the face of complete relativism. Today, however, it is agreed that certain sorts of "mental" reaction may be considered abnormal in every society. Also, as Dr. Wegrocki points out, it is possible to have pan-

---

NOTE: Reprinted, with slight abridgment, from the *Journal of Abnormal and Social Psychology*, Vol. 34 (1939), pp. 166–78, by permission of the author and the American Psychological Association.

*human standards of normality not merely in terms of the universality of
certain cultural judgments but also in terms of the part a set of symptoms
plays in the total economy of the personality.[1]*

*Many other areas of common interest to psychiatry and anthropology
have been explored in the recent literature.[2]*

ONE of the most significant contributions to a prope· orientation and
envisagement of human behavior has been the body of data coming in
the past few decades from the field of ethnological research. Human behavior had so long been seen in terms of the categories of Western
civilization that a critical evaluation of cultures other than our own
could not help exercising a salutary effect on the ever-present tendency
to view a situation in terms of familiar classifications. The achievement
of a realization that the categories of social structure and function are
ever plastic and dynamic, that they differ with varying cultures and that
one culture cannot be interpreted or evaluated in terms of the categories
of another, represents as tremendous an advance in the study of social
behavior as did the brilliant insight of Freud in the field of depth psychology.

In connection with this modern ethnological conception of the relativity of interpretations and standards, there has arisen the problem of
whether the standard of what constitutes abnormality is a relative or an
absolute one. Foley [7] infers from the ethnological material at hand that
abnormality is a relative concept and criticizes Benedict [1], who seems to

---

[1] Other recent contributions to this subject are the following: A. L. Kroeber,
"Psychosis or Social Sanction," *Character and Personality*, Vol. 8 (1940), pp. 204–15;
Ernest Beaglehole, "Culture and Psychosis in Hawaii," in *Some Modern Hawaiians*,
University of Hawaii Res. Public. No. 19 (1939); Ernest Beaglehole, "Culture and Psy-
chosis in New Zealand," *Journal of the Polynesian Society*, Vol. 48 (1939), pp. 144–55;
Erwin H. Ackerknecht, "Psychopathology, Primitive Medicine, and Primitive Cul-
ture," *Bulletin of the History of Medicine*, Vol. 14 (1943), pp. 30–67; N. J. Demerath,
"Schizophrenia among Primitives," *American Journal of Psychiatry*, Vol. 98 (1942),
pp. 703–7; and Arnold W. Green, "Social Values and Psychotherapy," *Journal of
Personality*, Vol. 14 (1946), pp. 199–228.

[2] See, for example, Robert E. L. Faris, "Reflections of Social Disorganization in
the Behavior of a Schizophrenic Patient," *American Journal of Sociology*, Vol. 50
(1944), pp. 134–41; Norman Cameron, "The Paranoid Pseudo-Community," *American Journal of Sociology*, Vol. 49 (1943), pp. 32–8; Paul Schilder, "The Sociological
Implications of Neuroses," *Journal of Social Psychology*, Vol. 15 (1942), pp. 3–21;
and Morris E. Opler, "Some Points of Comparison and Contrast between the Treatment of Functional Disorders by Apache Shamans and Modern Psychiatric Practice,"
American Journal of Psychiatry, Vol. 92 (1936), pp. 1371–87. The influence of
psychiatry on anthropology has been surveyed, with an extensive bibliography, by
C. Kluckhohn, "The Influence of Psychiatry on Anthropology in America During
the Past One Hundred Years," in *One Hundred Years of American Psychiatry*
(American Psychiatric Association, Columbia University Press, New York, 1944),
pp. 589–617.

present evidence for the "statistical or relativity theory" yet "at times appears inconsistent in seeking for an absolute and universal criterion of abnormality." Briefly, the evidence from Benedict can be subsumed under three headings: (1) behavior considered abnormal in our culture but normal in other societal configurations; (2) types of abnormalities not occurring in Western civilization; and (3) behavior considered normal in our society but abnormal in others.

1. Of "our" type of abnormal behavior considered normal in other cultures, Benedict gives, as an example, that of the Northwest Coast Indians whom Boas has studied at first hand. "All existence is seen in this culture in terms of an insult-complex." This complex is not only condoned but culturally reinforced. When the self-esteem and prestige of the chief is injured, he either arranges a "potlatch" ceremony or goes head-hunting. The injury to his prestige is a function of the prevalent insult-complex. Almost anything is an insult. It may be the victory of a rival chief in a potlatch competition; it may be the accidental death of a wife, or a score of other situations, all of which are interpreted as having a reference to the individual.

If, on the other hand, he has been bested in competition with a rival chief, he will arrange a potlatch ceremony in which he gives away property to his rival, at the same time declaiming a recitative in which there is "an uncensored self-glorification and ridicule of the opponent that is hard to equal outside of the monologue of the abnormal"; "either of the two mentioned above procedures are meaningless without the fundamental paranoid reading of bereavement."

Among other "abnormal" traits which are an integral part of some culture patterns, ethnological literature mentions the Dobuans, who exhibit an "unnatural" degree of fear and suspicion; the Polynesians, who regard their chief as taboo to touch, allegedly because of a prevalent *défense de toucher* neurosis; the Plains Indians with their religiously colored visual and auditory hallucinations; the Yogis with their trance states; and the frequent institutionalizations of homosexuality, whether in the religions of different cultures (e.g., shamans of North Siberia or Borneo) or in their social structures (e.g., the berdache of American Indian tribes or the homosexual youth of Grecian-Spartan antiquity).

2. The second argument in favor of the cultural envisagement of abnormality is the existence of "styles" of abnormalities which presumably do not occur in our Western type of cultures. The "arctic hysteria" noted by Czaplicka [5] and its tropical correlative *lâttah* [Clifford, 3, 4], with their picture of echolalia, echopraxia, and uncontrolled expression of obscenities, as well as the "amok" seizure of the Malayan world, are given as examples.

3. Of normal behavior in our culture considered abnormal in others, Benedict mentions as most conspicuous the role of personal initiative and drive in our own as compared with the Zuñi culture. Among the Zuñi Indians, for example, "the individual with undisguised initiative and

greater drive than his fellows is apt to be branded a witch and hung up by his thumbs." Similarly, what seems to us a perfectly normal pattern of behavior—acquisitiveness, for example—would be looked upon by the potlatch celebrants as just a little "queer." For them, possession of property is secondary to the prestige they acquire when they distribute it.

These are, then, briefly, the bases for the assertion that abnormality is a relative concept, differing from culture to culture, that no particular efficacy attaches to the expression, and that it gains meaning only in terms of the social milieu in which it is considered.

The question of what are the differentia of normality and abnormality is of course the crux of the problem. Is the concept of abnormality culturally defined? Is that which is *regarded* as abnormal or normal in a particular culture the *only* criterion for calling a behavior pattern such? Foley, for whom abnormality is a purely statistical concept, would answer in a positive manner and points to Benedict's example of the Northwest Coast Indians who institutionalize "paranoia." Let us, however, consider this "institutionalization of paranoia," as well as some of the other bits of evidence, critically.

It is not stretching the point to call the Indians' megalomaniac activities and beliefs "delusions" in the sense that the paranoiac in the psychopathic institution has his beliefs called delusions? Macfie Campbell states that "the delusions of the ill-balanced and the beliefs of the orthodox are more closely akin than is usually recognized," and we cannot separate the two as sharply as we would wish to do. The abnormal delusion proper is, however, an attempt of the personality to deal with a conflict-producing situation, and the delusion "like fever, becomes an attempt by nature at cure." The patient's delusion is an internal resolution of a problem; it is his way of meeting the intolerable situation. That is why it is abnormal. It represents a spontaneous protective device of the personality. It is a crystallization of something which hitherto had been pre-potent. The individual's personality thereafter refracts and reflects in terms of a distorted slant.

The Haida chief, upon the death of a member of his family, also experiences a certain tension. He resolves this tension, however, in a way which is not only socially sanctioned but socially determined. His reaction is not something spontaneous, arising out of the nuclear substrate of his instinctive life. It is not a crystallization in a certain direction of some previously unrealized protective potentialities of the psyche. His reaction is pre-determined socially. Since his milieu expects that reaction of him, he acts upon that expectation when the situation arises. Of course it is possible that historically the behavior may have had and must have had some spontaneous protective significance—most likely imbedded deeply in a web of primitive beliefs about magic practices. Yet the modern Alaskan chief, unlike his distant prototype, has no conflicts of doubt about the likelihood that malignant forces have caused the death of some member of his family; he *knows,* and he acts upon that knowledge by venting his emotions. There is no permanent change in his personality when a tension-producing situation arises. Emotions are aroused and appeased with no change in the personality profile.

In the personality of one who is labelled "abnormal," this change is, however, to be found. There is always "the way he was before" and "the way he is now," regardless of the fact that a present symptomatological picture had its roots in a pre-potent substrate which would make for a particular personality outline. This is not true of the Haida chief, in whom the "delusions" of reference and grandeur are externally imposed patterns. A Northwest Coast Indian, if given the opportunity for a naturalistic investigation of the situations that provoke his "paranoid" reactions —as, for example, through an education—could unlearn his previous emotional habits or at the least modify them. He is capable of insight; the true paranoiac is usually beyond it. The latter, if he kills the person who he thinks is persecuting him, only temporarily resolves his difficulty; the Indian chief who kills another family "to avenge the insult of his wife's death" achieves a permanent affective equilibration with regard to that incident. His prestige restored, he once more enjoys his self-respect. The Haida defends imagined assaults against his personal integrity only when some violent extra-personal event occurs. The paranoid psychotic defends himself against imagined assaults even though there is no objective evidence of any.

The point that the writer would then emphasize is that the delusions of the psychotic and the delusions of the Northwest Coast Indian cannot by any means be equated. Mechanisms like the conviction of grandeur are abnormal not by virtue of unique, abnormal qualia but by virtue of their *function in the total economy of the personality*. The true paranoiac reaction represents a *choice of the abnormal;* the reaction of the Haida chief represents no such choice. There is but one path for him to follow. If one of the chief's men showed paranoid symptoms by proclaiming that *he* really was the chief of the tribe and that his lawful place was being usurped, the institutionalization of paranoid symptoms within that culture would not, I am sure, prevent the rest of the tribe from thinking him abnormal.

Fundamentally the same criticism might be applied to the *défense de toucher* neurosis supposedly exemplified by the Polynesian taboo on touching the chief. Here, as in paranoia, we must consider whether the mechanism is a cultural habit reinforced by emotional associations or whether it is a true morbid reaction. Obviously it is the former. Thus by reason of that fact it *is not* abnormal. As in the paranoia of the Haida, there is no choice here and consequently no conflict. If a person is brought up with the idea that to touch the chief means death, his acting upon that idea in adulthood is not a neurosis but simply a habit. There is, in other words, the genetic aspect to true abnormality which cannot be evaded but which is overlooked when we speak of "abnormal" symptoms. The explanation of the delusion of persecution of the Dobu is, of course, subject to the same criticism, which would hold likewise for trances, visions, the hearing of voices, and hysterical seizures. When these are simply culturally reinforced pattern-suggestions, they are not abnormal in the true sense of the word. When the Plains Indian by a rigid physical regimen of exhaustion and fatigue plus a liberal dose of suggestion achieves a vision, that achievement is not an abnormal reaction in the same sense that the visual hallucination of the psychotic is.

A similar example presented as an argument for the cultural definition of abnormality is the supposed institutionalization of homosexuality among different cultures. The difficulty with all discussions of this enormously complex topic is the lack of agreement among investigators as to the sense in which the homosexual is abnormal. Obviously, homosexuality is not the same type of morbid mental reaction as paranoia; in fact, it is not a morbid mental reaction at all. The abnormality of homosexuality exists at a different level; it is social and biological rather than psychological. By this it is meant that homosexuality is not a compromise symptom due to a conflict, as is the case with paranoid manifestations. Whatever abnormality attaches to it is a secondary function due to the conflicts it creates in a social milieu. In short, it creates conflicts; it is not created by them. Only in those cases where it is used as an escape mechanism can it truly be called abnormal. Homosexuality as an abnormal form of behavior cannot be spoken of in the same sense in which one speaks of visual and auditory hallucinations or grandiose delusions. Sex inversion is rather a statistical type of abnormality. It represents extreme deviations from the norm and makes for the non-conformity which engenders social antagonism and ostracism in certain societies. We are not justified, then, in saying that certain cultures institutionalize this abnormality; because when we call homosexuality an abnormality in the same sense in which we speak of a delusion as abnormal, we are misusing the term and being inconsistent about its application.

The second point of view from which the cultural definition is sometimes argued is that there occur among different cultures abnormalities which are peculiar to them, as, for example, "arctic hysteria," "amok," *lâttah, et al.* The untenability of this typothesis becomes evident when a little analytic insight is applied to the phenomena considered. In his masterly analysis of the *lâttah* reaction, Van Loon [12] has also, the writer thinks, given a good explanation of "arctic hysteria." "*Lâttah*," he writes, "is chiefly a woman's complaint. The symptoms appear in consequence of a fright or some other sudden emotion; the startled patient screams" and exhibits echolalia, echopraxia, shouting of obscenities, and a strong feeling of fear and timidity. "The immediate cause of becoming *lâttah*, the patients report to be a dream of a highly sexual nature which ends in the waking up of the dreamer with a start. The waking up is here a substitute for the dream activity, protecting the dreamer's consciousness against the repressed complex." The same analysis might be applied to "arctic hysteria," which seems to show all the *lâttah* symptoms, although the occurrence of a sudden waking from sleep as the beginning of the complaint is nowhere reported.

The amok type of seizure, Van Loon explains, is due to hallucinations of being attacked by men or animals and seems to be confined to men. Clifford [3], on the other hand, considers amok from a genetic standpoint as having a background of anger, grudge, excitement, and mental irritation, what the Malay calls *sâkit hâti* (sickness of the liver). "A Malay loses something he values, his father dies, he has a quarrel—any of these things cause him 'sickness of the liver.' The state of feeling which drives the European to suicide makes the Malay go amok." In the heat of the moment "he may strike his father, and the hatred of self which results causes him to long for death and to seek it in the only way which occurs to a Malay, viz., by running amok." The psychoanalyst, with his theories of the introjective process in melancholia, would doubtless find "amok" a fertile field for interesting speculative analogies.

From the above discussion it is evident that the various "unique types of mental disorder" probably would yield readily to an analysis in terms of the categories of psychopathology. The mental disturbance can, to be sure, be understood only in terms of the cultural and social pattern within which it occurs, but the form that it does take is a secondary function of the abnormality. Only in this sense does culture condition abnormality. The paranoid reaction can occur in almost any culture, but the form that it takes is culturally modifiable. Although the psychotic can feel himself persecuted in almost any culture, in the one the persecutor may be the sorcerer, in another a usurping chief, and in still another the president of the United States.

The third argument for the cultural definition of abnormality is one which infers that inasmuch as traits considered normal in our society are considered abnormal in other cultures, abnormality can be looked upon as simply that form of behavior which a group considers aberrant.

As previously mentioned with reference to homosexuality, the term "abnormality" cannot with exactitude be applied to *all* those forms of behavior which fail to meet with social sanction. That would be making the term meaningless. The same criticism might be applied to this third argument. There is no element of internal conflict in the Zuñi, for instance, who, feeling full of energy, gives vent to that energy. His behavior is aberrant because it conflicts with the prevailing pattern. That, however, *does not* constitute abnormality. Such a Zuñi is not abnormal; he is delinquent [9]. He is maladjusted to the demands of his culture and comes into conflict with his group, not because he adheres to a different standard but because he violates the group standards which are also his own (to paraphrase Mead). Edwards [6] pointed out this very frequent confusion concerning abnormality when he differentiated four standard types of individuals: "The *average* individual is the fictitious individual who ranks at the midpoint in all distributions of test results. The *normal* individual is one who is integrated, healthy, and without any great variation from the average. The *adjusted* individual is one who is reasonably well-fitted into his environment and to its demands. The *effective* individual is one who, whether he is adjusted or normal, accomplishes his purposes."

From the above criticisms we can readily see that a relativity or statistical theory of abnormality, which argues from the ethnological material at hand, cannot stand a close analysis. From Benedict's writings, one might get the impression that she also, like Foley, believes that what is abnormal depends simply on whether or not it is regarded as such by the greater majority of individuals in a specific culture. In various parts of her book [2], though, as well as in her articles, there are statements which run counter to any such belief. That is what Foley had in mind when he stated that Benedict seemed "inconsistent in seeking for an absolute and universal criterion of abnormality." In a private communication, Benedict explains this seeming paradox as follows: "In 1930–31, when I wrote the article you refer to and the bulk of the book, writers in abnormal psychology constantly confused adequate personal adjustment and certain fixed symptoms. I wanted to break down the confusion, to show that interculturally adequate functioning and fixed symptoms could not be equated." When Benedict showed that the Northwest Coast Indians exhibited paranoid-like symptoms, she did not wish to prove that what the psychopathologist would call "abnormal" has no universal validity, but rather that, in spite of the fact that Northwest Coast Indians acted *like* paranoids, they were actually well-adjusted and "adult" individuals. What she was really arguing against was the confusion between fixed symptoms and adequate personal adjustment.

There is, of course, one way in which culture *does* determine abnormality, and that is in the number of possible conflicts it can present to its component individuals. (In this sense "determine" has, however, a different meaning from that used above.) In a civilization like that of Samoa, for example, there is a minimum of possible aberrant behavior because of the rarity of situations which can produce conflicts in individuals. Even there, however, as Mead points out [9], Christianity with its introduction of a different set of standards is bringing to the islands that choice which is the forerunner of conflict and neurosis. A similar source of abnormality is the destruction of native culture and the production of new stresses. Profound depression and the absence of the will-to-live are among some of the abnormalities produced [10]. McDougall [8] suggests that there may be temperamental differences in ability to adapt oneself to varying environments. He mentions, for example, the American Indian as being unadaptable because of his introverted disposition, as opposed to the care-free adaptable Negro with his extravert temperament. Seligman [11] voices the same opinion when he speaks of the extraverted Papuan and the introverted Malayan.

Benedict speculates that possibly the aberrant may represent "that arc of human capacities that is not capitalized in his culture" and that "the misfit is one whose disposition is not capitalized in his culture." She concludes that "the problem of understanding abnormal human behavior in any absolute sense independent of causal factors is still far in the future." "When data are available in psychiatry, this minimum definition of abnormal human tendencies will be probably quite unlike our culturally conditioned, highly elaborated psychoses such as those described, e.g., under the terms 'schizophrenic' and 'manic-depressive.'" Keeping away from any committal to a relativity theory of abnormality, she is criticized therefor by Foley [7] for whom "it is obvious that deviation implies relative variability of behavior; the responses of the individual must be considered in relation to the responses of other individuals." For Foley, therefore, "deviation from normative mean" and "abnormality" are synonymous. In that sense, then, abnormality is for him a statistical concept.

There are, however, many objections to this aspect of the statistico-relative formulation. The most important one is probably the fact that a statistical theory considers only the actual observable behavior of an individual, without delving into its meaning. Thus an erroneous identification is established between behavior patterns which are similar but do not have the same causal background. For example, the paranoid behavior of the schizophrenic and the "paranoid" behavior of the Northwest Coast Indian are equated; because the latter does not represent a deviation from the norm of that culture's behavior pattern while the former does, they are accepted as substantiating the statistical relativity theory. Of course we may arbitrarily define "abnormality" in such a way that it will mean the same as "deviation"; if, however, we proceed from an empirical point of view, it is obvious that we are not justified in equating such abnormalities as the psychoneuroses, the psychoses, sex inversion and amentia, and in speaking of them as deviations from a norm.

A statistical norm implies a graduated scale in which the items can be ranged on the basis of the possession of a "more" or "less" of a certain property. In that sense we can see that only sex inversion and amentia can be spoken of as deviations from a norm, inasmuch as it would have some meaning to speak of a "more" or "less" of development of cortical neurones or a "more" or "less" of a sex-determining hormone. It would, however, have only a qualitative significance if we spoke of one person's being "more" deluded than another, or "less" paranoid. In this sense, therefore, we are not justified in speaking of abnormality as a statistical concept. There are, rather, certain abnormalities which, because of their nature, can be ranged on graduated scales, and others which, because of a different substrate, cannot be similarly measured. A proximate graduation scale similar to one used in attitude testing can of course be utilized, but its use is bound up with all the prejudice that subjective judgments embody when topics of wide personal opinion-variance are considered. Besides, there is no real basis for comparative evaluation. Should, for example, a paranoid trait, such as the conviction of persecution, be measured by judges with respect to the degree in which it inhibits the satisfactory functioning of the total personality and causes personal unhappiness, or with respect to the degree in which it interferes with an adequate adjustment to the social group and creates opposition within the environment? Should the frequency with which it manifests itself in life situations be the determining criterion, or the intensity with which it is adhered to? Finally, should the degree of insight a person has into it be the standard? The bases for judging the "more" or "less" of a paranoid trait are, as is obvious, very divergent. No single criterion is any more justifiable than any other. A "paranoid scale" would, therefore, be of slight operational significance.

Of the three types of possible definitions of abnormality mentioned by Morgan—(1) the normative, (2) the pathological, and (3) the statistical [10]—we can see that when we take all those types of behavior which are referred to as abnormal, there will be some which will fall more readily into one category, others which will more easily fit into another. Thus all the symptoms associated with psychoneuroses seem best defined by a normative approach which arbitrarily postulates, on the basis of the best psychiatric opinion, a theoretical integrated, balanced personality, wide deviations from which would be looked upon as "abnormal." The functional psychoses also fit into this group, shading, however, into the pathological, where the organic psychoses, behavior disturbances due to cerebral lesions or malfunctionings, and extreme amentias belong; the statistical can really claim only the less decidedly pathological aments and those sex-inverts where the presumption of constitutional involvement is strong.

The confusion arises, of course, from subsuming different types of abnormalities under one heading, "abnormality," and speaking of them as if they were homogeneous entities. Obviously abnormal behavior is *called* abnormal because it deviates from the behavior of the general group. It is not, however, the *fact* of deviation which makes it abnormal but its causal background. That is why the hallucinations of the Plains Indians are not abnormal, while those of the schizophrenic are. It is not the *fact* of social sanction in Plains Indian society which makes that bit of behavior normal, but the fact that it does not have the background of a symptomatic resolution of an inner conflict such as produces that phenomenon in the schizophrenic. The "abnormal" behavior of the Indian is analogous to the behavior of the psychotic *but not homologous*. Just because it is analogous, the confusion has arisen of identifying the two.

If, therefore, behavior anomalies which are at bottom constitutionally or pathologically conditioned be excluded or subsumed as a different group under the category of the non-normal, we could state the quintessence of abnormality as *the tendency to choose a type of reaction which represents an escape from a conflict-producing situation instead of a facing of the problem*. An essential element in this type of problem-resolution is that the conflict does not seem to be on a conscious level, so that the strange bit of resulting behavior is looked upon as an abnormal intruder and, at least in its incipient stage, is felt as something which is not ego-determined. Inasmuch as pathological and, above all, constitutional factors cannot be partialled out in the etiology of a behavior anomaly, however, the above definition has only an ideal value of slight practical significance.

It does clarify, though, to some extent, the confusion which arises from labelling any bit of behavior "abnormal." It is obvious, for example, that masturbation *per se* is not abnormal and represents a quite normal, i.e., usual, growing-up phenomenon. In certain instances, however, its great frequency or its inappropriateness point to the use of it as an escape mechanism. What holds true for masturbation is true of all "abnormal mechanisms." *It is not the mechanism that is abnormal; it is its function which determines its abnormality.* It is precisely for this reason that the institutionalized "abnormal" traits in various cultures are not properly called "abnormal" entities. Because this distinction is not kept in mind and because a primarily statistico-relative conception of abnormality is adhered to, the unwarranted conclusion is drawn that standards of "abnormality" differ with cultures and are culturally determined.

REFERENCES

1. Benedict, R.: "Anthropology and the Abnormal," *J. Gen. Psychol.*, Vol. 10 (1934), pp. 59–82.
2. Benedict, R.: *Patterns of Culture* (Boston, Houghton Mifflin, 1934).
3. Clifford, H.: *In Court and Kampong* (London, G. Richards, 1897).
4. Clifford, H.: *Studies in Brown Humanity* (London, G. Richards, 1898).
5. Czaplicka, M. A.: *Aboriginal Siberia: A Study in Social Anthropology* (Oxford, Clarendon Press, 1914).
6. Edwards, A. S.: "A Theoretical and Clinical Study of So-Called Normality," *J. Abn. and Soc. Psychol.*, Vol. 28 (1934), pp. 366–76.
7. Foley, J. P., Jr.: "The Criterion of Abnormality," *J. Abn. and Soc. Psychol.*, Vol. 30 (1935), pp. 279–90.
8. McDougall, W.: *Outline of Abnormal Psychology* (New York, Scribners, 1926).
9. Mead, M.: *Coming of Age in Samoa* (New York, Morrow, 1928).
10. Morgan, J. J. B.: *The Psychology of Abnormal People* (New York, Longmans Green, 1928).
11. Seligman, C. G.: "Temperament, Conflict, and Psychosis in a Stone Age Population," *Brit. J. Psychol.*, Vol. 9 (1929), pp. 187–202.
12. Van Loon, F. H. G.: "Amok and Lâttah," *J. Abn. and Soc. Psychol.*, Vol. 21 (1926), pp. 434–44.

INDEX

Abbey, Edward A., 230

Abel, T. M., 258

Abnormality, critique of concepts, 691–701

Aborigines, North American, 260; *see also* American Indians

Abraham, Karl, 149, 182

Acculturation: and alcoholism, 683; of Hopi Indians, 277, 286; of immigrants, 132; resistance to, 134; and sorcery (Saulteaux), 266

Achievement demands, on American youth, 665

Ackernecht, Erwin H., 692

Action, conjunctivity of, 25

Activity: forms of, 13–15; process, 37

Adams, James Truslow, 236

Adjustment: and cue concept, 25; among Hopi Indians, 276–91; to military life, 387; occupational, 367; to social change, 132

Adland, Charlotte, 671

Administrative forms, and personality, 663–70

Adolescence: in American culture, 365–6; conformity during, 44; delinquency in, 218; and democratic character formation, 663; intimacy in, 221; in Kwoma culture, 139; problems of, 216; sex roles in, 573; *Sturm und Drang* period, 530; superego development in, 27; *see also* Childhood, Infancy, Puberty rites

Adulthood, stages, 221

Affectlessness, study of, 618–32

Affection, source, 127

Age grading: in American culture, 363–75; comparative cultural concept, 529; and parental standards, 658–9, 666; protest against, 539; and subordination, 570

Aggression: and alcoholism, 682; and breastfeeding, 170; displaced, under Nazism, 452; displaced oral, 152; in-family, 536; as frustration reaction, 437; intra-group, 396, 565; as guilt reaction, 432; in Hopi culture, 289; infantile, 498–513; in Kwoma culture, 138; and man-

Aggression (*continued*)
ner of birth, 414; Negro, 597–608; oral, 161; in Pedranos culture, 324; physical source, 494; rating scale, 178; repressed (Saulteaux), 270; repression, in slavery, 600; in Saulteaux Indians, 260–74; sexual, in childhood, 206; sexual (Pilagá), 297–8; and sorcery, 267–74; substitution for (Hopi), 279; value of, 22; *see also* Hostility

Albanian culture, infant swaddling, 508

Alcoholism, 131; patterns, 685–6; in primitive societies, 680–9; and psychosis, 458–63

Alexander, Franz, 421–35

Alexander the Great, 228

Algonkians, 260

Allen, Frederick H., 361

Allison, Davis, 320

Allport, Gordon W., 21, 142, 144, 147, 183, 341, 379, 380, 436–55

Aloofness, rating scale, 178

Ambition, rating scale, 178

Ambivalence: and bureaucracy, 379; definition, 11; in Hopi culture, 284; and mental health, 611; in Saulteaux, 262

American culture: age and sex factor, 363–75; and autocratic atmosphere, 669; behavior orientation, 351; "chance" belief in, 194; class distinction in, 131, 354; community study, 532–44; core of, 133; democratic personality in, 663–70; dynamic nature, 430; environmental concept in, 424–5; feeding regulation, 523; individualism in, 339; male and female role, 356; minorities in, 217; occupational structure, 585; old age in, 374; parental disturbance in, 196; problems of, 219; social drives in, 135; social stratification in, 309; status systems in, 567–76; time concept in, 349; and toilet training, 198; variants in, 353

American Indian culture, 658; alcoholism in, 686–7; infant swaddling, 508

"Amok" seizure, 693, 697

Anabolism, 37

A NOTE ON THE TYPE IN WHICH
THIS BOOK IS SET

This book was set on the Linotype in Janson, a recutting made direct from the type cast from matrices made by Anton Janson some time between 1660 and 1687.

Of Janson's origin nothing is known. He may have been a relative of Justus Janson, a printer of Danish birth who practised in Leipzig from 1614 to 1635. Some time between 1657 and 1668 Anton Janson, a punch-cutter and typefounder, bought from the Leipzig printer Johann Erich Hahn the typefoundry which had formerly been a part of the printing house of M. Friedrich Lankisch. Janson's types were first shown in a specimen sheet issued at Leipzig about 1675. Janson's successor, and perhaps his son-in-law, Johann Karl Edling, issued a specimen sheet of Janson types in 1689. His heirs sold the Janson matrices in Holland to Wolffgang Dietrich Erhardt, of Leipzig.

The book was composed, printed, and bound by Kingsport Press, Inc., Kingsport, Tennessee.